Joe H Jones

BA 470

College of Business

BASIC STATISTICS
A Real World Approach

Second Edition

BASIC STATISTICS
A Real World Approach

VINCENT E. CANGELOSI
Louisiana State University

PHILLIP H. TAYLOR
University of Arkansas

PHILIP F. RICE
Louisiana Tech University

WEST PUBLISHING COMPANY
St. Paul · New York · Los Angeles · San Francisco

A study guide has been developed to assist you in mastering concepts presented in this text. The study guide reinforces concepts by presenting them in condensed, concise form. Additional illustrations and examples are also included. The study guide is available from your local bookstore under the title, *Study Guide/Workbook to Accompany Cangelosi, Taylor and Rice's Basic Statistics: A Real World Approach, 2nd Edition* (prepared by Satterfield and Huffman).

COPYRIGHT © 1976 by West Publishing Company
COPYRIGHT © 1979 by West Publishing Company
50 West Kellogg Boulevard
P.O. Box 3526
St. Paul, Minnesota 55165

Printed in the United States of America

Library of Congress Cataloging in Publication Data

Cangelosi, Vincent E
    Basic statistics.

    Includes index.
    1. Statistics. 2. Social sciences—Statistical methods. 3. Commercial statistics. I. Taylor, Phillip H., joint author. II. Rice, Philip F., joint author. III. Title.
HA29.C27 1979   519.5   78-24592
ISBN 0-8299-0194-9

3rd Reprint—1980

To Jean, Gloria, and Jane

Contents

Measures of Dispersion, Skewness, and Kurtosis   44

3

Elementary Probability Theory   68

4

The Random Variable and Probability Distributions   99

5

Random Sampling and the Sampling Distribution of Sample Means and Proportions   130

6

Estimation: Means and Proportions   156

7

Statistical Testing: One Sample   169

8

Confidence Intervals and Hypothesis
Testing: Two Samples   192

9

Some Practical Aspects of Sampling   221

10

Nonparametric Statistical Tests   248

11

Correlation and Regression Analysis: The Simple Linear Case   276

12

Correlation and Regression Analysis: The Nonlinear and Multivariate Cases   315

13

Index Numbers   345

14

Time Series Analysis: Trend   363

15

Time Series Analysis: Seasonality   396

16

Expected Values and Decision Making   415

17

Appendixes   437

Answers to Odd-Numbered Problems   511

Index   525

Preface

In writing this text for students in business administration, economics, and other social sciences, we attempted to accomplish the following objectives:
1. To write a book which is understandable to students—one from which students can learn statistics.
2. To make learning statistical techniques interesting.
3. To present the subject of statistics in such a way that it has relevance to *real-world* applications.
4. To design a text that gives the instructor maximum flexibility in presenting a course suitable to student needs.

The first objective is accomplished by making a realistic evaluation of the mathematics background of the types of students who will most likely take a statistics course for which this text will be used. Not only do we consider that most of these students will probably have not taken any more than one course in algebra, but we also acknowledge that the mathematics retention factor is reasonably low for many students. Moreover, we feel that this book is written in a comprehensible, explanatory, and personal fashion.

We also feel that learning statistics does not have to be dull and monotonous. Since most everyday, real-world affairs involve statistics of some sort, a text dealing with statistical techniques can draw upon many real-world experiences for applications of these techniques. We have attempted to use the real world as our laboratory to help accomplish objectives two and three.

While the text may serve as the basis for a one-semester course in statistics, the topical coverage is rather extensive. Therefore, the text is designed so that the instructor may eliminate chapters without disrupting the continuity of the course or running out of material. For example, one instructor may not want to include the multiple regression analysis or the nonparametric chapter in a course for business students; yet another may want to exclude time-series analysis and index numbers for other social science students. In either case these eliminations are possible with sufficient material in the remainder of the book to cover a semester's work.

This book was conceived several years ago when the authors felt the need for a text in statistics which would help students with real-world problems. It is based upon what we have learned in our many years as classroom teachers and practitioners, as well as on the education and experience provided by our many friends, colleagues, and professors.

The first edition of this text was published in 1976. Its acceptance by our colleagues, and their students, encouraged us to continue to work with the material in the book for its further improvement. Thus, in this revised edition you will find the topical coverage of the material to be slightly expanded; but most importantly, some of the sections have been rewritten to enhance the presentation of the topics. We have also greatly expanded the problem sets at the end of each chapter. In this regard, we have also included three cases, each covering the concepts in several chapters, at strategic points in the text.

While we recognize that the final responsibility for this text, good or bad, rests with us, we wish to express our appreciation to Professors Marie Burkhead (Northwestern State University), James Calloway (Louisiana Tech University), Jay Craven (University of Miami), William Kornegay (Miami Dade Community College), Evelyn Rosenkranz (Monmouth College), Betty Thorne (University of Tennessee in Chattanooga), and Gregory Ulferts (University of Southern Colorado). Also, we express appreciation to Professors Donald Satterfield and Douglas Huffman (Memphis State University) who contributed many helpful suggestions while developing the study guide/workbook. However, we are most indebted to the members of our respective families for their encouragement when we were discouraged; for their patience during the many days and nights we were unable to be with them; and foremost of all, their continued love when we were unlovable.

BASIC STATISTICS
A Real World Approach

Data Presentation

CHAPTER

1

This text focuses on statistics as a research tool for decision making. Accordingly, we shall define the term *statistics* as the collection, presentation, analysis, and interpretation of quantitative information. In collecting such information, one usually searches for existing data, takes a sample survey, or runs an experiment. Almost anyone who collects data does so for a purpose; usually because it helps to measure and evaluate. Through analyzing and evaluating statistical data, public opinion has been determined; young athletes have been tendered enormous contracts; and economic programs have been initiated.

Evaluating or analyzing data, however, usually does not complete the researcher's task. Frequently the researcher must communicate this knowledge to others. To provide this communication the researcher may incorporate tables, charts, or empirical frequency distributions in the presentation of the research report. The purpose of this chapter is to introduce you to some of these tools of data presentation. Subsequent chapters will deal with the collection, analysis, and interpretation techniques.

Tables

A *table* is an orderly presentation of data. It should have a *title* that accurately and clearly identifies the contents and should have *column headings* that categorize the data within the table. The first column usually identifies for whom, what, when, or where the other columns of data were collected. That is, the first column usually represents years, months, states, countries, nationality, religion, or some such

1

breakdown for which the same data are to be presented for comparison. And finally, a good table will identify the *source* of the original data.

As an illustration of a well-prepared table, take a look at Table 1-1. You should notice that the table has a title and a source. The first column identifies what the information in the other columns is about. Notice, also, that the other columns have headings so that the data underneath are understandable. This table exhibits how data can be differentiated using what is called qualitative classes (see column one.) Table 1-2, on the other hand, presents its data on the basis of quantitative classes (see column one).

TABLE 1-1. *OPEC Crude Oil Capacity and Production (millions of barrels per day)*

	ESTIMATED CAPACITY DEC. 1976	ESTIMATED PRODUCTION 1977	PROJECTED CAPACITY 1980	PROJECTED PRODUCTION 1980[a]	PROJECTED PERCENT SHUT-IN CAPACITY 1980
Total OPEC	38.4	31.0	45.5	28.5	37.4%
Arabian Peninsula	18.1	13.5	22.7	8.0	65.0%
Saudi Arabia	11.5	9.2	15.0	5.2	65.0%
Kuwait	3.5	1.8	3.5	1.3	65.0%
United Arab Emirates	2.4	2.1	3.2	1.1	65.0%
Qatar	0.7	0.4	1.0	0.4	65.0%
Other OPEC	20.3	17.5	22.8	20.5	10.0%
Iran	6.7	5.6	7.2	6.5	10.0%
Iraq	3.0	2.1	4.0	3.6	10.0%
Venezuela	2.6	2.3	2.5	2.2	10.0%
Libya	2.5	2.1	2.5	2.2	10.0%
Nigeria	2.3	2.2	3.0	2.7	10.0%
Indonesia	1.7	1.7	2.0	1.8	10.0%
Algeria	1.0	1.1	0.9	0.8	10.0%
Gabon	0.3	0.2	0.3	0.3	10.0%
Ecuador	0.2	0.2	0.4	0.4	10.0%

[a]Allocation of total OPEC production for 1980 assumes non-Arabian Peninsula countries produce at 90% of available capacity, and that the difference is allocated proportionately among the four Arabian Peninsula producers.
Source: Across the Board, Volume XV, Number 1, January 1978, The Conference Board, Inc., p. 19.

If you examine Table 1-2, you will find that the table is based on information from one of the *Current Population Reports.* It is a good example of a complete table for it is titled, sourced, and in a clear manner provides data for the income groups delineated. Study this table carefully to see if you can read it.

TABLE 1-2. *Percent Distribution of Families and Unattached Individuals by Total Money Income Levels, 1955 and 1975*

TOTAL MONEY INCOME	1955	1975
Under $2,000	25.8	5.3
$2,000 and under $4,000	25.4	11.7
$4,000 and under $6,000	24.7	10.2
$6,000 and under $8,000	13.6	9.3
$8,000 and under $10,000	5.2	8.7
$10,000 and under $15,000	4.0	19.8
$15,000 and under $25,000	0.8	24.0
$25,000 and over	0.4	10.8

Source: Current Population Reports, Series P-60, Number 105, June 1977, U.S. Bureau of the Census.

Charts

Charts, as opposed to tables, resort to a *pictorial* presentation of data. And, because of the sense of familiarity with pictures, a reader usually feels more at ease with charts.   Consequently, charts are very popular data presentation aids.   Two widely used charts are the *bar chart* and the *pie chart*.

Bar Charts

Just as you would expect, *bar charts* communicate information through the use of bars.   Two examples of bar charts are shown in Figures 1-1 and 1-2.   Notice that in one of the charts the bars are vertical and in the other, they are horizontal.   In both cases the width of the bars is constant and the length of each bar indicates the magnitude of the data.

Pie Charts

As you probably guessed, a *pie chart* is simply a division of a circle (a pie) into "pie slices."   This circular shaped chart is particularly useful when data are stated in percentage terms.   The "pie" represents the whole (100 percent) and is subdivided into "slices" that represent the respective parts.   Of course, the "slices" expressed as percentages must total 100 percent.   Figure 1-3 illustrates the use of a pie chart.

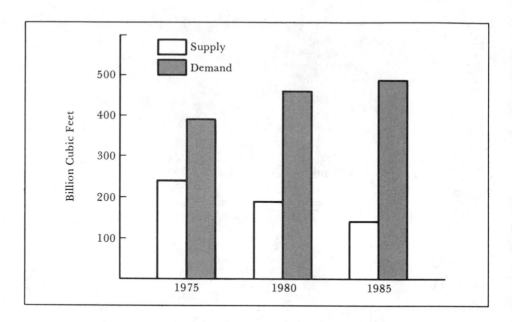

FIGURE 1-1. *Arkansas Gas Supply/Demand Balance Continued Growth-Medium Supply   Source: The Ozarks Regional Commission Regional Energy Alternatives Study: Arkansas Summary,* August 1977, Ozarks Regional Commission, p. 16.

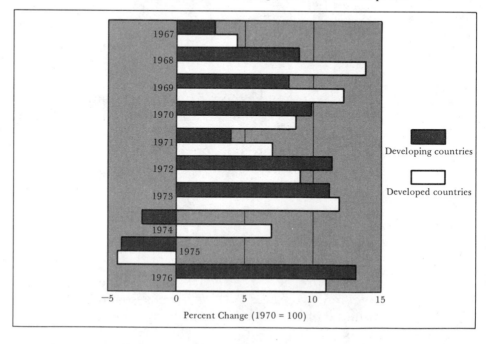

FIGURE 1-2. *Volume of Trade   Source:* "International Trade," *Road Maps of Industry,* Number 1819, November, 1977, The Conference Board, Inc.

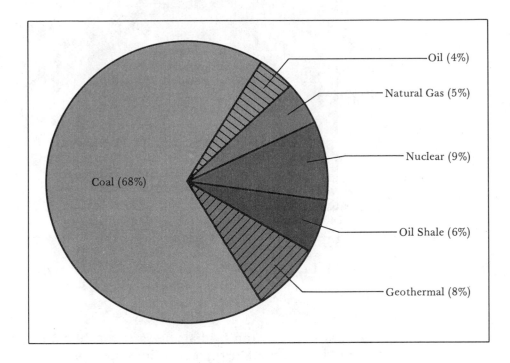

FIGURE 1-3. *U.S. Energy Reserves, 1975   Source: EXXON USA,*
Second Quarter, 1977, Exxon Company, p. 29.

Graphs

A *graph* is a line connecting points representing the coordinated values
of two variables. One variable is represented on the vertical axis (the
ordinate). Because it does relate the variables in a precise way, the
graph form is frequently utilized to present statistical data when the
main objective is to stress the relationship between the two variables.
Figures 1-4 and 1-5 are examples of graphs using two different methods of scaling the vertical axis.

In Figure 1-4 an arithmetic scale is used along the vertical axis.
This means that the distance used to represent $100 billion (or any
amount) is constant up and down the axis. If data are displayed in this
manner, the graph is called *arithmetic*.

Using an arithmetic scale, however, can create a problem if the
data to be displayed are very different in size. For example, you
might find it necessary to plot the numbers 1, 10, 150, 1,000, and
10,000 on the ordinate scale of a graph. To show this set of five numbers using an arithmetic scale would require a graph that would be
either very tall or one that would not show much, if any, difference in
the numbers 1, 10, and 150. The problem can be avoided by using a
graph similar to that shown in Figure 1-5. This type of graph differs

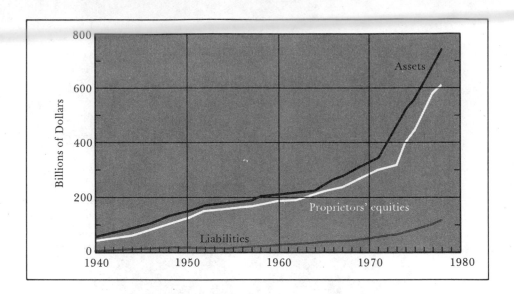

FIGURE 1-4. *Balance Sheet of the Farming Sector   Source: Review,* Volume 60, Number 1, January 1978, Federal Reserve Bank of St. Louis, p. 18.

from an arithmetic graph in that the scale on the vertical axis is logarithmic rather than an arithmetic. Since a logarithmic scale is used only on the ordinate, this type of graph is sometimes called a *semi-log graph.*

Note that in Figure 1-5 the distance used to represent 10 million barrels of oil diminishes as you move away from the intersection of the axes. This occurs because distances used to represent values are determined by the logarithms of the values. The logarithm of 1 is 0.00000; that of 2 is 0.30103; and the logarithmic value for 10 is 1.00000. Thus, on a log scale the value 2 is shown as 30.103 percent of the way between 1 and 10. The logarithmic scale in Figure 1-5 is called a "four-phase" scale. So long as the largest number to be plotted on an axis does not exceed 10 times the smallest number, only a single phase is needed. However, an additional phase must be added to accommodate numbers larger than 10 times the minimum number. That is, for the numbers 1 and 12, two phases are required; 1 and 150, three phases; etc. Finally, one disadvantage of this type of graph is that neither 0 nor negative numbers can be plotted on a logarithmic scale.

Frequency Distributions

Take a look at Table 1-3. What do you see? At best, you see 50 disorganized numbers. In fact, if the table were attempting to present any information, we would have to chastise ourselves and say it is a bad

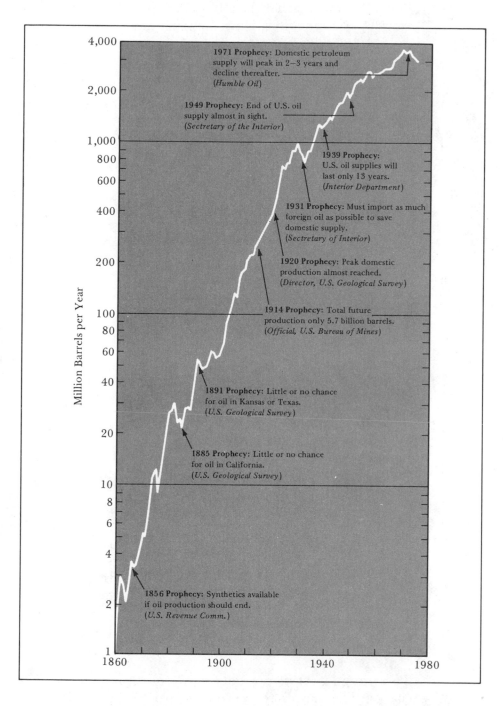

FIGURE 1-5. *Prophecy vs. Reality, Forecasts of U.S. Crude Oil Production*
Source: Across the Board, Volume XV, Number 1, January
1978, The Conference Board, Inc., p. 20.

table.   However, since the table was used to present only the raw data, we will not condemn its construction, but we will say that the presentation of raw data is not too helpful if one is expected to derive some meaning from the data.

TABLE 1-3.   *Per Capita State and Local Taxes for Selected States* and the District of Columbia in Fiscal Year 1976*

	667	749	778
935	596	588	793
728	658	549	768
703	651	493	814
964	530	486	924
820	581	586	584
590	823	684	609
709	701	1140	527
847	570	742	489
593	454	571	455
728	610	671	549
731	791	903	566
598	769	711	

*Does not include Alaska, where the per capita figure in 1976 was $1,896.
Source: Monthly Tax Features, Volume 21, Number 9, December 1977, Tax Foundation, Inc., p. 1.

Now suppose it has become your job to present this set of data in an orderly and meaningful manner.   A tabular listing of the numbers in descending or ascending order (an ordered array) would help somewhat, but it would certainly be a long table.   So let's investigate a way to aggregate the data.

A *frequency distribution* is developed when the data are divided into two or more *groups* called *classes,* with some identification given to each class.   The identifying nature of the classes can be either qualitative or quantitative, just the same as the first column of a complete table.   In order to construct a frequency distribution, however, we need to examine the concept of grouping data.

Grouping the Data

To arrange a set of raw data into groups, it is necessary to establish classes. The classes represent the groups (intervals for quantitative classes and categories or attributes for qualitative classes).   Each class that is established must be unique.   That is, there must be no overlap in the classes in order that each individual item in the data set will belong to only one class. The number of drownings classified according to activity, as shown in Table 1-4, is an example of *qualitative* classes. Similar data, except classified according to age, are presented in Table 1-5 as an illustration of *quantitative* classes.

TABLE 1-4. *Drownings by Selected Type of Activity, United States, 1971*

ACTIVITY	NUMBER OF DEATHS
Swimming	1,238
Playing on Ice	55
Fishing from Boat	343
Boating	349
Surfing	13
Hunting	48
Scuba Diving	58
Attempting a Rescue	143
Taking a Bath	100

Source: Statistical Bulletin, Volume 58, May 1977, Metropolitan Life Insurance Co., p. 4.

TABLE 1-5. *Drownings by Age of Victim, United States, 1971*

AGE	NUMBER OF DEATHS
Under 1	55
1–4	736
5–9	549
10–14	624
15–19	943
20–24	396
25–34	464
35–44	395
45–54	331
55–64	217
65 and over	252

Source: Statistical Bulletin, Volume 58, May 1977, Metropolitan Life Insurance Co., p. 4.

Table 1-5 needs some explanation, however, regarding its construction. The first column, age, in the table represents the classes. All but one of the classes have *limits*, upper and lower, that make the classes unique, but the placing of the data into classes may still require some interpretation. Consider a person who was 19 years and 8 months old. Into which class would 19⅔ go?

Where 19⅔ resides is decided in the following manner. If the data require rounding, more definite *boundaries* are needed than the class limits given in Table 1-5. That is, it may be decided to include all items less than 20 but equal to or more than 15 in the fifth class. As a matter of fact, when dealing with ages this is the usual procedure. However, had the data involved weights, dollars, inches or other similar units, another rounding procedure might be used. For example, the boundary of a class might be halfway between the upper limit of one class and the lower limit of the next class. The point is that a decision has to be made about the boundaries so that you will know where to place values that fall between the upper limit of one class and the lower limit of another.

Now notice in Table 1-5 that the class limits are unique. That is, the upper limit of one class must be less than the lower limit of the next class. Another way to present the class limits for this table is as follows:

1 and under 5

5 and under 10

10 and under 15

Stating limits in this fashion will sometimes produce a slightly different grouping, but it will also eliminate any doubt as to where a particular item belongs.

So far we have looked at class limits without discussing the class interval. The *interval* is the width of the class or the difference between the upper boundary and the lower boundary of a class. Thus, in Table 1-5 the boundaries, and not the class limits, must be used to establish the class interval. Consequently, using the boundaries for the second class (1 and 5), subtraction yields a class interval of four.[1]

The next item in Table 1-5 that we need to examine is the *class midpoint*. The class midpoint is the value represented by the point equidistant between the upper and lower boundaries of the class. The midpoint of any class can be established by adding the two boundaries and dividing their sum by two. Thus, the midpoint of the second class of Table 1-5 is $(1 + 5)/2 = 3$.

Finally, it should be pointed out that not all the class intervals in Table 1-5 are equal. While equal intervals for all classes are frequently desired, they are not essential. In this instance, there were also some extreme values that required the last class (65 and over) to be left without an upper limit. (Otherwise we might have had several classes in which no observations fell.) When a class is missing a limit, either upper or lower, it is called an *open end* class.

---

1.   Remember that in this instance the upper boundary of the second class approaches the lower limit of the third class. Since this depends on the rounding method used, sometimes the upper boundary and the lower limit will be different.

Frequency Distribution Construction

Now that you know something about grouping data, let us proceed to
the construction of a frequency distribution for the data shown in Table 1-3. To facilitate the grouping, first arrange the per capita taxes
in order of magnitude. (It makes no difference whether you begin
with the smallest and proceed to the largest or vice versa.) Data arranged in this way are called an *ordered array*. Table 1-6 shows the per
capita tax data arrayed in descending order.

TABLE 1-6. *Per Capita State and Local Tax Data Array for Fiscal 1976*

	769	667	571
1140	768	658	570
964	749	651	566
935	742	610	549
924	731	609	549
903	728	598	530
847	728	596	527
823	711	593	493
820	709	590	489
814	703	588	486
793	701	586	455
791	684	584	454
778	671	581	

Source: Table 1-3.

It is now necessary to face the question: "How many classes are
needed?" Unfortunately, there are no rules that will fix the number
of classes that should be in a given frequency distribution. There are,
however, two guidelines that will aid you in frequency distribution construction. They are (1) Keep the number of intervals between 5 and
20, and (2) avoid empty classes.

With these two guidelines in mind, a good starting place in constructing a frequency distribution is to examine the difference in the
extreme values in our data set. That is, subtract the lowest data item
from the highest ($1140 − $454 = $686). To assure that there is a
place in the frequency distribution for every item in the data set, the
product of the number of classes and the class interval must be equal to
or greater than this difference ($686).

At this point, either experience or experimentation is the only help available.[2]   In our particular case, class intervals of $75, $100, or $150 might do the job adequately.   Selecting the interval of $100 indicates the need for seven classes.   That is, dividing the selected class interval into the difference between the largest and smallest values in the data set yields $686/$100 = 6.86.   Then, the number of classes is rounded to seven.   Table 1-7 shows the per capita tax data grouped into seven classes with $450 chosen as the lower limit of the first class.

TABLE 1-7.   *Frequency Distribution for Fiscal 1976 per Capita State and Local Taxes*

CLASS INTERVAL	FREQUENCY
$450–$549	9
550– 649	13
650– 749	14
750– 849	9
850– 949	3
950–1049	1
1050–1149	1
	50

Source: Table 1-6.

It should be noted that the boundaries of the first class are $449.50 and $549.50, which gives the class interval of $100.   The midpoint for the first class is ($449.50 + 549.50)/2 = $499.50.   This method of stating the class limits, however, has the disadvantage that the class limits shown are not the actual boundaries of the classes.   Table 1-8 presents the data grouped using class limits that correspond to the boundaries.   As stated before, this type of grouping eliminates uncertainty regarding the placement into classes of any observations.   Consequently, this is the method of stating class intervals that will be used throughout this text.   Since it is not intended that this is the only frequency distribution that might be established from the data set— nor is it necessarily the "best"—why not take a few minutes and make your own!

The frequency distribution in Table 1-8 is an *absolute frequency distribution*.   That is, it shows the actual number of data items falling within each class.   Thus, in the first class there are nine items, in the

---

2.   There is a method, called Sturges' Rule, that can assist you in determining the number of classes.   Sturges' Rule is stated as follows:

$$k = 1 + 3.3 \log n$$

where k = the number of classes when rounded to the nearest whole number,
          n = the number of observations in the data set.

However, it should be remembered that this method is a general guide and not a substitute for experienced judgment.

second class there are 13 items, and so on. In contrast to this absolute reporting is the type of distribution known as a *relative frequency distribution*. The relative frequency distribution states the frequency of each class as a percent of the total frequencies. Table 1-9 presents the per capita tax data as a relative frequency distribution. The first class contains 18 percent of the observations in the distribution or $(9/50)100 = 18\%$.

TABLE 1-8. *Frequency Distribution for Fiscal 1976 per Capita State and Local Taxes*

CLASS INTERVAL	FREQUENCY
$450 and under $550	9
550 and under 650	13
650 and under 750	14
750 and under 850	9
850 and under 950	3
950 and under 1050	1
1050 and under 1150	1
	50

Source: Table 1-6.

TABLE 1-9. *Relative Frequency Distribution, per Capita State and Local Taxes in Fiscal 1976*

CLASS INTERVAL	RELATIVE FREQUENCY
$450 and under $550	18%
550 and under 650	26
650 and under 750	28
750 and under 850	18
850 and under 950	6
950 and under 1050	2
1050 and under 1150	2
	100%

Source: Table 1-8.

Histograms and Frequency Polygons

The data given in Table 1-8 could be displayed graphically to present the information pictorially. One form this presentation could take is known as a histogram. A *histogram* is a bar chart that shows a frequency distribution with data presented along both axes. The class interval is represented by the width of the bar on one axis, and the frequency of the class is given by the height of the bar on the other axis. As with other bar charts, this conveys to anyone viewing the histogram both the absolute frequency of each class and the size of the one class frequency relative to the frequencies of all other classes. The data in Table 1-8 are shown as a histogram in Figure 1-6.

In presenting a frequency distribution as a histogram, there is an implicit assumption that the data items in each class are evenly distributed throughout the class. If this even distribution is not a realistic presentation of the data, it may be more accurate to display the data using a *frequency polygon*. To construct a frequency polygon, a series of points, whose coordinates are the midpoints and frequencies of each class is plotted. A frequency polygon, then, is a line joining these points. The dashed line in Figure 1-6 is a frequency polygon showing the same information as the histogram and also having the same meaning. That is, it should provide pictorially both a medium for viewing the shape of the frequency distribution and a look at the frequency of occurrence for each class.

There is one final point that you should notice concerning the frequency polygon in Figure 1-6. The dashed line does not terminate

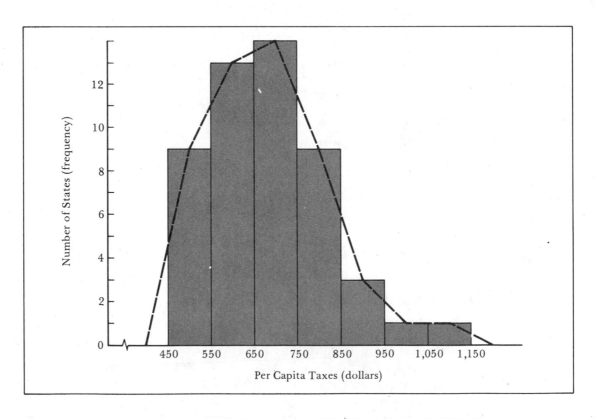

FIGURE 1-6. *Per Capita State and Local Tax Data for Fiscal 1976 Displayed as a Histogram and a Frequency Polygon*

at the midpoint of the last class, nor does it originate at the midpoint of the first class. The polygon is "closed in" by extending the dashed line to the abscissa. In practice, the line is made to intersect the axis at points that correspond to the midpoints of what would be the next class at each end of the frequency distribution.

Cumulative Frequency Distributions

At times you may find it convenient to present the data in a manner that will show the number of values that are greater than or less than some quantity. Such a presentation is called a *cumulative frequency distribution*. For an example of a "More Than" distribution, take a look at Table 1-10. In that table, you will notice that the class limits have been changed to allow the frequencies to accumulate. That is, the data are shown as "$450 or more," "$550 or more," and so on, rather than each class interval being exclusive. The cumulative frequencies are obtained by summing down the frequency column in Table 1-8. That is, the first cumulative frequency is 50 because all items are "$450 or more." For the second cumulative frequency, the summation begins with 13 and takes in all frequencies down the column (13 + 14 + 9 + 3 + 1 + 1 = 41). This procedure is continued until all data items are accounted for.

TABLE 1-10. *"More Than" Cumulative Frequency Distribution*

PER CAPITA TAXES	FREQUENCY
$450 or more	50
550 or more	41
650 or more	28
750 or more	14
850 or more	5
950 or more	2
1050 or more	1
1150 or more	0

Source: Table 1-8.

Reversing the manner in which the frequencies are accumulated produces a "Less Than" cumulative frequency distribution as shown in Table 1-11. Notice that the frequency begins with 0 and ends with 50 in the "Less Than" distribution, and that the frequency begins with 50 and ends with 0 in the "More Than" distribution.

TABLE 1-11. *"Less Than" Cumulative Frequency Distribution*

PER CAPITA TAXES	FREQUENCY
Less than $450	0
Less than 550	9
Less than 650	22
Less than 750	36
Less than 850	45
Less than 950	48
Less than 1050	49
Less than 1150	50

Source: Table 1-8.

Now, these cumulative frequency distributions can be used to establish additional information about the data set. For example, Table 1-10 indicates that all 50 selected areas have per capita taxes of $450 or more, and 14 of them have per capita taxes of $750 or more. According to these data, no area had per capita taxes in excess of $1150 in fiscal 1976. Similarly, Table 1-11 points out that none of the selected areas has per capita taxes below $450 and that all are below $1150.

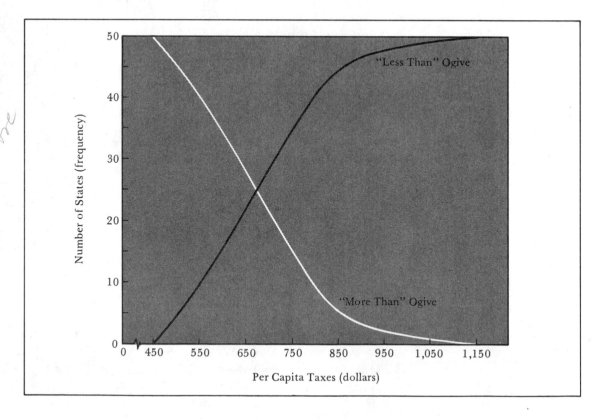

FIGURE 1-7. *Per Capita State and Local Tax Data for Fiscal 1976 Displayed as Ogives* Source: Tables 1-10 and 1-11.

Finally, the information shown in Table 1-10, as well as that shown in Table 1-11, can be displayed graphically. Such a graph is called an *ogive*. In the case of an ogive using the data in the form shown in Table 1-10, a series of points with coordinates of the lower limits and respective frequencies is plotted. These points are then connected to form the ogive. Figure 1-7 presents, in ogive form, the cumulative distributions of the per capita tax data as shown in Tables 1-10 and 1-11.

Frequency Array

Sometimes arranging data into classes is not convenient because you may wish to retain the individual identity of each item of data. Other times, the data set may not contain enough different values to justify constructing the type of frequency distribution shown in Table 1-8, or, still another possibility is that there are only a few different values in the data set and these values occur in whole number form. In this latter case, we simply mean that no fractional units occur.

Look at Table 1-12. This is an example of a *frequency array*. Rather than tabulating the data by classes, a frequency array displays the data as a tabulation of the occurrence of specific values. In the example in Table 1-12, the number of banks in any county is discrete (a whole number). Thus, the table is simply a tabulation of the number of counties with 1, 2, and so on up to 11 banks.

TABLE 1-12. *Frequency Array of the Number of Banks in Arkansas Counties in 1977*

NUMBER OF BANKS	NUMBER OF COUNTIES (FREQUENCY)
1	9
2	22
3	12
4	16
5	5
6	3
7	5
8	1
9	1
10	0
11	1
	75

Source: 1977 Report of the Bank Commissioner of Arkansas, Office of State Bank Department, Little Rock.

1. Define
    (a) statistics
    (b) frequency distribution
    (c) histogram
    (d) pie chart
    (e) class interval
    (f) class midpoint
    (g) class limits
    (h) class boundaries
    (i) ogive
2. List the components of a good table.
3. Distinguish between qualitative and quantitative classes.

—————————— PROBLEMS ——————————

1. In its *1977 Annual Report,* the Commercial National Bank provided the following information on its assets:

ITEM	1976	1977
Cash and Due from Banks	$ 48,531,000	$ 45,707,000
Investments	21,393,000	31,091,000
Funds Loaned	58,700,000	61,575,000
Net Loans	115,853,000	142,708,000
Other Assets	9,906,000	10,542,000
Total Assets	$254,382,000	$291,623,000

   a. Present these data in bar chart form for each year.
   b. Present these data in pie chart form for each year.

2. Several alternative projections of natural gas demand and supply for the Ozarks region have been made. The following one for 1985 assumes a high economic growth rate and low levels of supply. Demand will be: Oklahoma—710, Missouri—564, Louisiana—2176, Kansas—530, Arkansas—530. The supply available in each state will be: Louisiana—941, Kansas—192, Missouri—147, Arkansas—114, Oklahoma—561. (All demand and supply data are in billions of cubic feet of natural gas.) The source of this projection is *The Ozarks Regional Commission Regional Energy Alternatives Study: Arkansas Summary* published in August 1977.
   a. Present these data in table form. Include in your table the shortage or surplus expected in each state and the region as a whole.
   b. Present these data in bar chart form for each state.
   c. Present these data in pie chart form, using one chart for demand and one for supply.

3. One of the breakdowns for gross national product is by major type of product: durable goods, nondurable goods, services, and structures. Durable goods were $258.2 billion in 1975 and $303.4 billion in 1976; nondurable goods were evaluated at $428.0 billion in 1975 and $460.9 billion in 1976; services were $699.2 billion in 1975 and $782.0 billion in 1976; and structures were $143.5 billion in 1975 and $160.2 billion in 1976. The source of these data is Table 2 on page 8 of the October 1977 issue of the *Survey of Current Business.*
a. Present these data in table form.
b. Present these data in a bar chart for each year.
c. Present these data in a pie chart for each year.

4. On page 18 of the October 1977 issue of the *Survey of Current Business,* the following personal income data are given:

STATE	PERSONAL INCOME
Connecticut	$24,880
Maine	6,298
Massachusetts	41,486
New Hampshire	5,459
Rhode Island	6,526
Vermont	2,869

Actually, the above data are for the second quarter of 1977 and are stated in millions of dollars, seasonally adjusted at annual rates. The states shown are the New England region as shown in the publication.
a. Present these data in a bar chart.
b. Use a pie chart to present the information.
c. Comment on the two methods of presenting the data.

5. A small bread company recorded the following information concerning last year's operating expenses:

ITEM	EXPENDITURES
Cost of Products Sold	$500,000
Administrative Expenses	25,000
Commissions	75,000
Interest	9,000
Depreciation	60,000
Income Taxes	40,000

a. Present these data in a bar chart.
b. Present these data in a pie chart.
c. Comment on the two presentations.

6. Construct a bar chart of the following employment data for Washington County in 1978.

OCCUPATION	NUMBER
Professional and Technical	8200
Farm Workers	6800
Clerical and Kindred	4500
Sales Personnel	6000
Craftsmen and Foremen	14000
Operative Workers	9700
Household Workers	200
Service Workers	7300
Laborers	5200
Other	2500

7. S and S News Service has recorded the following 5-year earnings pattern. Present the data in a bar chart:

YEAR	EARNINGS PER SHARE
1971	$0.90
1972	0.85
1973	0.70
1974	0.75
1975	0.95

8. Convert the data in Problem 6 to percentages and present these percentages in a pie chart.

9. Present the data in Problem 7 on a graph.

10. The following data are for Madison County for 1978:

AGE	NUMBER OF FEMALES	NUMBER OF MALES
15	100	95
16	125	130
17	130	120
18	160	150
19	190	180
20	205	200
21	200	210
22	240	225
23	250	260

a. Construct a graph for females.
b. Construct a graph for males.

11. The following data are federal government receipts and outlays (in $ millions) for 1971 through 1978 (estimated):

YEAR	RECEIPTS	OUTLAYS
1971	$188,392	$211,425
1972	208,649	232,021
1973	232,225	247,074
1974	264,932	269,621
1975	280,997	326,105
1976	299,197	365,648
1977	356,861	401,896
1978	401,372	462,882

a. Construct a graph showing both receipts and outlays.
b. Construct a bar chart with the bars for receipts and outlays for each year adjacent to one another.

12. The Commercial National Bank lists the following information concerning borrowed funds in its *1977 Annual Report.*

YEAR	AMOUNT	YEAR	AMOUNT
1969	$ 591,000	1974	$34,007,000
1970	1,740,000	1975	21,501,000
1971	9,650,000	1976	44,716,000
1972	12,110,000	1977	51,655,000
1973	42,665,000		

Present this information in semi-log graph form.

13. The Sunflower Nursing Home began operations in 1969 and the following is its record of the number of residents the home has had:

YEAR	NUMBER
1969	5
1970	12
1971	55
1972	75
1973	120
1974	125
1975	140
1976	242
1977	136
1978	248

a. Construct an arithmetic graph with these data.
b. Construct a semi-log graph with these data.
c. Comment on your results using the two graph forms.

14. Use the following frequency distribution to answer the questions given below:

DATA	FREQUENCY
50 and under 60	5
60 and under 70	11
70 and under 80	18
80 and under 90	14
90 and under 100	6

   a. How many classes are there?
   b. What is the class interval?
   c. What are the class midpoints?
   d. Convert the above frequency distribution to a relative frequency distribution.
   e. Construct a histogram.
   f. Construct a frequency polygon.
   g. What percent of the data are 70 or above?
   h. What percent of the data are less than 80?

15. At the end of its first year of operations, a new credit union has 60 accounts as given below:

$ 50	$300	$510	$720	$ 950	$1250
50	350	550	750	1000	1250
70	370	560	750	1000	1260
120	380	560	800	1000	1270
150	390	570	800	1000	1280
150	400	580	800	1010	1290
200	500	580	810	1050	1300
250	500	600	820	1080	1350
260	500	600	820	1090	1380
280	500	630	830	1090	1400

   a. Using the above data, construct a frequency distribution.
   b. Convert the frequency distribution to a "More Than" cumulative frequency distribution.

16. Use the data on ages of women working at the Northwest Egg Breaking Plant to complete this problem.

43	58	21	24	31	49	40	51	55	28
50	33	62	30	25	39	59	29	36	42
38	46	42	18	50	41	37	35	40	52
47	35	57	55	50	36	45	32	45	42
36	52	64	34	32	47	41	44	39	38

   a. Present these data as an array.
   b. Prepare a frequency distribution with about seven or eight classes. (Save your work for use later.)
   c. Plot the data in part (b) as a histogram.

    d. Convert the frequency distribution to a relative frequency distribution.

    e. Convert the histogram of part (c) to a frequency polygon.

17. Use your frequency distribution in Problem 16 to answer the following questions:

    a. Convert the frequency distribution to a "More Than" cumulative frequency distribution.

    b. Construct a "More Than" ogive.

    c. Convert the frequency distribution to a "Less Than" cumulative frequency distribution.

    d. Construct a "Less Than" ogive.

18. A service station owner has the following information on customer purchases of gasoline:

GALLONS PURCHASED	FREQUENCY
5 and under 8	17
8 and under 11	25
11 and under 14	29
14 and under 17	52
17 and under 20	39
20 and under 23	28
	190

    a. What is the class interval?

    b. What are the midpoints of the first and last class?

    c. Construct a "More Than" and "Less Than" cumulative frequency distribution.

    d. How many of the customers purchased 14 or more gallons? What percentage of the customers is this?

    e. How many customers purchased at least 8 but less than 17 gallons?

19. Use the following charts and answer the questions on page 24.

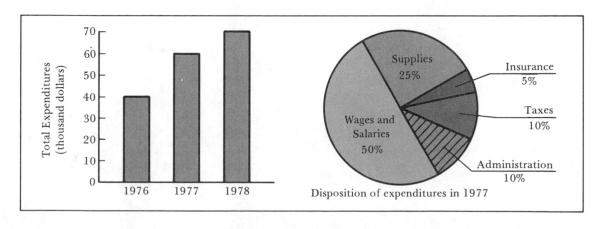

Disposition of expenditures in 1977

a. What were the administrative expenses in 1977?

b. How much was the insurance in 1977?

c. What was the dollar increase in expenditures from 1976 to 1978?

d. What was the percentage increase in expenditures from 1976 to 1978?

20. Table 1-3 contains per capita state and local tax data for every state and the District of Columbia except Alaska. In fiscal year 1976, the per capita tax figure for Alaska was $1896.

a. Construct a frequency distribution of the per capita tax data for all 50 states plus the District of Columbia.

b. Comment on how the frequency distribution you constructed differs from that in Table 1-8 and why.

Measures of Central Tendency

CHAPTER
2

In Chapter 1 we concentrated on various ways of presenting quantitative data. As you might have guessed, however, data presentation alone does not provide much explicit information about the data under investigation. The data must be described and analyzed. It is time, then, to introduce methods that can be used to measure and to describe certain features of the data. In this chapter, we are going to become familiar with certain *measures of central tendency,* which are often referred to as *averages.* In Chapter 3, we will examine measures for describing the dispersion or spread of data.

When describing a set of data, whether the set is individual data items or a frequency distribution such as the one constructed for per capita taxes in Table 1-8, a *measure of central tendency* is often given to identify the center of the data. In the following discussion, you will be introduced to five measures of central tendency sometimes referred to as averages. They are: the *arithmetic mean,* the *weighted arithmetic mean,* the *median,* the *mode,* and the *geometric mean.* As you will discover, we have more than one measure of central tendency because there is more than one way to describe the "middle." Their meanings are different, and often there is a need to have more than one way to describe the "middle."

The Arithmetic Mean

Perhaps the best known average (measure of central tendency) is the *arithmetic mean,* or simply the "mean," which is the "center" of the values of a set of observations. That is, the *arithmetic mean* of a set of observations is the sum of their values divided by the number of observations. Algebraically, the equation for determining the arithmetic mean of n observations is

$$\overline{X} = \frac{\Sigma X}{n} = \frac{X_1 + X_2 + X_3 + \cdots + X_n}{n} \tag{2.1}$$

As a matter of accepted practice, the symbol $\overline{X}$ (pronounced X bar or bar X) will be used in this text to refer to the mean of sample data. A sample consists of n observations drawn from a larger group called the population whose number of observations is represented by the symbol N. For the population, the symbol representing the arithmetic mean is the Greek letter μ (mu). A value such as $\overline{X}$, which is a sample mean, is referred to as a *statistic* while a value such as μ, which is a population mean, is referred to as a *parameter*. Thus, for a population we have

$$\mu = \frac{\Sigma X}{N} = \frac{X_1 + X_2 + X_3 + \cdots + X_N}{N} \tag{2.2}$$

To illustrate the computation, suppose that the ages of five students in a particular class are 19, 20, 21, 18, and 22. If these observations represent a population of five (the whole class) then the arithmetic mean (parameter) is

$$\mu = \frac{\Sigma X}{N} = \frac{\Sigma X}{5} = \frac{19 + 20 + 21 + 18 + 22}{5} = 20$$

Now, let's suppose that a sample of three students is drawn from the class of five. The three ages drawn are 20, 21, and 22. Since the three ages represent a sample from the class, the arithmetic mean (statistic) is

$$\overline{X} = \frac{\Sigma X}{n} = \frac{\Sigma X}{3} = \frac{20 + 21 + 22}{3} = 21$$

This value, $\overline{X}$, is an estimate of the population mean.

Grouped Data—Calculation of the Arithmetic Mean

You have learned how the arithmetic mean is computed by summing the values of a set of observations and dividing by the number of observations. Also, you have seen the arithmetic mean calculated for ungrouped data by using Equations 2.1 and 2.2. At times, however, the data you must work with will be grouped into class intervals. Consequently, Equations 2.1 and 2.2 cannot be used to determine the mean. We must investigate a procedure for calculating the mean of grouped data. Table 1-8 is reproduced in Table 2-1 to facilitate our grouped data discussion.

As you can see in Table 2-1, we no longer have knowledge about the exact value of the individual observations because of the grouping process. We only know that a certain number of observations belong in a certain class interval.

TABLE 2-1. *Per Capita State and Local Tax Data*

CLASS INTERVAL	FREQUENCY
$450 and under $550	9
550 and under 650	13
650 and under 750	14
750 and under 850	9
850 and under 950	3
950 and under 1050	1
1050 and under 1150	1
	50

Source: Table 1-8.

To calculate the mean, it is necessary that a term equivalent to ΣX of Equation 2.1 be estimated. In order to make this estimate, *the midpoint of each class is assumed to represent the unknown observations in each class.* This assumption (known as the midpoint assumption) allows us to estimate the sum of the observations in each class by multiplying each class midpoint (M) by the frequency (f) of the respective class. Then the mean can be calculated by summing the multiplication done for each class and dividing the sum by the number of observations such that:

$$\overline{X} = \frac{\Sigma f M}{\Sigma f} = \frac{\Sigma f M}{n} \qquad (2.3)$$

where $\overline{X}$ = the arithmetic mean
f = frequency for a class
M = midpoint of a class
n = total number of observations

To illustrate, study Table 2-2 to see how the calculation is performed:

TABLE 2-2. *Calculation of the Arithmetic Mean for Grouped Data*

CLASS INTERVAL	CLASS MIDPOINT (M)	FREQUENCY (f)	fM
$450 and under $550	$500	9	$4500
550 and under 650	600	13	7800
650 and under 750	700	14	9800
750 and under 850	800	9	7200
850 and under 950	900	3	2700
950 and under 1050	1000	1	1000
1050 and under 1150	1100	1	1100
		50	$34100

$$\overline{X} = \frac{\Sigma fM}{\Sigma f} \qquad \overline{X} = \frac{\$34100}{50} = \$682$$

From Table 2-2 you can see that the arithmetic mean computed from the grouped data is $682.[1] This compares favorably with the mean that results from using Equation 2.1 and the data in Table 1-6, which is $681.52. The difference ($0.48 or less than 0.1 percent) is the result of grouping, which caused the individual identity of each observation to be lost. Thus, how well the mean computed from a frequency distribution compares with the mean computed from ungrouped data depends on how close fM represents the sum of the actual observations in each class. The extent to which this is true, or to which there are offsetting errors, will determine how closely the two means compare.

Another Approach for Calculating the Arithmetic Mean

Now that you have the idea of how to handle grouped data computations, you are ready to examine a quicker method for calculating the arithmetic mean from grouped data. The answer is the same as the one from the previous method; it's just easier to compute.

The method demonstrated will use an assumed mean and deviations of midpoints from the assumed mean. Keep your eye on Table 2-3 as you read the following explanation.

From Table 2-3, you can see that the equation used to calculate the mean was

---

1.   This is the arithmetic mean of the per capita state and local taxes for the selected areas. The actual per capita state and local taxes differ due to population differences among the states. You will understand why this difference occurs when you study the weighted arithmetic mean.

$$\overline{X} = \overline{X}_a + \left(\frac{\Sigma fd}{n}\right) i \qquad \textit{Skip} \qquad \textbf{(2.4)}$$

where $\overline{X}$ = the arithmetic mean
$\overline{X}_a$ = an assumed mean
f = frequency of each class
i = class interval
$n = \Sigma f$
d = deviation of each midpoint (M) from the assumed mean in class interval units $[(M - \overline{X}_a)/i]$

TABLE 2-3. *Another Approach to Calculating the Arithmetic Mean of Grouped Data*

CLASS INTERVAL	CLASS MIDPOINT (M)	FREQUENCY (f)	$M - \overline{X}_a$	d	fd
$450 and under $550	$500	9	−200	−2	−18
550 and under 650	600	13	−100	−1	−13
650 and under 750	700	14	0	0	0
750 and under 850	800	9	+100	+1	+ 9
850 and under 950	900	3	+200	+2	+ 6
950 and under 1050	1000	1	+300	+3	+ 3
1050 and under 1150	1100	1	+400	+4	+ 4
		50			− 9

$$\overline{X} = \overline{X}_a + \left(\frac{\Sigma fd}{n}\right) i \qquad\qquad \overline{X} = 700 + \left(\frac{-9}{50}\right) 100$$

$$\overline{X}_a = \$700 \qquad\qquad\qquad = 700 + (-0.18)\, 100$$

$$= 700 - 18$$

$$\overline{X} = \$682$$

The reason we can use Equation 2.4 is that an assumed mean can be adjusted to the true mean by an accounting of the deviation of each midpoint from the assumed mean. Since each midpoint represents several observations, the first step is to multiply the deviation, in class interval units, of each midpoint from the assumed mean by the frequency of the corresponding class. These products are then added (Σfd) to find the total deviation in class interval units from the assumed mean and are divided by the number of observations to provide the average deviation. The average deviation is then multiplied by the class interval to convert the deviation to the proper dimensions ("dollars" in our example), and this product is added to the assumed mean. The result is that if the assumed mean is too high, it will be adjusted downward; if the assumed mean is too low, it will be adjusted upward.   It should be noticed that the arithmetic mean calculated in Table 2-3 is the same as the one in Table 2-2.

More About the Arithmetic Mean

The arithmetic mean may be thought of as the point of balance with regard to the values of the data items.   That is, when the mean is subtracted from each observation, the sum of these differences will equal $0, \Sigma (X - \overline{X}) = 0$.   In other words, the negative differences equal the positive ones.   (Actually, Equation 2.4 is based on this characteristic of the arithmetic mean.)   Additionally, if these differences are squared and then summed, $\Sigma (X - \overline{X})^2$, this sum is the *minimum value* that can be attained from summing the squared differences.   No other number can be used in place of the mean to give a smaller sum of squared differences.   For any given data set, no other number demonstrates these two properties and, in this respect, the arithmetic mean is the most representative measure of central tendency.

As a method of locating the middle, the arithmetic mean has both advantages and disadvantages.   Among the advantages are

1. A mean exists for every set of data. *(ungrouped)*
2. Every value of the variable in the set of data influences the value of the arithmetic mean.

Some of the disadvantages are

1. The mean can be overly influenced by extreme values of the variable.   For example, the mean of 2, 4, 6, 8, 10, and 900 is 155, although five of the six observations are values of 10 or less.
2. The mean cannot be determined for open-ended frequency distributions without additional information.

Perhaps the most significant advantages of the arithmetic mean in relation to other measures of central tendency are those related to other statistical measures and analysis techniques.   As you cover the material in the following chapters, you will see repeated reference to the arithmetic mean.   Overall, it is used in more ways than the other measures of central tendency, and thus it may be thought of as being more popular than either the median or the mode.

The Median

A second average (measure of central tendency) is the median. It is the *positional middle* of the data. In other words, in an ordered array of the data, one-half of the values of the variable precede the median and one-half follow it.

Procedurally, the first step in calculating the median is to locate its position in the ordered array or frequency distribution. For ungrouped data the median is located at the $(n + 1)/2$ position in the array where n is the number of observations of the variable. As an example, in the ordered array 10, 9, 8, 7, 6, there are 5 values. Thus, the median is located at $(5 + 1)/2 = 3$. Referring to the array, the third number is 8. Therefore 8 is the median.

If the data set happens to have an even number of observations, the median is assumed to be halfway between the two central values in the order array. Thus, in Table 2-4 the median is the $(50 + 1)/2 = 25.5$th number in the array. Since the 25th value is 671 and the 26th is 667, the median is $(671 + 667)/2 = 669$.

TABLE 2-4. *Per Capita Tax Data Arrayed*

	769	667	571
1140	768	658	570
964	749	651	566
935	742	610	549
924	731	609	549
903	728	598	530
847	728	596	527
823	711	593	493
820	709	590	489
814	703	588	486
793	701	586	455
791	684	584	454
778	671	581	

Source: Table 1-6.

Grouped Data—Calculation of the Median

If the data have been grouped, as stated before, the identity of the individual observations has been lost. Consequently, the median must be an estimated value. Specifically, the estimate depends on locating the median class. Then, under the assumption that the items in the median class are evenly spaced, the median is determined using the following equation:

$$Md = L + i\left(\frac{n/2 - F}{f_{md}}\right) \tag{2.5}$$

where L = the lower boundary of the class in which the median is located

i = the class interval

n = the number of values of the variable in the distribution

F = the cumulated frequency of the classes preceding the median class

f_{md} = the frequency of the median class (the median class is the class where the cumulative frequency is equal to, or exceeds for the first time, $n/2$).

Table 2-5 demonstrates the median calculation using the per capita tax data. The median class is located by finding the class containing the median. This is determined by identifying the class containing the 25.5th observation. The first and second classes contain the first 22 observations, but this does not carry us to the 25.5th. Going to the third class, however, we see that this class contains the 23rd through the 36th observations. Here, then, is the median class.

TABLE 2-5. *Calculation of the Median of the Per Capita State and Local Tax Data*

CLASS INTERVAL	FREQUENCY
$450 and under $550	9
550 and under 650	13
650 and under 750	14
750 and under 850	9
850 and under 950	3
950 and under 1050	1
1050 and under 1150	1
	50

$n/2 = 50/2$
$F = 22$
$f_{md} = 14$
$L = 650$
$i = 100$

$$Md = L + i\left(\frac{n/2 - F}{f_{md}}\right)$$

$$= 650 + 100\left(\frac{50/2 - 22}{14}\right)$$

$$= 650 + 100\left(\frac{25 - 22}{14}\right)$$

$$= 650 + 100\left(\frac{3}{14}\right)$$

$$= 650 + 21.43$$

$$Md = \$671.43$$

Utilization of Equation 2.5 yields a median of $671.43. This differs from the true median of $669 that we found using the data in Table 2-4. The explanation of the difference must be that the assumption of evenly spaced observations in the median class does not hold exactly.

More About the Median

As an average, the median is not affected by extreme values of the variable, only by the position of values. This characteristic of the median provides its primary advantage. For instance, if Charlie Harris's sales are exceptionally high compared to those of other salesmen, it is probably best to use a median in computing the average sales of the salesmen. In that way, Charlie's sales will not influence the average because only those values of the variable that are near the positional middle of an ordered array influence the median.

The median is also used when you desire to know the positional middle. An illustration of this might be the average number of employees absent each day. In this case, a positional average would indicate that a certain number, or more, of employees were absent every day. Furthermore, when the median is subtracted from each of the values of the variable, the sum of the absolute difference is the minimum that can be obtained. That is, $(\Sigma |X - Md| = \text{minimum})$.[2]

Among the disadvantages of this measure of central tendency are:

1. It cannot be used with many statistical analytical procedures.
2. Like the arithmetic mean, the median is sometimes an artificial number. That is, *no value* in the set of observations might actually be the median.

The Mode

A third measure of central tendency is the *mode*. By definition the mode is the observed value that occurs most frequently. It is a measure of central tendency in that it locates the point where the variable values occur with the greatest density. For ungrouped data, the mode is determined by counting the frequency of each value and defining the mode as the value with the highest frequency of occurrence.

To illustrate, consider the following eight values: 2, 4, 5, 7, 7, 9,

---

2.   Notice that for the median the sum of the *absolute differences* is a minimum, and for the arithmetic mean the sum of the differences, *paying attention to the sign*, is 0.

10, 12. All values occur once except the 7 which occurs twice. Thus, 7 is the mode of these data.

In Example 2.1, a real world problem is presented that illustrates a situation in which the mode is a more meaningful measure of central tendency than either the arithmetic mean or the median.

▶ **Example 2.1**   An architect has been employed to design a series of model homes for a large housing development. One of the questions the architect must resolve is the size of the garage. Should it hold one, two, or more cars? After doing some research, he discovered the following averages with respect to the automobile ownership of the families that are in the potential market for houses of this type.

$$\overline{X} = 2.25$$
$$Md = 1.5$$
$$Mo = 2$$

Of these three averages, the mode is the one the architect decided to use. Why? First, it is rather obvious that the mean and median may not be actual numbers of cars owned by any family. But more important, the mode indicates the garage size that would meet the needs of the largest numbers of families. ◀

For ungrouped data, the mode is always equal to a value of one of the observations. This characteristic is not necessarily true for the mean or the median. Thus, at times the mode is the best of these three measures of central tendency.

But what about grouped data? As you probably have already noticed, no equation has been given to determine the mode from grouped data. This omission was intentional because the identities of the observations are lost with grouped data. Practically, if the data have been grouped, identification of the modal class is possible but not the mode.

Considering the data in Table 2-5, the modal class is the class with 14 observations; that is, the class with the most observations. If we tried to make any further calculations and determine a single value for the mode, it could be very misleading. Looking at the data in Table 2-4 helps explain why. If you recall, Table 2-4 is the arrayed data from which Table 2-5 was constructed. A close look at these data will reveal that there is no modal value. The values 728 and 549 occur twice, and the other 48 values occur only once. Thus, the data set does not have a single value that occurs more frequently than all others.

Finally, the mode is not influenced by extreme values, but rather by the frequency of occurrence. However, the mode can be located at one extreme of the data. Consequently, its use could result in misleading conclusions.

The Weighted Arithmetic Mean

nothing new

When calculating the arithmetic mean using Equation 2.1, each value of the variable was assumed to be of equal importance. In some situations, however, such an assumption might be misleading. In fact, if the values are not of the same importance, it is better to calculate a *weighted arithmetic mean* using Equation 2.6.

$$\overline{X}_w = \frac{\Sigma\,(WX)}{\Sigma\,W} \tag{2.6}$$

where $\overline{X}_w$ = weighted arithmetic mean
W = weight for each value
X = the individual values

To illustrate, let us consider calculating the semester grade point average for a college student taking a 16-hour load. This sixteen-hour load is composed of four three-hour courses and two two-hour courses. Now, let us assume that our student earns the grade of A in two three-hour courses and one two-hour course, the grade of B in one two-hour course, the grade of C in a three-hour course, and a grade of D in a three-hour course. What is his semester grade point average? As you already know, it would be incorrect to just assign $A = 4$, $B = 3$, $C = 2$, and $D = 1$ to the grades and divide by the number of courses, $(4 + 4 + 4 + 3 + 2 + 1)/6 = 3.000$, because each course does not carry the same hour credit. Consequently, the correct procedure is to calculate a weighted arithmetic mean as follows:

$$\overline{X}_w = \frac{\Sigma\,(WX)}{\Sigma\,W}$$

$$\overline{X}_w = \frac{3\,(4) + 3\,(4) + 2\,(4) + 2\,(3) + 3\,(2) + 3\,(1)}{3 + 3 + 2 + 2 + 3 + 3}$$

$$\overline{X}_w = 2.9375$$

Thus, the correct grade point average is 2.9375 rather than 3.000.

Equation 2.3 is also an equation for calculating a weighted arithmetic mean. If you examine it closely, you will see that in using Equation 2.3 to calculate the arithmetic mean of a frequency distribution, the result is simply a weighted arithmetic mean of the class midpoints. As a matter of fact, the arithmetic mean of a frequency array such as that shown in Table 1-12 can be found in the same way. Table 2-6 contains an illustration of how Equation 2.6 is used to determine the arithmetic mean of a frequency array.

TABLE 2-6. *Computing the Arithmetic Mean of a Frequency Array*

NUMBER OF BANKS (X)	NUMBER OF COUNTIES (W)	WX
1	9	9
2	22	44
3	12	36
4	16	64
5	5	25
6	3	18
7	5	35
8	1	8
9	1	9
10	0	0
11	1	11
	75	259

$$\overline{X}_w = \frac{\Sigma\,(WX)}{\Sigma W}$$

$$= \frac{259}{75}$$

$$\overline{X}_w = 3.453$$

The Geometric Mean

Another average is the *geometric mean* which is the n th root of the product of n numbers. It is calculated as follows:

$$G = \sqrt[n]{X_1 \cdot X_2 \cdot X_3 \cdot \ldots \cdot X_n} \tag{2.7}$$

Actually, this mean has a special use in business and economic problems, for frequently the situation arises where there is a need for an *average percentage change.*

▶ **Example 2.2**   To illustrate, let us consider the sales of the Mom's Red Hot shown in Table 2-7. In the recent annual report, the president claimed that the sales had shown an average annual increase of 106.37 percent per year. This percentage was determined as follows:

$$\overline{X} = \frac{\Sigma X}{n} = \frac{104 + 203.8 + 311.3}{3} = 206.37$$

$$206.37 - 100 = 106.37$$

The 100 was subtracted to convert to an annual percentage increase.

TABLE 2-7. *Sales of Mom's Red Hot*

YEAR	SALES ($1,000)	PERCENT OF PREVIOUS YEAR
1972	50	
1973	52	104
1974	106	203.8
1975	330	311.3

However, if the average increase of 106.37 percent is applied to the original sales figure, the resulting sales at the end of the four year period would be $439,450.23.[3] You should note that this is much larger than the actual sales figure of $330,000. So, what happened?

What happened is that we used the wrong measure of central tendency. We overlooked the effect of compounding. That is, we forgot that the increase also increases.   If we had used the geometric mean, the problem would have been solved as follows using Equation 2.7:

$$G = \sqrt[3]{104 \cdot 203.8 \cdot 311.3}$$
$$\log G = 1/3 \ (\log 104 + \log 203.8 + \log 311.3)$$
$$= 1/3 \ (2.01703 + 2.30920 + 2.49318)$$
$$= 1/3 \ (6.81941)$$
$$\log G = 2.27314$$
$$G = 187.6\%$$

And if this percentage is applied to the original sales data, the sales increase is as follows:

$$50 \times 1.876 = \$ \ 93.8$$
$$93.8 \times 1.876 = 175.97$$
$$175.97 \times 1.876 = 330.12 \cong \$330 \text{ thousand}$$

Thus, the sales averaged an annual increase of 87.6 percent. ◄

---

3.   ($50)(2.0637) = $103.1850;   ($103.1650)(2.0637) = $212.9428845;   (212.9428845)(2.0637) = $439.19474; ($439.45023)(1000) = $439,450.23.

───────────── **SUMMARY OF EQUATIONS** ─────────────

2.1 $\overline{X} = \dfrac{\Sigma X}{n}$,   *page 26*

2.2 $\mu = \dfrac{\Sigma X}{N}$,   *page 26*

2.3 $\overline{X} = \dfrac{\Sigma fM}{\Sigma f} = \dfrac{\Sigma fM}{n}$,   *page 27*

2.4 $\overline{X} = \overline{X}_a + \left(\dfrac{\Sigma fd}{n}\right) i$,   *page 29*

2.5 $Md = L + i \left(\dfrac{n/2 - F}{f_{md}}\right)$,   *page 32*

2.6 $\overline{X}_w = \dfrac{\Sigma (WX)}{\Sigma W}$,   *page 35*

2.7 $G = \sqrt[n]{X_1 \cdot X_2 \cdot X_3 \cdot \ldots \cdot X_n}$,   *page 36*

REVIEW QUESTIONS
---

1. Identify what each equation shown at the end of the chapter determines and when to use each one.
2. Define
   (a) arithmetic mean
   (b) median
   (c) mode
   (d) weighted arithmetic mean
   (e) geometric mean
3. Comment on the use of the geometric mean in business situations.
4. Distinguish between a statistic and a parameter.
5. In order to calculate the arithmetic mean from grouped data, it is necessary to estimate the sum of the individual observations. How is this accomplished?
6. Explain the significance of the assumption that the items in the median class are evenly spaced when calculating the median from grouped data.

PROBLEMS
---

1. Micro Motors test drove 8 of its Microstats in order to gather data for its advertising campaign. The data given below are the miles per gallon achieved by the 8 Microstats. Compute the arithmetic mean of these data for the Advertising Manager.
   24.3, 25.8, 29.2, 25.6, 27.2, 28.0, 24.9, 27.6

2. Using the data in Problem 1, compute the median miles per gallon for the 8 test cars. Based upon this median value, what statement can Micro Motors make about the cars?

3. Listed below are the number of overtime hours per worker last week at Rooster's Heaven Poultry Company. Determine the arithmetic mean, median, and mode of the number of overtime hours. What can you tell management about the average amount of overtime?

2	6	8	10	5
5	6	2	3	3
5	8	2	3	6
9	5	4	6	10
2	6	6	5	1
1	6	9	3	6

4. The Dean of Students wants to include a statement in the University catalog concerning the cost of off-campus housing. Since he cannot get this information from every student living off-campus, he decided to ask 25 students how much they spend and use their average expenditure in his statement. The following data are the monthly expenditures reported by this group of students. Determine the arithmetic mean, median, and mode of the expenditures for the Dean.

13	33	76	90	38
25	40	61	92	54
27	59	45	63	71
27	73	51	52	93
29	93	62	37	34

5. Mr. Rent owns seven houses, which are rented on a monthly basis. If he receives monthly rents of $125, $150, $275, $100, $110, $200, and $80, what is
   a. the mean monthly rent.
   b. the median monthly rent.
   c. the modal monthly rent.

6. Sun Dial Manufacturing Co. has determined that the average number of overtime hours per employee per year was 300 for the previous year. The company currently has 25 employees. Based on a 40-hour week and a 50-week operating year, does Sun Dial need any additional employees? If so, how many? What assumption did you make?

7. Fast Sam is a senior finance major at Quick Buck College. Sam is working his way through his last semester by selling Sanskrit dictionaries between classes. In order to stay in school, Sam must average 24 sales per day. His last 24 days' sales are listed below. Do you think he will make it? Find the mean, median, and mode to help you make up your mind.

34	50	32	42	14	26	33	34	11	14	5
38	5	39	7	13	26	26	23	1	37	2
37	47									

8. The Loosecane Yacht Company manufacturers a limited number of high-quality sailboats. The company's accountant, Jane Loosecane, has maintained accurate records for the cost of each sailboat. They are $475, $510, $490, $470, $475, $500, $505, and $995.
   a. Calculate the mean cost per sailboat.
   b. Is the mean the best measure of central tendency to use in this situation?

9. In the small community of Robinsonville, there are eleven families—Mr. Robinson, and his ten employees. The following is a list of the family incomes in the community last year: $5,000, $7,600, $5,500, $9,000, $7,000, $100,000, $7,800, $6,250, $10,000, $15,000, and $8,500.
   a. Determine the best measure of central tendency to represent the average yearly family income of Robinsonville.
   b. Calculate the value of this best measure.

10. The Fully-Ripe Fruit Market is considering wrapping its onions in packages. The management has decided to make the average number purchased by each customer the standard package size. Fifteen customers purchased the following number of onions: 2, 4, 3, 34, 10, 4, 5, 4, 4, 1, 2, 1, 3, 4, and 6. What should be the number of onions in each package at the Market?

11. The manager of a food store has been receiving complaints concerning the time customers must wait at the checkout counter. He feels that if he knew the arithmetic mean of the waiting times, he could then decide if he needs to expand his capacity. From the following data, calculate the arithmetic mean.

WAITING TIME (MINUTES)	FREQUENCY
0 but less than 5	850
5 but less than 10	620
10 but less than 15	510
15 but less than 20	30
20 but less than 25	10

$n = 2020$

12. In an academic grading system where A = 4, B = 3, C = 2, D = 1, and F = 0, a group of college juniors have grade point averages as indicated by the following frequency distributions.

GRADE POINT AVERAGE	NUMBER OF STUDENTS
1.0 and under 1.5	1
1.5 and under 2.0	2
2.0 and under 2.5	10
2.5 and under 3.0	9
3.0 and under 3.5	5
3.5 and under 4.0	3

$n = 30$

a. Calculate the arithmetic mean using Equation 2.3.
b. Calculate the arithmetic mean using Equation 2.4.
c. Compare your answers in parts (a) and (b).
d. Calculate the median using Equation 2.5.

13. A sample of 25 charge customers of Deferred Payments Unlimited revealed the following frequency distribution of unpaid balances.

UNPAID BALANCE	FREQUENCY
$30 and under 40	3
40 and under 50	8
50 and under 60	7
60 and under 70	4
70 and under 80	3

$n = 25$

a. Determine the mean unpaid balance.
b. Determine the median unpaid balance.

14. At the annual convention of Life Insurance, Inc., the following distribution of whole life insurance policies was presented:

DOLLAR VALUE OF POLICY	NUMBER SOLD
$ 0 and under 10,000	15
10,000 and under 20,000	18
20,000 and under 30,000	30
30,000 and under 40,000	10
40,000 and under 50,000	7
50,000 and over*	20

$n = 100$

*The average value of each policy in this group is $90,000.

a. Estimate the total dollar value of whole life policies sold.
b. Calculate the mean value of each policy.
c. Calculate the median value of the policies.

15. Childrens Books, Ltd., has records showing the following sales:

BOOK PRICE	SALES (THOUSANDS)
$ 1.00 and under 3.00	100
3.00 and under 5.00	150
5.00 and under 7.00	200
7.00 and under 9.00	250
9.00 and under 11.00	300
11.00 and under 13.00	200

a. Determine the mean book price.
b. Determine the median book price.
c. What can you say about the modal book price?

16. The following is a tabulation of the number of drownings, by age of the victim, in hotel swimming pools in 1977.

AGE OF VICTIM	NUMBER OF DEATHS
Less than 5	16
5 but less than 10	9
10 but less than 15	11
15 but less than 20	7
20 but less than 25	5
25 but less than 30	8
30 but less than 35	4

    a. Determine the median age of the victims.

    b. Compute the arithmetic mean of the ages.

17. Mr. and Mrs. Savings have four savings accounts. In account A, they have $4,000; in account B, they have $6,000; in account C, they have $3,000; and in account D, they have $10,000. If account A earns 5½ percent per year, account B earns 4 percent per year, account C earns 6 percent per year, and accound D earns 7 percent per year, what is the average annual percent earnings of the Savings?

18. During the past five years, State University has recorded the enrollments given below:

YEAR	ENROLLMENT
1974	20,000
1975	22,500
1976	24,600
1977	27,500
1978	29,000

    a. What has been the average annual rate of increase in enrollment?

    b. If the percentage increase had been maintained exactly each year, what would have been the enrollments for each year.

19. Hazel bought 10 pounds of sugar for 40¢ per pound, 5 pounds of potatoes for 15¢ per pound, 3 pounds of tomatoes for 60¢ per pound, and 8 pounds of bananas for 30¢ per pound. What is the average price per pound of her purchases?

20. Mr. Logan has a large apartment complex which has one-, two-, and three-bedroom apartments. The one-bedroom rents for $135 per month; the two bedroom rents for $205 per month; and the three bedroom rents for $265 per month. If there are 50 one-bedroom apartments, 40 two-bedroom apartments, and 10 three-bedroom apartments, what is the mean monthly rental of the apartments?

21. Ms. Investor began an investment program with $1,000 in 1973. Sometimes she invested in stocks, and sometimes in bonds. From the data below, determine the average annual rate of return.

YEAR	AMOUNT
1973	$1000
1974	1052
1975	1114
1976	1129
1977	1152
1978	1194

22. The faculty retirement fund has purchased 10 shares of Red Inc. at $50 per share, 25 shares of Blue Co. at $15 per share, and 20 shares of Green Ltd. at $40 per share. What is the average price per share of stock?

Measures of Dispersion, Skewness, and Kurtosis

CHAPTER
3

In the previous chapter, measures of central tendency were discussed in an effort to familiarize you with averages used to describe the "middle" of the data. But central tendency doesn't tell the whole story. In describing and analyzing data we must also consider its dispersion, skewness, and kurtosis. *Dispersion* relates to the *spread* or *variability* of the data. *Skewness* refers to the *symmetry* of the data, and *kurtosis* describes its *peakedness*. Therefore, this chapter is concerned with measuring the extent to which the values of a set of observations are equal and how they are spread about a central value. To illustrate dispersion, consider the sets of data A, B, and C shown in Table 3-1. Each set has the same arithmetic mean (50) and median (50), but they are not alike in dispersion or variability.

TABLE 3-1. *Three Data Sets with Different Amounts of Dispersion*

A	B	C
50	25	0
50	25	0
50	50	0
50	50	100
50	75	100
50	75	100

In set A the values of the six items are identical. There is no dispersion; they are perfectly homogeneous. In set B the items vary over three values; and in set C the items vary over two values, but there is greater spread in the third set. No item in set C is close to the mean or the median, so you can see that dispersion (spread) is an important characteristic of data description.

Our discussion of dispersion will include the following measures: the *range,* the *quartiles,* the *standard deviation,* and the *coefficient of variation.* First, the range and the quartiles will be discussed.

The Range and Quartiles

The *range* of a set of data, as you probably already realize, is the difference between the highest and lowest values in the data set. In Table 3-1, the range of set A is 0; the range of set B is 50; and the range of set C is 100. Thus, the range is a very easy measure of dispersion to compute. It has the disadvantage, however, that it can be influenced by extreme values, thereby providing misleading information about the true spread of the majority of the data.

Consequently, the spread of data is sometimes expressed in terms of quartiles so that segments of the data can be observed. In Figure 3-1 the lower or first quartile is the value Q_1 below which 25 percent of data can be found and above which 75 percent of the data lie. The second quartile (Q_2 and also the median) is the value above which and below which can be found 50 percent of the data. And finally, the third quartile is the value Q_3 above which 25 percent of the data lie and below which 75 percent of the data occur.

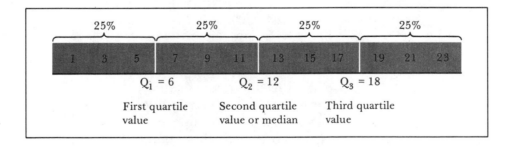

FIGURE 3-1. *Presentation of Quartiles*

When using ungrouped data, the quartile positions can be determined using the following equations:

$$Q_1 \text{ position} = \frac{n+2}{4} \qquad (3.1)$$

$$Q_2 \text{ position} = \frac{n+1}{2} \qquad (3.2)$$

$$Q_3 \text{ position} = \frac{3n+2}{4} \qquad (3.3)$$

Remember, though, that these equations identify the location and not the value of the quartile. Thus, in Figure 3-1, the first quartile position is

$$Q_1 \text{ position} = \frac{12+2}{4} = \frac{14}{4} = 3.5$$

Numerically, Q_1 is the average of the third and fourth values or $(5 + 7)/2 = 6 = Q_1$.

When using grouped data, the quartile *values* are determined using the following equations (Table 3-2):[1]

$$Q_1 = L + \left(\frac{n/4 - F}{f_{Q_1}} \right) i \qquad (3.4)$$

$$Q_2 = L + \left(\frac{n/2 - F}{f_{Q_2}} \right) i \qquad (3.5)$$

$$Q_3 = L + \left(\frac{3n/4 - F}{f_{Q_3}} \right) i \qquad (3.6)$$

where L = the lower boundary of the class in which the quartile is located
n = number of observations
F = the cumulative frequencies of the classes preceding the quartile class
f_Q = the frequency of the quartile class
i = the class interval

Use of either the range or quartiles results in representing dispersion in terms of a few selected values. Thus, this type of dispersion measure fails to make use of most of the values of the observations in the data set. To overcome this, another type of dispersion measure is available. This second type deals with measuring dispersion from one of the measures of central tendency. There are several of this second type of dispersion measure, but we will investigate only one, the *standard deviation*.

---

1.   Notice that Equation 3.5 is the same as the equation to determine the median.

TABLE 3-2. *Finding Quartile Values of Grouped Data Per Capita State and Local Taxes*

CLASS INTERVAL	FREQUENCY
$450 and under $550	9
550 and under 650	13
650 and under 750	14
750 and under 850	9
850 and under 950	3
950 and under 1050	1
1050 and under 1150	1

$$Q_1 = L + \left(\frac{n/4 - F}{f_{Q_1}}\right)i \qquad Q_2 = L + \left(\frac{n/2 - F}{f_{Q_2}}\right)i \qquad Q_3 = L + \left(\frac{3n/4 - F}{f_{Q_3}}\right)i$$

$$n/4 = 50/4 = 12.5 \qquad n/2 = 50/2 = 25 \qquad \frac{3n}{4} = \frac{3(50)}{4} = 37.5$$

$$Q_1 = 550 + \left(\frac{12.5 - 9}{13}\right)100 \qquad Q_2 = 650 + \left(\frac{25 - 22}{14}\right)100$$

$$Q_3 = 750 + \left(\frac{37.5 - 36}{9}\right)100$$

$$= 550 + \left(\frac{3.5}{13}\right)100 \qquad = 650 + (3/14)\,100$$

$$= 650 + (0.2143)\,100 \qquad = 750 + \left(\frac{1.5}{9}\right)100$$

$$= 550 + (0.2692)\,100$$

$$= 650 + 21.43 \qquad = 750 + (0.1667)\,100$$

$$= 550 + 26.92$$

$$Q_2 = 671.43 \qquad = 750 + 16.67$$

$$Q_1 = 576.92$$

$$Q_3 = 766.67$$

The Standard Deviation

In order to measure dispersion from some measure of central tendency, it is, of course, necessary to choose one of the averages. To compute the standard deviation, the average used is the arithmetic mean. If you recall our earlier discussion, one property of the arithmetic mean is that $\Sigma (X - \overline{X})^2$ is the minimum obtainable from summing the squared differences about any of the averages. And if we are dealing with a population, $\Sigma (X - \mu)^2$ is the minimum also. Now if we take this minimum sum of squared differences, divide it by the number of observations, and then take the square root, we have calculated the standard deviation.

For a population the standard deviation σ (pronounced sigma) is

$$\sigma = \sqrt{\frac{\Sigma (X - \mu)^2}{N}} \tag{3.7}$$

where μ = the population mean
          N = the number of items in the population
          X = the individual values

If you should happen to be working with sample data, and you want the sample standard deviation to provide a good estimate of the population standard deviation, use Equation 3.8.

$$s = \sqrt{\frac{\Sigma (X - \overline{X})^2}{n - 1}} \qquad (3.8)$$

There are two differences in Equation 3.7 and 3.8. First, $\overline{X}$ is used rather than μ. Second, the denominator in the equation for s is $n - 1$, while for σ it is N. This second difference looks like a small one, but it warrants an additional explanation.

The discussion of the arithmetic mean in Chapter 2 mentions that the sum of the variations about the mean is 0 or $\Sigma (X - \overline{X}) = 0$. With this in mind, suppose you have five numbers, 2, 4, 6, 8, and 10. Their arithmetic mean is 6. If you select any four of them, the value of the fifth number is fixed. In other words it is not free to vary. For example, if you select the first four numbers (2, 4, 6, and 8), the fifth number *must* be 10 if $\Sigma (X - \overline{X}) = 0$.

$$(2 - 6) + (4 - 6) + (6 - 6) + (8 - 6) + (X - 6) = 0$$
$$-4 \quad - \quad 2 \quad + \quad 0 \quad + \quad 2 \quad + (X - 6) = 0$$
$$-4 \quad + (X - 6) = 0$$
$$X - 10 = 0$$
$$X = 10$$

In computing s from a sample containing n values of the variable X, you have only $n - 1$ *degrees of freedom*. Put in a more general form, *degrees of freedom is the number of variables that can vary freely in a given set of variables.*

As you can see from Equations 3.7 and 3.8, the standard deviation is the square root of the mean of the squared deviations. Thus, it is sometimes referred to as the root mean square deviation. If we remove the radical, that is, leave the results as σ^2 or s^2, we have what is known as the *variance*. The *variance* is the *standard deviation squared*, or the *standard deviation* is the *square root* of the *variance*.

The Variance and Standard Deviation: Calculation from Ungrouped Data

Shown in Table 3-3 are the necessary computations for the determination of s. Notice that the arithmetic mean $\overline{X}$ is first determined to be

$500, and then $500 is subtracted from each sales observation X to find the individual deviations from the mean $X - \overline{X}$ shown in the second column. These deviations are then squared and summed and placed in Equation 3.8, which yields $s = \$63.25$.

TABLE 3-3. *Computation of the Standard Deviation from Ungrouped Data (Long Method)*

DAILY SALES FOR MARKET'S MARKET

SALES (X)	($X - \overline{X}$)	($X - \overline{X}$)²
$ 525	+ 25	625
475	− 25	625
425	− 75	5,625
450	− 50	2,500
525	+ 25	625
600	+100	10,000
$3,000	0	20,000

$$\overline{X} = \frac{\Sigma X}{n} \qquad\qquad s = \sqrt{\frac{\Sigma (X - \overline{X})^2}{n - 1}}$$

$$\overline{X} = \frac{\$3000}{6} \qquad\qquad s = \sqrt{\frac{20000}{6 - 1}}$$

$$\overline{X} = \$500 \qquad\qquad s = \sqrt{4000}$$

$$s = \$63.25$$

Consistent with the definition of the arithmetic mean, the sum of these variations is 0. These deviations are then squared and summed. Next, the sum of the squared deviations is divided by the degrees of freedom $(n - 1)$ to determine the variance (4000). The square root of the variance is $63.25 and is the sample standard deviation s. Notice that the variance is in terms of "dollars squared" or, in other words, it is not in the same units as the original data. The standard deviation, however, is in the same units as the original data and is thus a more practical measure of variation than is the variance. *— for descriptive purposes.*

Standard Deviation:
Shortcut Method for Ungrouped Data

As you studied Table 3-3, it probably occurred to you that if there were 100 sales figures rather than six, then a much greater computational effort would be necessary to calculate the standard deviation. Your observation would be correct, and you would have identified a need for a shortcut procedure to use on ungrouped data. Equation 3.9 is such a shortcut method for entire populations, and Equation 3.10 is the shortcut method when using sample data.[2]

$$\sigma = \sqrt{\frac{\Sigma X^2}{N} - \left(\frac{\Sigma X}{N}\right)^2}$$ (3.9)

$$s = \sqrt{\frac{\Sigma X^2 - \frac{(\Sigma X)^2}{n}}{n - 1}}$$ (3.10)

To illustrate how much easier it is to use Equation 3.10 rather than Equation 3.8, take a look at Table 3.4. Now compare the work involved in Table 3-4 against the effort necessary in Table 3-3. Don't you agree that the shortcut method is easier?

TABLE 3-4. *Computation of the Standard Deviation from Ungrouped Data (Short Method)*

MARKET'S MARKET

SALES (X)	X^2
$ 525	275,625
475	225,625
425	180,625
450	202,500
525	275,625
600	360,000
$3,000	1,520,000

$$s = \sqrt{\frac{\Sigma X^2 - \frac{(\Sigma X)^2}{n}}{n - 1}}$$

$$s = \sqrt{\frac{1,520,000 - \frac{(3000)^2}{6}}{6 - 1}}$$

$$s = \sqrt{\frac{1,520,000 - 1,500,000}{5}}$$

$$s = \sqrt{\frac{20000}{5}}$$

$$s = \sqrt{4000}$$

$$s = \$63.25$$

---

2.   Both of these so-called shortcut equations are really mathematical derivations of the so-called long-method equations. Therefore, the solutions from Equations 3.7 and 3.9 are equal. The same is true of Equations 3.8 and 3.10.

Standard Deviation: Grouped Data

At times, as discussed earlier, you will find it necessary to work with grouped data. Also, as before, since the identities of the individual observations are no longer known, it will be necessary to make use of the midpoint as the representative of the observations in each class.

The Long Method

Suppose we have data that have already been grouped as shown in Table 3-5. If we must begin with grouped data, Equations 3.11 and 3.12 can be used to determine the standard deviation.

$$s = \sqrt{\frac{\Sigma f (M - \overline{X})^2}{n - 1}} \qquad (3.11)$$

$$\sigma = \sqrt{\frac{\Sigma f (M - \mu)^2}{N}} \qquad (3.12)$$

where M = midpoint of each class
f = frequency of each class
n = total number of observations in the sample
N = total number of observations in the population
$\overline{X}$ = arithmetic mean of the sample
μ = arithmetic mean of the population

Table 3-5 presents the computations necessary to implement Equation 3.11.

Equations 3.11 and 3.12 require that the arithmetic mean be calculated first and that the difference, $M - \overline{X}$ or $M - \mu$, be squared. As you worked through the illustration in Table 3-5, you probably wished for a shorter method. Fortunately, again, there is a short method.

The Shortcut Method

A shortcut approach for calculation of the standard deviation from grouped data follows the same concept as the shortcut approach for the arithmetic mean that was demonstrated in Table 2-3. That is, deviations from the assumed mean are expressed in terms of class interval units. The equation is shown below.

$$s = i \sqrt{\dfrac{\Sigma f\,(d)^2 - \dfrac{(\Sigma fd)^2}{n}}{n-1}} \qquad (3.13)$$

Table 3-6 illustrates the implementation of Equation 3.13 using the Big Pickle Company data. As a refresher, the arithmetic mean was calculated using Equation 2.4. Then the standard deviation was calculated using Equation 3.13. Notice that only one additional column, the last one, was needed to determine the standard deviation. A comparison of Tables 3-5 and 3-6 shows that the values of $\overline{X}$ and s are the same using either set of equations, but that less effort is entailed in Table 3-6.

TABLE 3-5. *Computation of the Standard Deviation from Grouped Data (Long Method)*

BIG PICKLE COMPANY MONTHLY PICKLE PRODUCTION

CASES	MIDPOINT M	FREQUENCY f	fM	$(M - \overline{X})$	$(M - \overline{X})^2$	$f(M - \overline{X})^2$
95 and under 105	100	4	400	−26	676	2,704
105 and under 115	110	14	1540	−16	256	3,584
115 and under 125	120	18	2160	− 6	36	648
125 and under 135	130	15	1950	+ 4	16	240
135 and under 145	140	10	1400	+14	196	1,960
145 and under 155	150	7	1050	+24	576	4,032
155 and under 165	160	2	320	+34	1156	2,312
		70	8820			15,480

$$\overline{X} = \frac{\Sigma fM}{\Sigma f}$$

$$\overline{X} = \frac{8820}{70}$$

$$\overline{X} = 126 \text{ cases/month}$$

$$s = \sqrt{\frac{\Sigma f(M - \overline{X})^2}{n - 1}}$$

$$s = \sqrt{\frac{15480}{70 - 1}}$$

$$s = \sqrt{224}$$

$$s = 14.97 \text{ cases/month}$$

TABLE 3-6. *Computation of the Standard Deviation from Grouped Data (Shortcut Approach)*

BIG PICKLE COMPANY MONTHLY PICKLE PRODUCTION

CASES	CLASS MIDPOINT (M)	f	d	fd	$f(d)^2$
95 and under 105	100	4	−3	−12	36
105 and under 115	110	14	−2	−28	56
115 and under 125	120	18	−1	−18	18
125 and under 135	130	15	0	0	0
135 and under 145	140	10	1	10	10
145 and under 155	150	7	2	14	28
155 and under 165	160	2	3	6	18
		70		−28	166

$$\overline{X} = \overline{X}_a + \left(\frac{\Sigma fd}{n}\right) i$$

$$= 130 + \left(\frac{-28}{70}\right) 10$$

$$= 130 - 4$$

$$\overline{X} = 126 \text{ cases/month}$$

$$s = i \sqrt{\frac{\Sigma f(d)^2 - \frac{(\Sigma fd)^2}{n}}{n-1}}$$

$$s = 10 \sqrt{\frac{166 - \frac{(-28)^2}{70}}{69}}$$

$$= 10 \sqrt{\frac{166 - \frac{784}{70}}{69}}$$

$$= 10 \sqrt{\frac{166 - 11.2}{69}}$$

$$= 10 \sqrt{\frac{154.8}{69}}$$

$$= 10 \sqrt{2.24}$$

$$= 10 \, (1.497)$$

$$s = 14.97 \text{ cases/month}$$

Coefficient of Variation

Up until now, two types of measures of dispersion have been discussed. The first type was of the nature of the range and quartiles where only a few observations are used, and the second type was of the nature of the standard deviation (or variance) where all observations are used to measure dispersion about the arithmetic mean. The third and last type of measure of dispersion we will discuss shows *relative dispersion;* the particular one we are concerned with is called the *coefficient of variation.*

An *absolute measure* of dispersion (for example, the standard deviation) is converted to a relative measure (the coefficient of variation) by stating the absolute measure as a percent of a measure of central tendency, usually the arithmetic mean. Equations 3.14 and 3.15 are used to calculate the coefficient of variation, which is the ratio of the standard deviation to the mean.

$$V = \left(\frac{s}{\overline{X}}\right) 100 \tag{3.14}$$

$$V = \left(\frac{\sigma}{\mu}\right) 100 \tag{3.15}$$

To illustrate the use of the coefficient of variation, consider that the results calculated in Table 3-6 represent the first 70 months of operation of the Big Pickle Company, and that an additional 12 months of operation have yielded production data of $\overline{X} = 150$ and $s = 16$. Calculation of the coefficient of variation for each period is shown in Table 3-7.

TABLE 3-7. *Calculation of Coefficient of Variation*

PERIOD	$\overline{X}$	s	$V = \left(\dfrac{s}{\overline{X}}\right) 100$
70 months	126	14.97	11.9
12 months	150	16	10.7

As you can see, the 12-month period had the highest average and the lowest coefficient of variation. This would be good news to the production manager who would know that output had been increased, and that the *relative variation* had actually decreased. Thus, even though the standard deviation increased from 14.97 cases to 16.0 cases, relative to the change in average output, this increase would not be considered very large. In fact, the later period actually has a relatively more homogeneous output.

Finally, since the coefficient of variation is expressed as a percentage, it can be used to compare measures of dispersion expressed in different units. This, in fact, is one of the important characteristics of any measure of relative dispersion.

Skewness and the Relationship of the Mean, Median, and Mode

Skewness refers to the shape of a frequency distribution; specifically, skewness can be related to symmetry. If a frequency distribution is bell shaped as shown in Figure 3-2 it is said to be symmetrical.   That is, 50 percent of the values lie on each side of the mode.   Figure 3-2 is also a *unimodal* distribution. That is, it has a single mode.

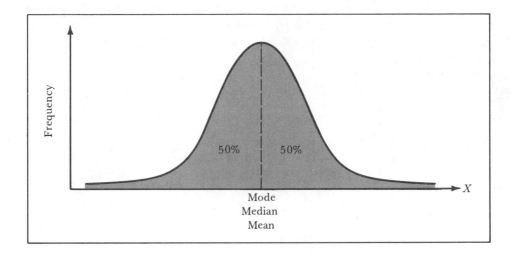

FIGURE 3-2.   *Symmetrical Frequency Distribution*

Figure 3-2 demonstrates another property of the symmetrical distribution.   In addition to being unimodal, the median and the mean are equal to the mode.   Now turning to Figure 3-3, we see the relationship of the mean, median, and mode for *skewed* distributions. If the distribution is skewed right, the value of the mode is less than the median, which is less than the mean.   But if the distribution is skewed left, the value of the mode is greater than the median, which is greater than the mean.   In both situations, the median value lies between the value of the mean and the mode.   Also the mean is always

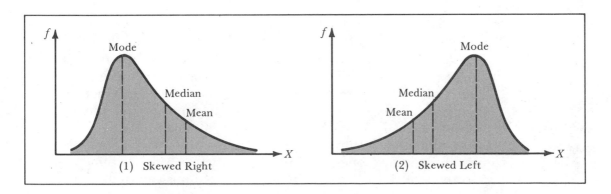

FIGURE 3-3.   *Skewed Frequency Distributions*

in the direction of the skewness relative to the mode. That is, if the data are skewed right, the extreme values are to the right, so the mean is to the right of, or greater than, the mode. The converse is true for data that are skewed left.

There are several methods available for measuring skewness. The one presented here, in Equation 3.16, is known as the *Pearsonian coefficient of skewness.*

$$Sk = \frac{3\,(\overline{X} - Md)}{s} \tag{3.16}$$

where Sk = Pearsonian coefficient of skewness
           $\overline{X}$ = arithmetic mean
           Md = median
           s = standard deviation

You can find the values of the arithmetic mean and the standard deviation for the Big Pickle Company data in Table 3-5. The median for those data is 114.44 cases per month. Thus, using Equation 3.16, you can calculate the Pearsonian coefficient of skewness as follows:

$$Sk = \frac{3\,(126 - 124.44)}{14.97}$$

$$= \frac{3\,(1.56)}{14.97}$$

$$= \frac{4.68}{14.97}$$

$$Sk = +0.31$$

But what does $Sk = +0.31$ mean? First of all, the plus (+) indicates the data are *positively,* or right, *skewed* similar to the top distribution in Figure 3-3. A negative sign (−) would indicate a left, or *negatively, skewed* distribution. The Pearsonian coefficient of skewness ranges from −3 to +3, with $Sk = 0$ meaning a perfectly symmetrical distribution. Thus, when $Sk = +0.31$ we can conclude that the distribution has a relatively small amount of positive skewness.

Equation 3.16 is based on the observation that the median will lie approximately one-third of the way between the arithmetic mean and the mode as is the case in both distributions shown in Figure 3-3. Since this is not always true, you should refer to a more advanced text if you need a more precise measure of skewness.

Kurtosis

[handwritten in margin: Cannot be determined graphically]

The extent to which the data contained in a distribution tend to be concentrated at its midpoint is referred to as the *kurtosis* of the distribution. Thus, kurtosis is another term for *peakedness*.

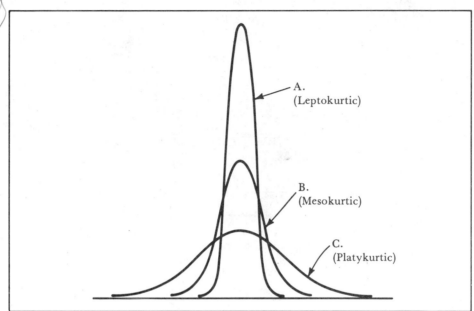

FIGURE 3-4. *Three Distributions with Differing Amounts of Kurtosis*

In Figure 3-4, three distributions are shown, each with different amounts of kurtosis. The curve labeled A is called *leptokurtic* and represents a distribution with values concentrated around its midpoint. The second, curve B, is *mesokurtic*. It has intermediate kurtosis but less concentration at the midpoint than does curve A. Finally, curve C is relatively flat and is called *platykurtic*. In other words, the values in the distribution are relatively uniformly scattered so that there is little peakedness.

─────────────── **SUMMARY OF EQUATIONS** ───────────────

3.1 Q_1 position $= \dfrac{n + 2}{4}$,   *page 46*

3.2 Q_2 position $= \dfrac{n + 1}{2}$,   *page 46*

3.3 Q_3 position $= \dfrac{3n + 2}{4}$,   *page 46*

3.4 $Q_1 = L + \left(\dfrac{n/4 - F}{f_{Q_1}}\right) i$,   *page 46*

3.5 $Q_2 = L + \left(\dfrac{n/2 - F}{f_{Q_2}}\right) i$,   *page 46*

3.6 $Q_3 = L + \left(\dfrac{3n/4 - F}{f_{Q_3}}\right) i$,   *page 46*

3.7 $\sigma = \sqrt{\dfrac{\Sigma(X - \mu)^2}{N}}$,   *page 47*

3.8 $s = \sqrt{\dfrac{\Sigma (X - \overline{X})^2}{n - 1}}$ *page 48*

3.9 $\sigma = \sqrt{\dfrac{\Sigma X^2}{N} - \left(\dfrac{\Sigma X}{N}\right)^2}$,   *page 50*

3.10 $s = \sqrt{\dfrac{\Sigma X^2 - \dfrac{(\Sigma X)^2}{n}}{n - 1}}$,   *page 50*

3.11 $s = \sqrt{\dfrac{\Sigma f (M - \overline{X})^2}{n - 1}}$,   *page 51*

3.12 $\sigma = \sqrt{\dfrac{\Sigma f (M - \mu)^2}{N}}$,   *page 51*

3.13 $s = i \sqrt{\dfrac{\Sigma f (d)^2 - \dfrac{(\Sigma fd)^2}{n}}{n - 1}}$,   *page 52*

3.14 $V = \left(\dfrac{s}{\overline{X}}\right) 100$,   *page 54*

3.15 $V = \left(\dfrac{\sigma}{\mu}\right) 100$,   *page 54*

3.16 $Sk = \dfrac{3 (\overline{X} - Md)}{s}$ *page 56*

REVIEW QUESTIONS

1. Identify what each equation shown at the end of the chapter calcuiates and when to use each one.

2. Define
   (a) range
   (b) first quartile
   (c) second quartile or median
   (d) third quartile
   (e) standard deviation
   (f) variance
3. Discuss the use of the coefficient of variation.

─────────────────── **PROBLEMS** ───────────────────

1. A sample of ten workers at the Ace plant in Walnut Ridge was asked how far they drove to work each day. The following represents their responses (in miles):
   25, 8, 1, 2, 3, 7, 4, 3, 9, 5
   a. What is the range for the number of miles driven?
   b. Compute both Q_1 and Q_3. In this instance, what do these values tell the management of the Ace plant?
   c. Compute the standard deviation for the above data.

2. The following values represent the average number of Medicaid patients seen by a group of doctors each day.
   15, 8, 11, 9, 13, 16, 19, 20, 6, 38
   a. Determine the range.
   b. Determine the first quartile value.
   c. Determine the second quartile value. What measure of central tendency is this value?
   d. Determine the third quartile value.

3. Use the data in Problem 2 and determine the following:
   a. The arithmetic mean
   b. The standard deviation using Equation 3.8. $(x - \bar{x})^2 / n - 1$
   c. The standard deviation using Equation 3.10. $\left(\Sigma x^2 - \frac{(\Sigma x)^2}{n} \right) (n-1))$
   d. Compare your answers in parts (b) and (c).
   e. The variance.

   Treat as sample.

4. At the end of its first year of operations, a new credit union has 60 accounts as given below.

5	30	51	72	95	125
5	35	55	75	100	125
7	37	56	75	100	126
12	38	56	80	100	127
15	39	57	80	100	128
15	40	58	80	101	129
20	50	58	81	105	130
25	50	60	82	108	130
26	50	60	82	109	130
28	50	63	83	109	130

   a. Compute the arithmetic mean.
   b. Determine the median.
   c. Does this data set have a mode?
   d. Compute Q_1 and Q_3.
   e. What is the range?
   f. Compute the standard deviation.
   g. Which equation should be used in part (f)?   Why?

5. Construct a frequency distribution using the credit union account data in Problem 4.   Using your frequency distribution
   a. Compute the arithmetic mean.
   b. Determine the median.
   c. Compute the standard deviation.
   d. Determine the variance of the data.

6. Use the data from Problem 3, Chapter 2 and determine the following:
   a. Range.
   b. Standard deviation using Equation 3.7.
   c. Standard deviation using Equation 3.9.
   d. Variance.
   e. First quartile value.
   f. Second quartile value.
   g. Third quartile value.
   h. Coefficient of variation.

7. Use the data in Problem 4, Chapter 2 and determine
   a. the arithmetic mean.
   b. the standard deviation.
   c. the variance.

8. For the following ungrouped data, determine the mean, the standard deviation, and the coefficient of variation:

   $$5, 8, 11, 14, 19, 22$$

9. The Company has three plants.   Certain information about these three plants is given below:

PLANT	NUMBER OF EMPLOYEES	MEAN ANNUAL SALARY	STANDARD DEVIATION
No. 1	100	$ 8,300	$ 750
No. 2	180	11,200	1200
No. 3	250	9,000	900

   a. Which plant has the largest payroll?
   b. Which plant appears to have the most homogeneous salaries?
   c. If the range can be approximated by the expression, $\mu \pm 3\sigma$, which plant has the widest range of salaries?
   d. Approximate the lowest and highest salary of The Company.

10. Use the data in Problem 1, and determine the coefficient of variation. What does your calculated value imply?

11.

CLASSES	FREQUENCY
5 and under 15	6
15 and under 25	10
25 and under 35	24
35 and under 45	12
45 and under 55	8
	60

a. Determine the arithmetic mean using Equation 2.4. 2,3 $\frac{\Sigma f m}{n}$
b. Determine the standard deviation using Equation 3.11
c. Determine the standard deviation using Equation 3.13.
d. Compare your answers from parts (b) and (c).
e. Determine the variance.
f. Determine the first quartile value.
g. Determine the second quartile value.
h. Determine the third quartile value.
i. Determine the coefficient of variation.
j. Compute the Pearsonian coefficient of skewness.

12. Determine the standard deviation for the data of Problem 12, Chapter 2.
a. Use Equation 3.11.
b. Use Equation 3.13.
c. Comment on the use of these two equations.

13. At a major state university, assistant professors have an arithmetic mean salary of $15,000 with a standard deviation of $5,000, while professors have an arithmetic mean salary of $40,000 with a standard deviation of $10,000. Which group has the greater relative dispersion? What is it?

14. Two organizations, A and B, have the same mean salary of $10,000. Organization A has a standard deviation of $500, and Organization B has a standard deviation of $200.
a. From which organization would you probably get the best starting salary? B - less variation
b. In 10 years, from which organization would you probably get the best salary? A - greater dispersion

15. Two growers of grapefruit have determined the following statistics for their current crops.

GROWER M *more uniform*	GROWER N *larger*
$\overline{X} = 15$ ounces	$\overline{X} = 14$ ounces
$s = 1$ ounce	$s = 2$ ounces

drop

a. Which grower has grapefruit that are more uniform in weight?
b. Which grower probably has the larger grapefruit? uses 3σ

16. An organization has the following distribution of ages for its current employees:

AGES	NUMBER OF EMPLOYEES
18 and under 28	2
28 and under 38	10
38 and under 48	30
48 and under 58	25
58 and under 68	17

a. Determine the first quartile value.
b. Determine the third quartile value.
c. Comment on the age situation in this organization.
d. Compute the Pearsonian coefficient of skewness.

17. An automobile manufacturer is considering two brands of batteries, I and II, as standard equipment on its 1980 model. Information on the two batteries is given below:

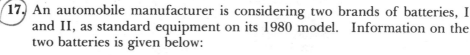

BATTERY I	BATTERY II
$\overline{X}$ = 40 months	$\overline{X}$ = 36 months
s = 4 months	s = 2 months

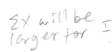
ΣX will be larger for I

a. If the decision criterion is to maximize total months, which battery should be chosen? *Battery I*
b. Which battery should be chosen if uniformity of service on each car is the decision criterion? *II $V_1 = 10\%$, $V_{II} = 5.6\%$*

18. Use the data from Problem 7, Chapter 2 to determine the following: *(drop)*
a. The standard deviation.
b. The first quartile.
c. The third quartile.

19. A recent study in Unknown Country revealed the following statistics on per capita cigarette consumption:

1975	1976
$\overline{X}$ = 0.45 packages	$\overline{X}$ + 0.40 packages
s = 0.10 packages	s = 0.05 packages

Comment on the results. *minimum is higher*

20. For the 50 states, the following is a frequency distribution of the labor force participation rates of their populations.

PARTIPATION RATE (CLASS MIDPOINT)	FREQUENCY
0.70	2
0.71	8
0.72	15
0.73	10
0.74	8
0.75	5
0.76	2
	50

a. Determine the variance of the participation rates.
b. Compute the Pearsonian coefficient of skewness.
c. Are the data positively or negatively skewed?

CASE STUDY

Main State University

For the past 100 years, Main State has had an open admissions policy for all students graduating from an accredited high school in the United States. However, members of the Faculty Senate of the University have recently become concerned about the policy and have appointed a committee to investigate the advisability, advantages, and disadvantages of revoking it. The committee is to report back to the Faculty Senate with its findings and recommendations.

Members of the committee requested from the Office of Institutional Research summary information on ACT scores and other data for incoming freshmen classes for the past three years. Specifically, the following information was requested and received by the committee:

1. Distribution of ACT score frequencies for all incoming freshmen, women and men, for Fall semester, 1976-78.
2. The grade point averages of freshmen (based on maximum 4.0), for men and women, for Fall semester, 1976-78.
3. Distribution of the averages of high school graduates for all incoming freshmen, women and men, for Fall semester, 1976-78.

The data actually supplied the committee are shown in Tables 1 through 5.

Based on a preliminary analysis of the data given to the committee, one of its members placed a motion before the committee (for discussion) that it consider a selective admissions policy based on a cutoff of the composite ACT score at 13. That is, no one with a composite score of 13 or less would be admitted as a freshman to MSU.

While several arguments and different positions were presented in the course of the discussion, the following points summarize the major concern of various members of the committee holding different views.

1. This policy would cause too high a reduction in enrollment.
2. It definitely discriminates against women students.
3. It doesn't really eliminate the bottom quartile of the poorer students.
4. MSU would be better off using high school grades instead of ACT scores.

Prepare a report, including tables, charts, and graphs, for presentation to the Faculty Senate. Include in your report an analysis that covers each of the four points noted above.

TABLE 1. *Distribution of ACT Score Frequencies for All Incoming Freshmen, 1976–1978*

STANDARD SCORE	ACT ENGLISH			ACT MATHEMATICS			ACT SOCIAL STU			ACT NATURAL SCI			ACT COMPOSITE		
	1976	'77	'78	1976	'77	'78	1976	'77	'78	1976	'77	'78	1976	'77	'78
36	0	0	0	5	4	8	0	0	0	0	0	0	0	0	0
35	0	0	0	16	13	12	0	0	0	5	13	2	0	0	0
34	0	0	0	21	22	13	10	6	8	24	16	16	4	0	0
33	12	7	2	24	15	27	19	16	9	51	53	62	4	3	2
32	9	4	2	37	24	37	40	31	22	128	120	131	11	11	7
31	17	8	18	49	60	41	42	41	45	126	183	144	37	27	37
30	21	28	27	97	58	69	81	76	76	228	177	189	60	55	51
29	38	17	34	190	131	115	109	85	86	204	189	197	104	89	91
28	35	35	53	203	152	177	170	143	126	197	181	173	141	111	133
27	83	67	52	212	222	244	186	163	140	157	141	156	183	192	142
26	124	92	79	202	282	235	215	208	207	171	200	167	210	201	202
25	170	149	141	246	250	227	287	248	219	195	171	159	247	218	201
24	183	200	187	245	201	171	253	213	201	169	163	207	260	247	216
23	300	240	261	184	165	172	324	295	215	209	246	163	235	257	243
22	403	338	329	159	140	122	208	195	170	183	189	236	297	256	218
21	349	370	297	168	176	157	240	221	124	252	164	214	257	246	217
20	339	314	249	166	160	160	185	144	169	218	221	190	264	221	225
19	351	299	297	185	168	171	137	97	143	184	185	138	272	217	240
18	325	335	295	255	184	187	63	63	113	314	167	152	262	220	210
17	285	251	252	257	240	198	128	147	110	234	201	161	221	219	213
16	240	180	193	311	172	174	81	59	113	175	188	183	212	197	201
15	156	155	189	165	161	155	136	147	108	245	157	209	191	145	169
14	121	112	147	144	128	132	133	140	171	131	152	108	170	142	165
13	117	119	100	49	73	113	163	134	125	115	109	107	126	115	121
12	111	112	126	84	92	86	165	143	142	69	47	59	111	121	94
11	97	98	97	63	50	44	122	106	201	56	33	44	113	93	91
10	89	72	73	119	83	95	125	149	117	41	19	32	75	56	64
9	76	47	56	61	68	56	127	110	164	28	12	19	44	33	40
8	51	39	41	57	28	37	99	100	118	22	16	23	28	16	40
7	28	17	29	76	73	65	122	88	81	14	10	6	21	8	17
6	24	10	23	27	38	67	53	63	63	14	6	9	11	9	13
5	15	7	11	36	23	11	63	46	34	2	0	7	4	2	3
4	5	4	2	17	34	38	32	20	19	10	2	1	3	4	2
3	2	2	6	19	7	17	26	15	13	5	0	3	0	0	1
2	1	3	1	15	18	15	18	12	10	1	0	2	0	0	1
1	1	0	1	14	16	22	16	8	8	1	0	1	0	0	0

TABLE 2. *Distribution of ACT Score Frequencies for Women, 1976–78*

STANDARD SCORE	ACT ENGLISH			ACT MATHEMATICS			ACT SOCIAL STU			ACT NATURAL SCI			ACT COMPOSITE		
	1976	'77	'78	1976	'77	'78	1976	'77	'78	1976	'77	'78	1976	'77	'78
36	0	0	0	0	1	1	0	0	0	0	0	0	0	0	0
35	0	0	0	3	3	5	0	0	0	0	3	0	0	0	0
34	0	0	0	5	2	0	2	3	2	3	3	2	0	0	0
33	5	4	2	6	4	4	2	3	4	10	14	14	0	0	1
32	4	4	1	17	6	9	11	9	4	35	34	34	0	5	3
31	10	7	10	15	17	16	11	10	12	27	67	54	13	11	11
30	14	19	16	39	18	22	27	26	19	78	68	59	23	18	12
29	23	11	23	60	45	39	43	31	23	60	57	76	46	39	35
28	17	23	33	67	52	64	58	58	52	85	69	68	47	38	48
27	43	36	33	89	86	102	72	64	59	53	62	69	68	68	66
26	73	66	39	74	111	89	102	96	97	79	77	76	80	76	83
25	105	85	78	109	107	90	130	98	87	91	78	79	99	101	84
24	105	114	106	100	88	75	120	95	88	79	80	99	116	105	94
23	159	137	144	85	77	77	148	123	99	93	101	79	109	116	103
22	211	175	177	74	72	59	101	86	77	79	98	116	135	120	97
21	184	186	159	69	77	82	101	115	72	130	87	114	119	116	106
20	162	146	122	92	87	73	98	63	77	119	109	103	136	107	114
19	163	140	131	89	80	79	71	45	69	96	95	71	126	105	120
18	143	150	144	144	97	110	35	31	67	178	98	88	141	115	108
17	131	107	114	132	131	98	60	80	52	127	123	86	115	108	101
16	103	68	77	171	91	93	42	25	56	103	120	96	120	109	113
15	66	51	78	101	96	87	62	75	54	158	90	121	112	82	94
14	46	52	49	83	73	73	75	86	95	86	100	67	98	81	92
13	49	46	43	19	49	61	99	79	67	81	67	71	73	65	72
12	40	40	48	52	51	50	92	78	82	38	27	30	63	65	50
11	37	41	39	34	31	28	68	60	102	34	22	28	71	61	53
10	41	28	28	75	51	58	71	81	67	25	7	14	41	35	37
9	31	19	23	34	40	44	71	60	94	20	9	13	26	19	20
8	22	13	12	43	18	21	65	54	62	13	13	15	16	10	25
7	11	8	15	50	38	48	63	56	44	11	5	4	8	4	8
6	7	4	11	13	26	40	37	43	39	11	5	5	9	7	7
5	5	6	4	24	14	8	31	28	20	1	0	5	1	1	1
4	2	1	0	10	21	22	15	9	9	7	0	1	3	1	2
3	1	0	1	14	4	12	11	8	4	4	0	3	0	0	1
2	1	1	0	12	12	11	13	5	3	0	0	0	0	0	0
1	0	0	1	10	12	11	7	5	3	0	0	1	0	0	0

TABLE 3. *Distribution of ACT Score Frequencies for Men, 1976–1978*

STANDARD SCORE	ACT ENGLISH			ACT MATHEMATICS			ACT SOCIAL STU			ACT NATURAL SCI			ACT COMPOSITE		
	1976	'77	'78	1976	'77	'78	1976	'77	'78	1976	'77	'78	1976	'77	'78
36	0	0	0	5	3	7	0	0	0	0	0	0	0	0	0
35	0	0	0	13	10	7	0	0	0	5	10	2	0	0	0
34	0	0	0	16	20	13	8	3	6	21	13	14	4	0	0
33	7	3	0	18	11	23	17	13	5	41	39	48	4	3	1
32	5	0	1	20	18	28	29	22	18	93	86	97	11	6	4
31	7	1	8	34	43	25	31	31	33	99	116	90	24	16	26
30	7	9	11	58	40	47	54	50	57	150	109	130	37	37	39
29	15	6	11	130	86	76	66	54	63	144	132	121	58	50	56
28	18	12	20	136	100	113	112	85	74	112	112	105	94	73	85
27	40	31	19	123	136	142	114	99	81	104	79	87	115	124	76
26	51	26	40	128	171	146	113	112	110	92	123	91	130	125	119
25	65	64	63	137	143	137	157	150	132	104	93	80	148	117	117
24	78	86	81	145	113	96	133	118	113	90	83	108	144	142	122
23	141	103	117	99	88	95	176	172	116	116	145	84	126	141	140
22	192	163	152	85	68	63	107	109	93	104	91	120	162	136	121
21	165	184	138	99	99	75	139	106	52	122	77	100	138	130	111
20	177	168	127	74	73	87	87	81	92	99	112	87	128	114	111
19	188	159	166	96	88	92	66	52	74	88	90	67	146	112	120
18	182	185	151	111	87	77	28	32	46	136	69	64	121	105	102
17	154	144	138	125	109	100	68	67	58	107	78	75	106	111	112
16	137	112	116	140	81	81	39	34	57	72	68	87	92	88	88
15	90	104	111	64	65	68	74	71	54	87	67	88	79	63	75
14	75	60	98	61	55	59	58	54	76	45	52	41	72	61	73
13	68	73	57	30	24	52	64	55	58	34	42	36	53	50	49
12	71	72	78	32	41	36	73	65	60	31	20	29	48	56	44
11	60	57	58	29	19	16	54	46	99	22	11	16	42	32	38
10	48	44	45	44	32	37	54	68	50	16	12	18	34	21	27
9	45	28	33	27	28	12	56	50	70	8	3	6	18	14	20
8	29	26	29	14	10	16	34	46	56	9	3	8	12	6	15
7	17	9	14	26	35	17	59	32	37	3	5	2	13	4	9
6	17	6	12	14	12	27	16	20	24	3	1	4	2	2	6
5	10	1	7	12	9	3	32	18	14	1	0	2	3	1	2
4	3	3	2	7	13	16	17	11	10	3	2	0	0	3	0
3	1	2	5	5	3	5	15	7	9	1	0	0	0	0	0
2	0	2	1	3	6	4	5	7	7	1	0	2	0	0	1
1	1	0	0	4	4	11	9	3	5	1	0	0	0	0	0

TABLE 4. *Arithmetic Mean Grade Point Averages for Fall Semester of Enrolled Freshmen Men and Women, Main State University, 1976–1978*

FALL SEMESTER	WOMEN	MEN
1976	2.211	2.114
1977	2.324	2.045
1978	2.327	2.137

TABLE 5. *Frequency Distributions of the High School Grade Averages of Enrolled Freshmen (Women and Men), of Main State University, 1976–1978*

HIGH SCHOOL AVERAGES	MEN			WOMEN			TOTAL		
	1976	'77	'78	1976	'77	'78	1976	'77	'78
3.76-4.00	147	141	184	185	235	228	332	376	412
3.51-3.75	133	120	143	153	156	166	286	276	309
3.26-3.50	175	179	170	208	182	179	383	361	349
3.01-3.25	204	199	230	231	193	207	435	392	437
2.76-3.00	275	231	230	304	231	237	579	462	467
2.51-2.75	233	180	159	192	154	150	425	334	309
2.26-2.50	269	234	205	245	189	175	514	423	380
2.01-2.25	202	211	171	176	136	118	378	347	289
1.76-2.00	213	169	150	137	105	103	350	274	253
1.51-1.75	99	67	69	51	31	25	150	98	94
1.26-1.50	83	69	42	40	31	23	123	100	65
1.01-1.25	41	27	30	15	15	7	56	42	37
0.76-1.00	13	14	9	5	4	3	18	18	12
0.51-0.75	1	1	3	1	0	0	2	1	3
0.26-0.50	0	2	2	1	2	0	1	4	2
0.01-0.25	0	0	3	0	0	1	0	0	4
TOTAL	2088	1844	1800	1944	1664	1622	4032	3508	3422

Elementary Probability Theory

CHAPTER
4

Before a football game, a coin is flipped to determine which team will kick off and which team will receive the football. What is going to turn up—a head, a tail? Neither we nor the teams involved can be sure until the coin is flipped and the outcome is observed. We do know, however, that it will be a head or a tail, assuming the coin cannot stand on edge. Moreover, if the coin can be assumed to be perfectly balanced in weight such that neither side is more likely to be the up side, it would seem reasonable to conclude that either side is equally likely to appear. That is, the head should occur half of the time, and the tail should occur the other half of the time, thereby giving each team an equal chance of winning the coin toss.

In the preceding paragraph, the word *likely* appeared twice and the word *chance* once with reference to the possible occurrence of a future event. Such situations where chance and likelihood are discussed represent situations where the conclusions are often stated in terms of *probability*. In fact, probability stands as the foundation upon which statistical analysis is built. Later we will define probability more specifically, but for the moment let us accept "likelihood" as a reasonable synonym for it.

For instance, one might assert, "There is a strong likelihood that it will rain" or, "There is a better chance that it will rain than that it will not rain." Yet either statement could have been phrased, "There is a higher *probability* that it will rain than that it will not."

More specifically, the statement might have assigned some numerical value to describe the likelihood of rain. Now such a statement might read, "There is a 0.60 probability of rain." To convey the same information, the television weatherman would say, "There is a 60 percent chance of rain today." So let us see what this 0.60 probability, or 60 percent chance, connotes.

Consider a scale from 0 to 1.0, where zero means there is no possibility and 1.0 means there is absolute certainty that an event will occur. Thus, if the probability of rain is zero, there is no possibility of rain, whereas, if the probability of rain is 1.0, *it will rain.* Moreover, if the probability of rain is 0.5, then it is equally likely that it will or will not rain.

PROBABILITY OF RAIN	INTERPRETATION
0	No Possibility
0.5	Equally as Likely
1.0	Absolute Certainty

A probability value can be any value between and including the two extremes of zero and one. It is equivalent to expressing a relationship on a numerical scale. Thus, as a working definition, let us define probability as the prospect that an event will occur at some future time measured on a scale from zero (the event will not occur) to one (the event will occur with certainty).

Sources of Probability Numbers

So far we have established that *probability numbers* range from zero to one inclusive. Zero probability implies that there is no possibility that an event can occur (it is impossible) whereas a probability of one implies that an event is certain to occur. The next question that needs to be answered is "Where do we get probability numbers?" The answer can be uncovered by examining three concepts of a probability number.

Classical Concept

The classical concept of probability is that probability numbers are known or can be determined *a priori* (before the fact). Thus, the exact probability that a particular event might occur would be known before the fact. Illustrations of this type of probability number are

1. the probability of drawing an ace from a deck containing 52 cards;
2. the probability of drawing a red ball from a jar containing 3 red balls and 7 green balls.

To determine each of the preceding *a priori* probabilities, the following formula can be used:

~A alternate notation

$$P(A) = \frac{N(A)}{N(A) + N(\overline{A})} \qquad (4.1)$$

where $P(A)$ = the probability of defined event A occurring

$N(A)$ = the number of possible ways A can occur in a situation

$N(\overline{A})$ = the number of possible ways A cannot occur in the same situation

$N = N(A) + N(\overline{A})$ = the total number of possibilities of A and not-A involved in the situation.

Considering the illustration of drawing an ace from a deck of 52 cards, we see that be defining an ace as the event (A) of interest the following substitution results.

$$N(A) = \ 4 \ (\text{aces})$$

$$N(\overline{A}) = 48 \ (\text{all other cards})$$

$$N = 52 \ (\text{all cards})$$

$$P(A) = \frac{N(A)}{N(A) + N(\overline{A})} = \frac{N(A)}{N} = \frac{4}{4 + 48} = \frac{4}{52} = \frac{1}{13} = 0.077$$

Thus, the probability of drawing an ace on a single draw is 0.077. Similarly, if a ball is drawn from the aforementioned jar, the following substitution results when A, the event of interest, is defined as drawing a red ball.

$$N(A) = \ 3 \ (\text{red balls})$$

$$N(\overline{A}) = \ 7 \ (\text{green balls})$$

$$N = 10 \ (\text{all balls})$$

$$P(A) = \frac{N(A)}{N(A) + N(\overline{A})} = \frac{N(A)}{N} = \frac{3}{3 + 7} = \frac{3}{10} = 0.30$$

And so, the probability of a red ball being drawn is 0.30.

Now, it should be remembered that these are examples of *a priori* probabilities. *A priori* probabilities are those that can be determined before any event occurs; however, *a priori* probabilities are subjected to a particular criticism that we should discuss.

As you may have realized when you worked through the previous two illustrations, for the $P(A)$ calculation to be correct, it is necessary to assume that all cards or all balls had an equal probability of being selected. Now, notice what has happened. We have described a type of probability *(a priori)* that relies on the assumption of equal probability. Thus, we have used probability to define probability, and that is the criticism.

As this is a rather technical criticism, let us not concern ourselves with the ambiguity that results from the definition, but simply remember that these are situations where the possibilities (number of ways) can be determined before the occurrence of the event. And, from these possibilities, we can determine the classical probability *(a priori)* of an event before it happens.

Relative Frequency Concept

Another concept of probability is based on the relative frequency of occurrence. This concept utilizes the relationship of the number of times a particular event actually occurred to the total number of trials observed. To illustrate, consider that when flipping a coin 1000 times (1000 trials), the head turned up 450 times, we would determine a probability of the occurrence of a head based on relative frequency of $450/1000 = 0.45$. Since this probability is determined after the fact *(a posteriori)* and is based on observed data, it is referred to frequently as an empirical probability. The empirical probability of an event A can be determined as follows:

$$P\,(A) = \frac{a}{n}$$

where a is the number of times event A occurred out of n total observations or trials.

As the number of trials, n, increases infinitely, then the relative frequency probability becomes stable. That is, the probability value a/n becomes the limit of the relative frequency as n approaches infinity.[1] As a further illustration of the concept of relative frequency, consider the following example.

▶ **Example 4.1**   The Tireless Manufacturing Company has maintained records on the number of defective tires returned over the past year. Of the 200,000 tires produced, 3,000 were defective. In order to determine the probability of any one tire being defective, the relative frequency concept is utilized, and the empirical probability is:

---

1.   If you are familiar with limits, then $P\,(A) = \lim\limits_{n \to \infty} \dfrac{a}{n}$

$$P \text{ (defective tire)} = \frac{3{,}000 \text{ (number of defectives)}}{200{,}000 \text{ (total number produced)}} = 0.015$$

Thus, the implication is that if all aspects of the manufacturing process remain unchanged, future production of defective tires will occur at a rate of 15 defectives for each 1,000 tires produced. ◄

Here again, we must be reminded to take care in interpreting this probability. Remember, the concept of relative frequency is based on an infinite number (or at least a large number) of trials. Obviously, we never use an infinite number of observations, so the empirical probability can be erroneous if too few trials are observed. Certainly 10 trials would not be enough to determine the probability of a defective tire, whereas with our 200,000 trials we have a right to feel pretty secure about our estimate of the probability of a defective tire.

Finally, you might like to know that empirical probability has been found to be quite useful in setting premium rates in the insurance industry. These rates are based on the probability of the occurrence of events against which a company may issue an insurance policy—e.g., insurance on your car. Since there are few, if any, cases in the insurance industry where the possible number of ways events may occur can be determined, the probabilities upon which insurance rates are based are determined by the relative frequency of occurrence. Thus, relative frequency probability explains, in part, why automobile insurance premiums differ in some parts of the country and for different age-groups of drivers.

Subjective Probability

At times, there are situations (certainly in business) when probabilities cannot be determined by either the *a priori* or the empirical methods. In these cases, one may have to resort to a probability based on intuition or judgment. This type of probability estimate is called *subjective probability*. However, do not confuse intuitive judgment with empirical observation. The latter is based on observed data, while the former relies on personal evaluation (something like shooting from the hip). Since subjective probability is based on experience, the subjective probability estimates should also get better as one's experience improves. And from time to time the probability may undergo revision based on additional experience.

Methods of Counting

In using the classical (*a priori*) concept of probability the equation is dependent on our being able to determine the values that go into the numerator and denominator—the possibilities.

$$P(A) = \frac{N(A)}{N(A) + N(\overline{A})} \qquad (4.1)$$

skip

where $N(A) + N(\overline{A}) = N$ (total number of possibilities)

That is, *a priori* probabilities are based on the number of possibilities. Sometimes these possibilities are reasonably easy to count; other times they are difficult to determine. When they are difficult to determine, we frequently need assistance in counting the possibilities. Four methods used to help count the possibilities are presented in this chapter.

The Multiplication Method

give example of number of schedules from 3 stat sections, 4 finance, 5 accounting, 6 english + 7 government

If a particular experiment is determined to have *m* possible outcomes on the first trial and *n* possible outcomes on the second trial, the total number of possible outcomes for the two trials is the product $m \times n$. Let us examine this method of counting using a tree diagram.[2]

Suppose you have four required and five elective courses that you could take. Further suppose that required courses are available only in the Fall semester and that only electives can be taken in the Spring. If you are planning to take one required course in the Fall and one elective course in the Spring, how many different pairs of courses are available to you?

Notice that we have two distinct groups, required courses (4 items) followed by electives (5 items). Then the solution is $4 \times 5 = 20$ as demonstrated by the tree diagram in Figure 4-1. This solution was obtained by setting $m = 4$ and $n = 5$. The examples can be expanded to any number of groups, and the total number of outcomes equals the product of the number of elements in each group.

Multiple Choice Arrangement

give example of examination

Same as multiplication method / special case / what all choices are equal size

Notice that this section deals with a type of arrangement. Such is the case for the next three methods of counting. The first method (the Multiplication Method) dealt with two or more groups, whereas we will now examine arrangements within one of the groups.

Suppose we identify the four required courses as A, B, C, and D; and we want to arrange them in pairs such that one is taken during one semester and the other is taken in the next semester. How many

---

2.   Tree diagrams can be very helpful and will be used frequently in this chapter. They are pictorial representations of the situation we are trying to evaluate.

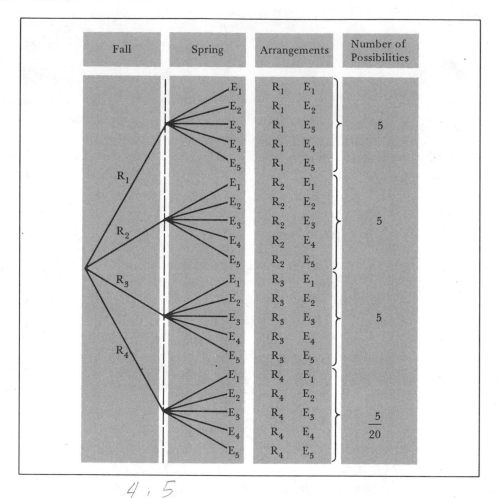

FIGURE 4-1.   *Tree Diagram for the Multiplication Method*

distinctly different arrangements of two courses can we get from the required four courses?   At this point, the question cannot be answered until we make the two assumptions required of *multiple choice arrangements.*

1. Duplication is permissible.   That is, a single item may be duplicated in a given pair.   (You might fail a given course and have to repeat it.)
2. Order is important.   That is, AB is considered a different arrangement from BA.

   The possible multiple choice arrangements of the courses are shown in Figure 4-2.

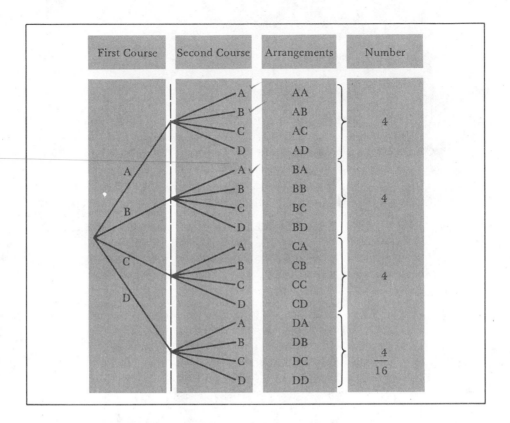

FIGURE 4-2. *Tree Diagram for Multiple Choice Arrangements*

4 · 4

From the tree diagram, we see that there are 16 arrangements under the assumptions of multiple choice. It is obvious, however, that for larger problems use of the tree diagram would prove cumbersome. Thus, the easier way is to use the following formula:

$$M_x^n = n^x \qquad\qquad (4.2)$$

where n = the size of the set
x = the number of places in the arrangement

▶ **Example 4.2**   Now let us repeat the illustration and demonstrate the use of the preceding equation. There are four courses ($n = 4$) and you are going to take one each semester for two semesters ($x = 2$). How many possible pairs do you have? If we assume that you might have to repeat a course (duplication is permissible) and that sequence AB is different from BA (order is important), then our number of pairs is $M_2^4 = 4^2 = 16$. ◀

*give example of
5 persons filling
hierarchical
5 positions, then
6 filling 3 positions*

Permutation Arrangement

Suppose we continue with your schedule plans for subsequent semesters and that we decide to change the condition regarding duplication. That is, let us now assume:

1. Duplication is not permissible.   (You do not expect to fail any courses.)
2. Order is important.   That is, AB is considered a different arrangement from BA.

Then, Figure 4-3 shows the possible pairs of courses.

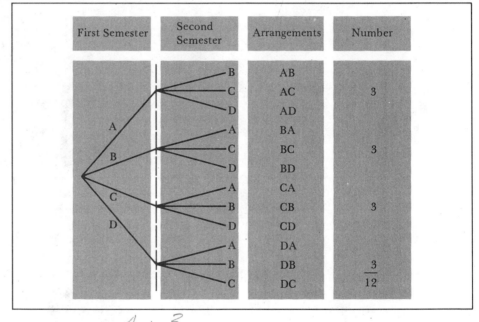

4 · 3

FIGURE 4-3.   *Tree Diagram for Permutation Arrangements*

Now, instead of 16 arrangements, we have 12 arrangements. These are the *permutation arrangements*.   The permutation formula for finding the number of possible arrangements under the assumptions of (1) no duplication, and (2) order is important, is:

$$P_x^n = \frac{n!}{(n-x)!}$$ (4.3)

$\frac{4 \cdot 3 \cdot 2 \cdot 1}{2 \cdot 1} = 4 \cdot 3$

This equation states that the number of permutations of *n* items taken *x* at a time is equal to *n* factorial (*n*!) divided by (*n* − *x*) factorial,

$(n - x)!.$[3]   As another example of counting with permutations, consider the following situation.

▶ **Example 4.3**   Let us use Equation 4.3 to determine the number of arrangements shown in Figure 4-3.   Remember, the situation is to schedule one of four courses (A, B, C, D) in each of the next two semesters under the assumptions of (1) no duplication, and (2) order is important.   Setting $n = 4$ and $x = 2$:

$$P_2^4 = \frac{4!}{(4 - 2)!} = \frac{4 \cdot 3 \cdot 2 \cdot 1}{2 \cdot 1} = 12 ◀$$

Combination Arrangement

Suppose we go back to planning your schedule of courses and decide that in addition to the restriction of no duplication we will consider the arrangement AB to be indistinguishable from the arrangement BA That is, we are now assuming that:

1. Duplication is not permissible.
2. Order is not important.

As far as your schedule is concerned, this means that no course may be taken twice; and that if A is taken first and B second, the reverse order will not be considered.   The tree diagram of Figure 4-4 depicts the possible combinatorial arrangements.

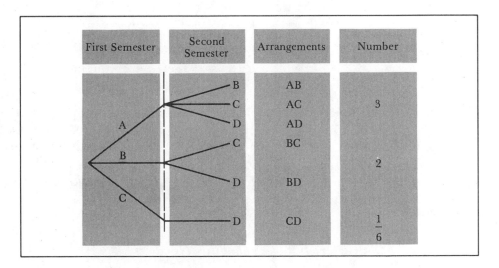

FIGURE 4-4.   *Tree Diagram for Combinatorial Arrangements*

---

3.   The symbol, !, is the notation for factorial. $n! = n \cdot (n - 1) \cdot (n - 2) \cdot \ldots \cdot (1)$.   Thus $5! = 5 \cdot 4 \cdot 3 \cdot 2 \cdot 1 = 120$; and by definition $0! = 1$.

Thus, under the conditions of (1) no duplication and (2) order is not important, we get only six distinct arrangements. The combination of n items taken x at a time is

$$C_x^n = \frac{n!}{x! \, (n - x)!} \qquad \text{(4.4)}$$

This formula is called the *combinatorial formula*. As an example of the use of this formula, consider the following situation.

▶ **Example 4.4**   Let us use Equation 4.4 to determine the number of arrangements shown in Figure 4-4. The situation is to schedule one of four courses (A, B, C, D) in each of the next two semesters under the assumptions of (1) no duplication, and (2) order is not important. Setting $n = 4$ and $x = 2$:

$$C_2^4 = \frac{4!}{2! \, (4 - 2)!} = \frac{4 \cdot 3 \cdot 2 \cdot 1}{2 \cdot 1 \cdot 2 \cdot 1} = 6 \blacktriangleleft$$

Probability: Characteristics and Terms

The general formulas,

$$P(A) = \frac{N(A)}{N} \text{ (a priori) and}$$

$$P(A) = \frac{a}{n} \text{ (relative frequency)},$$

can be used to determine probabilities of simple events. There are many times, however, when probabilities will be determined by combining simple probabilities. In this type of circumstance, we will be investigating compound events; and to do this we need to become familiar with certain characteristics and terms associated with probability numbers.

Characteristics of Probability Numbers

Let's suppose that an experiment has several possible outcomes which we'll call events—$A_1, A_2, A_3, \ldots, A_k$. If a group of numbers is to be identified as probability numbers they must possess the following characteristics:

1. The probability associated with any event cannot be any smaller than 0 or greater than 1.   That is, $0 \leq P(A_i) \leq 1$.
2. The sum of the event probabilities in the experiment $(A_1 + A_2 + A_3 + \ldots + A_k)$ must equal 1.

Thus,

$$\sum_{i=1}^{k} P(A_i) = 1$$

The first characteristic states that for all outcomes $(A_1, A_2, A_3, \ldots, A_k)$ of an experiment, the probability of any particular outcome, denoted $P(A_i)$, must be within the range 0 and 1.   The second characteristic states that the probability numbers assigned to all outcomes $[P(A_1), P(A_2), \ldots, P(A_k)]$ must sum to one, $P(A_1) + P(A_2) + \ldots + P(A_k) = 1$.   Since all possible events are considered, one of them *must* occur.

Consider, as an illustration, the flipping of a coin.   The possible outcomes are head and tail.   If the coin is fair, and each outcome is equally likely, then using Equation 4.1, $P(H) = N(H)/[N(H) + N(T)] = 1/(1 + 1) = 1/2$; and similarly $P(T) = 1/2$.   Now, notice that each probability is in the 0-1 range, and that the sum of the probabilities is 1.   Thus, the numbers have the necessary characteristics of probability numbers.

Now that we know something about the characteristics of probability numbers, we are almost ready to consider the so-called *laws of probability*.   Before we do this, though, let us take a quick look at some terms we need to know.

Complement

Let us understand the term *complement*.   The complement of heads is tails; or the complement of tails is heads.   More generally, the complement of any outcome A is all outcomes other than A or simply "not-A".   Symbolically, the expression denoting not-A is often written as $\overline{A}$.   And for any outcome A, the probability of not-A is

$$P(\overline{A}) = 1 - P(A) \tag{4.5}$$

In our example, if the probability of heads is $P(A)$, then $P(\overline{A}) = P(T) = 1 - P(H)$.   When the experiment has several possible outcomes, $P(\overline{A})$ represents the probability that some outcome other than A will occur.

Mutually Exclusive Events.

What does it mean to say that two events (outcomes), *A* and *B,* are *mutually exclusive?* Events *A* and *B* are mutually exclusive if they cannot occur simultaneously. Figure 4-5 diagrams an example of mutually exclusive events. In the figure, the area designated *A* refers to the number of possible ways event *A* might occur, and the area marked *B* refers to the number of possible ways event *B* might occur. The area outside *A* and *B* (call it *C*) depicts the possible ways for neither event *A* nor *B* to occur. You should notice that there is no area indicating possible ways for events *A* and *B* to occur together. Then, events *A* and *B* are mutually exclusive.

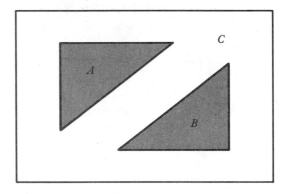

FIGURE 4-5. *Mutually Exclusive Events*

Independent and Dependent Events

In addition to mutually exclusive events, it is important that we understand the difference between independent and dependent events. If two events are such that the occurrence of one event does not affect or change the probability of the other event occurring, these events are said to be *independent.* But if the events are so related that if one event occurs the probability of the other event occurring is changed, these events are *dependent.* The probability of the second event would be considered *conditional* on the occurrence of the first event.

▶ **Example 4.5**   Consider a jar that contains four red balls and six green balls. Then $P(R) = 4/10$ and $P(G) = 6/10$. Now if a red ball is drawn and not replaced, the probability that a green ball would be drawn on the second draw is no longer 6/10. It has been changed to 6/9 conditional that a red ball was drawn on the first draw. Conditional probabilities are written as follows:

$$P\,(G\backslash R) = 6/9$$

In this case it is read as "the probability of a green ball, given that a red has occurred, equals 6/9." If the events had been independent (the red ball was replaced after the first draw) rather than dependent, then the probability of a green ball on the second draw would have remained 6/10. ◀

You should notice that the conditional probability was determined from a reduced population. That is, we started with 10 balls but considered only 9 when defining the conditional probability. This is always the case when dealing with dependent events. If the population from which the second event could occur was not reduced after the occurrence of the first event, then the events would have to be independent because one event is not affected by the occurrence of the other event.

Laws of Probability

Now that we have learned certain terms in probability, we are ready to examine and use the *laws of probability* as they relate to *compound events.*

Multiplication Law

One particular compound event involves the occurrence of *all* the individual events under consideration. These events are considered either to occur all at the same time or successively. Such a compound event requires the calculation of a *joint* probability using the following general expression:

$$P\,(A \text{ and } B) = P\,(A) \cdot P\,(B\backslash A) \qquad\qquad (4.6)$$

Notice that the joint probability is a product of the probabilities of each event required to occur. The equation given above is for *dependent* events; if the events are *independent,* then the expression for joint probabilities becomes:

$$P\,(A \text{ and } B) = P\,(A) \cdot P\,(B) \qquad\qquad (4.7)$$

▶ **Example 4.6**   Consider tossing a fair coin and rolling a fair die. What would be the probability of a head occurring on the coin toss and a six occurring on the die roll?   Since the events are independent, then:

$$P \text{ (Head and Six)} = P\,(H) \cdot P\,(S) = 1/2 \cdot 1/6 = 1/12 \blacktriangleleft$$

The outcomes and probabilities of the experiment described in Example 4.6 are shown in the tree diagram, Figure 4-6.   The figure shows all four possible outcomes when a coin is tossed and a die rolled. They are: Head and Six; Head and No Six; No Head and Six; No Head and No Six; and the sum of all of these joint occurrences is one, just as we expected.   The occurrence we are interested in is "Head and Six" with a probability of 1/12.

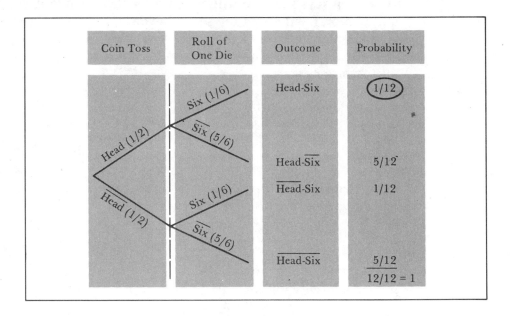

FIGURE 4-6.   *Tree Diagram for Coin Toss and Roll of One Die*

▶ **Example 4.7**   Now let us apply the *multiplication law* to *dependent events*.   Suppose we return to our jar that has four red balls and six green balls.   What would be the joint probability of a red on the first of two draws and a green on the second draw, given that the red ball is not replaced?

$$P \text{ (Red and Green)} = P \text{ (Red)} \cdot P \text{ (Green\textbackslash Red)}$$

$$P \text{ (Red and Green)} = 4/10 \cdot 6/9$$

$$P \text{ (Red and Green)} = 4/15 \blacktriangleleft$$

This example also can be illustrated using a tree diagram as shown in Figure 4-7. Again there are four possible outcomes: Red and Green; Red and Not Green; Not Red and Green; Not Red and Not Green, and the probabilities sum to one as expected. The outcome of interest to us is "Red and Green" which is equal to 24/90 or 4/15 as demonstrated earlier. This also illustrates that the use of the tree diagram is very helpful when calculating joint probabilities.

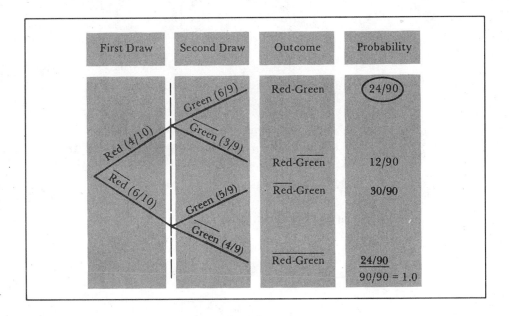

FIGURE 4-7. *Tree Diagram for Red and Green Balls*

Remember that joint probabilities require the multiplication law which is

$$P(A \text{ and } B) = P(A) \cdot P(B \backslash A) \tag{4.6}$$

and if A and B are independent events, the law is

$$P(A \text{ and } B) = P(A) \cdot P(B) \tag{4.7}$$

Addition Law

Suppose you wanted to know the probability of A or B occurring. In this case, if either A or B occurs, then we say the event occurs. Thus, for two events, A and B, the probability of either A or B occurring is determined by the following equation:

$$P (A \text{ or } B) = P (A) + P (B) - P (A \text{ and } B) \qquad \textbf{(4.8)}$$

In Equation 4.8, if events A and B are mutually exclusive, then the last term on the right side, $P (A \text{ and } B)$, is equal to zero and the addition law for mutually exclusive events reduces to:

$$P (A \text{ or } B) = P (A) + P (B) \qquad \textbf{(4.9)}$$

special case only

▶ **Example 4.8**   Let us determine the probability of drawing a red or green ball from a jar containing three red, four green, and five blue balls.   Since the events are mutually exclusive, then

P (Red or Green) $= P$ (Red) $+ P$ (Green) $= 3/12 + 4/12 = 7/12$

Also, notice that the probability of drawing a red or a green or a blue ball is:

P (R or G or B) $= 3/12 + 4/12 + 5/12 = 12/12 = 1.0$

The above conforms to our understanding that something will occur with a probability of one if all possible outcomes are defined. ◄

Now, let us move to an illustration where the events are *not mutually exclusive*.   If the events are not mutually exclusive, they have a common or like area of occurrence as depicted by the shaded area in Figure 4-8.

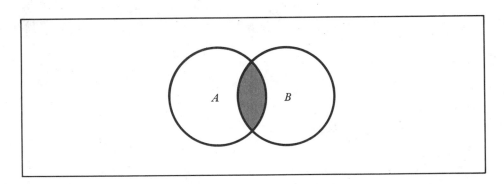

FIGURE 4-8. *Diagram of Events That Are Not Mutually Exclusive*

For instance, if we define event A as the drawing of a heart (there are 13 hearts) from a deck of 52 cards and event B as the drawing of a king (there are four kings) from the same deck, then the shaded area of Figure 4-8 represents the king of hearts. Thus, the king of hearts is an outcome that would mean both events occurred. This identifies them as events that are not mutually exclusive. Consequently, if you want to determine the probability of a heart or king on one draw, you must be careful to avoid double counting the king of hearts. That is, if you consider that there are thirteen hearts (including the king of hearts), you must pretend that there are only three kings; or conversely if you consider that there are four kings, you must pretend that there are only twelve hearts. Either way, it should be apparent that the number of cards available to make your draw successful is sixteen—either $12 + 4$ or $13 + 3$. The probability of drawing a heart or king is $16/52$.

Now, utilizing the addition law:

$$P\ (H \text{ or } K) = P\ (H) + P\ (K) - P\ (H \text{ and } K)$$

$$= 13/52 + 4/52 - 1/52$$

$$P\ (H \text{ or } K) = 16/52$$

Generalizing the Laws

In the preceding discussion of the multiplication and addition laws, two events were considered. These laws, however, are valid for as many events as you care to consider as long as you are careful.

To illustrate, if the addition law is applied to three events (A, B, C), then:

$$P\ (A \text{ or } B \text{ or } C) = P\ (A) + P\ (B) + P\ (C) - P\ (A \text{ and } B) -$$
$$P\ (A \text{ and } C) - P\ (B \text{ and } C) + P\ (A \text{ and } B \text{ and } C)$$

Figure 4-9 can be used to show how the above equation accomplishes the desired result. Each numbered area corresponds to a specific probability such that $P\ (A) = \text{area } 1 + \text{area } 2 + \text{area } 6 + \text{area } 7$.

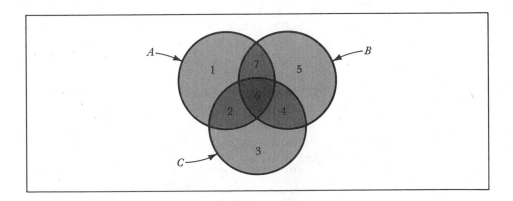

FIGURE 4-9. *Compound Events With Three Events*

Similarly,

$$P(B) = 4 + 5 + 6 + 7$$
$$P(C) = 2 + 3 + 4 + 6$$
$$P(A \text{ and } B) = 6 + 7$$
$$P(A \text{ and } C) = 2 + 6$$
$$P(B \text{ and } C) = 4 + 6$$
$$P(A \text{ and } B \text{ and } C) = 6$$

Then,

$$P(A \text{ or } B \text{ or } C) = (1 + 2 + 6 + 7) + (4 + 5 + 6 + 7) + (2 + 3 + 4 + 6) - (6 + 7) - (2 + 6) - (4 + 6) + 6$$
$$P(A \text{ or } B \text{ or } C) = 1 + 2 + 3 + 4 + 5 + 6 + 7,$$

which are the actual areas within the three circles.

SKIP

Examples Using the Multiplication and Addition Laws

There are instances where the determination of a desired probability involves combining the laws of multiplication and addition as illustrated in the following examples.

▶ **Example 4.9**   The Go-Save Shopping Center, in celebrating its first anniversary, is giving away two automobiles by a drawing.   The rules are simple. There are two barrels—one red and one green—in the shopping center mall.   Every time a purchase is made, a green ticket is given to a male customer for deposit in the green barrel, and a red ticket is given to a female customer for deposit in the red barrel.

Mr. and Mrs. Armstrong are not regular customers of the shopping center, but they have been shopping there lately because they both would like to win a new automobile.   Mr. Armstrong has deposited 10 tickets in the green barrel, and Mrs. Armstrong has deposited 20 tickets in the red barrel.   On the night of the drawing, there are 1,000 tickets in the green barrel and 5,000 tickets in the red barrel.

Now let us determine the probability that Mr. and Mrs. Armstrong win cars.   First let us caution you that there are three ways they might win a car; either Mr. Armstrong or Mrs. Armstrong or both might win a car.   Perhaps the easiest approach to this type of problem is through a tree diagram as shown in Figure 4-10.

Assuming that a ticket is drawn from the green barrel first and the red barrel second (the order can be reversed if you desire), the four outcomes are:

$P\,(HS)$ (Both win) $= 200/5{,}000{,}000$
$P\,(H\overline{S})$ (He wins, she loses) $= 49{,}800/5{,}000{,}000$
$P\,(\overline{H}S)$ (He loses, she wins) $= 19{,}800/5{,}000{,}000$
$P\,(\overline{HS})$ (Neither wins) $= 4{,}930{,}200/5{,}000{,}000$

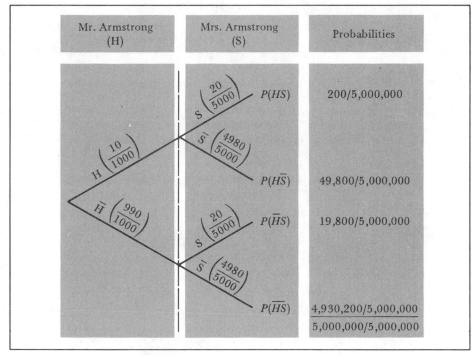

FIGURE 4-10. *Tree Diagram for the Armstrongs Winning a Car*

Now, notice the outcomes are joint probabilities and that the *multiplication law* was used to determine the probability of each; but to determine the probability of winning a car, it is necessary to use the addition law.   Thus

$P\,(\text{car}) = P\,(HS) + P\,(H\overline{S}) + P\,(\overline{H}S)$
$\qquad\quad = 200/5{,}000{,}000 + 49{,}800/5{,}000{,}000 + 19{,}800/5{,}000{,}000$
$\qquad\quad = 69{,}800/5{,}000{,}000$
$P\,(\text{car}) = 349/25{,}000 = 0.01396$ ◀

▶ **Example 4.10**   Suppose there are 100 stores that stock the same product.   Table 4-1 presents a cross-classification of the stores according to product sales and the height of shelf display.

TABLE 4-1. *Hypothetical Product Sales for 100 Stores*

PRODUCT SALES	SHELF DISPLAY			
	High	Eye-Level	Low	Total
High	4	32	5	41
Average	10	15	11	36
Low	6	4	13	23
	20	51	29	100

Then the following probabilities, among others, might be determined:

P (Average Sales) $= 36/100 = 0.36$
P (Eye-Level Display) $= 51/100 = 0.51$
P (High Sales\Eye-Level Display) $= 32/51 = 0.63$

And, the question of independence can be answered. Notice that the probability of High Sales and High Shelf Display is $4/100$. If Product Sales and Shelf Display are independent, then

P (High Sales and High Display) $= P$ $(H.S.) \cdot P$ $(H.D.)$
That is, $4/100$ $= (41/100)(20/100)$
But $4/100$ $\neq 820/10000$

Thus, they are not independent. There is a relationship between product sales and shelf display, and the actual relationship becomes

P (High Sales and High Display) $= P$ $(H.S.) \cdot P$ $(H.D.\H.S.)$
                $= (41/100)(4/41)$
                $= 4/100$

And, this result agrees with our earlier determination. ◀

Revised Probabilities and Bayes' Formula

From time to time a situation arises in which the probability of an event can be reevaluated, or revised, based on information obtained about a subsequent event. This situation is an example of another type of *a posteriori* (after the fact) probability. It differs from the empirical probability that was also determined after the fact in that empirical probability, in a sense, attempts to estimate the *a priori* (before the fact) or classical probability.

The type of posterior probability that we are now interested in is a revision of the original priors, whether they be *a priori* or *empirical* or *subjective* probabilities. To determine these revised possibilities we will employ Bayes' formula.

Around two hundred years ago the Reverend Thomas Bayes discovered an interesting relationship that permits the calculation of posterior (revised) probabilities from prior probabilities. This discovery is called Bayes' Theorem; in essence, Bayes' Theorem states: If there are two events A and B, such that the $P(A)$ and $P(B\backslash A)$ are known, it is possible to determine $P(A\backslash B)$ from the original priors. Notice that $P(A\backslash B)$ is a revision of the original prior $P(A)$. To demonstrate how this can be done, it is necessary to observe that

$$P(A \text{ and } B) = P(A) \cdot P(B\backslash A)$$

and

$$P(A \text{ and } B) = P(B) \cdot P(A\backslash B)$$

and then equating the right sides such that

$$P(B)\,\dot{P}(A\backslash B) = P(A)\,P(B\backslash A)$$

and thus

$$P(A\backslash B) = \frac{P(A) \cdot P(B\backslash A)}{P(B)} \qquad (4.10)$$

As you can see, the numerator of the right side contains $P(A)$ and $P(B\backslash A)$, the two original known probabilities. The denominator, however, still must be determined. As the denominator is the probability of the event B occurring, it is necessary that *all* possible ways that B might occur be included in this probability, $P(B)$. Thus $P(B) = P(A) \cdot P(B\backslash A) + P(\overline{A}) \cdot P(B\backslash\overline{A})$; and there may be more than one $P(\overline{A}) \cdot P(B\backslash\overline{A})$ since there may be several events identified as not-A. Consider the following example.

▶ **Example 4.11** A plant has two machines. Machine A produces 60 percent of the output with a fraction defective of 0.02. Machine B produces 40 percent of the output with a fraction defective of 0.04. If a single unit of output is observed to be defective, determine the revised probabilities that should be assigned to each machine (See Table 4-2).

TABLE 4-2. *Probability Revision Example Calculations*

EVENT (Ei)	PRIOR $P(Ei)$	CONDITIONAL $P(D\backslash Ei)$	JOINT $P(Ei) \cdot P(D\backslash Ei)$	POSTERIOR $P(Ei\backslash D)$
Machine A	0.60	0.02	0.012	0.012/0.028 = 0.429
Machine B	0.40	0.04	0.016	0.016/0.028 = 0.571
			0.028	

Even though Machine A produces a larger proportion of the output, the revised probability of 0.571 indicates that the quality control investigation should begin with Machine B. ◀

Another approach to the preceding problem could have been the use of a tree diagram as shown in Figure 4-11. Notice again in Figure 4-11 that the sum of the probabilities for all outcomes is 1.00. Now from Bayes' formula, we know that:

$$P(A \backslash D) = \frac{P(A \text{ and } D)}{P(D)} = \frac{P(A \text{ and } D)}{P(A) \cdot P(D \backslash A) + P(B) \cdot P(B \backslash A)}$$

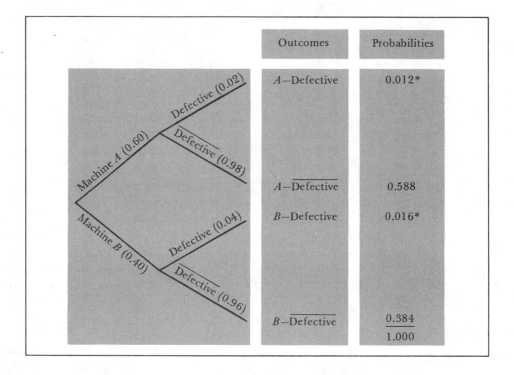

FIGURE 4-11. *Tree Diagram for Bayes' Formula (Example 4.11)*

All we need to do is select the probabilities from the tree diagram needed to make the revision calculation. The ones we need are denoted by asterisks. They are the outcomes involving a defective. Each of these probabilities appears in the denominator, and the probability involving Machine A and a defective appears in the numerator such that:

$$P(A \backslash D) = \frac{0.012}{0.012 + 0.016}$$

$$= \frac{0.012}{0.028}$$

$$P(A \backslash D) = 0.429$$

And, similarly, the probability associated with Machine B can be revised as follows:

$$P\ (B\backslash D) = \frac{P\ (B\ and\ D)}{P\ (D)}$$

$$= \frac{0.016}{0.012 + 0.016}$$

$$= \frac{0.016}{0.028}$$

$$P\ (B\backslash D) = 0.571$$

Notice on the tree diagram (Figure 4-11) that the first branches represent outcomes for the original event, and the second group of branches represent the observable outcomes. Now let us look at another example where the original probabilities are more subjective than empirical.

▶ **Example 4.12**   A marketing manager believes the market demand potential of a new product to be high with a probability of 0.30, or to be average with a probability of 0.50, or to be low with a probability of 0.20. From a sample of 20 employees, 14 indicated a very favorable reception to the new product. In the past, such an employee response (14 of 20 favorable) has occurred with the following probabilities; if the actual demand is high, the probability of a favorable reception is 0.80; if the actual demand is average, the probability of a favorable reception is 0.55; and if the actual demand is low, the probability of the favorable reception is 0.30. Thus, given a favorable reception, revise the prior probabilities estimated by the manager (Table 4-3).

TABLE 4-3.   *Probability Revision Example Calculations*

EVENT (Ei)	PRIOR $P\ (Ei)$	CONDITIONAL $P\ (F\backslash Ei)$	JOINT $P\ (Ei) \cdot P\ (F\backslash Ei)$	POSTERIOR $P\ (Ei\backslash F)$
High Demand	0.3	0.80	0.240	0.24/0.575 = 0.4174
Average Demand	0.5	0.55	0.275	0.275/0.575 = 0.4783
Low Demand	0.2	0.30	0.060	0.06/0.575 = 0.1043
			0.575	

Thus, the probability of high demand is greater (0.4174), whereas the probability of low demand is reduced to 0.1043. ◀

As in Example 4.11, another approach to Example 4.12 could have been the use of a tree diagram as shown in Figure 4-12.

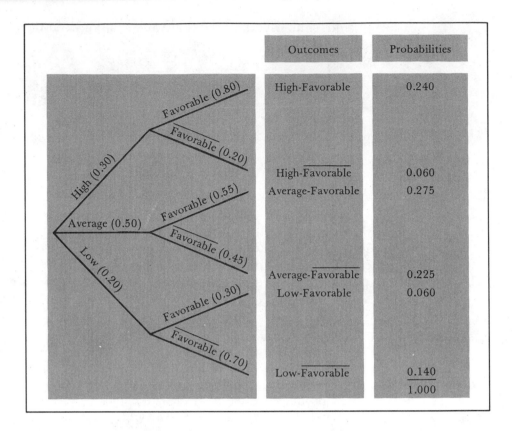

FIGURE 4-12. *Tree Diagram for Bayes' Formula*

As before, we select the probabilities from the tree to make the desired revision. For instance if we want the revised probability associated with high demand, then:

$$P(H \backslash F) = \frac{P(H \text{ and } F)}{P(F)}$$

$$P(H \backslash F) = \frac{P(H \text{ and } F)}{P(H) \cdot P(F \backslash H) + P(A) \cdot P(F \backslash A) + P(L) \cdot P(F \backslash L)}$$

$$P(H \backslash F) = \frac{0.24}{0.24 + 0.275 + 0.060}$$

$$= \frac{0.24}{0.575}$$

$$P(H \backslash F) = 0.4174$$

In a like manner, $P(A \backslash F)$ and $P(L \backslash F)$ can be determined. Why don't you try to calculate them using the answers in Table 4-3 to check your work?

4.1 $\quad P(A) = \dfrac{N(A)}{N(A) + N(\overline{A})},\quad$ *page 70*

4.2 $\quad M_x^n = n^x,\quad$ *page 75*

4.3 $\quad P_x^n = \dfrac{n!}{(n-x)!},\quad$ *page 76*

4.4 $\quad C_x^n = \dfrac{n!}{x!\,(n-x)!},\quad$ *page 78*

4.5 $\quad P(\overline{A}) = 1 - P(A),\quad$ *page 79*

4.6 $\quad P(A \text{ and } B) = P(A) \cdot P(B\backslash A),\quad$ *page 81*

4.7 $\quad P(A \text{ and } B) = P(A) \cdot P(B),\quad$ *page 81*

4.8 $\quad P(A \text{ or } B) = P(A) + P(B) - P(A \text{ and } B),\quad$ *page 84*

4.9 $\quad P(A \text{ or } B) = P(A) + P(B),\quad$ *page 84*

4.10 $\quad P(A\backslash B) = \dfrac{P(A) \cdot P(B\backslash A)}{P(B)},\quad$ *page 89*

REVIEW QUESTIONS

1. Identify what each equation at the end of this chapter calculates.
2. Define:
   (a) probability number
   (b) *a priori* probability
   (c) relative frequency concept of probability
   (d) subjective probability
   (e) combination
   (f) permutation
   (g) mutually exclusive events.
   (h) complement
3. State the characteristics by which a group of numbers can be identified as probability numbers.
4. Explain the purpose of the multiplication law.
5. Explain the purpose of the addition law.
6. Discuss Bayes' Theorem.

1. Phil Traveler, president and owner of Hotel Chain, Inc., always visits one of his hotels in Florida in the summer and one of his hotels in Virginia in the winter. If he owns seven hotels in Florida and three hotels in Virginia, how many different pairs does he have from which to choose?

2. Statistics students at Big U. were given an examination consisting of ten questions. If three questions may be omitted, in how many ways can a student select the problems he will answer?

3. Hotel Chain, Inc., is seeking a new president and two senior vice-presidents. Eight people are serious candidates. In how many ways can a president, a first vice-president, and a second vice-president be chosen?

4. In how many ways can six different economics texts be arranged on a shelf?

5. The Arkansas license plate consists of exactly three letters then three digits. How many distinct license plates can be made using this scheme?

6. If a three-volume set of books is placed on a shelf at random, what is the probability that
   (a) the books will be in the correct order?
   (b) volume 2 will be in the first position?

7. From a container that has 10 identical units, 3 of which are defective, 2 units are drawn in succession without replacement; determine
   (a) the probability that both units are not defective.
   (b) the probability that the second unit drawn is defective.

8. On a slot machine there are three reels with the ten digits, 0, 1, 2, 3, 4, 5, 6, 7, 8, 9, plus a star on each reel. When a coin is inserted and the handle is pulled, the three reels revolve independently several times and come to rest. What is the probability of
   (a) two stars?
   (b) one star?
   (c) no stars?

9. An advertising agency is preparing to bid on a very large account. The potential customer is willing to review two presentations from the agency. Over the past six weeks the agency has prepared 5 possible presentations from which must be selected 2. The manager feels that the order of the presentation might be important in acquiring the contract. To be sure he doesn't overlook any possible arrangement, you are asked to determine how many possible arrangements of the 5 presentations exist.

10. In the Two-Jack Emerald Company, there are five economists and seven engineers. A committee of two economists and three engineers is to be formed. In how many ways can this be done if
    (a) one particular engineer must be on the committee.
    (b) two particular economists cannot be on the same committee.

11. A florist has two show windows at the front of her store. As the Christmas season approaches, she must decide on the window decorations to

use. If there are six possible window decorations, how many possible arrangements are available? (Assume no duplication and order is not important.)

12. If a die is rolled twice, what is the probability that the first roll yields a 5 or 6 and the second roll anything but a 3?

13. Three people are working independently at solving a statistics problem. The respective probabilities that they will solve the problem are 1/4, 1/2, and 1/3. What is the probability that the problem will be solved?

14. Ann, Mary, and Joan Smith are all partners in the law firm of Smith, Smith, and Smith. While they are supposed to be in their office at 9 o'clock in the morning, each Smith has the reputation of not getting to the office on time six out of ten mornings. One morning, Mr. Frank P. Mogul, one of the Smiths' most important clients, enters the offices of the law firm at 9 o'clock sharp. Mr. Mogul is known for his respect for punctuality. His coming is unannounced and he finds none of the three Smiths at the offices. Enraged, he severs his business connections with Smith, Smith, and Smith. When the Smiths hear about his reasons for this, they are upset and ask what the probability is for a client to arrive at 9 o'clock sharp and find none of the three there. What is the answer?

15. Jean and Phyllis practice shooting at a target. If, on the average, Jean hits the target 3 times out of 5, and Phyllis hits it 4 times out of 8, what is the probability, when both shoot simultaneously, that the target is not hit?

16. Four people in turn each draw a card from a deck of 52 cards without replacement. What is the probability that the first person draws the ace of spades, the second the king of diamonds, the third a king, and the fourth an ace?

17. One urn contains 6 red discs and 3 blue discs, and a second urn contains 7 red discs and 4 blue ones. A disc is selected from each urn, and they are placed in a bag containing 4 red and 4 blue discs. What is the probability of drawing a red disc from the bag?

18. In how many ways can 5 identical jobs be filled by selections from 12 different people?

19. Big Al, Susan, and Steve are applying for three different jobs at Smash, Inc. The probability is 1/2 that Big Al will get his job, 1/3 that Susan will get hers, and 3/4 that Steve will be employed.
    (a) What is the probability that all three get their respective jobs?
    (b) None will be employed?
    (c) Only one will get his job?

20. A couple owns an apartment in the city and a cabin at the lake. In any one year the probability of the apartment being burglarized is 0.02, and the probability of the cabin being burglarized is 0.05. For any one year what is the probability that
    (a) both will be burglarized?
    (b) one or the other (but not both) will be burglarized?
    (c) neither will be burglarized?

21. An urn contains 7 red discs and 3 blue discs. After two draws with replacement, what is the probability of:
    (a) 2 red discs.
    (b) only 1 blue disc.

22. Repeat Problem 21 without replacement after the first draw.

23. Two cards are drawn successively at random from a 52-card deck, without replacement, and reshuffling after each draw. What is the probability that one or the other, but not both, will be a heart?

24. From 7 men and 4 women, how many committees can be selected consisting of
    (a) 3 men and 2 women;
    (b) 5 people of which at least 3 are men?

25. If a machine produces to specification 94 percent of the time, what is the probability of two defective items in a row? (Assume independent events.)

26. There is a new form of Greek roulette at Bank U. A student flips 2 honest coins; if 2 heads occur, he must draw a chip from a bag containing 30 black chips and 20 white ones. If 2 heads occur and he selects a white chip, he must shave his beard. What is the probability that he will have to shave?

27. An airline has a daily flight between two cities that is on schedule 80 percent of the time. When the flight is on schedule, the weather has been rated good 90 percent of the time; and when the flight is not on schedule the weather has been rated poor 70 percent of the time. If on a given day, the weather is rated good, what is the revised probability that the flight will be on schedule?

28. In a particular city 40 percent of the households watch BROKEN SET on TV, a program sponsored by Product A. Of those households watching BROKEN SET, 15 percent use Product A. In the city, however, 25 percent of the households use Product A.
    (a) Determine the probability of selecting a household that does not watch BROKEN SET and does not use Product A.
    (b) Determine the probability of selecting a household that does watch BROKEN SET if the household is known to use Product A.

29. The Safety Council did a traffic survey of 1,000 drivers to determine whether age made a difference in the usage of safety belts. Based on the results given below:

	SAFETY BELTS	NO SAFETY BELTS	
Under 40	250	177	427
40 and Over	325	248	573
	575	425	1000

(a) What is the probability that a driver under 40 uses the safety belt?

(b) What is the probability of a driver over 40 not using a safety belt?

(c) Is the use of a safety belt independent of age?

30. Use the information concerning 300 individuals in the table below to answer the following questions:

	CHECKING ACCOUNT	NO CHECKING ACCOUNT	
Savings Account	150	25	175
No Savings Account	50	75	125
	200	100	300

(a) What is the probability of selecting an individual who has a checking account?

(b) What is the probability of selecting an individual who has a savings account?

(c) What is the probabiilty of selecting an individual who has both a savings account and a checking account?

(d) Is having a checking account independent of having a savings account?

31. Suppose that 100 stores have been cross-classified according to management and sales.

	MANAGEMENT			
Sales	A	B	C	TOTAL
High	15	3	2	20
Average	5	50	15	70
Low	1	2	7	10
	21	55	24	100

Use the data to determine

(a) the probability of high sales.

(b) the probability of B management.

(c) the probability of average sales given A management.

(d) the probability of B management given high sales.

(e) if sales and management are independent.

32. The owner of a ladies' speciality shop has cross-classified her customers in the following manner:

	PURCHASE	NO PURCHASE	
Female	70	140	210
Male	50	40	90
	120	180	300

(a) What is the probability that a customer is a female?

(b) If a purchase is made, what is the probability that the purchaser was a male?

(c) Is the likelihood of a purchase independent of who makes the purchase?

33. A candidate for political office runs spot announcements on both TV and radio. A recent survey of registered voters revealed that 50 percent of those interviewed had seen at least one of the TV announcements; of this 50 percent, 60 percent had heard the candidate on radio. Of all the people interviewed, however, 70 percent had heard the candidate on radio.

(a) What is the probability that a registered voter has seen the candidate on TV and heard him on radio?

(b) What is the probability that a voter has seen or heard the candidate?

34. If in Problem 33 a voter is selected who has heard the candidate on radio, what is the probability that the voter has also seen the candidate on TV?

35. It has been determined that 60 percent of all new car sales are made to previous customers. Former customers buy the Deluxe Model 40 percent of the time, and new customers buy the Deluxe Model 30 percent of the time. If a Deluxe Model was purchased, what is the probability that it was sold to a former customer.

36. Empirical evidence indicates that 80 percent of all applicants are good auto insurance risks. A certain test identifies a good risk as such 90 percent of the time but identifies a bad risk incorrectly 20 percent of the time. If an applicant passed the test, what is the probability he is a good risk?

The Random Variable and Probability Distributions

CHAPTER
5

In the previous chapter we were concerned with defining and calculating probability numbers. In this chapter we define the term *random variable,* and show that there are various ways for the occurrence of a random variable to be described. These ways to describe random variables are known as *probability distributions.*

The Random Variable

A random variable is a variable whose value is determined by the chance outcome of an experiment. To illustrate, let us consider a box that contains a red and a green ball. We shall define the outcomes of interest to be the results of two draws, with replacement, from the box. The possible outcomes of the experiment are shown in Figure 5-1. Now if we are interested in how many times the green ball occurs for each outcome, then let the random variable be the *number* of green balls per outcome.

Table 5-1 shows the possible values of the random variable. Notice that the first column in the table represents the outcomes of the experiment, and that the second column is the value of the random variable resulting from each outcome.

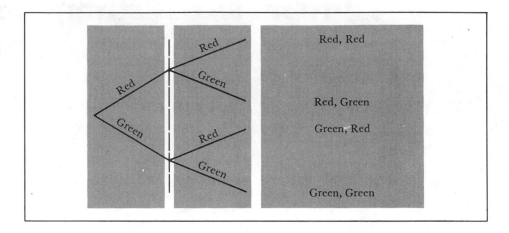

FIGURE 5-1. *Outcomes from Two Draws with Replacement*

TABLE 5-1. *Random Variable: The Number of Green Balls*

OUTCOMES	NUMBER OF GREEN BALLS (THE RANDOM VARIABLE)
Red	0
Red Green	1
Green Red	1
Green Green	2

Classification of Random Variables

A random variable will display one of two classifications. It will either be a *discrete* random variable or a *continuous* random variable.

A discrete random variable is a random variable which may assume a countable or limited number of quantitative values. Table 5-1 is an example of a discrete random variable. The numerical values 0, 1, 1, 2 are certainly countable as there are only three values possible. Thus, these numerical values represent a discrete random variable.

In contrast to the discrete random variable is the continuous random variable. *If, over a range, the random value may assume any numerical value in the range, the random variable is called a continuous random variable.* Notice that the number of values that the random variable might assume would not be countable or finite. That is, the random variable might take on any one of an infinite number of numerical values; thus, both the scale and the random variable are said to be continuous.

Probability Distributions

If we know all possible outcomes of a random experiment, and if we know all values that the random variable can assume, the probability that a particular random variable might occur can be determined. That is, we could determine the probability that the random variable X would take on one of the possible values, namely x_i. Symbolically, this would be written $P\ (X = x_i)$. Further if we determine all the probability numbers for the random variable such that

$$0 \le P\ (X = x_i) \le 1$$

and

$$\sum_{i=1}^{n} P\ (X = x_i) = 1, \text{ then}$$

we have established a probability distribution for the random variable X. Thus, very generally, we can think of a probability distribution as the *listing* of the probabilities for each numerical value that the random variable might assume. There are differences, however, between *discrete* and *continuous* probability distributions. We shall discuss discrete probability distributions first.

Discrete Probability Distributions

To illustrate the general concepts of a probability distribution and of a cumulative probability distribution, we shall employ the discrete random variable. Consider, if you will, the data presented in Table 5-1. Looking at the table, we observe that there are four outcomes (Red Red, Red Green, Green Red, Green Green) that determine the values of the random variable of interest. Further, we observe that each outcome occurs only once. Since each outcome occurs only once and there are four outcomes, the probability of any outcome can be determined *a priori* to be 0.25; thus, a third column concerning probability might be added to Table 5-1 yielding Table 5-2.

TABLE 5-2. *Outcomes, Random Variables, and Probabilities*

OUTCOMES	NUMBER OF GREEN BALLS	PROBABILITIES
Red Red	0	1/4
Red Green	1	1/4 ⎫
Green Red	1	1/4 ⎬ ½
Green Green	2	1/4 ⎭

In Table 5-2 we see that probabilities have been assigned for each value of the random variable (number of green balls). This column of probabilities coincides with our definition of a probability distribution, which is the listing of the probability numbers for each numerical value that the random variable might assume. And if the sum of these probabilities is 1, then we know that we have a complete probability distribution. In our illustration, the probabilities do sum to 1, so we have constructed the probability distribution for the number of green balls that occur in two draws, with replacement, from a container with a red and green ball.

Notice in Table 5-2 that the probability associated with one green ball is really 1/2. This is obtained by adding the probability associated with Red Green to the probability associated with Green Red. Since we are interested in only the random variable (the number of green balls), there are only three numerical values of the random variable in this case.

Table 5-2 can be converted easily into a cumulative probability distribution such as that displayed in Table 5-3. In this table, the probability column is interpreted as the probability that the random variable will be equal to or less than a given numerical value x. Symbolically, the probability column is

$$P\ (X \leq x)$$

Thus, Table 5-3 can be developed in the following manner.

(1) $$P\ (X \leq 0) = P\ (X = 0)$$
           $$P\ (X \leq 0) = \tfrac{1}{4}$$

(2) $$P\ (X \leq 1) = P\ (X = 0) + P\ (X = 1)$$
           $$= \tfrac{1}{4} + \tfrac{1}{2}$$
           $$P\ (X \leq 1) = \tfrac{3}{4}$$

(3) $$P\ (X \leq 2) = P\ (X = 0) + P\ (X = 1) + P\ (X = 2)$$
           $$P\ (X \leq 2) = \tfrac{1}{4} + \tfrac{1}{2} + \tfrac{1}{4}$$
           $$P\ (X \leq 2) = 1.0$$

These results are presented in Table 5-3.

TABLE 5-3. *Cumulative Probability Distribution*

NUMBER OF GREEN BALLS (x)	PROBABILITY $P\ (X \leq x)$
0	¼
1	¾
2	1.00

Now that we have developed the concepts of a probability distribution and of a cumulative probability distribution, we are ready to investigate certain specific probability distributions. First, we will examine four well-known discrete probability distributions.

The Uniform Distribution

If all possible numerical values that a random variable may assume have an equal probability of occurring, then the resulting probability distribution is said to be a *uniform* distribution. The general equation[1] for a uniform distribution is

$$P(x) = 1/N \tag{5.1}$$

This equation calculates the probability that the random variable X will assume the numerical value x where there are N values.

A common illustration of the uniform distribution is the single roll of a fair die. Since there are six surfaces on a die and each side has an equal chance to occur, the probability mass function would be

$$P(x) = 1/6 \text{ for } x = 1, 2, \ldots, 6$$

By defining $N = 6$, we have described a specific uniform distribution. $P(x) = 1/6$ tells us that the random variable X is uniformly distributed and the probability that X equals any of the possible values is $1/6$. Such a distribution as this is called a *one* parameter distribution, because there is only one parameter, N, in the probability equation.

The Binomial Distribution

Suppose that, in tossing three fair coins, we want to determine the probability that exactly two heads will show. The question that arises is how to proceed.

One approach would be to use the counting schemes learned in Chapter 4. Using the multiple choice procedure, we can determine that there are eight different arrangements possible when a coin is tossed three times. $M_x^n = n^x = M_3^2 = 2^3 = 8$. These eight arrangements are shown in Figure 5-2.

---

1.   For discrete distributions, such equations are called probability mass functions (*pmf*). Notice we are using the notation $P(x)$.

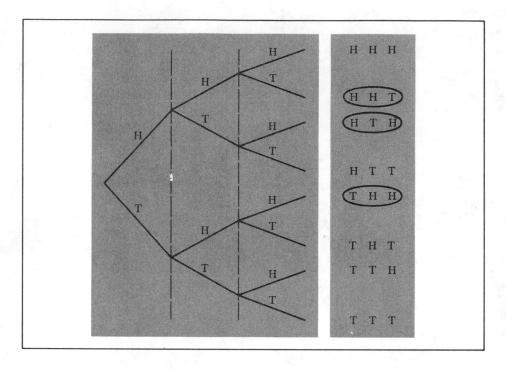

FIGURE 5-2.   *Possible Arrangements of Heads and Tails in Three Tosses*

Next, we need to determine the number of ways that two heads might occur.   In Figure 5-2, these are the circled outcomes.   Thus, the probability of two heads in three tosses of a fair coin is

$$P \text{ (two heads, three tosses)} = \frac{\text{Number of ways to get two heads}}{\text{Number of outcomes from three tosses}} = 3/8 = 0.375$$

Obviously, for large experiments with many outcomes, the preceding procedure would not be very effective.   Fortunately, however, there is an expression for the binomial distribution just as there was for the uniform distribution.   The general equation for a binomial distribution is

$$P(x) = C_x^n \, \pi^x \, (1 - \pi)^{n-x} \tag{5.2}$$

which calculates the probability of x occurrences in n trials. π is the probability that x occurs on any one trial.   Thus, for our preceding probability of two heads in three tosses of a fair coin, Equation 5.2 would yield

$$P(2) = C_2^3 (0.50)^2 (0.50)^1$$

$$= \frac{3!}{2!(3-2)!}(0.50)^2 (0.50)$$

$$= \frac{3!}{2!}(0.25)(0.50)$$

$$= \frac{3 \cdot 2 \cdot 1}{2 \cdot 1}(0.125)$$

$$P(2) = 0.375$$

Equation 5.2, the binomial equation, is valid under three conditions. They are

1. *Each trial is dichotomous; there are only two possible outcomes such as head-tail or success-failure.*
2. *On each trial, the probability of success (head or tail) is equal to π and remains constant.*
3. *There are n independent trials.*

$$P(x) = C_x^n \pi^x (1 - \pi)^{n-x} \tag{5.2}$$

▶ **Example 5.1**   Consider a jar containing four green chips and six red ones. Suppose four chips are drawn *with replacement.*
(a) What is the probability that exactly three of the four chips drawn are green?
(b) What is the probability that at least one chip is green?
(c) What is the probability that at most two of the four chips are green?
      The solution to this probability problem can be found easily by using the binomial formula where:

$$\pi = 4/10 = 0.40$$

$$1 - \pi = 6/10 = 0.60$$

$$n = 4$$

Thus, the specific binomial equation for this situation is:

$$P(X = x) = C_x^4 (0.4)^x (0.6)^{4-x}$$

(a) To answer the first part of the question, we simply substitute 3 for *x*. Therefore,

$$P(X = 3) = C_3^4 (0.4)^3 (0.6)^1$$

$$= (4)(0.064)(0.6)$$

$$P(X = 3) = 0.1536$$

(b) The solution to the second part of the question requires that we find the probabilities for $X = 1, X = 2, X = 3$, and $X = 4$ and sum them. That is,

$$P\,(X > 0) = P\,(1) + P\,(2) + P\,(3) + P\,(4)$$

However, since X can be only 0, 1, 2, 3, or 4, we may simplify the calculations somewhat by taking advantage of the complementarity characteristic of probability numbers. Specifically,

$$P\,(X > 0) = 1 - P\,(X = 0)$$

$$= 1 - C_0^4\,(0.40)^0\,(0.60)^4$$

$$= 1 - 0.1296$$

$$P\,(X > 0) = 0.8704$$

(c) To say "at most two chips are green" is to say there can be no more than two green chips. In other words, there can be 0, 1, or 2 green ones. Thus, we want $P\,(X = 0, 1,\text{ or } 2) = P\,(0) + P\,(1) + P\,(2)$, or

$$P\,(X \leq 2) = C_0^4\,(0.4)^0\,(0.6)^4 + C_1^4\,(0.4)^1\,(0.6)^3 + C_2^4\,(0.4)^2\,(0.6)^2$$

$$= 0.1296 + 0.3456 + 0.3456$$

$$P\,(X \leq 2) = 0.8208 \blacktriangleleft$$

▶ **Example 5.2**   A particular telephone system for Public Surety Insurance Agency is busy 30 percent of the time. Suppose five separate calls are made to Public Surety. What is the probability that one of the five calls gets a busy signal? What assumption are you making regarding the different calls?

Here again we are faced with a probability problem that suggests the binomial. Let X equal the number of busy calls; X can possibly be 0, 1, 2, 3, 4, or 5 in this case. Also $\pi = 0.3$ and $1 - \pi = 0.7$. Using the binomial function, we find that

$$P\,(X = 1) = C_1^5\,(0.3)^1\,(0.7)^4$$

$$P\,(X = 1) = 0.3602$$

The underlying assumption made here is that each call coming to Public Surety is independent of all other calls, which may not really be a good assumption. ◀

After studying the preceding examples, you have probably realized that computations using Equation 5.2 could become quite time consuming. For this reason, tables have been developed to make using the binomial much easier. Appendix D and Appendix E are provided to assist in obtaining binomial probabilities for various values of π, n, and x. Let us demonstrate the use of these tables by determining the answers for Example 5.1.

In Example 5.1(a), Appendix D (additional discussion on the use of the appendix is provided at the beginning of Appendix D) can be used to determine $P(X = 3)$. Across the top of each page in Appendix D are the various values of π that are provided in the table; and down the left side of each page are the various values of n and x. Locating $n = 4$ and then $x = 3$, we go across the row until we are under the column headed $\pi = 0.40$. At the intersection of the row and column is the value, 0.1536, which we previously calculated as $P(X = 3)$.

In Example 5.1(c), it is possible to determine $P(X \leq 2)$ from Appendix E (additional discussion on the use of the appendix is provided at the beginning of Appendix E). In Appendix E, across the top of each group of numbers is the value of n. Locating $n = 4$, we then locate the intersection of $\pi = 0.40$ and $x = 2$ resulting in the value, 0.8208. Thus, $P(X \leq 2) = 0.8208$.

Finally, in Example 5.1(b), we have to be a little more careful in our use of Appendix E. Appendix E determines $P(X \leq x)$, so we will have to subtract as was done in the example to determine $P(X > 0)$. Specifically, we locate the value for $P(X \leq 0)$ in the table and subtract the value from 1. Thus, using Appendix E, when $n = 4$, $\pi = 0.40$, and $x = 0$, then $P(X \leq 0) = 0.1296$; and $P(X > 0) = 1.000 - 0.1296 = 0.8704$.

The Poisson Distribution

Whereas the binomial distribution can be used to determine the probability of a certain number of occurrences in a given number of trials, it is sometimes desirable to be able to determine the probability of the number of occurrences per unit of time or space. When the frame of reference is a unit of time or a unit of space rather than the number of trials, it is appropriate to use the Poisson distribution. The Poisson probability function is shown by the following equation:

$$P(x) = \frac{\mu^x e^{-\mu}}{x!} \tag{5.3}$$

where x = value of the random variable X
μ = the average number of occurrences per unit time or space
e = the base of the natural log system (2.71828)

Examples of situations where the Poisson distribution is particularly useful for computing probabilities include: within a specified time frame, how many machines break down, how many vehicles arrive at a toll booth, how many customers need service, or how many calls come in at a switchboard. Other examples that do not involve time include: number of defects per yard of woven material, number of typeset errors per page of print, and number of fatalities per passenger mile of driving. In each of the preceding situations, the Poisson distribution can be employed to calculate the probabilities associated with the particular number of occurrences per unit of measurement if the following conditions are met.

1. μ, the average number of occurrences per unit of time or space, remains constant.
2. the number of occurrences in one time period is independent of previous occurrences.
3. μ is proportional to the length of time or space and decreases as the length decreases such that the probability of more than one occurrence in a very short length is practically zero.

The following examples demonstrate the use of the Poisson distribution.

▶ **Example 5.3**   The Key Insurance Agency receives an average of 2 calls per minute through its switchboard. Use the Poisson distribution and determine:
(a) the probability of three calls in any one minute.
(b) the probability of five calls in two minutes.

(a) Since $\mu = 2$, the solution is:

$$P(x) = \frac{\mu^x e^{-\mu}}{x!}$$

$$P(X = 3) = \frac{2^3 e^{-2}}{3!}$$

$$= \frac{(8)(2.71828)^{-2}}{6}$$

$$P(X = 3) = 0.1804$$

(b) In this case, $\mu = 2 \cdot 2 = 4$ (2 calls per minute and there are two minutes) since the time measurement associated with μ must be the same as the time measurement associated with x.

$$P(X = 5) = \frac{4^5 e^{-4}}{5!}$$

$$= \frac{(1024)(2.71828)^{-4}}{120}$$

$$P(X = 5) = 0.1563 \ ◀$$

Solution note: The preceding probabilities could have been obtained directly from Appendix F.   Use of the Appendix is explained in a section at its beginning.   Why don't you rework Example 5.3 using Appendix F now!

▶ **Example 5.4**   On a particular thruway there has been an average of three accidents per day. Use Appendix F to determine:
(a) the probability of no accidents on a given day.
(b) the probability of three or four accidents on a given day.

(a) Using Appendix F and $\mu = 3$, the column heading for $\mu = 3$ is located, and the row heading for $x = 0$ is located.   The intersection of these particular row and column yields the answer. Thus:

$$P\,(X = 0) = 0.0498$$

(b) Again using Appendix F and $\mu = 3$, the column heading for $\mu = 3$ is located, and then the row headings for $x = 3$ and $x = 4$ are located.   From the Appendix, the two required probabilities are read as:

$$P\,(X = 3) = 0.2240$$

$$P\,(X = 4) = 0.1680$$

And the solution is:

$$P\,(X = 3 \text{ or } 4) = 0.2240 + 0.1680 = 0.3920 \blacktriangleleft$$

The Poisson distribution can also be used as an approximation for the binomial when μ is small.   In this situation, the mean of the Poisson distribution μ is determined as $n\pi$.

▶ **Example 5.5**   Consider the situation of a very large box of electrical fuses, 4 percent of which are known to be defective.   Determine the probability of selecting 5 fuses at random and obtaining 1 defective fuse.   The problem clearly suggests use of the binomial distribution which yields:

$$P\,(X = 1) = C_1^5\,(0.04)^1\,(0.96)^4$$

$$= 5\,(0.04)\,(0.8493)$$

$$P\,(X = 1) = 0.1699$$

However, using the Poisson approximation we let

$$\mu = n\pi = 5\,(0.04) = 0.20$$

and using Appendix F and $x = 1$, then,

$$P\,(X = 1) = 0.1637$$

Thus, the approximation is accurate through the second decimal place and should be considered quite satisfactory. ◀

The Hypergeometric Distribution

In the cases of both the binomial and the Poisson distributions, the events relating to the random variable were independent. Recall that in the binomial example, the chips were drawn *with replacement.*

The *hypergeometric* mass function is an appropriate model for estimating probabilities of a random variable when selection is made *without replacement* from a finite number of items. In other words, if it is *not reasonable* to assume that the events are independent, the hypergeometric distribution is appropriate for the random variable. As in the case of the binomial, the dichotomy (either-or situation) condition applies to the hypergeometric distribution.

Stated alternatively, the hypergeometric distribution is appropriate when a sample of size n is drawn, without replacement, from a finite population of size N. Under these conditions, the probability that the sample contains exactly x items of the first type and $n - x$ items of the second type can be determined given that the population N contains N_1 items of the first type and N_2 items of the second type.

The formula for the hypergeometric probability mass function pmf is shown as:

$$P(x) = \frac{C_x^{N_1} C_{n-x}^{N_2}}{C_n^N} \tag{5.4}$$

where x = value of the random variable, X
N_1 = the number of one kind of outcome (say a success) from which the selection is made
N_2 = the number of the other kind of outcome (say failure) from which the selection is made
n = the number of items in the random sample
$N = N_1 + N_2$

▶ **Example 5.6**   Consider a jar containing four green chips and six red ones. Suppose four chips are drawn *without* replacement. (They may be drawn all at once.)
(a) What is the probability that exactly three of the four chips drawn are red?
(b) What is the probability that at least one chip is red?

A hypergeometric pmf is the appropriate one for making these probability estimates. We have a dichotomy, but the events are not independent, because the color of the chips drawn each time influences the probabilities of obtaining a red or green chip in the next draw. To answer both questions, let us first define and assign values to the terms of the hypergeometric pmf:

N_1 = number of red chips in the jar = 6
N_2 = number of green chips in the jar = 4
n = number of chips to be selected or drawn = 4
x = value of the random variable, the number of red chips
drawn = 3
$N = N_1 + N_2 = 10$

(a)
$$P(X = 3) = \frac{C_3^6 C_1^4}{C_4^{10}}$$

$$= \frac{\frac{6!}{3!\,3!}\,\frac{4!}{1!\,3!}}{\frac{10!}{4!\,6!}}$$

$$P(X = 3) = \frac{(20)(4)}{210} = \frac{8}{21} = 0.381$$

(b) Here we can find the probability of getting 0 red chips and subtract this probability from 1. This difference will give us the
probability of getting 1, 2, 3, or 4 red chips. Thus, let $x = 0$, and
use Equation 5.4.

$$P(X = 0) = \frac{C_0^6 C_4^4}{C_4^{10}}$$

$$= \frac{\frac{6!}{0!\,6!}\,\frac{4!}{4!\,0!}}{\frac{10!}{4!\,6!}}$$

$$P(X = 0) = \frac{1}{210}$$

Therefore,

$$P(X \geq 1) = 1 - \frac{1}{210} = \frac{209}{210} = 0.9952 \blacktriangleleft$$

Continuous Probability Distributions

The distributions discussed so far have all been discrete distributions.
Before ending our discussion of probability distributions, however, it
is important to introduce continuous distributions. In particular we
need to become familiar with the normal distribution and the exponential distribution. First let us discuss the normal distribution.

The Normal Distribution

Distributions of data are somewhat like people—they come in many shapes. Figure 5-3 contains a few of the more common ones, but the one that is of particular significance is the normal distribution. Why? Because the normal distribution can be used to approximate the distribution of many physical phenomena, and forthcoming chapters deal with several topics based on the knowledge of the normal distribution.

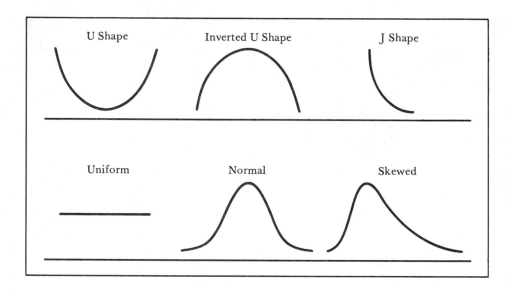

FIGURE 5-3. *Various Distribution Shapes*

With all the possible shapes—both human and statistical—how do you recognize a normal distribution when you see one? A few characteristics of normal distributions might help you to recognize them.

1. Continuous—the values of the variable are infinitely divisible and infinite in number;
2. Symmetrical—the mean, median and mode are equal;
3. Asymptotic—the left- and right-hand tails of the distribution approach the abscissa, but reach it only at infinity.
4. Unimodal—there is only one mode.

The above characteristics will help you to recognize many distributions as being non-normal, but they will not let you be certain that a particular distribution is a normal distribution. The precise shape of a normal distribution is outlined by the following equation:

$$f(x) = \frac{1}{\sqrt{2\pi\sigma^2}} e^{-(x-\mu)^2/2\sigma^2} \tag{5.5}$$

where $f(x)$ = the ordinate value (height) when $X = x$
   x = any value of the random variable, X
   μ = the mean of X
   σ^2 = the variance of X
   π = 3.1416 (a constant)
   e = 2.71828 (base of natural log system—a constant)

Using Equation 5.5 you can determine the coordinates $[x, f(x)]$ for a sufficiently large set of points to produce a normal curve (a graph of a normal distribution) for any arithmetic mean and variance.[2]

The statement of the exact shape of a normal curve enables the statistician to rely upon the existence of a definite relationship between the arithmetic mean, the variance (or standard deviation) and the area under the curve. Specifically, for any normal distribution, the arithmetic mean plus one standard deviation ($\mu + 1\sigma$) will always encompass 34.13 percent of the area under the curve as shown in Figure 5–4. The mean plus two standard deviations ($\mu + 2\sigma$) includes 47.72 percent of the area, and the mean plus three standard deviations ($\mu + 3\sigma$) bounds 49.86 percent of the area. And, since the normal

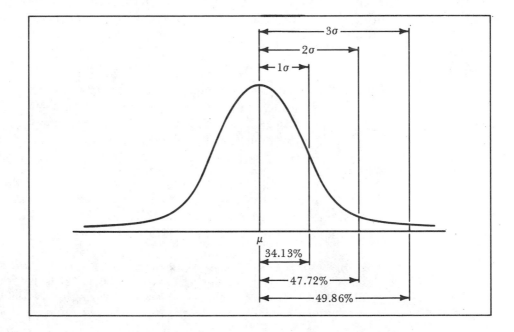

FIGURE 5-4.   *Areas Under a Normal Curve*

---

2.   For continuous variables, the equations are called probability density functions (*pdf*). Notice for the continuous distribution we are using $f(x)$.

curve is symmetrical, these values also hold when the standard deviation is subtracted from the mean.

There are an infinite number of possible normal distributions, each dependent upon the values of μ and σ. However, we can convert any normal distribution (with parameters μ and σ) into a standard form. This standardized distribution is referred to as the *standard normal distribution*.

The standard normal distribution is a normal distribution with parameters, $\mu = 0$ *and* $\sigma = 1$. Appendix H contains the areas under the curve for the standard normal distribution. In Appendix H we see that areas under the normal curve are listed for various values of Z where Z represents the number of standard deviations away from μ. That is if $Z = 1$, then the area between μ and $(\mu + 1\sigma)$ can be read from Appendix H. Thus, we can read in the table the area between the mean and one standard deviation from the mean, which is 0.3413. Look back at Figure 5-4 and you can see the area that we are discussing. Consequently, we use Appendix H to determine areas under the normal curve for any normally distributed random variable.

Fortunately, it is not necessary to convert an entire normal distribution to a standard normal distribution in order to use Appendix H. Rather, it requires only that we measure the number of standard deviations from the mean associated with that portion of the distribution that is of interest. This can be accomplished using Equation 5.6.

$$Z = \frac{X - \mu}{\sigma} \tag{5.6}$$

where   Z = the number of standard deviates
X = some value of the random variable—a point on the abscissa
μ = the mean of the normal distribution
σ = the standard deviation of the normal distribution

Thus, we express the distance from the mean μ along the abscissa in standard deviates (Z's) rather than in units of the variable (dollars, bushels, etc.). Look at the following examples in order to see how Appendix H is used.

▶ **Example 5.7**   A small town has maintained records showing that the daily high temperature during the Spring has averaged 15° Celsius with a standard deviation of 4° Celsius. Assume the random variable (high temperature) is normally distributed and determine the following:

(a) the percentage of days that will have a high temperature between 15°C and 19°C.

(b) the percentage of days that the high temperature will exceed 21°C.

(c) the probability that the high temperature will be less than 12°C.
(d) the high temperature that will be exceeded only 10 percent of the
     time.

(a) In order to determine the area under a normal curve (expressed
     as a percentage) between the mean (15) and a value of $X(19)$, it is
     necessary to determine the appropriate Z and then use Appendix
     H. Look at Figure 5–5 to see the area that is being determined.

$$Z = \frac{X - \mu}{\sigma}$$

$$= \frac{19 - 15}{4}$$

$$Z = 1.0$$

Referring to Appendix H, when $Z = 1$, the area is 0.3413.
Thus, 34.13 percent of the days have temperatures between 15° and
19°.

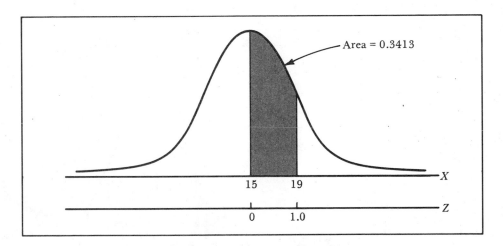

FIGURE 5-5. *Area Under Normal Curve in Example 5.7(a)*

(b) In order to determine the area under a normal curve that is above
     a particular value of X, such as above 21 as shown in Figure 5-6,
     it is first necessary to determine the area between 15 and 21.
     Then the area between 15 and 21 must be subtracted from
     0.5000.

$$Z = \frac{X - \mu}{\sigma}$$

$$= \frac{21 - 15}{4}$$

$$Z = 1.5$$

Using Appendix H, when $Z = 1.5$, the area $= 0.4332$. Thus, the area above 21 is $0.5000 - 0.4332 = 0.0668$; or the percentage of days with high temperatures above 21° is 6.68 percent.

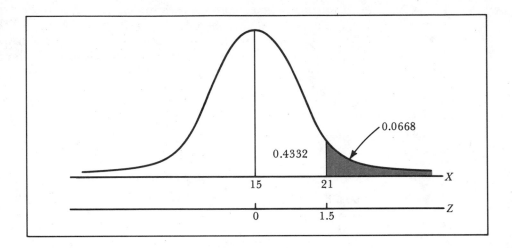

FIGURE 5-6. *Area Under Normal Curve in Example 5.7(b)*

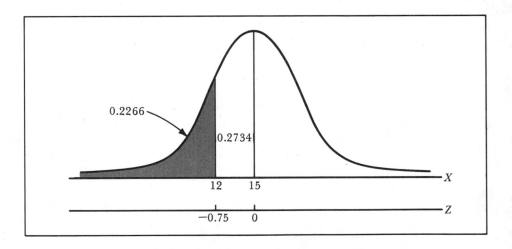

FIGURE 5-7. *Area Under Normal Curve in Example 5.7(c)*

(c) To determine the area under a normal curve below a particular value of X, such as below 12 as shown in Figure 5-7, initially we determine the area between 12 and 15. Then the area between 12 and 15 must be subtracted from 0.5000.

$$Z = \frac{X - \mu}{\sigma}$$

$$= \frac{12 - 15}{4}$$

$$Z = -0.75$$

From Appendix H, if $Z = -0.75$, the area $= 0.2734$. When using Appendix H, ignore the sign (plus or minus) of Z because the sign identifies only whether we are looking on the left (minus sign) side of the mean or the right (plus sign) side of the mean. Thus, the area below 12 is $0.5000 - 0.2734 = 0.2266$; and the probability that the high temperature will be less than 12°C is 0.2266.

(d) Sometimes we know the probability and must determine the value of X associated with that probability. Look at Figure 5-8 to get the idea. First we must determine the value of Z then solve Equation 5-6 for X. Looking again at Figure 5-8, we can see that the area between 15 and X is 0.4000 $(0.5000 - 0.1000)$. Thus, we use Appendix H in a reverse procedure. That is, we look up areas rather than Z values. Looking in Appendix H, the area closest to 0.4000 is 0.3997 which corresponds to $Z = 1.28$. That is, the unknown value of X is 1.28 standard deviations to the right of μ. Now we can solve for X.

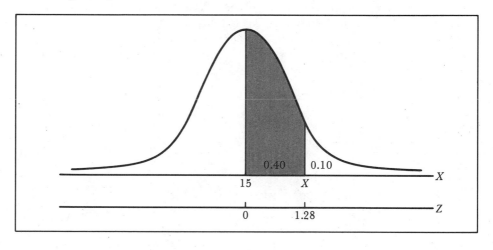

FIGURE 5-8. *Area Under Normal Curve in Example 5.7(d)*

$$Z = \frac{X - \mu}{\sigma}$$

$$1.28 = \frac{X - 15}{4}$$

$$X = 15 + (1.28)\, 4 = 15 + 5.12$$

$$X = 20.12$$

Thus, the temperature that should be exceeded only 10 percent of the time is 20.12°C. ◀

At this point, let us emphasize a few things about the normal distribution.

1. The area under a normal curve contains all (100 percent) of the values of the variable in the normal distribution.
2. One-half of the area of the curve, and thus one-half (50 percent) of the values of the variable, are less than, and one-half are more than, the arithmetic mean.
3. The ranges along the abscissa are infinitely divisible, and the area under the curve at any specific point is 0.
4. Standard deviates are not additive; that is, it is necessary to add or subtract areas (not deviates) to determine the area under the curve over some specific range of the variable.   Let us consider another example.

▶ **Example 5.8**   A gasoline distributor has maintained records that indicate sales of Special gasoline average 3,500 gallons per week with a standard deviation of 250 gallons. Assuming that sales are normally distributed, determine the probability of weekly sales between 3,300 and 3,800 gallons. Look at Figure 5-9 to see the areas that represent the solution to the problem.

Area 1 is determined as follows:

$$Z_1 = \frac{X - \mu}{\sigma}$$

$$Z_1 = \frac{3300 - 3500}{250} = -0.80$$

From Appendix H, Area 1 = 0.2881.

Area 2 is determined in the same manner as Area 1.

$$Z_2 = \frac{X - \mu}{\sigma}$$

$$Z_2 = \frac{3800 - 3500}{250} = 1.2$$

From Appendix H, Area 2 = 0.3849

Now, to obtain the area between 3,300 and 3,800, it is necessary to add Area 1 and Area 2 together. Thus the probability of sales between 3,300 and 3,800 gallons is:

$$P(3300 < X < 3800) = 0.2881 + 0.3849$$

$$P(3300 < X < 3800) = 0.6730 \blacktriangleleft$$

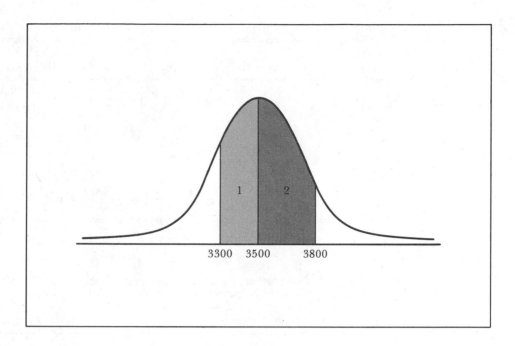

FIGURE 5-9. *Areas Under Normal Curve in Example 5.8*

Try some of the problems at the end of this chapter that involve normal distributions.   If you cannot work them, or do not understand why a particular procedure should be used, reread the normal distribution section of this chapter.   An understanding of the normal distribution will help you comprehend more clearly some of the material presented in subsequent chapters.

Normal Approximation to the Binomial

The expression, $P(x) = C_x^n \pi^x (1 - \pi)^{n-x}$, is increasingly cumbersome to work with as n becomes larger.   Fortunately, though, when n is large and π is neither very large nor very small, the binomial distribution can be approximated by the normal distribution.

In order to use the normal distribution as an approximation to the binomial, it is necessary to calculate the mean and standard deviation of the binomial distribution.   To illustrate this calculation, suppose we investigate the probabilities associated with obtaining either 0, 1, 2, or 3 heads on the three tosses of a fair coin ($\pi = 0.50$). Using Appendix D to determine $P(X = x)$, Tables 5-4 and 5-5 demonstrate the calculation of the mean and standard deviation, respectively.   Thus, from these two tables we see that $\mu = 1.5$ and $\sigma = 0.866$. (This procedure is analogous to calculating the mean and standard deviation using grouped data).

M   *f*   *M·f*

TABLE 5-4. *Possible Outcomes when Tossing Three Coins*

x NUMBER OF HEADS	*P* (*X* = *x*) PROBABILITY	*x* · *P* (*X* = *x*) (NO. OF HEADS) (PROB)
0	0.1250	0.0000
1	0.3750	0.3750
2	0.3750	0.7500
3	0.1250	0.3750
	1.0000	1.5000

$$\mu = \frac{\Sigma [x \cdot P (X = x)]}{\Sigma P (X = x)} = \frac{1.5000}{1.0000} = 1.5 \ \text{heads}$$

M   *f*   *M-μ* $f(M-μ)^2$

TABLE 5-5. *Calculation of Standard Deviation for Binomial Distribution*

x NUMBER OF HEADS	*P* (*X* = *x*) PROBABILITY	(*x* − *μ*)	(*x* − *μ*)²	*P* (*X* = *x*) · (*x* − *μ*)²
0	0.1250	−1.5	2.25	0.28125
1	0.3750	−0.5	0.25	0.09375
2	0.3750	0.5	0.25	0.09375
3	0.1250	1.5	2.25	0.28125
	1.0000			0.75000

$$\sigma = \sqrt{\frac{\Sigma [P (X = x) \cdot (x - \mu)^2]}{\Sigma P (X = x)}}$$

$$\sigma = \sqrt{\frac{0.75}{1.00}}$$

$$\sigma = 0.866$$

The computations in Tables 5-4 and 5-5 are rather time consuming for large values of *n*. Fortunately, μ and σ can be determined much easier using Equations 5.7 and 5.8 as shown below. Then this mean and standard deviation can be used to provide a normal approximation to binomial probabilities. Generally, the normal approximation is reasonably satisfactory when both $n\pi$ and $n (1 - \pi) > 5$.

$$\mu = n\pi \tag{5.7}$$

$$\mu = 3 (0.5)$$

$$\mu = 1.5$$

what order?

$$\sigma = \sqrt{n\pi\,(1-\pi)} \qquad\qquad (5.8)$$

$$\sigma = \sqrt{3\,(0.50)\,(0.50)}$$

$$\sigma = \sqrt{0.75}$$

$$\sigma = 0.866$$

what units?

▶ **Example 5.9**   Suppose that 45 percent of a telephone operator's service requests are long distance calls. What is the probability of having to place 11 long distance calls in 20 service requests? Using the binomial table in Appendix D, and setting $\pi = 0.45$ and $n = 20$, $P\,(X = 11) = 0.1185$.

Now let us approximate this binomial situation by using the normal distribution. First, we must calculate the mean and standard deviation as follows:

$$\mu = n\pi \qquad\qquad \sigma = \sqrt{n\pi\,(1-\pi)}$$

$$= 20\,(0.45) \qquad\quad = \sqrt{20\,(0.45)\,(0.55)}$$

$$\mu = 9.0 \qquad\qquad\quad = \sqrt{9\,(0.55)}$$

$$\sigma = \sqrt{4.95} = 2.22$$

Now look at Figure 5-10 to see how the normal distribution (a continuous distribution) is divided up to approximate the binomial distribution (a discrete distribution). As you know, we would ordinarily calculate only the probability of more than, or less than 11. Thus, in order to calculate the probability of exactly 11, we must make adjustments so that the normal distribution can be used to determine exact probabilities. As you can see in Figure 5-10, the adjustment is to place boundaries (10.5 and 11.5) around the value of interest (11). The purpose is to create an area under the normal curve that can serve as the approximation of $P\,(X = 11)$. Once the boundaries are established, we then calculate the approximate value for $P\,(X = 11)$ by determining $P\,(9 < X < 10.5)$ and subtracting it from $P\,(9 < X < 11.5)$ as shown below.

$$Z = \frac{X - \mu}{\sigma} \qquad\qquad Z = \frac{X - \mu}{\sigma}$$

$$= \frac{10.5 - 9}{2.22} \qquad\qquad = \frac{11.5 - 9}{2.22}$$

$$= \frac{1.5}{2.22} \qquad\qquad\quad = \frac{2.5}{2.22}$$

$$Z = 0.68 \qquad\qquad\quad Z = 1.13$$

$$\text{Area} = 0.2517 \qquad\qquad \text{Area} = 0.3708$$

Thus, $P\,(X = 11) = 0.3708 - 0.2517 = 0.1191$, which is very close to the value (0.1185) from Appendix D. ◀

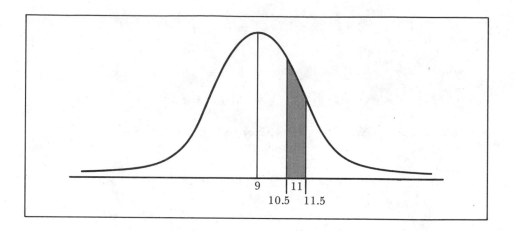

FIGURE 5-10. *Use of the Normal Distribution to Approximate the Binomial*

The Exponential Distribution

Earlier in this chapter you were introduced to the Poisson distribution, which can be used to calculate the probability of the number of occurrences per unit of time or space. Suppose, however, that rather than determining the number of occurrences per unit of time, it is desirable to determine the length of time or space between occurrences. We can do this with the exponential distribution. In particular, if a random process follows the Poisson distribution, the probabilities of the length of time between two occurrences can be determined using the exponential distribution.

The exponential distribution is a continuous distribution, and the function which determines the shape is:

$$f(x) = \frac{1}{\lambda} e^{-x/\lambda} \qquad (5.9)$$

where $f(x)$ = the ordinate value (height) when $X = x$
λ = average time between occurrences (the reciprocal of the Poisson parameter μ)
x = any value of the random variable X
e = 2.71828, the base of the natural log system

Figure 5-11 shows an exponential curve developed using Equation 5.9. In Figure 5.11, λ was assumed to be 0.2 minutes between occurrences, which would correspond to $\mu = 5$ in a Poisson distribution. That is, $\lambda = 0.2$ minutes (12 seconds) is associated with a Poisson process that averaged 5 occurrences per minute.

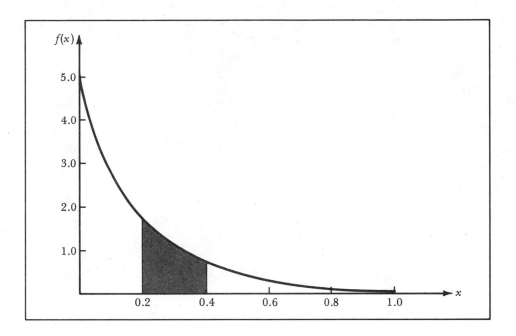

FIGURE 5-11. *Exponential Curve* ($\lambda = 0.2$)

Equation 5.10 can be used to calculate the cumulative probability in the right end of the exponential curve. Equation 5.10 will be employed in Examples 5.10 and 5.11.

$$P\,(X > x) = e^{-x/\lambda} \tag{5.10}$$

where x = time before an occurrence

▶ **Example 5.10**   An electronic part has a performance history that follows the exponential curve such that average time between failures is 150 hours. Thus, for any part, what is the probability that it will operate more than 300 hours.

$$P\,(X > x) = e^{-x/\lambda}$$

$$P\,(X > 300) = e^{-300/150} = e^{-2.0}$$

$$P\,(X > 300) = 0.13534 \text{ (From Appendix I)} \blacktriangleleft$$

▶ **Example 5.11**   At one of the large eastern airports, FAA records show that the average interval between flight arrivals is 4 minutes between 7 A.M. and 9 P.M. The FAA records also indicate that the intervals between arrivals are exponentially distributed. From time to time, the control tower suffers equipment failure. When this happens, it takes no less than 4 but not more than 8 minutes to repair. What is the probability that a plane will arrive during any given equipment failure?

Using Equation 5.10, we can answer this question by determining $P(X > 4)$ and $P(X > 8)$ and then subtracting the two probabilities.

$$P(X > 4) = e^{-4/4}$$

$$P(X > 4) = 0.36788 \text{ (from Appendix I)}$$

$$P(X > 8) = e^{-8/4}$$

$$P(X > 8) = 0.13534 \text{ (from Appendix I)}$$

Thus,

$$P(4 < X < 8) = 0.36788 - 0.13534$$

$$P(4 < X < 8) = 0.23254 \blacktriangleleft$$

SUMMARY OF EQUATIONS

5.1 $\quad P(x) = 1/N, \quad$ *page 103*

5.2 $\quad P(x) = C_x^n \, \pi^x \, (1 - \pi)^{n-x}, \quad$ *page 104*

5.3 $\quad P(x) = \dfrac{\mu^x e^{-\mu}}{x!}, \quad$ *page 107*

5.4 $\quad P(x) = \dfrac{C_x^{N_1} C_{n-x}^{N_2}}{C_n^N}, \quad$ *page 110*

5.5 $\quad f(x) + \dfrac{1}{\sqrt{2\pi\sigma^2}} e^{-(x-\mu)^2/2\sigma^2}, \quad$ *page 113*

5.6 $\quad Z = \dfrac{X - \mu}{\sigma}, \quad$ *page 114*

5.7 $\quad \mu = n\pi, \quad$ *page 120*

5.8 $\quad \sigma = \sqrt{n\pi(1 - \pi)}, \quad$ *page 121*

5.9 $\quad f(x) = \dfrac{1}{\lambda} e^{-x/\lambda}, \quad$ *page 122*

5.10 $\quad P(X > x) = e^{-x/\lambda}, \quad$ *page 123*

REVIEW QUESTIONS

1. Identify the probability distributions represented by the Equations above.
2. Define
   (a) random variable
   (b) discrete random variable

(c) continuous random variable
(d) standard normal distribution
(e) probability distribution

3. What are the two characteristics that identify a group of numbers as being a probability distribution?

4. Name the continuous probability distributions presented in this chapter.

5. Name the discrete probability distributions presented in this chapter.

6. State the three characteristics of a binomial distribution.

7. Identify when the Poisson distribution is appropriate.

8. Identify when the exponential distribution is appropriate.

---------------------------------- **PROBLEMS** ----------------------------------

1. A company's records yielded the following information.

SALES ($1,000)	FREQUENCY
50	2
100	15
150	17
200	13
250	3

(a) Determine the probability distribution for sales.
(b) Construct a cumulative probability distribution for sales.

2. Determine the outcomes that could result if a coin is flipped three times.

3. Assuming the coin in Problem 2 is a fair coin, develop the probability distribution that results if the random variable is defined as the number of heads that occur.

4. Repeat Problem 3 if $P(H) = 0.6$.

5. Develop the cumulative probability distribution for
(a) Problem 3 above
(b) Problem 4 above

6. Identify which of the following are probability distributions:
(a) $P(X = x) = x$ for $x = 1/6, 1/3, 1/2$
(b) $P(X = x) = 2/x$ for $x = 3, 4, 5$

(c) $P(X = x) = \dfrac{x - 2}{12}$ for $x = 1, 2, 3, 4$

(d) $P(X = x) = \dfrac{x^2}{14}$ for $x = 1, 2, 3$

7. Identify which of the following are probability distributions:

(a) $P(X = x) = \dfrac{x - 3}{9}$ for $x = 3, 4, 5, 6, 7$

(b) $P(X = x) = \dfrac{x}{3}$ for $x = \dfrac{1}{2}, 1, \dfrac{3}{2}$

(c) $P(X = x) = \dfrac{x^2 - 1}{50}$ for $x = 2, 3, 4, 5$

(d) $P(X = x) = \dfrac{1}{x}$ for $x = 2, 4, 6, 12$

8. Historically, 20 percent of the inquiries concerning a newspaper ad result in a sale. If 8 inquiries result, what is the probability of no more than 2 sales? (Appendix D or E will help.)

9. If 40 percent of all consumers use product CTR, what is the probability that in a sample of 10 consumers,
(a) exactly 4 will be CTR users?
(b) at least 4 will be CTR users?
(c) 8 or more will be CTR users?
(d) no more than 1 will be a CTR user?

10. In the manufacture of no-deposit-no-return bottles, 5 percent of the bottles are defective. What is the probability in a sample of 7 bottles that there are:
(a) no defectives.
(b) 3 defectives.
(c) more than 3 defectives.

11. If actuarial tables indicate that 85 percent of 10-year olds will survive 40 years, what is the probability that only one of three 10-year olds will survive 40 years?

12. Use Appendix D or Appendix E to determine the following probabilities:
(a) $P(X = 4)$ if $n = 15$ and $\pi = 0.40$
(b) $P(X \le 4)$ if $n = 12$ and $\pi = 0.10$
(c) $P(5 \le X \le 11)$ if $n = 20$ and $\pi = 0.35$
(d) $P(X = 6)$ if $n = 14$ and $\pi = 0.65$
(e) $P(X \ge 9)$ if $n = 19$ and $\pi = 0.60$
(f) $P(5 \le X < 8)$ of $n = 14$ and $\pi = 0.75$

13. Use Appendix D or Appendix E to determine the following probabilities:
(a) $P(X = 6)$ if $n = 14$ and $\pi = 0.25$
(b) $P(X \le 5)$ if $n = 11$ and $\pi = 0.20$
(c) $P(X \ge 4)$ if $n = 11$ and $\pi = 0.40$
(d) $P(X = 7)$ if $n = 13$ and $\pi = 0.75$
(e) $P(2 \le X \le 5)$ if $n = 18$ and $\pi = 0.30$
(f) $P(4 < X \le 9)$ if $n = 17$ and $\pi = 0.40$

14. If vehicles arrive at the toll booth at an average of 2 per minute:
(a) What is the probability of no arrivals in one minute?
(b) What is the probability of 1 arrival in one minute?
(c) What is the probability of 5 arrivals in three minutes?

15. The long distance operator receives an average of 3 calls per minute.
   (a) What is the probability of no calls in one minute?
   (b) What is the probability of 2 calls in one minute?
   (c) What is the probability of 4 calls in three minutes?

16. A weaving machine has a history of causing a defect in the cloth on the average of one defect per 10 yards of material produced.
   (a) What is the probability of no defects in 10 yards of cloth?
   (b) What is the probability of no defects in 1 yard of cloth?
   (c) What is the probability of two defects in 5 yards of cloth?

17. If computer requests from terminals are Poisson distributed with an average of 4 requests per minute, what is the probability of
   (a) 2 requests in one minute?
   (b) 6 requests in two minutes?
   (c) no more than five requests in one minute?

18. Candy machines break down on the average of 2 per week.
   (a) What is the probability of no breakdowns in a week?
   (b) What is the probability of 1 breakdown in ¼ of a week?
   (c) What is the probability of 6 breakdowns in 2 weeks?

19. A box contains 100 marbles. Five of the marbles are purple and the rest are green. If 8 marbles are drawn without replacement, what is the probability of obtaining exactly 2 purple marbles?

20. A retailer has 80 cans of vacuum packed tennis balls. If 3 of the cans contain yellow balls, what is the probability that a sample of 10 cans will contain one can of yellow balls?

21. A machine produces its output with a fraction defective of 0.06. If a lot consists of 100 items, what is the probability that a sample of 5 items will contain exactly 1 defective?

22. Assume a normal distribution with $\mu = 45$ and $\sigma = 10$, and determine the following probabilities.
   (a) $P(X < 65)$ (d) $P(X > 50)$
   (b) $P(X > 35)$ (e) $P(X < 42)$
   (c) $P(35 < X < 55)$ (f) $P(37 < X < 44)$

23. Assume a normal distribution with $\mu = 30$ and $\sigma = 5$, and determine the following probabilities.
   (a) $P(X > 34)$ (d) $P(X > 40)$
   (b) $P(X < 27)$ (e) $P(X < 24)$
   (c) $P(31 < X < 33)$ (f) $P(27 < X < 32)$
            (g) $P(25 < X < 29)$

24. The weights of grapefruit from The Valley average 15 ounces with a standard deviation of 1.6 ounces. If the weights are normally distributed, determine the percentage of grapefruit between:
   (a) 15 and 18 ounces.
   (b) 13 and 14.5 ounces.
   (c) 16 and 17.2 ounces.

25. From Problem 24, only 10 percent of the grapefruit would be larger than what weight?

26. Jane's average trip time from Elkstown to Newton is one hour and 45 minutes with a standard deviation of 10 minutes. Assuming a normal distribution, what percent of the trips are:
(a) between one hour and 30 minutes and one hour and 55 minutes?
(b) greater than two hours?
(c) less than one hour and 40 minutes?

27. Snow Cones, Ltd., has daily ice consumption that is normally distributed with $\mu = 600$ pounds and $\sigma = 25$ pounds. If it is desirable not to run out of ice more than 5 percent of the time, how many pounds of ice should be purchased for each day?

28. A manufacturer of batteries has determined the mean life of their battery in a smoke detector is 15 months with a standard deviation of 1.5 months. If battery lives are normally distributed, what should be the length of the guarantee if they want to replace a maximum of:
(a) 5 percent? (b) 2 percent?

29. If $\pi = 0.30$, use the normal approximation for the binomial to determine the probability of 140 defectives in a lot size of 500 items.

30. A manufacturing process ordinarily produces 35 defective bottles per 100 bottles manufactured. If a lot size of 1,000 bottles is manufactured, determine the probability of
(a) exactly 370 defectives. (b) more than 370 defectives.

31. A very expensive lightbulb has been designed such that the average time between failures is 1,000 hours. What is the probability a particular bulb will last more than 1,500 hours? (Assume an exponential distribution.)

32. A study has determined that long-distance, operator-assisted calls occur at the rate of 1 every 5 minutes. Assuming that the time interval between occurrences is exponentially distributed, determine the probability that more than 8 minutes will elapse before service is needed.

33. A recent study by a store manager established the following frequency distribution for customers checking out of the store during a very busy afternoon.

Assume the time interval between checkouts is exponentially distributed and determine
(a) the probability of more than 1 minute elapsing before the next customer arrives to check out.
(b) the probability of more than 3 minutes elapsing before the next customer arrives to check out.

NUMBER OF CUSTOMERS ARRIVING AT CHECKOUT PER MINUTE	FREQUENCY
0	2
1	7
2	15
3	12
4	11
5	10
6	6

Random Sampling and the Sampling Distribution of Sample Means and Proportions

CHAPTER
6

The method of *inductive reasoning* involves drawing a conclusion or making some generalization based upon limited or less than complete (specific) information. In a vague sense, it is often stated that inductive reasoning involves "reasoning from the specific to the general." In statistics, using "limited information," obtained through *sampling,* to make generalizations is a form of inductive reasoning.

Elements of Statistical Sampling

If you have ever had a large bag of jelly beans, you know how curious you can be about their various colors and flavors. You might want to know, for instance, how many beans, or what proportion of the total number, are licorice.

Suppose that each jelly bean has a number on it. In addition to increasing your curiosity about who would put numbers on jelly beans, you might want to determine the mean, the median, or the modal averages of these numbers. Besides merely satisfying your curiosity, it might be important to learn something about the different characteristics of the jelly beans in your large bag. You could examine (or eat) the entire bag of beans and could thus gain the information you desire. But you would have no more jelly beans. Sampling, using inductive reasoning, would be another way to determine some of the characteristics of the jelly beans in the bag.

The purpose of *sampling* is usually to make an *inference*. To infer, in this sense, means to use limited information from a sample to generalize about one or more characteristics of some larger body of data. In our example with the bag of jelly beans, then, we might draw a smaller number than the total number of beans in the bag (the sample) in order to learn something about the characteristics of all the beans in the entire bag (a population).

Now let us examine the jelly bean illustration in terms of the practical aspects of sampling. The entire collection of all jelly beans in the bag is called the *universe*. This universe, consisting of any number of jelly beans, may generate several variables. One variable might be the number of jelly beans of various colors; another variable might be the proportion of the jelly beans of various flavors; and another variable might be the numbers on each jelly bean. Each variable generates a *population* distribution. General characteristics of the population (color, flavor, numbers as in our illustration) are called *parameters*. The *population parameters* are what we try to make inferences about.

The components in the universe from which a selection is drawn are called the *sample units*. In our example, each individual jelly bean is a *sample unit* because that is the basic component from which the selection is made. The basic components providing the values of the variables or data generating the various populations are called *sample elements*. In other words, *sample units* actually selected make up the sample, and *sample elements* provide the quantitative data. A sample, then, is any number of components drawn or selected that does not include the entire group of the units in the universe. A characteristic measure of the sample is called an *obtained statistic*, or an *estimator*. *Estimators* (obtained from the sample) estimate *parameters* (characteristics of the entire population).

Random Sampling

As stated previously, the purpose of sampling is to make an inference or generalization about a population characteristic. In order to make valid generalizations, it is necessary to select random (probability) samples. The concept of random sampling implies that no external factors affect the selection of the sample units; that is, the person collecting the sample has no influence over the specific units that either appear or do not appear in the sample.

A *random* or *probability sample* is a sample in which the probability of any item in the population being selected is known. All samples used in this text will be assumed to be random samples. To be quite specific, in this text all samples will be assumed to be *simple random samples*.[1]   Since this is the case, it is necessary to investigate the concept of a *sample random sample* selected from a *finite population* and from an *infinite population*.

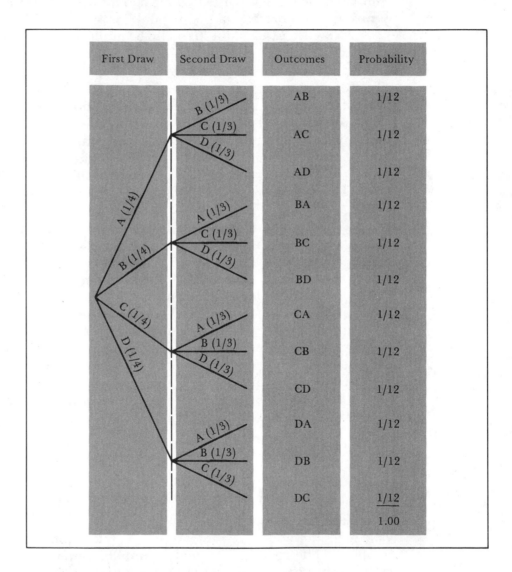

FIGURE 6-1.   *Simple Random Sample from a Finite Population*

---

1.   Other types of random samples (cluster, stratified, and systematic) are discussed in Chapter 10.

Simple Random Sampling: A Finite Population

If a sample containing n elements is selected from a finite population of N elements when $n < N$, it is a *simple random sample if each sample of size n has an equal probability of being selected.*

For example, let us consider a finite population containing four elements, A, B, C, and D. Suppose we select a sample of two elements from this finite population. Figure 6-1 shows the possible samples and the probability that any of the 12 possible samples will be selected is equal to 1/12.

A closer examination of Figure 6-1 reveals that there are only six individual samples if viewed from the eyes of the individual taking the sample. That is, the outcome AB is distinguishable from BA only by noting the order of selection. If the order of selection is ignored, then there are only six outcomes (AB or BA), (AC or CA), (AD or DA), (BC or CB), (BD or DB), and (CD or DC). And, in fact, as far as the sample is concerned, it is the elements in the sample and not the order of the elements that is important. Thus, any of the six outcomes has a probability of selection of $1/12 + 1/12 = 1/6$.

Now, realizing that the number of sample outcomes possible when order is not considered is 6, it is apparent that the total number of outcomes could have been determined using the *combinatorial equation*. That is, from a finite population of 4 elements, the number of possible samples of size 2 is

$$C_2^4 = \frac{4!}{2! \, (4 - 2)!} = \frac{4 \cdot 3 \cdot 2 \cdot 1}{2 \cdot 2} = 6$$

and extending our knowledge to generate the expressions for the probability of any sample of size two from a finite population of four, the probability becomes

$$\frac{1}{C_2^4} = 1/6$$

Thus, we see that from a finite population of size N, simple random samples of size n have equal probability which can be calculated as

$$\frac{1}{C_n^N}$$

Finally, you should notice that the preceding illustration relied on sampling *without replacement,* the most common type of sampling.

Simple Random Sampling:
An Infinite Population

Sampling from an infinite population is the same as sampling from a finite population *with replacement*. A simple random sample for an infinite population is a sample of size n selected so that each element of the universe has an equal chance of being chosen as a sample unit. This results in each sample of n items having an equal chance of being selected. Thus, it is apparent that this is the same definition of a simple random sample as that given for a finite population.

Under the infinite population, however, the assumption of independence is met. Obviously this is the case with replacement, but not the case without replacement.

The Sampling Distribution of Sample Means

As stated earlier, the usual purpose of sampling is to obtain information that can be used to describe a large body of data. The purpose might be to estimate the population mean of a particular universe. The procedure would be to select a sample of size n, determine $\overline{X} = \Sigma X/n$, and use $\overline{X}$ as a point estimate of μ.

Now, realizing that the sample from which the estimator, $\overline{X}$, would be obtained is only one of many possible samples (as demonstrated in Figure 6-1), it is reasonable to raise the question, "How good is the estimate of μ?" To answer such a question, we must investigate what is known as the sampling distribution of $\overline{X}$. That is, what does the sampling distribution of all possible $\overline{X}$'s using a sample of size n look like?

First, let us state (without proof) that if the random samples were selected from a population that was known to be normally distributed with a finite mean μ and standard deviation σ, then the sampling distribution of sample means ($\overline{X}$'s) will be normally distributed. We should recognize, however, that on many occasions the population may not be normally distributed. Then, we have to use the Central Limit Theorem to help determine the shape of the sampling distribution of sample means.

The *Central Limit Theorem,* as you will soon begin to realize, is a very important theorem in statistical sampling. In fact, the Central Limit Theorem will form the basis from which most of the interval estimation and inferences in this text will be made. So what does it say?

CENTRAL LIMIT THEOREM: The central limit theorem states that if all samples of size n are selected from a population with a finite mean μ and standard deviation σ, that as n is increased in size the distribution of sample means $\overline{X}$'s will tend toward a normal distribution with mean of μ (the population mean) and a standard deviation equal to $\sigma/\sqrt{n}$ (called the standard error of the mean) or $\sigma_{\overline{x}}$.

In other words, it says that sample means, $\overline{X}$'s, tend to have a frequency distribution that is normally distributed; and this occurrence is generally the case regardless of the shape of the population frequency distribution. Usually, a sample size of $n > 30$ is considered large enough for the Central Limit Theorem to be effective. As you can imagine, this theorem is quite powerful.

Sampling from Infinite Populations

[handwritten: example is from a finite population, with replacement]

Let us investigate further this prediction of normality by the Central Limit Theorem. To do so, we will use a hypothetical distribution having as members of the population the numbers 1, 2, 3, and 4. Notice in Figure 6-2 that this hypothetical distribution is a uniform distribution. Notice, also, that the hypothetical distribution is a finite distribution with the following parameters:

$$\mu = \frac{\Sigma X}{N} = \frac{10}{4} = 2.5, \text{ and } \sigma = \sqrt{\frac{\Sigma (X - \mu)^2}{N}} \quad \sqrt{\frac{5.00}{4}} = 1.12$$

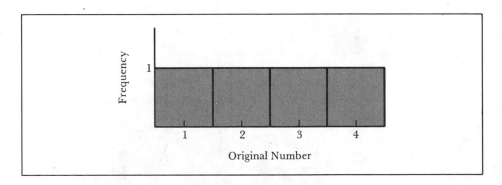

FIGURE 6-2. *Histogram of Original Data*

Now, let us take all possible samples of size 2 from this population under the assumption of replacement which, as stated previously, is equivalent to sampling from an infinite population. Table 6-1 shows all sixteen samples and their sample means.

TABLE 6-1. *Samples and Sample Means*

SAMPLE NUMBER	OBSERVED VALUES	MEAN
1	1, 1	1.0
2	1, 2	1.5
3	1, 3	2.0
4	1, 4	2.5
5	2, 1	1.5
6	2, 2	2.0
7	2, 3	2.5
8	2, 4	3.0
9	3, 1	2.0
10	3, 2	2.5
11	3, 3	3.0
12	3, 4	3.5
13	4, 1	2.5
14	4, 2	3.0
15	4, 3	3.5
16	4, 4	4.0

$$\text{number of samples} = n' = M_x^n = M_2^4 = 4^2 = 16$$

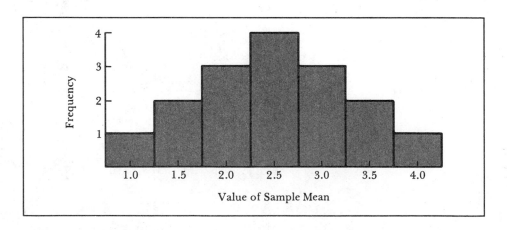

FIGURE 6-3. *Histogram for Sampling Distribution of Means when n = 2*

Using the data in Table 6-1 and plotting a histogram yields Figure 6-3. You should find it interesting to compare Figures 6-2 and 6-3. Why? Because from a uniform distribution of only four numbers (Figure 6-2) came a symmetrical distribution of sample means of size 2 (Figure 6-3). Thus, you can see that the distribution of sample means has started towards a normal one as the sample size increases. If you think of Figure 6-2 as a histogram of sample means when $n = 1$, it may help you to see the symmetry developing.

The Central Limit Theorem suggests that the parameters of the sampling distribution will be a mean equal to μ and a standard deviation equal to $\sigma/\sqrt{n}$. To make sure, we can calculate the mean and standard deviation of the sampling distribution.

First, let us determine the mean of the sampling distribution which we will denote as $\overline{\overline{X}}$. Now $\overline{\overline{X}}$ is equal to the sum of the sample means shown in Table 6-1 divided by n' where n' is the number of samples selected (16).

$$\overline{\overline{X}} = \frac{\Sigma \overline{X}}{n'} = \frac{40}{16} = 2.5$$

And we see that $\overline{\overline{X}} = 2.5$ which is also the value of μ, the population mean.[2]

Now, let us learn more about the standard deviation of the sampling distribution of $\overline{X}$'s. Call this measure of dispersion the *standard error of the mean* and denote it $\sigma_{\bar{x}}$. In fact, $\sigma_{\bar{x}}$, as the standard deviation of the distribution of sample means, is computed just like any other standard deviation. Compare the following formulas to those you learned in Chapter 3.

ungrouped data

$$\sigma_{\bar{x}} = \sqrt{\frac{\Sigma \, (\overline{X} - \overline{\overline{X}})^2}{n'}} \qquad \textbf{(6.1)}$$

grouped data

$$\sigma_{\bar{x}} = \sqrt{\frac{\Sigma f \, (\overline{X} - \overline{\overline{X}})^2}{n'}} \qquad \textbf{(6.2)}$$

---

2.   This computation will hold for any set of data. That is, $\dfrac{\Sigma \overline{X}}{\text{all samples of size } n} = \mu$

Try it for yourself.

Or, from the Central Limit Theorem we know that the standard deviation of the sampling distribution is

$$\sigma_{\bar{x}} = \frac{\sigma}{\sqrt{n}} \qquad (6.3)$$

The information in Table 6-1 is presented again in Table 6-2. But this time it is shown as a frequency distribution. The calculations included in the table show the same numerical result using either Equation 6.2 or 6.3.

TABLE 6-2. *Calculation of Standard Error of the Mean*

SAMPLE MEAN $(\bar{X})$	FREQUENCY	$X - \bar{\bar{X}}$	$f(\bar{X} - \bar{\bar{X}})^2$
1.0	1	-1.5	2.25
1.5	2	-1.0	2.00
2.0	3	-0.5	.75
2.5	4	0	0
3.0	3	$+0.5$	.75
3.5	2	1.00	2.00
4.0	1	1.5	2.25
	16		10.00

$$\bar{\bar{X}} = \mu = 2.5, \ \sigma_{\bar{x}} = \sqrt{\frac{\Sigma f(\bar{X} - \bar{\bar{X}})^2}{n'}} = \sqrt{\frac{10}{16}} = 0.79 \qquad \textbf{(Equation 6.2)}$$

$$\sigma_{\bar{x}} = \frac{\sigma}{\sqrt{n}} = \frac{1.12}{\sqrt{2}} = \frac{1.12}{1.41} = 0.79 \qquad \textbf{(Equation 6.3)}$$

To summarize, we have seen that the Central Limit Theorem is very important. Because of the Central Limit Theorem, we will expect the distribution of sample means to tend towards normality regardless of the shape of the original population distribution. We will expect the tendency towards normality to be stronger as the sample size, n, is increased. And, finally, the parameters of the sampling distribution are estimated as mean = μ, and standard deviation (called standard error of the mean) = $\sigma/\sqrt{n}$.

Sampling from Finite Populations

The preceding discussion was for sampling from an infinite population. However, sampling from finite populations is more likely to occur in business applications. Thus, we must investigate the effect

of the finite population on the estimate of the parameters for the
sampling distribution.   To avoid overcomplicating the situation, it
will suffice to say that there is an effect on the estimation procedure due to the finite population, but that this effect can be corrected.

First, to demonstrate the effect, let us repeat Table 6-1 in
Table 6-3.   Notice, however, that four of the sample means have
been eliminated.   That is because sampling without replacement
eliminates sample numbers 1, 6, 11, and 16.

TABLE 6-3.   *Sampling Without Replacement*

SAMPLE NUMBER	OBSERVED VALUES	MEAN
~~1~~	~~1, 1~~	~~1.0~~
2	1, 2	1.5
3	1, 3	2.0
4	1, 4	2.5
5	2, 1	1.5
~~6~~	~~2, 2~~	~~2.0~~
7	2, 3	2.5
8	2, 4	3.0
9	3, 1	2.0
10	3, 2	2.5
~~11~~	~~3, 3~~	~~3.0~~
12	3, 4	3.5
13	4, 1	2.5
14	4, 2	3.0
15	4, 3	3.5
~~16~~	~~4, 4~~	~~4.0~~

It may be a good idea to remind you that the original population is the same.   It is still the four numbers 1, 2, 3, and 4.   It is
still a uniform distribution as shown in Figure 6-2, and it still has
$\mu = 2.5$ and $\sigma = 1.12$.   Remember, it is the distribution of $\overline{X}$'s that
has changed, because the number of possible samples has been re-
duced from 16 to 12.   *Thus, the mean and standard error of the mean
must be determined for sampling distributions of finite populations.*   These
calculations are shown in Table 6-4.

As you can see from Table 6-4, the mean of the sampling dis-
tribution of $\overline{X}$'s has remained the same, $\overline{\overline{X}} = 2.5$; however, the standard
deviation of the sampling distribution has changed.   This is because
there are only 12 sample means, and the dispersion is not the same as
the previous 16 sample means (sampling with replacement).

TABLE 6-4. *Sampling Distribution from Finite Population*

SAMPLE MEAN	FREQUENCY	$f\overline{X}$	$\overline{X} - \overline{\overline{X}}$	$f(\overline{X} - \overline{\overline{X}})^2$
1.5	2	3.0	−1.0	2.00
2.0	2	4.0	−0.5	0.50
2.5	4	10.0	0	0
3.0	2	6.0	0.5	0.50
3.5	2	7.0	1.0	2.00
	12	30.0		5.00

$$\overline{\overline{X}} = \frac{30}{12} = 2.5 \qquad \sigma_{\overline{x}} = \sqrt{\frac{\Sigma f(\overline{X} - \overline{\overline{X}})^2}{n'}} = \sqrt{\frac{5}{12}} = 0.645$$

Direct calculation of the standard error of mean using the population standard deviation can be accomplished using Equation 6-4. Notice that Equation 6-4 requires use of the *finite population correction factor* (fpc), which is $\sqrt{(N - n)/N - 1)}$.

$$\sigma_{\overline{x}} = \frac{\sigma}{\sqrt{n}} \sqrt{\frac{N - n}{N - 1}} \tag{6.4}$$

For our example

$$\sigma_{\overline{x}} = \frac{1.12}{\sqrt{2}} \sqrt{\frac{4 - 2}{4 - 1}} = (0.79)(0.816) = 0.645$$

the same as shown in Table 6-4 where Equation 6.2 was used.

Using the Sampling Distribution of Sample Means

Let us reiterate the guidelines for establishing the shape of the sampling distribution of sample means obtained from a population with a finite mean μ and standard deviation σ. First, note that if the sample size is large ($n > 30$), the sampling distribution of sample means is normally distributed as explained by the Central Limit Theorem; and second, remember that if the random samples were taken from a normal population, the distribution of sample means will be normal regardless of the sample size and without use of the Central Limit Theorem. Under either of these situations, the sampling distribution will be as shown in Figure 6-4.

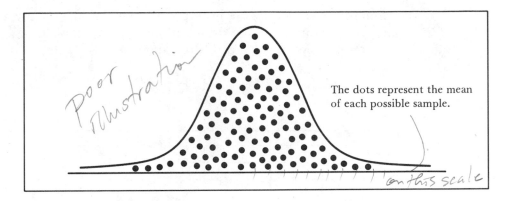

The dots represent the mean of each possible sample.

FIGURE 6-4. *Sampling Distribution of Sample Means*

Now that the shape of the sampling distribution of sample means has been established, we need to know how to make use of this information. Since the sampling distribution of sample means will be normally distributed under the conditions stated above, it is reasonable to suspect that we can make the same type of computations and probability statements for $\overline{X}$ that we did for a normally distributed random variable X in Chapter 5. Recall from Chapter 5 that any normal distribution could be converted to the standard normal distribution by using Equation 5.6 which is repeated below.

$$Z = \frac{X - \mu}{\sigma}$$

where Z = distance (measured in standard deviations) that the random variable X is from the true population mean,
          X = random variable
          μ = population mean
          σ = population standard deviation

The same approach can be extended to a sampling distribution of sample means to yield:

$$Z = \frac{\overline{X} - \mu}{\sigma_{\overline{x}}} \tag{6.5}$$

Equation 6.5 results from putting $\overline{X}$, which is also a random variable, in place of X, and putting $\sigma_{\overline{x}}$, the standard deviation of the sampling distribution (standard error of the mean), in place of σ. Figure 6-5 shows how the normally distributed $\overline{X}$'s are dispersed about the population mean. And, because the sampling distribution is normally distributed, we can relate the probabilities associated with any normal distribution to our sampling distribution.   For example 68.27 percent of the sample means lie within the range of $+1\sigma_{\bar{x}}$ and $-1\sigma_{\bar{x}}$ from the mean, μ.

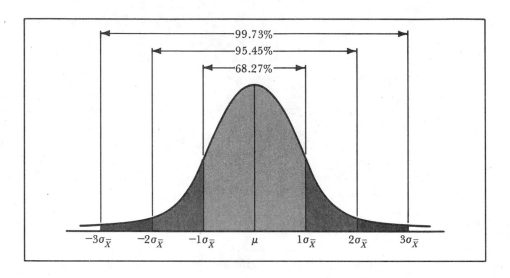

FIGURE 6-5.   *Sampling Distribution of Sample Means*

▶ **Example 6-1**   If a random sample of 25 items was selected from a normally distributed population with $\mu = 40$ and $\sigma = 5$, what is the probability that the sample mean will exceed 41.20?   (See Figure 6-6)

$$Z = \frac{\overline{X} - \mu}{\sigma_{\bar{x}}}$$

where

$$\sigma_{\bar{x}} = \frac{\sigma}{\sqrt{n}}$$

$$Z = \frac{41.2 - 40}{5/\sqrt{25}}$$

$$= \frac{1.2}{1}$$

$$Z = 1.2$$

Area = 0.3849 (From Appendix H)

$P\,(X > 41.20) = 0.5000 - 0.3849 = 0.1151$ ◀

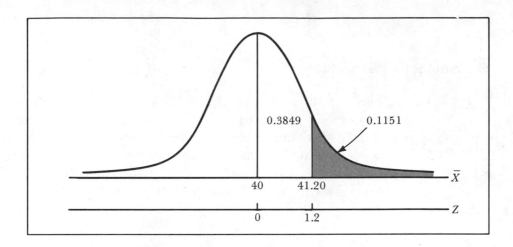

FIGURE 6-6. *Area Under Normal Curve in Example 6-1*

▶ **Example 6-2** Suppose the checking account balances at a large bank had $\mu = \$925$ and $\sigma = \$72$. What is the probability that a sample of 81 accounts would have an $\overline{X}$ less than 913? (See Figure 6-7)

$$Z = \frac{\overline{X} - \mu}{\sigma_{\overline{x}}}$$

where

$$\sigma_{\overline{x}} = \frac{\sigma}{\sqrt{n}}$$

$$= \frac{913 - 925}{72/\sqrt{81}} = \frac{-12}{8}$$

$$Z = -1.5$$

Area = 0.4332 (From Appendix H)

$$P\,(\overline{X} < \$913) = 0.5000 - 0.4332 = 0.0668 \blacktriangleleft$$

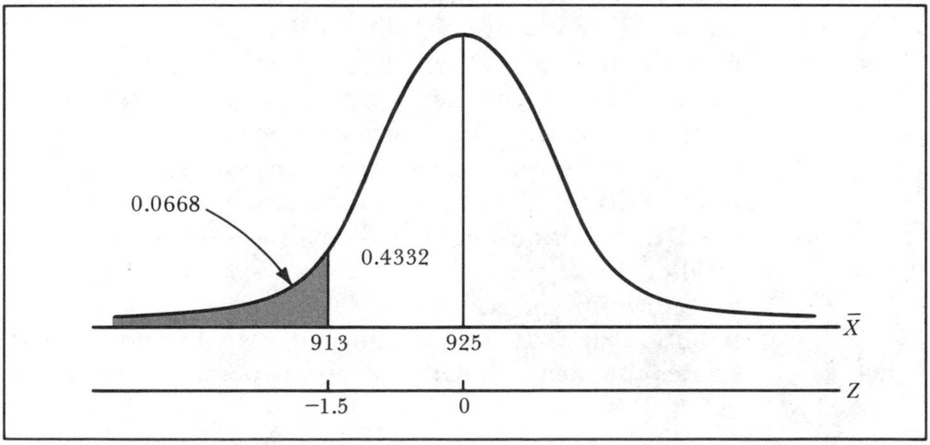

FIGURE 6-7. *Area Under Normal Curve in Example 6-2*

The Student t Distribution[3]

As you recall, the Central Limit Theorem states that the sampling distribution of $\overline{X}$'s tends to be normally distributed with a mean of μ and standard deviation (called standard error of the mean) of $\sigma/\sqrt{n}$. It has been pointed out, however, that the σ of a population is frequently unknown. Thus, in many sampling situations we estimate $\sigma_{\overline{x}}$ from sample data such that for an infinite population

$$s_{\overline{x}} = \frac{s}{\sqrt{n}} \tag{6.6}$$

And when the population is finite, the estimator of $\sigma_{\overline{x}}$ is

$$s_{\overline{x}} = \frac{s}{\sqrt{n}} \sqrt{\frac{N - n}{N - 1}} \tag{6.7}$$

Recognizing that this estimation of $\sigma_{\overline{x}}$ will take place whenever σ is unknown, the following question has to be raised: "What is the effect on the ratio $[(\overline{X} - \mu)/\sigma_{\overline{x}}]$ when it is changed to $[(\overline{X} - \mu)/s_{\overline{x}}]$?" And the answer is that the result is quite significant since $[(\overline{X} - \mu)/s_{\overline{x}}]$ is not normally distributed. In fact when σ is unknown, the distribution of $[(\overline{X} - \mu)/s_{\overline{x}}]$ is a sampling distribution known as the *Student t distribution,* providing we can assume that the *original population was normally distributed.*

The t *distribution* is similar to the standard normal distribution. Its curve resembles the physical characteristics of the standard normal in many ways. It is bell shaped, symmetrical, and universal. The two main differences are: (1) the t distribution has greater dispersion (the curve is flatter and its tails extend farther); (2) there is a particular t distribution for varying degrees of freedom (degrees of freedom, d.f., are one less than the sample size or $n - 1$).

This second difference requires a little explanation. Unlike the normal distribution, there is not a single t distribution. There is a specific t distribution for each value of $n - 1$ (degrees of freedom). The smaller the degrees of freedom $(n - 1)$, the flatter, or more dispersed, the curve becomes. Conversely, as $(n - 1)$ becomes large, the t distribution approaches the normal distribution. Figure 6-8 demonstrates the idea of a flatter, more dispersed curve as the t distribution.

---

3. Credit for developing the theoretical standard t distribution goes to William S. Gosset. This notion was first published in 1908 in a paper written by Gosset in which he used the pen name, "A. Student."

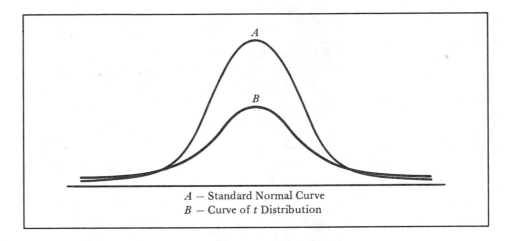

A — Standard Normal Curve
B — Curve of *t* Distribution

FIGURE 6-8. *Comparison of Normal Curve (t Distribution with ∞ Degrees of Freedom) with Curve of t Distribution Whose Degrees of Freedom Are Relatively Small*

Appendix J shows values of the *t* distribution for various degrees of freedom at selected probability levels. Notice that where the degrees of freedom (d.f.) = ∞ the *t distribution* is equivalent to the *standard normal distribution*. This table must be read for the appropriate degrees of freedom each time you want to determine a particular *t* value.

In converting to Student *t* distributions, the procedure is essentially the same as when converting to a standard normal distribution. Look at the formula below. We will make use of this relationship in the next chapter.

$$t = \frac{\overline{X} - \mu}{s_{\overline{x}}} \tag{6.8}$$

Except for using the values in the *t* table instead of those in the standard normal distribution table, Equation 6.8 is used just as you use the standard normal distribution (Equation 6.5).

The Sampling Distribution of Proportions

So far we have discussed sampling distributions of the arithmetic mean. There will be times, however, when the statistic of interest will not be the arithmetic mean. That is, the statistic might be a percentage sometimes referred to as a *proportion*. Since this is the case, we must investigate how *sample proportions* (p) are distributed where p is defined as the ratio of the number of successes to the total number of events in a binomial situation. Equation 6.9 shows how sample proportions are calculated.

$$p = \frac{X}{n} \tag{6.9}$$

where p = sample proportion
X = number of occurrences of the random variable
n = number of trials

Let us recall that proportions (when there are two responses such as yes-no) are actually distributed according to the *binomial distribution*. As the sample size increases sample proportions, however, will approach a normal distribution under the law of the Central Limit Theorem providing the population proportion, π, is not very large or very small.[4]

Figure 6-9 demonstrates the effect of the population proportion, π, on the distributions of sample proportions, p. As you can see, if the population proportion is quite small ($\pi < 0.5$), the p's tend to be skewed to the right, and if the population proportion is quite large

TABLE 6-5. *Samples of Size 2 and the Number of 3's Observed*

SAMPLE NUMBER	OBSERVED VALUES	NUMBER OF THREES	PROPORTION OF 3'S IN SAMPLE (p)
1	1, 1	0	0
2	1, 2	0	0
3	1, 3	1	1/2
4	1, 4	0	0
5	2, 1	0	0
6	2, 2	0	0
7	2, 3	1	1/2
8	2, 4	0	0
9	3, 1	1	1/2
10	3, 2	1	1/2
11	3, 3	2	1
12	3, 4	1	1/2
13	4, 1	0	0
14	4, 2	0	0
15	4, 3	1	1/2
16	4, 4	0	0

---

4.   A reasonable guide is that $n\pi$ and $n(1 - \pi)$ are both >5.

($\pi > 0.5$), the p's tend to be skewed left. Also, you must remember that for symmetry to develop, the sample size must be large.

Now that we know something of the sampling distribution of sample proportions, we need to know how to calculate the mean and standard deviation for this type of data. Let us consider the same finite population composed of the four numbers 1, 2, 3, and 4, which we used earlier in this chapter. Again let us take all possible samples of size 2, and let us count the number of 3's in each sample. Table 6-5 shows these results. (To help understand the entries in Table 6-5, consider sample number 7, which has observed values $(2, 3)$. The number of 3's in the sample is 1, and the sample size is 2; so the sample proportion is $p = X/n = \frac{1}{2}$).

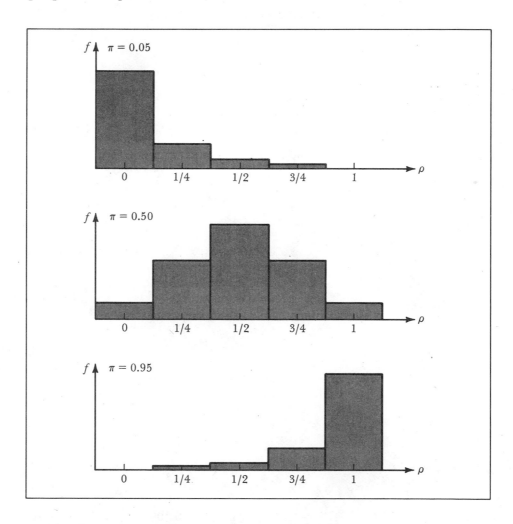

FIGURE 6-9. *Sampling Distributions of p's with Different Values of π and $n = 4$*

Now let us use the results in Table 6-5 to calculate the average of the sample proportions $(\bar{p})$ such that

$$\bar{p} = \frac{\text{sum of sample proportions}}{\text{number of samples}} = \frac{\Sigma p}{n'} \qquad (6.10)$$

$$= \frac{1/2 + 1/2 + 1/2 + 1/2 + 1 + 1/2 + 1/2}{16}$$

$$= 4/16$$

$$\bar{p} = 1/4$$

TABLE 6-6. *Calculation of Standard Error of the Proportion*

SAMPLE NUMBER	OBSERVED VALUES	PROPORTION OF 3'S IN SAMPLE p	$(p - \pi)$	$(p - \pi)^2$
1	1, 1	0	$-1/4$	1/16
2	1, 2	0	$-1/4$	1/16
3	1, 3	1/2	1/4	1/16
4	1, 4	0	$-1/4$	1/16
5	2, 1	0	$-1/4$	1/16
6	2, 2	0	$-1/4$	1/16
7	2, 3	1/2	1/4	1/16
8	2, 4	0	$-1/4$	1/16
9	3, 1	1/2	1/4	1/16
10	3, 2	1/2	1/4	1/16
11	3, 3	1	3/4	9/16
12	3, 4	1/2	1/4	1/16
13	4, 1	0	$-1/4$	1/16
14	4, 2	0	$-1/4$	1/16
15	4, 3	1/2	1/4	1/16
16	4, 4	0	$-1/4$	1/16

$$\Sigma (p - \pi)^2 = \frac{24}{16}$$

$$\sigma_p = \sqrt{\frac{\Sigma (p - \pi)^2}{n'}}$$

$$= \sqrt{\frac{24/16}{16}}$$

$$= \sqrt{\frac{24}{16 \cdot 16}}$$

$$\sigma_p = \sqrt{3/32} = 0.306$$

Notice that $\overline{p} = 1/4$ is equal to the true proportion of 3's in the population $(1, 2, 3, 4)$ which is $1/4$. The average of possible sample proportions (p's) is equal to the true population proportion, π.

Continuing on, let us calculate the standard deviation of the sample proportions, which is usually called the *standard error of the proportion.* The calculation in Table 6-6 shows the standard error of the proportion, σ_p, to be 0.306.

The calculations in Table 6-6, however, are rather long, particularly when the number of samples is large. Fortunately, a shorter way exists to determine the standard error of the proportion. Using Equation 6.11 with our example, we find that the answer in Table 6-6 could have been found much more readily.

$$\sigma_p = \sqrt{\frac{\pi(1-\pi)}{n}} \qquad (6.11)$$

$$\pi = 1/4$$

$$1 - \pi = 3/4$$

$$n = 2$$

$$\sigma_p = \sqrt{\frac{(1/4) \cdot (3/4)}{2}} = \sqrt{\frac{3/16}{2}}$$

$$\sigma_p = \sqrt{3/32}$$

$$\sigma_p = 0.306.$$

As before, if we sample from a finite population, we should use the finite correction factor as shown in Equation 6.12.

$$\sigma_p = \sqrt{\frac{\pi(1-\pi)}{n}} \sqrt{\frac{N-n}{N-1}} \qquad (6.12)$$

Now that you are familiar with the sampling distributions of sample proportions, you need to learn to make use of it. Since the *Central Limit Theorem applies to sample proportions,* we have an equation for converting the sampling distribution to a standard normal distribution. And Equation 6.13

$$Z = \frac{p - \pi}{\sigma_p} \qquad (6.13)$$

can be used just as you have done previously.

▶ **Example 6-3**   Suppose several hundred thousand (a large enough population not to need the finite population correction factor) greeting cards were manufactured and that 15 percent (π) of the cards were defective. What is the probability that a sample of 100 cards would contain a sample proportion exceeding 0.20 defective? (See Figure 6-10)

$$Z = \frac{p - \pi}{\sigma_p} \quad where \quad \sigma_p = \sqrt{\frac{\pi\,(1 - \pi)}{n}}$$

$$= \frac{0.20 - 0.15}{\sqrt{\dfrac{(0.15)\,(1 - 0.15)}{100}}}$$

$$= \frac{0.20 - 0.15}{\sqrt{\dfrac{(0.15)\,(0.85)}{100}}}$$

$$= \frac{0.05}{\sqrt{0.001275}}$$

$$= \frac{0.05}{0.03571}$$

$$Z = 1.40$$

Area = 0.4192 (From Appendix H)

$$P\,(p > 0.20) = 0.5000 - 0.4192 = 0.0808 \blacktriangleleft$$

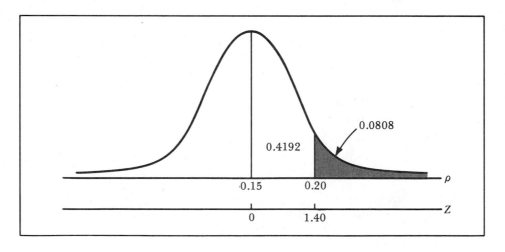

FIGURE 6-10.   *Area Under Normal Curve in Example 6.3*

You should recognize that the same situation exists for proportions that exists for arithmetic means; that is, we do not often know the population parameters, π and $1 - \pi$. Accordingly, we must frequently compute an estimate of σ_p (labeled s_p) as follows:

$$s_p = \sqrt{\frac{p\,(1 - p)}{n}} \qquad\qquad (6.14)$$

Equation 6.14 is appropriate if the sampling is from an infinite population. If the population is finite, the equation to use is:

$$s_p = \sqrt{\frac{p\,(1 - p)}{n}} \; \sqrt{\frac{N - n}{N - 1}} \qquad\qquad (6.15)$$

▶ **Example 6-4**   If a random sample of 200 individuals in a very large city showed 26 persons with the flu,
(a) estimate the proportion of people with the flu and
(b) estimate the standard error of the proportion.

(a) $p = X/n = \dfrac{26}{200} = 0.13$

(b) $s_p = \sqrt{\dfrac{p\,(1 - p)}{n}}$

$s_p = \sqrt{\dfrac{(0.13)\,(1 - 0.13)}{200}} = \sqrt{\dfrac{(0.13)\,(0.87)}{200}} = \sqrt{0.000566} = 0.0238$ ◀

———————————————— **SUMMARY OF EQUATIONS** ————————————————

(**6.1**) $\sigma_{\bar{x}} = \sqrt{\dfrac{\Sigma\,(\overline{X} - \overline{\overline{X}})^2}{n'}}, \qquad page\ 137$

(**6.2**) $\sigma_{\bar{x}} = \sqrt{\dfrac{\Sigma f\,(\overline{X} - \overline{\overline{X}})^2}{n'}}, \qquad page\ 137$

(**6.3**) $\sigma_{\bar{x}} = \dfrac{\sigma}{\sqrt{n}}, \qquad page\ 138$

(**6.4**) $\sigma_{\bar{x}} = \dfrac{\sigma}{\sqrt{n}}\,\sqrt{\dfrac{N - n}{N - 1}}, \qquad page\ 140$

(**6.5**) $Z = \dfrac{\overline{X} - \mu}{\sigma_{\bar{x}}}, \qquad page\ 141$

(6.6) $s_{\bar{x}} = \dfrac{s}{\sqrt{n}}$,   page 144

(6.7) $s_{\bar{x}} = \dfrac{s}{\sqrt{n}} \sqrt{\dfrac{N-n}{N-1}}$,   page 144

(6.8) $t = \dfrac{\bar{X} - \mu}{s_{\bar{x}}}$,   page 145

(6.9) $p = \dfrac{X}{n}$,   page 146

(6.10) $\bar{p} = \dfrac{\Sigma p}{n'}$,   page 148

(6.11) $\sigma_p = \sqrt{\dfrac{\pi(1-\pi)}{n}}$,   page 149

(6.12) $\sigma_p = \sqrt{\dfrac{\pi(1-\pi)}{n}} \sqrt{\dfrac{N-n}{N-1}}$,   page 149

(6.13) $Z = \dfrac{p - \pi}{\sigma_p}$,   page 149

(6.14) $s_p = \sqrt{\dfrac{p(1-p)}{n}}$,   page 151

(6.15) $s_p = \sqrt{\dfrac{p(1-p)}{n}} \sqrt{\dfrac{N-n}{N-1}}$,   page 151

REVIEW QUESTIONS

1. Identify what each equation at the end of this chapter calculates.
2. Define
   (a) inductive reasoning
   (b) inference
   (c) sample units
   (d) random sample
   (e) simple random sample
   (f) standard error of the mean
   (g) t distribution
   (h) universe
   (i) parameters
3. Distinguish between s and $s_{\bar{x}}$.
4. What is a sampling distribution?
5. State the Central Limit Theorem.
6. When is the finite population correction factor needed?

1. From a finite population of five items (A, B, C, D, E), determine how many samples of size two are possible
   (a) if duplication is allowed and order is observed.
   (b) if duplication is not allowed and order is observed.
   (c) if duplication is not allowed and order is not observed.

2. Write out the possible samples for each situation in Problem 1.

3. Using replacement, determine all possible samples of size 2 and the sample means from a population composed of the numbers 2, 4, 6, and 8.

4. Plot the sample means determined in Problem 3 in the form of a histogram.

5. Use your results in Problem 3 to
   (a) determine μ by calculating $\overline{\overline{X}}$.
   (b) determine the standard error of the mean using Equation 6.1 or 6.2.
   (c) duplicate the results of part (b) using Equation 6.3.

6. Use the population in Problem 3, and determine all possible samples of size 2 that can result when sampling without replacement. _class example_

7. Determine the sample means in Problem 6 and plot these sample means as a histogram.

8. Determine the standard error of the mean for Problem 6.
   (a) Use Equation 6.1 or 6.2.
   (b) Duplicate the results of part (a) using Equation 6.4.

9. Given a population composed of 9, 12, 6, and 15
   (a) Determine the number of samples of size 2 that could be obtained if the sampling were without replacement.
   (b) List all possible samples of size two obtained when sampling without replacement.
   (c) Determine μ by calculating $\overline{\overline{X}}$.
   (d) Determine the standard error of the mean using Equation 6.2.
   (e) Determine the standard error of the mean using Equation 6.4.

10. From the results of Problem 3, determine the sample proportions that result by counting the number of 4's in each sample. (Hint, see Table 6-5.)

11. Determine the population proportion π, using the sample proportions from Problem 10.

12. Determine the standard error of the proportion using the π from Problem 10.
    (a) Use Equation 6.11.
    (b) Duplicate this result by following the procedure shown in Table 6-6.

13. If random samples of size 36 are taken from a large population with $\mu = 80$ and $\sigma = 18$, what percent of the sample means
    (a) lie between 77 and 83?
    (b) would be greater than 82?
    (c) lie between 74 and 86?
    (d) would be less than 75?

14. If the population size is 3,149, the population mean is 185, the population standard deviation is 18, and the sample size is 144, determine the standard deviation of the sampling distribution of sample means.
    (a) Use Equation 6.3. *without fpc*
    (b) Use Equation 6.4. *with fpc*
    (c) Compare the two answers.

15. In a population of 21,000 management executives, the mean (μ) salary of the population is $39,000 with standard deviation (σ) of $4,500. Determine the standard deviation of the distribution of sample means for a sample size of 600.

16. A department store has 7,000 charge accounts. The comptroller takes a random sample of 36 of the account balances and calculates the standard deviation to be $42.00. If the actual mean (μ) of the account balances is $175.00, what is the probability that the sample mean would be
    (a) between $164.50 and $185.50?
    (b) greater than $180.00?
    (c) less than $168.00?

17. In a sample of 50 stocks, 20 declined over the period of a week. Estimate the
    (a) population proportion of stocks that declined.
    (b) standard error of the proportion of stocks that declined.

18. Determine the standard error of the proportion if $\pi = .60$ and $n = 196$. (Assume infinite population.)

19. Solve Problem 18 if $N = 8000$.

20. If a sample of size 100 yields a sample proportion of 0.20, estimate the standard error of the proportion.

21. If a sample of size 16 from a normal population produces a sample mean of 22 and a standard deviation of 4,
    (a) determine the standard error of the mean.
    (b) Are the sample means normally distributed?

22. If a sample of size 25 from a normal population yields $\overline{X} = 19$ and $s = 3.5$:
    (a) determine the standard error of the mean.
    (b) are the $\overline{X}$'s t-distributed?
    (c) explain your answer to part (b).

23. A distributor has 5,000 customers. If a random sample of 150 customers showed 25 delinquent accounts, estimate the
    (a) population proportion of delinquent accounts, and
    (b) standard error of the proportion of delinquent accounts.

24. An organization with 4,000 employees has established that Monday absenteeism has been 8 percent for several years. If a random sample of 80 employees was taken last Monday, what is the probability that the sample would contain more than 15 absentees?

25. If 45 percent of the homeowners in a very large city also own two or more

automobiles, what is the probability that a random sample of 300 homeowners will contain more than 150 homeowners who own at least two
automobiles?

26. The mean vase life of cut roses with preservative is 192 hours with a
standard deviation of 24 hours. If a random sample of 36 roses were
tested for vase life, determine the probability that the sample mean
would exceed 199.

Estimation: Means and Proportions

CHAPTER 7

As we have discussed earlier, one of the major reasons for taking a sample is to draw conclusions about population parameters. For instance, a political candidate may want to know what proportion of his constituents favors a political issue, or an advertiser may need to know the average income of the readers of a magazine. These conclusions about the population may be in the form of estimates and, short of taking a complete enumeration of the entire population, we depend on sample results to make these estimates.

In sampling, there are usually two types of estimates—*point estimates* and *interval estimates*. A point estimate is just what it implies, a single estimate of the *population parameter* made with a sample statistic. On the other hand, an interval estimate is a range constructed about the single point estimate by using the appropriate sampling distribution. You studied these sampling distributions in the previous chapter.

In this chapter, the emphasis will be on learning to construct interval estimates for certain population parameters. The reason for stressing interval estimates over point estimates is that point estimates are somewhat fragile. That is, it is rather unlikely that a single statistic (say the sample mean, $\overline{X}$) is going to equal exactly the population parameter under investigation. Thus, to avoid being wrong most of the time with a single point estimate, interval estimates are usually made.

The Confidence Interval

Interval estimates are referred to as *confidence intervals* and are constructed in such a way that a probability value can be associated with the interval estimates. More specifically, a confidence interval is a range, probabilistically constructed, about a point estimate to provide an interval estimate of a population parameter. The probability value indicates the likelihood that the *statement* concerning the population parameter is true.

Suppose a representative of a tire manufacturer has stated she is 95 percent confident that the average life of the company's tires is between 25,000 and 28,000 miles. Assuming that this statement is statistically based, we could conclude that the manufacturer has constructed a 95 percent confidence interval for average tire life. The *implication* is that the average tire life of all tires of this make is somewhere between 25,000 and 28,000 miles, with a probability of 0.95.

A more correct interpretation of the confidence interval, however, is to apply the probability level to the statement. That is, *the statement* that the true mean life of the tire is between 25,000 and 28,000 miles *is correct with a probability of 0.95.* And if you think about it, this makes better sense because the true mean is either in the interval or outside the interval. The probability is either 0 or 1.0, not 0.95. In other words, the true mean cannot jump in the interval 95 times out of 100 and out of the interval 5 times out of 100. Thus, the *statement* about the interval should be correct 95 times out of 100; the interval itself is either right or wrong.

Confidence Intervals for Means

In the previous chapter, two sampling distributions associated with $\overline{X}$ were introduced. These two sampling distributions were the normal distribution and the t distribution. Also, it was pointed out that when dealing with arithmetic means, sometimes the normal distribution is appropriate and sometimes the t distribution is appropriate. So, before we begin constructing confidence limits for the arithmetic mean, let us decide on the conditions that would dictate a certain sampling distribution.

Situations Under Which Samples Might Be Selected

When we are preparing to construct a confidence interval from information obtained from a random sample, we should raise three questions about the conditions under which the sample was collected. These three questions are as follows:

1. Is the population normally distributed?
2. Is the population standard deviation known or unknown?
3. Is the sample size large, $n > 30$, or small?

Each of these conditions—normality or lack of normality, known or unknown population standard deviation, and large or small sample—can affect the construction of the confidence interval. Table 7-1 shows eight different situations that you might be faced with, and it provides the expected sampling distribution for each situation.

TABLE 7-1. *Sampling Situations*

SITUATIONS	POPULATION DISTRIBUTION	POPULATION STANDARD DEVIATION	SAMPLE SIZE	APPROPRIATE $\overline{X}$ SAMPLING DISTRIBUTION
A	Normal[1]	Known	Large	Normal
B	Normal[1]	Known	Small	Normal
C	Normal[1]	Unknown	Large	Normal Via Central Limit Theorem
D	Normal[1]	Unknown	Small	t
E	Unknown	Known	Large	Normal Via Central Limit Theorem
F	Unknown	Known	Small	None[3]
G	Unknown[2]	Unknown	Large	Normal Via Central Limit Theorem
H	Unknown	Unknown	Small	None[3]

1.   Population distributions that differ slightly from normal are frequently treated as if they are normal distributions.

2.   Populations that differ a great deal from normality (such as a bimodal distribution) or that contain extreme values will probably require samples that seem very large relative to the size of the population.

3.   In Chapter 11, approaches to these situations will be discussed.

Inspection of the last column of Table 7-1 shows that four different categories are presented. The first category is comprised of situations A and B where the sampling was from a normal population with a known population standard deviation. In this situation, the appropriate sampling distribution is the normal distribution regardless of whether the sample size is large or small.

The second category is made up of those (C, E, G) where the appropriate sampling distribution is the normal distribution as a result of the Central Limit Theorem. In each of these cases, either the population distribution, the population standard deviation, or both, is unknown, and the sample size is large, $n > 30$. This large sample size allows the use of the Central Limit Theorem, thereby establishing the normality of the sampling distribution.

The third category is situation D in which the t distribution must be used. The t distribution is necessary because the unknown population standard deviation and the small sample prevent us from using the Central Limit Theorem. Notice, finally, that the population distribution is normal and that this condition is a necessity in order to use the t distribution.

The fourth category contains situations F and H for which there is no appropriate sampling distribution and thus no way to make interval estimates. What has happened in these two cases is that a small sample is combined with an unknown population distribution that prevents the use of the Central Limit Theorem or the t distribution.

Summarizing, we can say there are three categories of sampling situations that permit the construction of confidence intervals. They are (1) normal population and known population standard deviation, (2) large samples, and (3) small samples from a normal population with unknown population standard deviation. The first two categories employ the normal distribution, and the third category employs the t distribution. Now, knowing the sampling situations we might encounter, let us investigate the construction of confidence intervals for μ in each of the three categories.

Confidence Intervals for the Population Mean when Population Distribution is Normal and Population Standard Deviation is Known

As you might have already guessed, the situation of sampling from a normal population distribution with the population standard deviation being known is rather unlikely. It is, however, a situation where the sampling distribution of $\overline{X}$'s is always normally distributed regardless of sample size. And, because the sampling distribution is normal, we can relate the probabilities associated with any normal distribution to our sampling distribution.

For example, 68.27 percent of the sample means ($\overline{X}$'s) lie within the range of plus or minus one $\sigma_{\overline{x}}$ from the true mean, μ. Since this is true, there is a 0.6827 probability that a range of plus or minus one $\sigma_{\overline{x}}$ will contain the true population mean (μ). Using this same logic, it would be safe to say that there is a 0.95 probability μ will fall within the interval of any sample mean plus or minus 1.96 $\sigma_{\overline{x}}$. Figure 7-1 presents this relationship. Notice in Figure 7-1 that 95 percent of all the sample means in the sampling distribution are within a distance of $\pm 1.96\ \sigma_{\overline{x}}$ of μ. Thus, any sample mean that has 1.96 $\sigma_{\overline{x}}$ added to it and 1.96 $\sigma_{\overline{x}}$ subtracted from it forms an interval that has 0.95 probability of containing the true mean, μ.

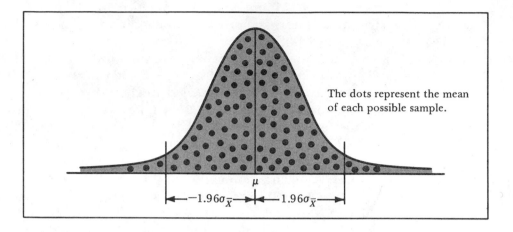

The dots represent the mean of each possible sample.

Figure 7-1. *A Representation of a Sampling Distribution Showing the Area Under the Normal Curve and 95 Percent of the Sample Means Within the Range $\pm 1.96\,\sigma_{\bar{x}}$*

This probability level is usually referred to as the level of confidence for the interval.   That is, confidence intervals may have confidence levels such as 0.80, 0.90, or 0.95, and the limits of the confidence interval are established from the expression

$$Z = \frac{\bar{X} - \mu}{\sigma_{\bar{x}}}$$

Solving the preceding expression for μ yields

$$Z\sigma_{\bar{x}} = \bar{X} - \mu$$

$$\mu = \bar{X} - Z\sigma_{\bar{x}},$$

and since Z can be positive or negative,

$$\mu = \bar{X} - Z\sigma_{\bar{x}} \text{ when } Z \text{ is positive, and}$$

$$\mu = \bar{X} + Z\sigma_{\bar{x}} \text{ when } Z \text{ is negative.}$$

Algebraically, this relationship may be expressed as follows:

$$(\bar{X} - Z\sigma_{\bar{x}}) \le \mu \le (\bar{X} + Z\sigma_{\bar{x}})$$

where $(\bar{X} - Z\sigma_{\bar{x}})$ is the lower confidence limit of the population mean, and $(\bar{X} + Z\sigma_{\bar{x}})$ is the upper confidence limit of the population mean. Now, we can write an expression for the construction of confidence intervals as follows:

$$(1 - \alpha)\% \text{ confidence interval for } \mu = \bar{X} \pm Z\sigma_{\bar{x}} \qquad (7.1)$$

Notice that Equation 7.1 is in terms of $(1 - \alpha)$ percentage. This alpha (α), which represents the probability that the interval is incorrect, takes as a numerical value the difference between 1.00 and the level of confidence such that for a 95 percent confidence interval, $\alpha = 0.05$. To demonstrate the construction of confidence intervals, let us consider the following example.

▶ **Example 7-1** Suppose that a random sample of 25 retail merchants from all of the merchants in Large City (assume a large population so the finite correction factor will not be needed) yielded a mean advertising expense for the past year of $1,250. Now, if the population of annual advertising expenditures in the city is *known* to be normally distributed and the population standard deviation is *known* to be $750, what would be the 95 percent confidence interval for the true mean advertising expense?

From Equation 7.1,

$$(1 - \alpha)\% \text{ confidence interval for } \mu = \overline{X} \pm Z\sigma_{\overline{x}}$$

$$95\% \text{ confidence interval for } \mu = \overline{X} \pm Z\sigma_{\overline{x}}$$

where Z must be determined from the normal curve table in Appendix H. To use Appendix H, divide the level of confidence by 2 ($0.95/2 = 0.4750$) and locate the associated Z value of 1.96 in Appendix H which leaves 2.5% of the area under the curve in each end of the distribution.

$$95\% \text{ confidence interval for } \mu = \$1,250 \pm (1.96)\left(\frac{\$750}{\sqrt{25}}\right)$$

$$= \$1,250 \pm (1.96)\,(150)$$

$$= \$1,250 \pm \$294$$

$$\$956 \leq \mu \leq \$1,544$$

Thus, we would state that the average advertising expenditure for retail merchants in Large City is in the range of $956 to $1544 with a probability of 0.95. You are cautioned to remember, however, that the probability is really attached to the statement about the range because the true mean is either in or out of the interval estimate. ◀

Confidence Interval for the Population Mean When Large Samples are Selected

Large samples ($n > 30$) are necessary in order for the Central Limit Theorem to be used as the basis of normality for a sampling distribution. The need for the large sample arises because, unlike the preceding example, either the population distribution, the population standard deviation, or both, happen to be unknown. If the population distribution is unknown and the population standard deviation is known, the confidence interval is

$$(1 - \alpha)\% \text{ confidence interval for } \mu = \overline{X} \pm Z\sigma_{\overline{x}} \qquad \textbf{(7.1)}$$

where $\sigma_{\overline{x}} = \sigma/\sqrt{n}$

However, if the standard deviation is unknown, the confidence interval is

$$(1 - \alpha)\% \text{ confidence interval for } \mu = \overline{X} \pm Zs_{\overline{x}} \qquad \textbf{(7.2)}$$

where $s_{\overline{x}} = s/\sqrt{n}$

and s is the sample estimate of the population standard deviation.

   Keep in mind that the circumstances are generally such that Equation 7.2 is used much more often than Equation 7.1.   Also, if the sampling is from a finite population, be sure to use the finite population correction factor when calculating $s_{\overline{x}}$ or $\sigma_{\overline{x}}$.

▶ **Example 7-2**   A machine designed to fill cereal boxes has been used on 4,000 boxes. Four hundred of these boxes were randomly selected and carefully weighed.   The mean weight of the boxes was found to be 16.1 ounces.

   The manufacturer of the machine, if he so desired, could base his sales claim on the point estimate $(\overline{X})$ of 16.1 ounces.   The problem, however, is that potential customers may be aware that $\overline{X} = 16.1$ ounces is only one of many possible sample means that might have occurred and they might be interested in having information about the dispersion of the fill.   Thus, the manufacturer calculated the sample standard deviation (s) to be 0.2 ounces, and he is going to provide his customers with a 90% confidence interval for the average fill of the machine using Equation 7.2 and $\alpha = 0.10$.   Then the

$$(1 - \alpha)\% \text{ confidence interval for } \mu = \overline{X} \pm Zs_{\overline{x}}$$

In addition $s_{\overline{x}} = (s/\sqrt{n}) \sqrt{(N - n)/(N - 1)}$ since the sampling was from a finite population.   Thus, the

90% confidence interval for $\mu = \overline{X} \pm Zs_{\overline{x}}$

$$= 16.1 \pm Z \left(\frac{0.2}{\sqrt{400}} \right) \sqrt{\frac{4000 - 400}{4000 - 1}}$$

$$= 16.1 \pm (1.64) \frac{0.2}{20} \sqrt{\frac{3600}{3999}}$$

$$= 16.1 \pm 1.64 \, (0.01) \sqrt{0.9002}$$

$$= 16.1 \pm 1.64 \, (0.01) \, (0.949)$$

$$= 16.1 \pm 0.0156$$

$$16.0844 \leq \mu \leq 16.1156 \text{ ounces}$$

Now the manufacturer has a range that he thinks includes the true average fill of his machine with a probability of 0.90. Remember, though, the probability is really attached to the statement about the range. ◀

Confidence Interval for the Population Mean When Small Samples Are Selected from a Normal Population with Unknown Population Standard Deviation

When small samples ($n < 31$) are selected, use of the Central Limit Theorem to imply normality of the sampling distribution is not valid. As you recall, this category was discussed in an earlier section of this chapter. It was explained that the t distribution was the proper sampling distribution for small samples taken from *normal* populations with *unknown* population standard deviation. Thus, beginning with the expression

$$t = \frac{\overline{X} - \mu}{s_{\overline{x}}}$$

and solving for μ yields

$$\mu = \overline{X} \pm t s_{\overline{x}}$$

and using our knowledge of confidence intervals, we can write the expression

$$(1 - \alpha)\% \text{ confidence interval for } \mu = \overline{X} \pm t s_{\overline{x}} \qquad (7.3)$$

where the value of t must be determined from Appendix J similar to the way we have used Z in the preceding confidence intervals, except we now have to consider the degrees of freedom ($n - 1$) and the value of α. Recall that the Chapter 6 discussion about the t distribution pointed out there is not a single t distribution; rather, a specific t distribution exists for each value of $n - 1$. Also, as $n - 1$ becomes large, say greater than 30, the t distribution can be approximated fairly well by the normal distribution.

▶ **Example 7-3**   A retailer was forced to buy large quantities of product A because the manufacturer of product A was faced with a prolonged strike. Three months have elapsed since the retailer made her purchase of product A, and she is concerned about the stability of the composition of the product, especially the loss of water and thus, weight. The net weight of the contents of a container of product A is supposed to be 8 ounces. The retailer has decided to take a sample of 16 containers and weigh the contents in order to construct a 98% confidence interval to see if the average is holding at 8 ounces. She has chosen a small sample because the product is expensive, and the test makes the sale of opened containers impossible. Further, she will use the t distribution because the sample is small and because she feels that the distribution of net weights of all containers of product A made by the manufacturer is normally distributed. If the sample results were $\overline{X} = 7.95$ and $s = 0.6$, then

$$(1 - \alpha)\% \text{ confidence interval for } \mu = \overline{X} \pm ts_{\overline{x}}$$

$$98\% \text{ confidence interval for } \mu = 7.95 \pm t\frac{0.6}{\sqrt{16}}$$

From Appendix J when $\alpha = 0.02$ and d.f. $= 15$, t is 2.602.

$$98\% \text{ confidence interval for } \mu = 7.95 \pm 2.602 \left(\frac{0.6}{4}\right)$$

$$= 7.95 \pm 2.602 \,(0.15)$$

$$= 7.95 \pm 0.39$$

$$7.56 \leq \mu \leq 8.34$$

Since this range does include 8 ounces, the retailer should not fear that much, if any, deterioration has taken place in the three months of storage. ◀

Confidence Interval for Proportions

In Chapter 6, it was pointed out that the statistic of interest will not always be the mean. At times we will be interested in percentages, usually referred to as proportions. Furthermore, it was pointed out in Chapter 6 that even though proportions (when there are two responses, like yes-no) are binomially distributed, the normal distribution can be used as the sampling distribution of the sample proportions when both $n\pi$ and $n\,(1 - \pi) > 5$.

We can start with the familiar standardized normal deviate expression,

$$Z = \frac{p - \pi}{\sigma_p}$$

and replace σ_p by the standard error of the proportion estimated from the sample as $s_p = \sqrt{p\,(1 - p)/n}$ to form

$$Z = \frac{p - \pi}{s_p}$$

Solving for π yields

$$(1 - \alpha)\% \text{ confidence interval for } \pi = p \pm Zs_p \tag{7.4}$$

▶ **Example 7-4**   A switchboard operator has been criticized by his chief for taking too long to answer incoming calls.   In defense of himself, the operator tells his chief that he is slow to answer incoming calls because he has to spend considerable time transferring a caller from one office to another office.   That is, many incoming callers talk to more than one person in the building.   The chief and the operator decide to check this idea by taking a sample of 200 incoming calls.   Of these two hundred calls, 80 calls, or 40 percent, required the additional service of being transferred to another office.   For this situation, a 95 percent confidence interval could be constructed using Equation 7.4. The sample proportion, p, is $80/200 = 0.40$.

$$(1 - \alpha)\% \text{ confidence interval for } \pi = p \pm Zs_p$$

$$95\% \text{ confidence interval for } \pi = 0.40 \pm 1.96 \sqrt{\frac{(0.40 \,(1 - 0.40)}{200}}$$

$$= 0.400 \pm (1.96) \sqrt{\frac{(0.40)\,(0.60)}{200}}$$

$$= 0.400 \pm 1.96 \sqrt{\frac{0.2400}{200}}$$

$$= 0.400 \pm 1.96 \sqrt{0.0012}$$

$$= 0.400 \pm 1.96 \,(0.0346)$$

$$= 0.400 \pm 0.068$$

$$0.332 \leq \pi \leq 0.468$$

Thus, the chief can see that the operator does have to transfer a heavy proportion of the incoming calls.   It looks as though at least 1 out of every 3 incoming calls will require additional service from the operator. ◀

SUMMARY OF EQUATIONS

7.1 $(1 - \alpha)\%$ confidence interval for $\mu = \overline{X} \pm Z\sigma_{\bar{x}}$,   *pages 160 and 162*

7.2 $(1 - \alpha)\%$ confidence interval for $\mu = \overline{X} \pm Zs_{\bar{x}}$,   *page 162*

7.3 $(1 - \alpha)\%$ confidence interval for $\mu = \overline{X} \pm ts_{\bar{x}}$,   *page 163*

7.4 $(1 - \alpha)\%$ confidence interval for $\pi = p \pm Zs_p$,   *page 164*

REVIEW QUESTIONS

1. Identify what each equation shown at the end of this chapter determines and when to use each one.

2. What is a point estimate?

3. What is an interval estimate?

4. What is a confidence interval?

5. Interpret the probability associated with confidence intervals.

6. List the three categories of sampling situations.

7. In Table 7-1, why is there no appropriate sampling distribution for Cases F and H?

8. What conditions are necessary to construct confidence intervals for proportions?

PROBLEMS

1. A recent random sample of 36 car owners revealed an average expenditure on gasoline of $39 with a standard deviation of $12. Use the sample information and construct a 95 percent confidence interval for average expenditures.

2. A random sample of 49 calculators yielded a mean life of 21 months with a standard deviation of 2.8 months. Construct a 90 percent confidence interval for the average life expectancy.

3. The tear strength of a particular paper product is known to be normally distributed. If a random sample from 9 rolls yielded a mean tear strength of 225 pounds per square inch with a standard deviation of 12 pounds per square inch, construct a 99 percent confidence interval for the average tear strength.

4. A manufacturer of a certain type of X-ray developing fluid has recently collected information on the life expectancy of the fluid from a random sample of 64 customers. If the random sample yielded a sample mean of 26 days and a standard deviation of 4 days, construct a 90 percent confidence interval for the average life expectancy of the fluid.

5. A consumer testing service is interested in estimating the mean weight of bags of a particular brand of peat moss. A random sample of size 100 produced an $\overline{X} = 43.5$ pounds. If the weight of the bags is known to be normally distributed with a standard deviation of 10 pounds, construct a 95 percent confidence interval for μ, the mean weight of the bags.

6. The number of typing mistakes per page that a new secretary makes is normally distributed with a standard deviation of 2. If a random sample of 25 pages of the secretary's work has a sample mean of 7 mistakes per page, construct a 99 percent confidence interval for μ.

7. If the secretary in Problem 6 had typed only 200 pages, construct a 90 percent confidence interval for μ.

8. The commuting times of a very large group of workers has a standard deviation of 21 minutes. If a random sample of 49 commuting times

had a mean of 52 minutes, construct a 98 percent confidence interval for the mean commuting time of all workers.

9. A random sample of the gasoline consumption of a particular make of car had a mean of 16 miles per gallon. It is known that the distribution of the gasoline consumption of all the cars of this type is slightly different from normal and has a standard deviation of 2.7 miles per gallon. If the sample size was 81, estimate with a level of confidence of 99 percent the mean gasoline consumption (μ) of this make of car.

10. The number of shirts finished per hour by a particular production line is normally distributed. A random sample of 25 hours output had a mean of 40 shirts per hour and a standard deviation of 9 shirts. Construct a 99 percent confidence interval for the mean number of shirts produced per hour.

11. Tests on 26 2 by 4 steel studs yielded an average breaking strength of 580 pounds with a standard deviation of 68 pounds. Assuming that breaking strengths are normally distributed, construct a 90 percent confidence interval for the average breaking strength.

12. A random sample of 64 sales slips yielded a sample mean of 79 cents and a standard deviation of 24 cents. Construct a 95 percent confidence interval for mean sales (μ).

13. A random sample of 25 motels in Vacation City revealed that the average room charge was $26 per day with a standard deviation of $6. Assuming that charges are normally distributed, construct a 90 percent confidence interval for μ.

14. If the sales slips in Problem 12 were for one salesperson who had made only 320 sales, construct a 95 percent confidence interval for this salesperson's mean sales (μ).

15. A random sample of 16 boxes of a popular fruit cereal contained a mean ($\overline{X}$) of 100 raisins per box with a standard deviation of 36 raisins per box. Assume that the number of raisins per box is normally distributed and construct a 95 percent confidence interval for μ.

16. Actually, the machine putting the raisins in the cereal boxes in Problem 15 was out of adjustment long enough for 400 boxes to be processed. If the sample of 16 boxes was taken from the batch of 400, estimate the mean number of raisins per box in this batch with a 95 percent confidence interval?

17. Land, Inc., wishes to estimate the average price of an acre of highway property in the country. A random sample of 100 recent land purchases showed a mean sale price of $4,000 per acre with a standard deviation of $360. Use the sample information to construct a 98 percent confidence interval for the average price.

18. A random sample of 200 accounts yielded an average unpaid balance of $75.00 with a standard deviation of $20. Use the sample information to construct a 90 percent confidence interval for the average unpaid balance.

19. In a physical fitness test, 240 out of 400 (60 percent) high school students could run a mile in less than 7 minutes. If the 400 students were a random sample, construct a 90 percent confidence interval of the proportion of all high school students that can run a mile in less than 7 minutes.

20. A concessionaire randomly sampled 200 people after the Big Game. If the sample indicated an average expenditure for concessions of $3.00 with a standard deviation of $0.50, construct a 95 percent confidence interval for μ.

21. A random sample of 100 students yielded a mean registration time of 90 minutes with a standard deviation of 12 minutes. Make an interval estimate for mean registration time of the student body. (Set $\alpha = 0.1$)

22. Repeat Problem 21 assuming there are 500 students in the college.

23. From a sample of 100 farm operators, 64 of whom owned their farms, make an interval estimate of the proportion of operators who own their farms. (Set $\alpha = 0.05$.)

24. A sample size of 225 managers of Barely Fried Chicken produced a mean salary of $22,000 with a standard deviation of $1,500. Assuming there is a very large number of the Barely Fried Chicken places, construct a 90% confidence interval for the average salary for all the managers.

25. In a large city, a research analysis organization wants to estimate the average rental expense for a particular group of merchants. If a sample of 100 merchants produced $\overline{X} = \$550$ and $s = \$75$, construct a 95 percent confidence interval for the average rental expense of all merchants.

26. Rework Problem 25 if there are only 500 merchants of this type in the city.

27. If a random sample of 500 registered voters had 310 respondents who planned to vote for the Last Candidate, what is your interval estimate of the proportion of voters that Last Candidate will receive from all the registered voters? (Set $\alpha = 0.01$).

28. A repairman is interested in the proportion of service calls that require repair time exeeding one hour. If a random sample of 50 calls from last year's calls yielded 10 service calls exceeding one hour, construct a 99 percent confidence interval for the proportion of service calls exceeding one hour.

29. The manufacturer of a high-precision product wishes to estimate the proportion of items manufactured that are defective. If a random sample of 200 items produced 10 defectives, construct a 95 percent confidence interval for π.

30. A random sample of 100 subscribers of a daily paper indicated 40 persons who had read a particular advertisement in the paper. Construct a 90 percent confidence interval for π.

Statistical Testing: One Sample

CHAPTER
8

In the preceding chapter, you were introduced to the concept of confidence intervals, and you learned to construct confidence intervals for the population mean and population proportion. In this chapter, we are going to investigate statistical testing with the objective of making inferences about a population mean or proportion. For the moment, however, let us limit our discussion to means in order to find out what statistical testing is all about.

Chance Difference Versus a Statistically Significant Difference

We may calculate a sample mean from sample data and compare it to a population mean; and we may find that the two values differ. The question we need to answer is whether the difference between the mean of the population (μ) and the sample mean ($\overline{X}$) is real or whether the difference is due to sampling variation that we know can exist.

To illustrate, look at Figure 8-1. The only $\overline{X}$'s that equal μ are the $\overline{X}$'s between the two arrows. All the other sample means, even though they are part of the sampling distribution with mean μ do not actually equal μ. Thus, on any given occasion in which we draw a sample, any one of the sample means in the sampling distribution might be selected; and most of the sample means would not equal the true population mean. This difference that could occur between the population mean and the sample means would be due to the sampling process. Such a difference is called a *chance difference* and is *not* considered *a statistically significant difference*.

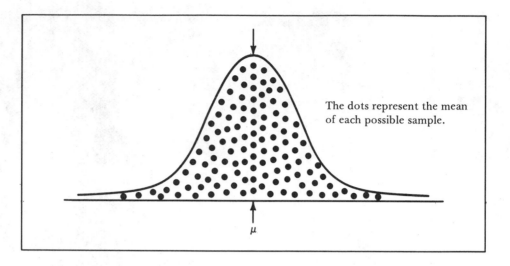

FIGURE 8-1. *A Sampling Distribution of $\overline{X}$'s With Mean of μ.*

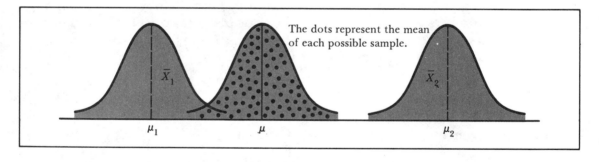

FIGURE 8-2. *Two Sample Means that Belong to a Population Mean Other Than μ.*

A *real* or *statistically significant difference* is due to a basic differ- ence not attributable to chance. Such a real difference between the sample mean and the population mean indicates that the sample mean $\overline{X}$ came from a population with a true mean other than μ. Figure 8-2 illustrates this possibility. Two sample means $\overline{X}_1$ and $\overline{X}_2$ are shown which, if drawn, would indicate that a significant difference exists. That is if $\overline{X}_1$ were selected, the distance $\overline{X}_1 - \mu$, is just too great to be attributed to chance. It would be correct to state $\overline{X}_1$ came from a population with another mean, μ_1; and μ_1 is different from μ. A similar statement can be made for $\bar{X}_2$. Let us turn to Example 8.1 to demonstrate the point.

▶ **Example 8.1**   The Old Deal Manufacturing Company has been manufacturing Part K for 15 years. During this time, the length of every Part K has been measured such that it is established that $\mu = 3.5$ inches and $\sigma = 0.10$ inches. Recently Old Deal installed a new machine, and the sample of 100 Part K's that was selected yielded a sample mean length of 3.65 inches. The question is whether the new machine is producing the same average size part as was previously produced. The distribution of Part K is normal.

An approach to answering the question would be to establish 99 percent confidence limits about μ and see if $\bar{X}$ falls in the confidence limits. Figure 8-3 shows the idea. If $2.58 \, \sigma_{\bar{x}}$ is added and subtracted from μ, upper and lower limits are established within which any sample mean should fall 99 times out of 100 if the population mean remains unchanged; thus, a chance error can occur 1 time in 100.

$$\text{Upper } \bar{X} \text{ Limit} = \mu + 2.58 \frac{\sigma}{\sqrt{n}} \qquad \text{Lower } \bar{X} \text{ Limit} = \mu - 2.58 \frac{\sigma}{\sqrt{n}}$$

$$= 3.5 + 2.58 \frac{0.10}{\sqrt{100}} \qquad \qquad = 3.5 - 0.0258$$

$$\text{Lower } \bar{X} \text{ Limit} = 3.4742$$

$$= 3.5 + 0.0258$$

$$\text{Upper } \bar{X} \text{ Limit} = 3.5258$$

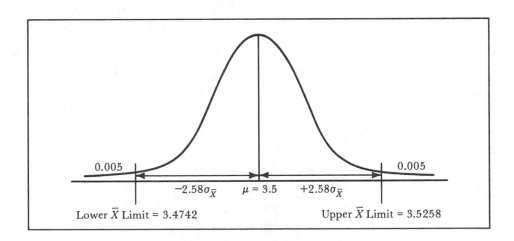

FIGURE 8-3. *99 Percent Confidence Limits.*

Since the limits, 3.4742 to 3.5258, do not contain $\overline{X} = 3.65$, the conclusion is drawn that the difference between $\overline{X}$ and μ is statistically significant. Thus, it is concluded that the new machine is not producing at the same average length as the old machine. (The difference is real, not chance, with a probability of being correct equal to 0.99.) ◄

The Test Procedure

The preceding example demonstrated the concept of a statistically significant difference. In this section, a more standardized approach to statistical testing will be given, and then the previous example will be reworked.

Formulating the Null and Alternate Hypothesis

In statistical testing, a null hypothesis is an assumption about a population. Sample information is used to make a decision about the null hypothesis. If the sample information supports the null hypothesis, then the null hypothesis is accepted. Acceptance of a null hypothesis implies that any difference observed between the sample information and the assumed population characteristic could be considered the result of chance variation. However, if the sample evidence does not support the null hypothesis, then the null hypothesis is rejected. Rejection of the null hypothesis implies that the sample result is significantly different from the assumption.

One additional comment on accepting the null hypothesis should be made. Some researchers prefer to say that a null hypothesis is *not rejected* rather than to accept the null hypothesis. Such a position is taken because it is difficult to prove, without any doubt, that a null hypothesis is true. However, whichever expression ("accept the null hypothesis" or "do not reject the null hypothesis") is used, just remember that such a decision is reached when the sample evidence does not justify rejecting the null hypothesis.

Symbolically, a null hypothesis about a population mean would be as follows:

$$H_0: \quad \mu = \mu_H$$

The symbol "H_0" means null hypothesis. The preceding null hypothesis states that the population mean μ has a value of μ_H (some number). Another way to state this is that there is no difference between μ and μ_H. Our goal is to test this null hypothesis.

Before actually testing the null hypothesis, it is appropriate to state an alternate hypothesis. The alternate hypothesis is a statement about the situation that should exist if the null hypothesis is rejected. An alternate hypothesis may take one of three forms as shown below:

(1) H_A: $\mu \neq \mu_H$ **(2)** H_A: $\mu < \mu_H$ **(3)** H_A: $\mu > \mu_H$

The first alternate states that if the population mean is not equal to μ_H, alternatively it may be greater than or less than μ_H. This type of alternate hypothesis is often referred to as a *two-sided test* of the null hypothesis.

Figure 8-4 demonstrates how the two-sided test works. First, a level of significance, or alpha level (α), has to be established. By establishing the level of significance, the boundaries for the regions of acceptance and rejection are defined. In Figure 8-4, an alpha of 0.05 is assumed. (The choice of an alpha value will be discussed at the end of the chapter.) Thus, the region of acceptance is represented by the area of the curve that would be expected to contain 95 percent of the $\overline{X}$'s; whereas, the region of rejection is represented in the two ends of the curve that could be expected to contain 5 percent of the $\overline{X}$'s.

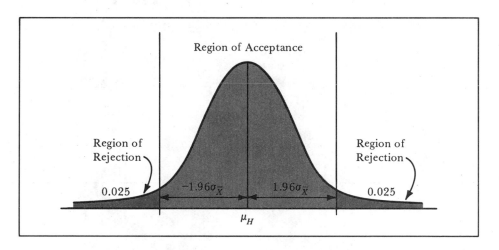

FIGURE 8-4. *A Two-sided Test of the Null Hypothesis With $\alpha = 0.05$.*

The purpose of the alpha level is to provide a way to decide whether the observed difference between the sample mean and hypothesized mean ($\overline{X} - \mu_H$) is a chance difference (sampling variation), or whether the difference ($\overline{X} - \mu_H$) should be declared a statistically significant difference. Using the level of significance shown in Figure 8-4, it is predetermined when H_0 is correct that 95 percent of the time any difference that is due to chance will be *attributed to chance* and 5 percent of the time any difference due to chance will be labeled incorrectly *a statistically significant difference.* Thus, an alpha level of 0.05 establishes that 5 percent of the time we would fail to identify that the true mean is equal to μ_H when in fact, it is.

The alternate hypotheses 2 and 3 are *one-sided tests* of the null hypothesis. That is, the researcher is interested only in detecting differences on one side of the hypothesized value of the mean. Hypothesis 2 states that if the null is rejected, it will be rejected because it is believed that the true mean is less than μ_H. Hypothesis 3, on the other hand, states that the null hypothesis will be rejected only if the true mean is greater than μ_H. Thus, these two alternate hypotheses are designed to detect differences on only one side of the null hypothesis.

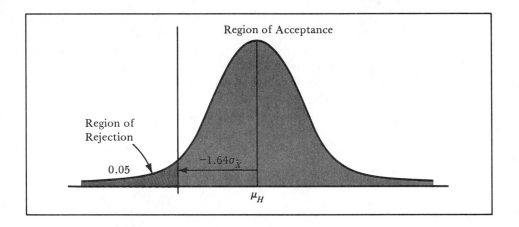

FIGURE 8-5. *A One-sided Test of the Null Hypothesis with* $\alpha = 0.05$.

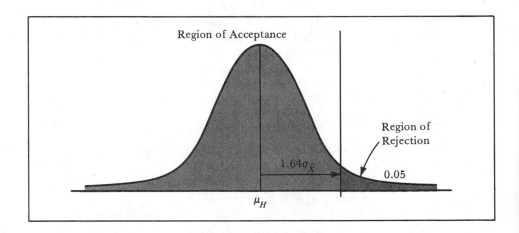

FIGURE 8-6. *A One-sided Test of the Null Hypothesis with* $\alpha = 0.05$.

Figures 8-5 and 8-6 show the regions of rejection and acceptance for the second and third alternate hypotheses, respectively. Notice that the rejection region is placed on one side of the curve. This is necessary since the alternate hypotheses were one sided; and in order to reject the null hypothesis and accept the alternate hypothesis, $\overline{X}$ must be less than $\mu_H - 1.64\sigma_{\bar{x}}$ (Figure 8-5) or $\overline{X}$ must be greater than $\mu_H + 1.64\sigma_{\bar{x}}$ (Figure 8-6).

Making the Decision

Given that appropriate null and alternate hypotheses are formulated, a method for making a decision is needed. The procedure is to identify the appropriate sampling distribution (normal or t distribution) and use that distribution to decide if the difference $(\overline{X} - \mu_H)$ *is too large to be considered a chance difference.*

For instance, in Figure 8-4, if the value $Z_{test} = (\overline{X} - \mu_H)/\sigma_{\bar{x}}$ is greater than $+1.96$ or less than -1.96, the difference $(\overline{X} - \mu_H)$, converted to a standard normal deviate, is too large to be a chance difference. The conclusion would have to be that the difference is a statistically significant difference. For Figure 8-4, the entire test procedure would be

$$H_0: \quad \mu = \mu_H$$

$$H_A: \quad \mu \neq \mu_H$$

$$\alpha = 0.05$$

Reject H_0 if $Z_{test} < -1.96$ or $Z_{test} > 1.96$

And for the one-sided tests, the test procedure would be:
For Figure 8-5:

$$H_0: \quad \mu = \mu_H$$

$$H_A: \quad \mu < \mu_H$$

$$\alpha = 0.05$$

Reject H_0 if $Z_{test} < -1.64$[1]

For Figure 8-6:

$$H_0: \quad \mu = \mu_H$$

$$H_A: \quad \mu > \mu_P$$

$$\alpha = 0.05$$

Reject H_0 if $Z_{test} > 1.64$

---

1.   Remember $-1.65 < -1.64$

Now that we have established a test procedure, let us rework Example 8.1.

1. Formulate the null hypothesis

   H_0: $\mu = 3.5$

2. Formulate the alternate hypothesis

   H_A: $\mu \neq 3.5$

3. Select the level of significance

   $\alpha = 0.01$; from Appendix H, the critical value of $Z = 2.58$

4. Establish rejection limits

   Reject H_0 if $Z_{test} < -2.58$ or $Z_{test} > 2.58$

5. Calculate Z_{test}

$$Z_{test} = \frac{\overline{X} - \mu_H}{\sigma_{\bar{x}}} \quad where \quad \sigma_{\bar{x}} = \frac{\sigma}{\sqrt{n}}$$

$$= \frac{3.65 - 3.5}{0.1/\sqrt{100}}$$

$$= \frac{3.65 - 3.5}{0.01}$$

$$= \frac{0.15}{0.01}$$

$$Z_{test} = 15$$

6. Make decision

   Since Z_{test} exceeds 2.58, reject H_0 and accept H_A; conclude that the population mean is now greater than 3.5 inches.

Arithmetic Mean: One Sample Test

Now that we have outlined a procedure for testing hypotheses using information from a single sample, we need to investigate the appropriate test for the sampling situations we encountered in chapter 7, which were

1. normal population and known population standard deviation
2. large samples in order to use the Central Limit Theorem, and
3. small samples from a normal population where the population standard deviation is unknown

Each of these situations has a particular test statistic for use in Step 5 shown above. For Situation 1, where the sample is drawn from a normal population with known population standard deviation, the test statistic is Equation 8.1.

$$Z_{test} = \frac{\overline{X} - \mu_H}{\sigma_{\overline{x}}} \qquad (8.1)$$

If Situation 2 exists where large samples are used in order to use the Central Limit Theorem, the test statistic is Equation 8.2.

$$Z_{test} = \frac{\overline{X} - \mu_H}{s_{\overline{x}}} \qquad (8.2)$$

And if Situation 3 exists, where the t distribution is appropriate, the test statistic is Equation 8.3.

$$t_{test} = \frac{\overline{X} - \mu_H}{s_{\overline{x}}} \qquad (8.3)$$

Now let us see an example for each of these situations.

▶ **Example 8.2**   *Test of Hypothesis About the Population Mean when Population Is Normal and Population Standard Deviation Is Known.* Suppose from the most recent data available that the Chamber of Commerce has established that family income is normally distributed with an average of $13,000 and a standard deviation of $4,500. However, because the data are several years old, the Chamber questions that the average has remained unchanged. To investigate the possibility of change, the Chamber took a random sample of 25 families which yielded a sample mean of $14,500. At the 0.05 level of significance, does it seem likely that change has occurred?

$$H_0 \quad : \quad \mu = \$13,000$$

$$H_A \quad : \quad \mu \neq \$13,000$$

$$\alpha = 0.05$$

Reject H_0 if $Z_{test} < -1.96$ or $Z_{test} > 1.96$

$$Z_{test} = \frac{\overline{X} - \mu_H}{\sigma_{\overline{x}}} \quad where \quad \sigma_{\overline{x}} = \frac{\sigma}{\sqrt{n}}$$

$$= \frac{14,500 - 13,000}{4,500/\sqrt{25}}$$

$$= \frac{1,500}{900}$$

$$Z_{test} = 1.67$$

Accept (do not reject) the null hypothesis. The difference between $\overline{X}$ and μ_H can be considered chance (sampling) variation. Thus, the Chamber of Commerce concludes that average family income has not changed. ◀

▶ **Example 8.3**

Test of Hypothesis About the Population Mean when Large Samples Are Selected. Suppose that a local advertising agency in Big Town believes the average advertising expenditure of a certain group of retail merchants is at least $1,500. To determine this expenditure, the agency collects a sample of expenditures from 64 merchants. The sample yielded an $\overline{X} = \$1,400$ and $s = \$128$. At the 0.05 level of significance, is the agency's belief correct? (Note: this will be a one-sided test because the agency believes the expenditure will be *at least* $1,500. The agency is wrong only if the population mean is actually less than $1,500, not if it is $1,500 or above.)

$$H_0: \quad \mu = \$1500 \quad \text{or} \quad \mu \geq \$1500$$

$$H_A: \quad \mu < \$1500$$

$$\alpha = 0.05$$

Reject H_0 if $Z_{\text{test}} < -1.64$ (See Figure 8-5)

$$Z_{\text{test}} = \frac{\overline{X} - \mu_H}{s_{\bar{x}}} \quad where \quad s_{\bar{x}} = \frac{s}{\sqrt{n}}$$

$$= \frac{1400 - 1500}{128/\sqrt{64}}$$

$$= \frac{-100}{128/8}$$

$$= \frac{-100}{16}$$

$$Z_{\text{test}} = -6.25$$

Reject the null hypothesis, the difference between $\overline{X}$ and μ_H is a statistically significant difference. The agency's belief is incorrect. ◀

▶ **Example 8.4**

Test of Hypothesis About the Population Mean when Small Samples Are Selected from a Normal Population with Unknown Population Standard Deviation. The Hogly Wogly Food Store has a special fall promotion every year when chickens are sold for a fixed amount rather than by the pound. The price this year is to be 70¢ per chicken.

In order not to lose money on this promotion, the Hogly Wogly management has determined that the average size chicken purchased from the supplier should not exceed 28 ounces. To determine the average size of the chicken, management has been allowed to weigh 25 of the chickens that are available. The sample mean was 30 ounces with a standard deviation of 2 ounces. If the weights of the chickens can be considered to be normally distributed, should management make the purchase? (Note: this will be a one-sided test because management is concerned only if the average size *exceeds* 28 ounces.)

$$H_0: \quad \mu = 28 \quad \text{or} \quad \mu \leq 28$$

$$H_A: \quad \mu > 28$$

$$\alpha = 0.05$$

Reject H_0 when $t_{\text{test}} > 1.711$ (obtain 1.711 from Appendix J, using $n - 1$ or $25 - 1 = 24$ degrees of freedom.)

$$t_{\text{test}} = \frac{\overline{X} - \mu_H}{s_{\bar{x}}} \qquad where \qquad s_{\bar{x}} = \frac{s}{\sqrt{n}}$$

$$t_{\text{test}} = \frac{30 - 28}{2/\sqrt{25}}$$

$$= \frac{2}{2/5}$$

$$t_{\text{test}} = 5$$

Reject H_0 and accept H_A. The chickens are too large for management to purchase. ◄

Proportions: One Sample Test

Just as we can construct confidence intervals for proportions, we can test hypotheses about the population proportion. The null hypothesis would be of the form,

$$H_0: \quad \pi = \pi_H$$

and the three possible alternatives would be of the form,

$$(1) \ H_A: \quad \pi \neq \pi_H \qquad (2) \ H_A: \quad \pi < \pi_H \qquad (3) \ H_A: \quad \pi > \pi_H$$

just as existed for the mean. The test procedure would also be the same as outlined for population means if a large sample is taken in order to use the Central Limit Theorem. The appropriate test statistic is shown in Equation 8.4

$$Z_{\text{test}} = \frac{p - \pi_H}{\sigma_p} \qquad\qquad \textbf{(8.4)}$$

$$where \qquad \sigma_p = \sqrt{\frac{\pi_H (1 - \pi_H)}{n}}$$

Notice that the standard error of the proportion σ_p is calculated using π_H, the hypothesized value of the population proportion rather than p, the sample proportion. The reasoning is that since we have hypothesized that $\pi = \pi_H$, it is appropriate to use π_H as an approximation for π when calculating σ_p. This approach differs, however, from the estimation method discussed in chapter 7 where, under the assumption of no knowledge about π, p was used as the approximation for π.

▶ **Example 8.5**   Mr. Candidate believes he is going to carry a particular city by a large margin. In fact, he believes he should get at least 60 percent of the vote. His campaign manager, who is not so optimistic, has employed the firm of Election Forecasters to check out Mr. Candidate's belief. If, in a sample of 400 voters, 232 indicated they would vote for Mr. Candidate, do you think his optimism is justified?[2]

$$p = \frac{X}{n} = \frac{232}{400} = 0.58$$

$$\sigma_p = \sqrt{\frac{\pi_H (1 - \pi_H)}{n}}$$

$$\sigma_p = \sqrt{\frac{(0.60)(0.40)}{400}}$$

$$= \sqrt{0.0006}$$

$$\sigma_p = 0.0245$$

$$H_0: \quad \pi = 0.60 \quad \text{or} \quad \pi \geq 0.60$$

$$H_A: \quad \pi < 0.60$$

$$\alpha = 0.05$$

Reject H_0 if $Z_{test} < -1.64$

$$Z_{test} = \frac{p - \pi_H}{\sigma_p}$$

$$= \frac{0.58 - 0.60}{0.0245}$$

$$Z_{test} = -0.816$$

Accept (do not reject) H_0, the difference between the sample proportion (0.58) and the hypothesized proportion (0.60) can be considered a chance difference. Therefore, Mr. Candidate's belief has not been disproved. ◀

---

2.   Remember that percentages are proportions multiplied by 100.

Type I and Type II Errors

In our discussion and examples in this chapter, there has not been any indication as to what possible errors the researcher might make when he is testing hypotheses. Table 8-1 shows the possible situations that exist when a hypothesis is tested. From the table you can see that there are four possibilities, two of which are the correct decision and two of which lead to erroneous decisions.

TABLE 8-1. *Possible Outcomes When Testing Hypotheses*

	H_0 IS ACCEPTED	H_A IS ACCEPTED
H_0 is correct	Correct Decision $1 - \alpha$	Type I Error α
H_A is correct	Type II Error β	Correct Decision $1 - \beta$

The first error in Table 8-1, denoted as *Type I Error,* occurs when the correct null hypothesis is rejected and the alternate, which is incorrect, is accepted. As you have probably noticed, the probability of Type I Error is α and results from the level of significance selected by the researcher. It is the probability that a researcher will fail to identify a *true* null hypothesis.

Figure 8-7 shows a two-sided situation where the α level is 0.10. This is a situation where the Type I Error has been set at 0.10. Figure 8-8 shows a one-sided situation where the α level is 0.10. Notice that in the one-sided case, the Type I Error occurs at only one end of the sampling distribution.

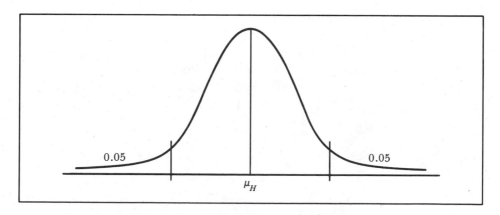

FIGURE 8-7. *Type I Error Set at 0.10 for Two-sided Test.*

181

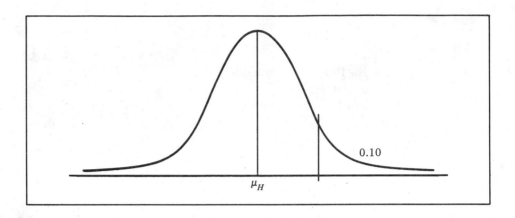

FIGURE 8-8. *Type I Error Set at 0.10 for One-sided Test.*

The choice of a value for α is a management decision rather than a statistical decision. The decision has to be made based on a judgment concerning the seriousness of committing Type I Error. The general guideline is to set α at a relatively low value (0.01 or 0.05) if rejecting a correct null hypothesis would be very costly and/or hazardous. Conversely, if accepting an incorrect null hypothesis (Type II Error) is costly or hazardous, the α value should be high (0.25 or higher).

Looking again at Table 8-1, the other error that appears is called *Type II Error,* and the probability of Type II Error is designated β. Type II Error occurs when the researcher accepts a null hypothesis that is incorrect. From Table 8-1, we see this indicates that the alternate hypothesis was the correct one, but the null hypothesis was accepted.

A Type II Error has a prerequisite in that this particular error can occur only if the hypothesized mean is incorrect (H_A is true). Or it is often said that Type II Error can occur only if the true mean changes. Figure 8-9 shows how this happens.

Figure 8-9 shows a situation where the null and alternate hypotheses would have been

$$H_0: \quad \mu = \mu_H$$

$$H_A: \quad \mu \neq \mu_H$$

where μ_H is the hypothesized or assumed value of the population mean. Using an α level of 0.10, we know that if a sample $\overline{X}$ were obtained that fell between the upper and lower $\overline{X}$ limits (region of acceptance), we would accept H_0. However, if the true mean is not μ_H

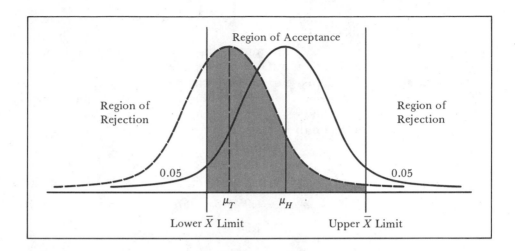

FIGURE 8-9. *Type II Error*

but is really μ_T, then the real sampling distribution is the dotted distribution. In this situation, the correct decision is to reject H_0 and accept H_A. Failure to do this results in Type II Error.

The probability of Type II Error is the shaded area of Figure 8-9. Type II Error can be determined as the area under the dotted sampling distribution curve that is between the upper and lower $\overline{X}$ limits. Example 8.6 demonstrates the calculation of Type II Error.

▶ **Example 8.6**   Suppose a random sample of 100 items yielded a sample mean of 50 and a standard deviation of 10. If the hypothesized mean had been 52, but the true mean is really 51, what is the probability of Type II Error if $\alpha = 0.05$?

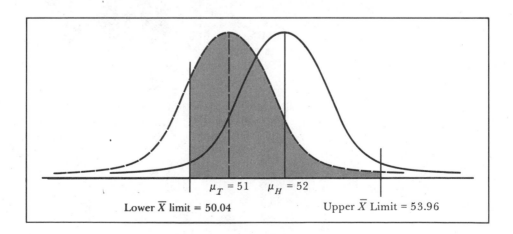

FIGURE 8-10a. *Location of Type II Error.*

Figure 8-10a shows the problem as it is perceived. The curve represented by the solid line is the sampling distribution that exists if $\mu_H = 52$. The upper $\overline{X}$ limit and the lower $\overline{X}$ limit are calculated using μ_{H_1} and the region between them is the region of acceptance. The dotted curve represents the sampling distribution when the actual mean is 51. The shaded area is the portion of the dotted curve within the region of acceptance and this shaded area is the likelihood of committing Type II Error.

$$LL_{\overline{X}} = \mu_H - Z_\alpha s_{\overline{x}} \qquad\qquad UL_{\overline{X}} = \mu_H + Z_\alpha s_{\overline{x}}$$

$$= 52 - 1.96\,(s/\sqrt{n}) \qquad\qquad = 52 + 1.96\,(s/\sqrt{n})$$

$$= 52 - 1.96\,(10/\sqrt{100}) \qquad = 52 + 1.96\,(10/\sqrt{100})$$

$$= 52 - 1.96\,(1) \qquad\qquad = 52 + 1.96\,(1)$$

$$= 52 - 1.96 \qquad\qquad = 52 + 1.96$$

$$LL_{\overline{X}} = 50.04 \qquad\qquad UL_{\overline{X}} = 53.96$$

Figure 8-10b demonstrates how to calculate Type II Error. The shaded area is divided into two regions, ① and ②. The areas of these two regions added together make up the probability of Type II Error or β.

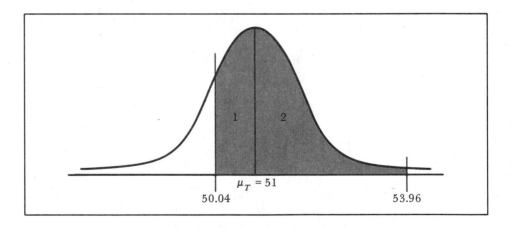

FIGURE 8-10b. *Computing Type II Error.*

$$Z_1 = \frac{LL_{\overline{X}} - \mu_T}{s_{\overline{x}}} = \frac{50.04 - 51}{1}$$

$$Z_1 = \frac{-0.96}{1} = -0.96$$

Area 1 = 0.3315

$$Z_2 = \frac{UL_{\bar{x}} - \mu_T}{s_{\bar{x}}} = \frac{53.96 - 51}{1}$$

$$= \frac{2.96}{1}$$

$$Z_2 = 2.96$$

$$\text{Area } 2 = 0.4985$$

Thus, the probability of committing Type II Error is 0.8300 (0.3315 + 0.4985). ◄

Operating Characteristic Curve and Power of the Test

An operating characteristic curve presents the *probability of accepting a null hypothesis* for various values of the population parameter at a given α level using a particular sample size. To construct an operating characteristic curve, different values of the population parameter μ or π are assumed, and the probability of accepting H_0 is determined for each assumed value. Then the assumed values of the parameter are plotted on the abscissa, while the probability of accepting H_0 is plotted on the ordinate. Thus, any point on the graph represents the probability of accepting the null hypothesis for a given value of the parameter.

Figure 8-11 is the operating characteristic curve for Example 8-6. Notice on Figure 8-11, when the population mean is 52, the

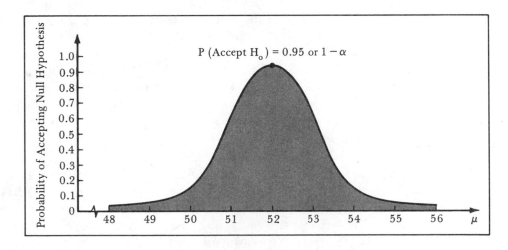

FIGURE 8-11. *Operating Characteristic Curve*
$H_0: \mu = 52, \alpha = 0.05, n = 100.$

probability of acceptance is 0.95, which is $1 - \alpha$. This occurrence is consistent with the original design of the test. That is, if $\mu = 52$, the null hypothesis should be accepted 95 percent of the time and rejected 5 percent of the time. Also notice on Figure 8-11 that when $\mu = 51$ the probability of accepting the null hypothesis is 0.8300. This value, 0.8300, is the Type II Error calculated in Example 8-6. Thus, all points on the operating characteristic curve are obtained by calculating β as demonstrated in Example 8-6. (All probabilities are β except at the peak of the curve which is $1 - \alpha$.)

Looking at Figure 8-11 again, if the actual value of the mean is either 51 or 53 (one unit from the hypothesized value of the mean) the probability of β error is 0.8300. Further, if the actual value of the mean is either 50 or 54 (two units from the hypothesized value of the mean), then the probability of β error is 0.4840. Consequently, when the actual value of the mean is two units above or below the hypothesized value of the mean, the test will fail to detect this fact 48.4 percent of the time.

Thus, use of the operating characteristic curve allows an evaluation of the decision (accept H_0) that might be reached for a given sample size and a given level of significance. The *OC* curve reveals that the test constructed for the situation described in Example 8-6 is prone towards β error if the actual value of the mean is close (one or two units) to the hypothesized value of the mean.

If these β errors are not acceptable risks to the decision maker, there are two courses of action that will reduce them. One action would be to increase the value of α, which would increase the probability of rejecting H_0, thereby reducing β error. The other action would be to increase the sample size. This action has the effect of reducing the region of acceptance by reducing $s_{\bar{x}} = s/\sqrt{n}$. Either of these actions would improve what is known as the *power of the test* $(1 - \beta)$.

The power of the test is *the inverse function of the operating characteristic curve*. That is, the power of the test is the probability of rejecting the null hypothesis for various possible values of the population parameter. Table 8-2 presents the power of the test values for the situation described in Example 8-6.

As you see from Table 8-2, the power of the test increases as the values of the mean get farther away from the hypothesized value of 52. For the values of μ, except 52, the power of the test $(1 - \beta)$ is the probability of making the correct decision. And, for this reason, it is often used instead of the operating characteristic curve.

Figure 8-12 is a graphic presentation of the power of the test and is usually referred to as the *power function*. As μ changes from 52, the probability values can be thought of as the probability of rejecting

H_0 when H_0 should be rejected. And from Figure 8-12, it can be seen that the probability of rejection does increase as the value of μ moves farther and farther away from 52.

TABLE 8-2. *Operating Characteristic Curve Values and Power of the Test;*
H_0: $\mu = 52$, $\alpha = 0.05$, $n = 100$.

POSSIBLE VALUES OF μ	PROBABILITY OF ACCEPTING H_0 (OPERATING CHARACTERISTIC CURVE)	PROBABILITY OF REJECTING H_0 (POWER OF THE TEST)	
48	0.0207	0.9793	
49	0.1492	0.8508	$1 - \beta$
50	0.4840	0.5160	
51	0.8300	0.1700	
52	0.9500	0.0500	α
53	0.8300	0.1700	
54	0.4840	0.5160	$1 - \beta$
55	0.1492	0.8508	
56	0.0207	0.9793	

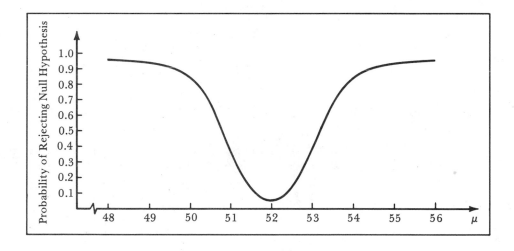

FIGURE 8-12. *Power Function*
H_0: $\mu = 52$, $\alpha = 0.05$, $n = 100$.

--- **SUMMARY OF EQUATIONS** ---

8.1 $Z_{\text{test}} = \dfrac{\overline{X} - \mu_H}{\sigma_{\bar{x}}}$, *page 177*

8.2 $Z_{test} = \dfrac{\overline{X} - \mu_H}{s_{\bar{x}}}$,   *page 177*

8.3 $t_{test} = \dfrac{\overline{X} - \mu_H}{s_{\bar{x}}}$,   *page 177*

8.4 $Z_{test} = \dfrac{p - \pi_H}{\sigma_p}$,   *page 179*

─────────── **REVIEW QUESTIONS** ───────────

1. Identify the use of each equation shown at the end of the chapter.
2. Define
   a. null hypothesis
   b. alternate hypothesis
   c. level of significance
   d. Type I Error
   e. Type II Error
   f. operating characteristic curve
   g. power of the test
3. Distinguish between a chance difference and a statistically significant difference.
4. List the steps in the statistical testing of a hypothesis.
5. Explain when Type II Error can occur.

─────────── **PROBLEMS** ───────────

1. Use the following to test the null hypothesis that $\mu = 45$ at the 0.05 level of significance.   Assume the population is normally distributed.

$$\sigma = 10$$
$$\overline{X} = 43$$
$$n = 100$$
$$H_A: \quad \mu \neq 45$$

2. Test the null hypothesis $\mu = 190$ against the alternate hypothesis $\mu \neq 190$ if a random sample of 49 items yielded an $\overline{X} = 182$.   If the distribution of the population is unknown, but the sample standard deviation is known to be 21, perform the test at an α level of 0.10.

3. Use the sample results given below to test the null hypothesis $\mu = 75$ against the alternate hypothesis $\mu > 75$.   Set $\alpha = 0.05$.

$$n = 64$$
$$s = 24$$
$$\overline{X} = 79$$

$\left[\dfrac{N-n}{N-1}\right]^{1/2}$

4. Work Problem 3 if $N = 500$. Does your conclusion change?

5. Use the sample results given below to test the null hypothesis $\mu = 120$ against the alternate hypothesis $\mu < 120$. Set $\alpha = 0.01$.

$$n = 81$$
$$s = 36$$
$$\overline{X} = 100$$

6. A sample of 25 machinists yielded a mean age of 38 years with a standard deviation of 8 years. If the industry ages for machinists are normally distributed with a mean of 43 years, what can you conclude about the machinists represented by the sample? Set $\alpha = 0.05$.

7. The manufacturer of a battery-operated toy car has made modifications on the motor of the car. One result of the modification seems to be an extended lifetime for the battery. From past data, it is known that the battery in the car had an average operating life of 30 hours. To investigate if the modification did improve battery life, a random sample of 100 cars with the new motor was obtained yielding a mean operating lifetime of 31 hours and a standard deviation of 0.75 hours. At the 0.05 level of significance can we infer that the battery lifetime was extended?

8. A chicken grower claims that the average weight of a particular group of chickens is at least 30 ounces. Before agreeing to a purchase, a particular customer selected a sample of 25 chickens, which yielded a sample mean of 29.75 ounces and a standard deviation of 1.5 ounces. If the weights can be considered to be normally distributed, should the claim be rejected at the 0.05 level of significance?

9. A gasoline retailer has maintained records indicating that weekly sales have averaged 1,600 gallons. In an attempt to increase sales, the retailer issues a small prize with purchases of 10 gallons or more. If after a 16-week period, the average weekly sales have increased to 1,650 gallons with a standard deviation of 150 gallons, is it wise to continue the prizes? Set $\alpha = 0.10$. Are any special assumptions necessary?

10. A machine used to fill cereal boxes has a setting of 12 ounces. A random sample of 49 boxes produced a sample mean of 11.9 ounces and a standard deviation of 0.35 ounces. At the 0.05 level of significance, what can you conclude about the machine setting?

11. Suppose the situation in Problem 10 had included only 16 boxes in the sample.
a. Would a *t*-test be appropriate?
b. If the *t*-test is appropriate, would your conclusion drawn in Problem 11 differ from Problem 10?

12. An automobile engine has been designed to have a trouble-free life of 20,000 miles. A random sample of 100 engines experienced malfunction at an average of 16,000 miles with a standard deviation of 4,000 miles. If $\alpha = 0.10$, what can you conclude about the trouble-free life?

13. A particular report claimed that the mean amount of tar per cigarette for a particular brand was at least 13 grams. If a random sample of 100 cigarettes yielded a mean of 12.85 grams with a standard deviation of 0.75 grams, can the company refute the claim? Set $\alpha = 0.01$.

14. A company maintains the time required for a service call does not exceed 25 minutes. If a random sample of 16 calls yields a sample mean of 26.25 minutes with a standard deviation of 3 minutes, at the 0.05 level of significance, what can be concluded?

15. In the preceding problem, if the sample size had been 36, would your conclusion change?

16. A company plans to market nationally a new product only if more than 70 percent of the test market responds favorably to a new product. If a random sample of 1,000 persons from the test market yielded 730 favorable responses, at the 0.05 level of significance, what can you conclude?

17. Test the null hypothesis $\pi = 0.70$ against the alternate hypothesis $\pi \neq 0.70$ if a sample proportion of 0.74 was obtained from a sample size of 300. Set $\alpha = 0.05$.

18. A union composed of several thousand employees is preparing to vote on a new contract. If a random sample of 400 members yielded 212 who planned to vote yes, should you conclude that the new contract will receive more than 50 percent yes votes? Set $\alpha = 0.05$.

19. A sample of 100 items from a large shipment of Part WK produced 6 defective parts. If the manufacturer guaranteed 97 percent acceptable parts, should the shipment be accepted? Set $\alpha = 0.01$.

20. Rework Problem 19 if the shipment contained 500 parts.

21. A store manager believes at least 40 percent of the purchases exceed $5.00. If a random sample of 60 shoppers yielded 22 persons with purchases in excess of $5.00, is the manager's claim substantiated? Set $\alpha = 0.05$.

22. New Products, Inc., has a standing rule for test marketing a new product. The rule necessitates a test market acceptance of greater than 80 percent before actual mass production of an item. If the latest product received 1,950 favorable responses from the 2,500 interviews, should mass production begin? Set $\alpha = 0.05$.

23. Smoke, Inc., has been manufacturing Smoker Fertilizer for years and has many distributors across the country. Recently, complaints that the 50-pound bag is light have been received. To investigate these complaints, Smoke has selected a sample of 144 bags and obtained $\overline{X} = 49.5$ and $s = 2$ pounds. At the 0.05 level, how do you react to the complaints?

24. A switchboard operator believes that at least 40 percent of the calls received are long distance. If a random sample of 200 calls produced 70 long distance calls, do you agree with the operator? Set $\alpha = 0.10$.

25. A chemical process is set to produce at the rate of 8.9 tons per hour. The

normal?

process, however, has produced at a higher and lower rate in the past. The most recent sample of 24 hours showed an average yield of 8.81 tons with a standard deviation of 0.20 tons. Does it appear at the 0.05 level of significance that the process average has changed?

26. A sample of 100 items from a normal population produced a sample mean of 30 and a standard deviation of 5. If the intent was to test the null hypothesis $\mu = 29$ against the alternate hypothesis $\mu \neq 29$, what is the probability of committing Type II Error when the true mean is 28.5? Set $\alpha = 0.05$.

27. A processing plant has a machine that places lids on all filled jars. The machine supposedly is set at a 30-pound rating with a known standard deviation of 8 pounds. At the 0.05 level of significance, what is the probability of committing Type II Error if a sample of 64 jars is taken and the true mean setting is 28 pounds?

28. For the preceding problem, construct an operating characteristic curve and a power curve.

29. A study was done on the daily cash balances of a bank to investigate the hypothesis that the average cash balance was $250,000. The sample of 100 yielded a mean of $248,000 and a standard deviation of $30,000. If the true mean was actually $246,000, what is the probability of Type II Error? Set $\alpha = 0.05$.

30. For the preceding problem, construct an operating characteristic curve and a power curve.

31. A certain developing solution is advertised to last at least 25 days under normal use. If a random sample of 36 customers yielded a sample mean of 23 with a standard deviation of 4 days, is the advertising claim supported? Set $\alpha = 0.01$.

32. An internal auditor claims that at least 90 percent of all discrepancies are corrected. If a random sample of 200 discrepancies yielded 175 corrections, is the claim supported? Set $\alpha = 0.05$.

Confidence Intervals and Hypothesis Testing: Two Samples

CHAPTER
9

In Chapter 6, you learned how to calculate the mean and standard deviation of the distribution of sample means ($\overline{X}$'s). You also learned several things about the shape of the distribution of sample means. There are situations, however, where we may select two samples and perform analysis based on the results of both. To be able to do this, we need to investigate the sampling distribution of two random samples.

The Sampling Distribution of the Difference Between Two Sample Means ($\overline{X}_1 - \overline{X}_2$)

Analogous to what we did in Chapter 6, if you take all possible samples of n size, from a population of N, and compare the mean of each sample with the mean of all the other possible samples, you will find that the arithmetic mean of the *difference* in sample means is zero (0). That is, the sampling distribution of differences in sample means will have a mean of 0 if the samples all came from the same population. Moreover, the differences will be normally distributed if the population is normally distributed or if the sample size is large (Central Limit Theorem). If the sample size is small, the differences in sample means selected from a normally distributed population will be Student t distributed.

But what if we have samples from two different populations? As a matter of fact, the mean of the distribution of differences in sample means will be equal to the difference in the two population means, or $\mu_1 - \mu_2$. The differences will be normally distributed if *both* populations are normal distributions with known variances or if the sample sizes are large.   The distribution will be Student t shaped if *the sample size is small and the standard deviations of both populations are equal and the populations are normal.*

The above should sound somewhat familiar to you since the sampling distribution of differences in two sample means is similar to the sampling distribution of sample means.   These similarities are summarized in Table 9-1.

TABLE 9-1.   *Comparisons of Sampling Distributions.*

SAMPLING DISTRIBUTION	NUMBER OF SAMPLES	
	ONE	TWO
Normal	Normal Population σ is known	Two Normal Populations σ_1 and σ_2 are known
Normal via Central Limit Theorem	Large Sample	Two Large Samples
Student t	Small Sample Normal Population σ is unknown	Two Small Samples Two Normal Populations $\sigma_1 = \sigma_2$ (unknown)

but equal

So far, we have mentioned the standard deviation of the population, or populations, from which the samples are taken, but we have not considered the standard deviation of the sampling distribution of differences.   This standard deviation is called the *standard error of the difference between two sample means* and can be calculated using Equation 9.1 if the standard deviations of the two populations are known.

$$\sigma_{\bar{x}_1 - \bar{x}_2} = \sqrt{\frac{\sigma_1^2}{n_1} + \frac{\sigma_2^2}{n_2}} \qquad (9.1)$$

where σ_1^2 = variance of the first population
σ_2^2 = variance of the second population
n_1 = the size of the first sample
n_2 = the size of the second sample

If the standard deviations of the populations are unknown but both samples are large (more than 30), then Equation 9.2 can be used to estimate the standard error of the difference between two sample means.

$$S_{\bar{x}_1-\bar{x}_2} = \sqrt{\frac{s_1^2}{n_1} + \frac{s_2^2}{n_2}} \qquad (9.2)$$

where $s_1^2 = \dfrac{\Sigma\,(X-\overline{X})^2}{n_1-1}$ = variance of the first sample

              $s_2^2 = \dfrac{\Sigma\,(X-\overline{X})^2}{n_2-1}$ = variance of the second sample

When the sample sizes are small (30 or less) *and* the standard deviations of the two populations are unknown but equal, Equation 9-3 can be used to calculate an estimate of the standard error.

$$S_{\bar{x}_1-\bar{x}_2} = \sqrt{\frac{s_1^2\,(n_1-1) + s_2^2\,(n_2-1)}{n_1+n_2-2}}\;\sqrt{\frac{1}{n_1} + \frac{1}{n_2}} \qquad (9.3)$$

Constructing Confidence Intervals for the Difference Between Two Population Means ($\mu_1 - \mu_2$)

In order to construct a confidence interval estimate of the difference between two population means we must first determine whether

1. The populations are normally distributed;
2. σ_1 and σ_2 are known;
3. $\sigma_1 = \sigma_2$; and
4. whether the sample sizes (n_1 and n_2) are greater than 30.

If both populations are normally distributed and σ_1 and σ_2 are known, then Equation 9.4 is appropriate to use.

$$(1-\alpha)\%\text{ confidence interval for }\mu_1-\mu_2 = (\overline{X}_1 - \overline{X}_2) \pm Z\sigma_{\bar{x}_1-\bar{x}_2} \quad (9.4)$$

where $\sigma_{\bar{x}_1-\bar{x}_2} = \sqrt{\dfrac{\sigma_1^2}{n_1} + \dfrac{\sigma_2^2}{n_2}}$

If, however, the population variances are unknown and the shapes of the population distributions are not necessarily normal, then Equation 9.5 can be used if both sample sizes are large (n_1 and $n_2 > 30$) so that the Central Limit Theorem is effective.

$$(1-\alpha)\%\text{ confidence interval for }\mu_1-\mu_2 = (\overline{X}_1 - \overline{X}_2) \pm Zs_{\bar{x}_1-\bar{x}_2} \quad (9.5)$$

where $s_{\bar{x}_1-\bar{x}_2} = \sqrt{\dfrac{s_1^2}{n_1} + \dfrac{s_2^2}{n_2}}$

And, if both populations are normally distributed, the population vari-ances are unknown but equal ($\sigma_1 = \sigma_2$), and the sample sizes are 30 or less, a confidence interval estimate for $\mu_1 - \mu_2$ can be made using Equation 9.6.

$$(1 - \alpha)\% \text{ confidence interval for } \mu_1 - \mu_2 = (\overline{X}_1 - \overline{X}_2) \pm t s_{\bar{x}_1 - \bar{x}_2} \qquad \textbf{(9.6)}$$

$$\text{where} \quad s_{\bar{x}_1 - \bar{x}_2} = \sqrt{\frac{s_1^2(n_1 - 1) + s_2^2(n_2 - 1)}{n_1 + n_2 - 2}} \sqrt{\frac{1}{n_1} + \frac{1}{n_2}}$$

$n_1 + n_2 - 2 = $ degrees of freedom

Now let us examine some example problems using each of the preceding confidence interval equations.

▶ **Example 9.1**   Two union representatives in a large company wonder if their respec-tive groups are paid the same hourly wage. In order to investigate this situation further, the first representative takes a sample of 25 em-ployees and determines that their average hourly wage is $4.65.   At the same time, the second representative has determined from a sample of size 20 that the average hourly wage of the second group is $4.52.   It is known that the two populations from which the samples were drawn are normally distributed and that $\sigma_1 = \$0.25$, and $\sigma_2 = \$0.20$.   Use this information to make a 90 percent confidence interval estimate for the difference between the two population means. Since $\alpha = 0.10$, then $Z = 1.64$, and

$$(1 - \alpha)\% \text{ confidence interval for } \mu_1 - \mu_2 = (\overline{X}_1 - \overline{X}_2) \pm Z\sigma_{\bar{x}_1 - \bar{x}_2}$$

$(1 - 0.10)$ confidence interval for

$$\mu_1 - \mu_2 = (\$4.65 - \$4.52) \pm 1.64 \sqrt{\frac{(0.25)^2}{25} + \frac{(0.20)^2}{20}}$$

$$= \$0.13 \pm 1.64 \sqrt{\frac{0.0625}{25} + \frac{0.0400}{20}}$$

$$= \$0.13 \pm 1.64 \sqrt{0.0025 + 0.0020}$$

$$= \$0.13 \pm 1.64 \sqrt{0.0045}$$

$$= \$0.13 \pm 1.64 \,(0.067)$$

$$= \$0.13 \pm \$0.11$$

$$\$0.02 \leq (\mu_1 - \mu_2) \leq \$0.24 \blacktriangleleft$$

▶ **Example 9.2**   A manufacturing company assembles its product on production lines known as Line A and Line B.   Line A is composed of employees with at least 2 years experience whereas Line B is composed of employees with less than six months experience.   A recent 35-day study of Line A revealed an average number of assemblies each day of 70 with a standard deviation of 5 units.   A similar study of Line B conducted for 40 days revealed an average of 64 units assembled per day with a standard deviation of 8 units. Construct a 95 percent confidence interval for the difference in the daily production of the two lines. Since $\alpha = 0.05$, then $Z = 1.96$, and

$(1 - \alpha)\%$ confidence interval for $\mu_1 - \mu_2 = (\overline{X}_1 - \overline{X}_2) \pm Z s_{\bar{x}_1 - \bar{x}_2}$

$(1 - 0.05)$ confidence interval for

$$\mu_A - \mu_B = (70 - 64) \pm 1.96 \sqrt{\frac{5^2}{35} + \frac{8^2}{40}}$$

$$= 6 \pm 1.96 \sqrt{\frac{25}{35} + \frac{64}{40}}$$

$$= 6 \pm 1.96 \sqrt{0.714 + 1.6}$$

$$= 6 \pm 1.96 \sqrt{2.314}$$

$$= 6 \pm 1.96 \, (1.52)$$

$$= 6 \pm 2.98$$

$$3.02 \le (\mu_A - \mu_B) \le 8.98 \; \blacktriangleleft$$

▶ **Example 9.3**   The manufacturer of two types of boots (A and B) is interested in the prices that are being charged for the boots. A random sample of 12 retailers who stock Type A yielded an average price of $13.00 with a standard deviation of $1.00, and a random sample of 10 retailers who stock Type B yielded an average price of $14.00 with a $2.00 standard deviation. Under the assumption that the populations of prices are normally distributed and that the population standard deviations, though unknown, are equal, construct a 98 percent confidence interval for the difference between the two average prices.[1] (Remember that the degrees of freedom (d.f.) are $n_1 + n_2 - 2$.) Since $\alpha = 0.02$ and $d.f. = 12 + 10 - 2$, then $t = 2.528$, and

$(1 - \alpha)\%$ confidence interval for $\mu_1 - \mu_2 = (\overline{X}_1 - \overline{X}_2) \pm t s_{\bar{x}_1 - \bar{x}_2}$

$(1 - 0.02)\%$ confidence interval for

$$\mu_A - \mu_B = (\$13 - \$14) \pm 2.528 \sqrt{\frac{1^2 (12 - 1) + 2^2 (10 - 1)}{12 + 10 - 2}} \sqrt{\frac{1}{12} + \frac{1}{10}}$$

$$= -\$1 \pm 2.528 \sqrt{\frac{1 (11) + 4 (9)}{20}} \sqrt{0.183}$$

$$= -\$1 \pm 2.528 \sqrt{2.35} \sqrt{0.183}$$

$$= -\$1 \pm 2.528 \, (1.53) \, (0.428)$$

$$= -\$1 \pm \$1.66$$

$$-\$2.66 \le (\mu_A - \mu_B) \le \$0.66 \; \blacktriangleleft$$

---

1.   The appendix to this chapter contains a statistical test to determine if two unknown variances are equal.

Testing Hypotheses About the Difference Between Two Population Means $(\mu_1 - \mu_2)$

When we test a hypothesis about the difference between two population means, we are testing the following null hypothesis,

$$H_0: \quad \mu_1 - \mu_2 = 0 \quad \text{or} \quad \mu_1 = \mu_2$$

against one of the three alternate hypotheses shown below

(1) $H_A: \mu_1 - \mu_2 \neq 0$ **(2)** $H_A: \mu_1 - \mu_2 < 0$ **(3)** $H_A: \mu_1 - \mu_2 > 0$

As was the case in Chapter 8, the first alternate is the two-sided test of the null hypothesis that there is no difference between the two population means; and the second and third alternates are the one-sided tests. Figure 9-1 shows the two-sided test; and Figures 9-2 and 9-3 show the one-sided tests for the second and third alternate hypotheses, respectively. Notice again, that on the one-sided tests, all of the Type I Error (α error) goes in a single tail of the sampling distribution.

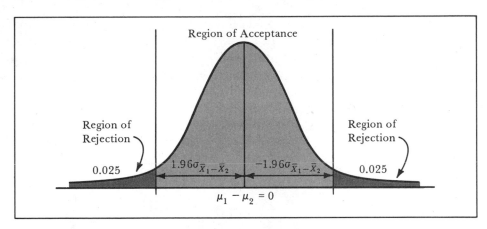

FIGURE 9-1. *A Two-sided Test of the Null Hypothesis with $\alpha = 0.05$.*

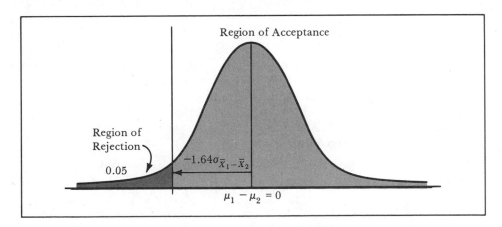

FIGURE 9-2. *A One-sided Test of the Null Hypotheses with $\alpha = 0.05$.*

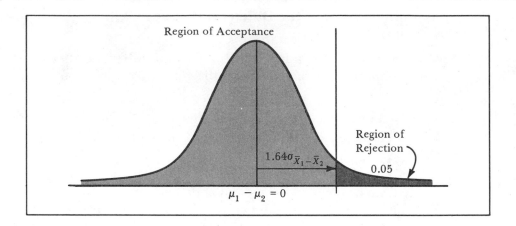

FIGURE 9-3. *A One-sided Test of the Null Hypothesis with $\alpha = 0.05$.*

In order to test the hypothesis, one of three following test equations will be used.   Equation 9.7 is used when both populations are normally distributed and σ_1 and σ_2 are known.   Equation 9.8 is used when n_1 and n_2 are large samples, which enables the Central Limit Theorem to be used, and Equation 9.9 is used when n_1 and n_2 are small samples selected from two normal populations that have the same standard deviation.

$$Z_{\text{test}} = \frac{(\overline{X}_1 - \overline{X}_2) - (\mu_1 - \mu_2)}{\sigma_{\bar{x}_1 - \bar{x}_2}} \tag{9.7}$$

$$Z_{\text{test}} = \frac{(\overline{X}_1 - \overline{X}_2) - (\mu_1 - \mu_2)}{S_{\bar{x}_1 - \bar{x}_2}} \tag{9.8}$$

$$t_{\text{test}} = \frac{(\overline{X}_1 - \overline{X}_2) - (\mu_1 - \mu_2)}{S_{\bar{x}_1 - \bar{x}_2}} \tag{9.9}$$

▶ **Example 9.4**   An insurance company has agents in two different states.   A random sample of 200 claims in state one produced an average claim of $750.00, and a random sample of 250 claims in state two produced an average claim of $720.00.   If it is known that the two populations from which the samples were drawn are normally distributed and that $\sigma_1 = \$100$ and $\sigma_2 = \$80$, at the 0.10 level of significance are the population means different?

$$H_0: \quad \mu_1 - \mu_2 = 0$$

$$H_A: \quad \mu_1 - \mu_2 \neq 0$$

$$\alpha = 0.10$$

Reject H_0 if $Z_{\text{test}} < -1.64$ or $Z_{\text{test}} > 1.64$

$$Z_{test} = \frac{(\overline{X}_1 - \overline{X}_2) - (\mu_1 - \mu_2)}{\sigma_{\bar{x}_1 - \bar{x}_2}}$$

$$= \frac{(750 - 720) - 0}{\sqrt{\dfrac{(100)^2}{200} + \dfrac{(80)^2}{250}}}$$

$$= \frac{30}{\sqrt{50 + 25.6}}$$

$$= \frac{30}{8.69}$$

$$Z_{test} = 3.45$$

Reject H_0, the two states do not have the same average claim. ◀

▶ **Example 9.5**   Stores Incorporated is investigating the average dollar sale of two stores in cities with trade areas that are about the same size. From store one, a random sample of 100 sales produced an average sale of $20.00 with a standard deviation of $4.00; and from the second store a random sample of 125 sales produced an average sale of $18.00 with a standard deviation of $5.00. At the 0.05 level of significance, does store one have a larger average dollar sale?

$$H_0: \quad \mu_1 - \mu_2 = 0$$

$$H_A: \quad \mu_1 - \mu_2 > 0$$

$$\alpha = 0.05$$

Reject H_0 if $Z_{test} > 1.64$ using Equation 9.8.

$$Z_{test} = \frac{(\overline{X}_1 - \overline{X}_2) - (\mu_1 - \mu_2)}{s_{\bar{x}_1 - \bar{x}_2}}$$

$$= \frac{(20 - 18) - 0}{\sqrt{\dfrac{4^2}{100} + \dfrac{5^2}{125}}}$$

$$Z_{test} = \frac{20 - 18}{\sqrt{0.16 + 0.20}} = \frac{2}{0.6} = 3.33$$

Reject H_0, store one has a higher average sale. ◀

▶ **Example 9.6**   The data in Example 9.3 can be used to test the hypothesis that the average price of Type A is less than the average price of Type B.

$$H_0: \quad \mu_A - \mu_B = 0$$

$$H_A: \quad \mu_A - \mu_B < 0$$

$$\alpha = 0.05$$

Reject H_0 if $t_{test} < -1.725$ using Equation 9.9.

$$t_{\text{test}} = \frac{(\overline{X}_A - \overline{X}_B) - (\mu_A - \mu_B)}{s_{\bar{x}_A - \bar{x}_B}}$$

$$= \frac{(13 - 14) - 0}{\sqrt{\dfrac{1^2 (12 - 1) + 2^2 (10 - 1)}{12 + 10 - 2}} \sqrt{\dfrac{1}{12} + \dfrac{1}{10}}}$$

$$= \frac{-1}{0.655}$$

$$t_{\text{test}} = -1.527$$

Accept (do not reject) H_0, the average prices are not statistically different. ◀

Dependent Samples

An unstated requirement for all of the previous discussion in this chapter is that the two random samples must be *independent*. For our purposes, independence can be taken to imply that *the individual observations of sample one have no known relationship to the individual observations in sample two*. There are times, however, when this assumption of independence is not valid; whenever this is the case, the preceding equations are not to be used. Instead, we must proceed with an analysis designed for *dependent samples*.

Dependent samples exist when the individual observations of one sample can be matched with individual observations of the other sample. Thus the samples form what can be referred to as *matched pairs*. This matching of pairs can be accomplished based on characteristics such as age, sex, professional skill, and other physical characteristics; or by having an observation on the same individual in both samples such as a pre- and post-test. When we do have these matched pairs, what we actually analyze is the difference in the two observations and treat this difference the same as the single samples discussed in chapters 7 and 8. Instead of working with sample means, we work with mean differences $(\overline{d})$.

$$\overline{d} = \frac{\Sigma d}{n} \tag{9.10}$$

where $\overline{d}$ = average of the differences (d) in all matched pairs from the two samples

d = difference between each matched pair of observations

n = number of matched pairs

and
$$s_d = \sqrt{\frac{\Sigma\,(d - \bar{d})^2}{n - 1}} \qquad\qquad (9.11)$$

where s_d = estimate of the standard deviation for all the matched pairs.

Figure 9-4 shows pictorially the type of sampling distribution we would be working with when we have matched pairs. $\bar{D}$, the mean of the sampling distribution, is actually $\mu_1 - \mu_2$. The measure of dispersion for this sampling distribution, standard error of the difference of the matched pairs, is

$$\sigma_{\bar{d}} = \frac{\sigma_d}{\sqrt{n}} \qquad\qquad (9.12)$$

and when σ_d is unknown we will estimate σ_d with Equation 9.11. If the samples are large so that the number of pairs is large, it is reasonable to assume that the $\bar{d}$'s will be normally distributed as shown in Figure 9-4. In this situation the test statistic is shown in Equation 9.13.

$$Z_{\text{test}} = \frac{\bar{d} - \bar{D}}{s_{\bar{d}}} \qquad\qquad (9.13)$$

If the number of pairs is small, however, we will consider the $\bar{d}$'s to be t distributed if it is reasonable to assume that the original populations were both normally distributed. For this situation, the test statistic is shown in Equation 9.14.

$$t_{\text{test}} = \frac{\bar{d} - \bar{D}}{s_{\bar{d}}} \qquad\qquad (9.14)$$

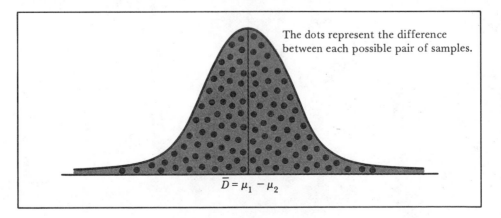

The dots represent the difference between each possible pair of samples.

$\bar{D} = \mu_1 - \mu_2$

FIGURE 9-4. *Sampling Distribution for Matched Pairs.*

▶ **Example 9.7**   Suppose we select 25 employees who have not completed high school and give them a reading test. After this first test, we give them some formal vocabulary training and then retest their reading skill. If Table 9-2 shows the results of the two reading tests, can we conclude that the vocabulary training was beneficial?

$$H_0: \quad \overline{D} = 0 \quad \text{or} \quad \mu_1 = \mu_2$$

$$H_A: \quad \overline{D} \neq 0 \quad \text{or} \quad \mu_1 \neq \mu_2$$

$$\alpha = 0.05$$

$$d.f. = 25 - 1 = 24$$

TABLE 9-2.   *Data for Example 9.7.*

FIRST TEST	SECOND TEST	DIFFERENCE (d)	$(d - \overline{d})$	$(d - \overline{d})^2$
65	67	2	-1	1
72	70	-2	-5	25
64	72	8	5	25
43	50	7	4	16
55	54	-1	-4	16
84	86	2	-1	1
72	80	8	5	25
52	50	-2	-5	25
49	62	13	10	100
80	81	1	-2	4
38	56	18	15	225
93	90	-3	-6	36
77	78	1	-2	4
62	64	2	-1	1
69	72	3	0	0
58	57	-1	-4	16
45	55	10	-7	49
90	88	-2	-5	25
60	62	2	-1	1
54	52	-2	-5	25
72	70	-2	-5	25
49	53	4	1	1
53	56	3	0	0
82	84	2	-1	1
66	70	4	1	1
		75		648

$$\overline{d} = \frac{\Sigma d}{n} = \frac{75}{25} = 3 \qquad\qquad s_d = \sqrt{\frac{\Sigma (d - \overline{d})^2}{n - 1}} = \sqrt{\frac{648}{25 - 1}}$$

$$s_d = \sqrt{27} = 5.20$$

Reject H_0 if $t_{\text{test}} < -2.064$ or $t_{\text{test}} > 2.064$.

$$t_{\text{test}} = \frac{\bar{d} - \bar{D}}{s_{\bar{d}}} \qquad \text{where} \qquad s_{\bar{d}} = \frac{s_d}{\sqrt{n}}$$

$$= \frac{3.0}{1.04} \qquad\qquad\qquad = \frac{5.20}{\sqrt{25}}$$

$$t_{\text{test}} = 2.88 \qquad\qquad\qquad s_{\bar{d}} = 1.04$$

Reject H_0, conclude that the vocabulary test was beneficial to the employees. ◀

The Difference Between Two Proportions: Test of Hypothesis

Just as we can test hypotheses about the difference between two population means, we can test hypotheses about the difference between two population proportions $(\pi_1 - \pi_2)$. In order to do this, we will select large samples so that we can use the Central Limit Theorem to assert that the differences between the sample proportions $(p_1 - p_2)$ are normally distributed. The mean of the sampling distribution of differences between sample proportions $(p_1 - p_2)$, is the difference between the population proportions $(\pi_1 - \pi_2)$, and the standard deviation *(standard error of the difference between sample proportions)* is

$$\sigma_{p_1 - p_2} = \sqrt{\frac{\pi_1 (1 - \pi_1)}{n_1} + \frac{\pi_2 (1 - \pi_2)}{n_2}} \tag{9.15}$$

As you can see, Equation 9.15 assumes that the population proportions are known. However, if π_1 and π_2 were known, there would be no reason to be sampling. Thus, when we are sampling, we will have to work with sample proportions p_1 and p_2. And under the assumption that $\pi_1 = \pi_2$, we estimate the common population proportion as

$$\hat{p} = \frac{n_1 p_1 + n_2 p_2}{n_1 + n_2} \tag{9.16}$$

$\hat{p}$, then, becomes the basis for estimating the standard error $s_{p_1 - p_2}$ as shown below:

$$s_{p_1 - p_2} = \sqrt{\frac{\hat{p} (1 - \hat{p})}{n_1} + \frac{\hat{p} (1 - \hat{p})}{n_2}} \tag{9.17}$$

and the appropriate test statistic is shown in Equation 9.18.

$$Z_{test} = \frac{(p_1 - p_2) - (\pi_1 - \pi_2)}{s_{p_1 - p_2}}$$

(9.18)

▶ **Example 9.8**   The manufacturer of Kornles believes that her munchy product being introduced in Region Y will be more popular than in Region X, where it is currently produced and distributed. In order to check this hypothesis, a random sample from each region was taken. The sample in Region X contained 700 people, 560 of whom claimed to prefer the taste of Kornles. In Region Y, 525 of the 750 people sampled responded favorably. Based on these results, does it appear that Kornles will be more popular in Region Y than they have been in Region X? (Set $\alpha = 0.05$.)

$$H_0: \quad \pi_y - \pi_x = 0 \quad \text{or} \quad \pi_y = \pi_x$$

$$H_A: \quad \pi_y - \pi_x > 0 \quad \text{or} \quad \pi_y > \pi_x$$

$$\alpha = 0.05$$

Reject H_0 if $Z_{test} > 1.64$.

$$Z_{test} = \frac{(p_y - p_x) - (\pi_y - \pi_x)}{s_{p_y - p_x}}$$

$$p_x = \frac{X_x}{n_x}$$

$$= \frac{(70 - 80) - 0}{\sqrt{\dfrac{\hat{p}(1 - \hat{p})}{n_y} + \dfrac{\hat{p}(1 - \hat{p})}{n_x}}}$$

$$= \frac{560}{700}(100)$$

$$p_x = 80\%$$

$$= \frac{70 - 80}{\sqrt{\hat{p}(1 - \hat{p})\left(\dfrac{1}{n_y} + \dfrac{1}{n_x}\right)}}$$

$$p_y = \frac{X_y}{n_y}$$

$$= \frac{525}{750}(100)$$

$$= \frac{70 - 80}{\sqrt{(74.8)(25.2)\left(\dfrac{1}{750} + \dfrac{1}{700}\right)}}$$

$$p_y = 70\%$$

$$= \frac{70 - 80}{\sqrt{(74.8)(25.2)(0.0013 + 0.0014)}}$$

$$\hat{p} = \frac{n_x p_x + n_y p_y}{n_x + n_y}$$

$$= \frac{-10}{2.26}$$

$$= \frac{700(80) + 750(70)}{700 + 750}$$

$$Z_{test} = -4.42$$

$$\hat{p} = 74.8$$

Accept (do not reject) H_0: The product is not doing better in the new market. In fact, based on the sample results, the product may be less popular in Region Y. ◀

—————————— **APPENDIX TO CHAPTER 9** ——————————

The F Distribution

Suppose we select all possible samples of size n_1 and size n_2 from a normal population. Then for each of these samples let us calculate $\overline{X}_1$ and $\overline{X}_2$ and use the statistics to calculate s_1^2 and s_2^2. Now suppose we form all possible ratios of s_1^2/s_2^2. The result is known as the *F distribution*. A graph of the *F* distribution is shown in Figure 9-5.

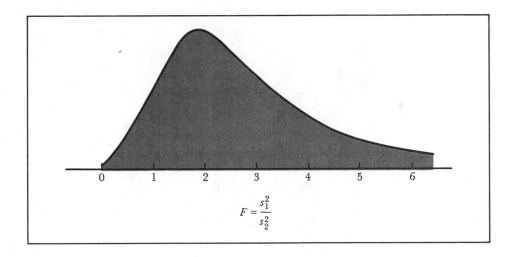

$$F = \frac{s_1^2}{s_2^2}$$

FIGURE 9-5. *The F Distribution.*

Actually, Figure 9-5 is only *one of many F* distributions since the *F* distribution is *a function of the degrees of freedom of the numerator and denominator of the ratio*. Thus, use of this distribution is similar to the use of the *t* distribution. Appendix K contains the *F* values associated with an 0.05 and the 0.01 probabilities in the right-hand end of the distribution.

Notice in Appendix K that the smallest *F* ratio is 1.00 (lower right corner of Appendix). This is because this Appendix contains only the right side values of *F* associated with the probabilities. Use of this distribution can best be demonstrated by an example.

▶ **Example 9.9**   A manufacturer believes he has two types of tires (assume tire wear to be normally distributed) whose variability in wear is the same. If a sample of 25 tires of the first type produced a variance of 2,000 miles and a sample of 20 tires of the second type produced a variance of 1,600 miles, is the manufacturer correct?

$$H_0: \quad \sigma_1^2 = \sigma_2^2$$

$$H_A: \quad \sigma_1^2 \neq \sigma_2^2$$

$$\alpha = 0.05$$

Reject H_0 if $F_{test} > 2.11$.

The F value is a function of the degrees of freedom of the numerator and denominator. Since Appendix K only contains F values of 1.00 or greater, it is necessary to use

$$F_{test} = \frac{s_{larger}^2}{s_{smaller}^2} \tag{9.19}$$

which means that the numerator degrees of freedom is associated with the larger sample variance; and the denominator degrees of freedom is associated with the smaller sample variance. In this example, the larger sample variance came from the sample of 25 tires, so the degrees of freedom for the numerator is $25 - 1 = 24$. The smaller sample variance came from the sample of 20 tires, so the degree of freedom for the denominator is $20 - 1 = 19$.

Thus, the F value is found by locating 24 across the top of the Appendix and 19 down the side of the Appendix. Now within Appendix K, when $\alpha = 0.05$, the F associated with this set of degrees of freedom is 2.11.

Now to complete the example, let us calculate the F_{test}.

$$F_{test} = \frac{s_{larger}^2}{s_{smaller}^2}$$

$$= \frac{2000}{1600}$$

$$F_{test} = 1.25$$

Accept (do not reject) H_0, conclude that the difference in the variances is a chance difference. ◄

Analysis of Variance

Earlier in this chapter, you were shown how to test the null hypothesis that $\mu_1 = \mu_2$ using either the normal or t distributions. Now suppose you wanted to test a null hypothesis concerning the equality of three or more means. An approach to testing more than two sample means is known as *analysis of variance*. Analysis of variance is based on the realization that there are two ways to calculate the population variance from sample data. Since there are two ways to estimate the same variance, *the value of the F ratio formed by the two estimates will vary from one (1) by chance alone if the data came from one normally distributed population with a mean and variance of μ and σ^2. If, however, the sample data came from two or more normal populations that have different means and a common variance, the F ratio formed by the two variance estimates will tend to be greater than 1, indicating more than one population.*

To demonstrate the computations, let us return to the population of four numbers, 1, 2, 3, and 4, which was used in Chapter 6. This particular population has a mean of 2.5 and a variance of 1.25. Table 9-3 shows all possible samples of size two. Using the sample data, we can demonstrate two methods of determining that $\sigma^2 = 1.25$.

TABLE 9-3. *All Possible Samples of Size Two.*

SAMPLE NUMBER	OBSERVED VALUES	SAMPLE MEAN
1	1, 1	1.0
2	1, 2	1.5
3	1, 3	2.0
4	1, 4	2.5
5	2, 1	1.5
6	2, 2	2.0
7	2, 3	2.5
8	2, 4	3.0
9	3, 1	2.0
10	3, 2	2.5
11	3, 3	3.0
12	3, 4	3.5
13	4, 1	2.5
14	4, 2	3.0
15	4, 3	3.5
16	4, 4	4.0

The first method is derived from our knowledge that $\sigma_{\bar{x}} = \sigma/\sqrt{n}$, which, when rewritten, shows that $\sigma^2 = n\sigma_{\bar{x}}^2$. Thus, if we calculate $\sigma_{\bar{x}}^2$ and multiply it by 2, we should have $\sigma^2 = 1.25$. Table 9-4 shows how it is done. First we calculate μ, which is the average of the sample means. Then we use μ to calculate the measure of dispersion for the sampling distribution, $\sigma_{\bar{x}}^2$. And to obtain the variance of the population, we multiply $\sigma_{\bar{x}}^2$ by 2 to obtain 1.25 as shown in Table 9-4.

TABLE 9-4.   *Determining the Population Variance from $\sigma_{\bar{x}}^2$.*

SAMPLE MEAN	FREQUENCY	$f(\bar{X})$	$(\bar{X} - \mu)$	$f(\bar{X} - \mu)^2$
1.0	1	1.0	-1.5	2.25
1.5	2	3.0	-1.0	2.00
2.0	3	6.0	-0.5	0.75
2.5	4	10.0	0	0
3.0	3	9.0	0.5	0.75
3.5	2	7.0	1.0	2.00
4.0	1	4.0	1.5	2.25
	16	40.0		10.00

$$\mu = \frac{\Sigma f(\bar{X})}{\Sigma f} \qquad\qquad \sigma_{\bar{x}}^2 = \frac{\Sigma[f(\bar{X} - \mu)^2]}{\Sigma f}$$

$$= \frac{40.0}{16} \qquad\qquad\qquad = \frac{10}{16}$$

$$\mu = 2.5 \qquad\qquad\qquad \sigma_{\bar{x}}^2 = 0.625$$

$$\sigma^2 = 2\,\sigma_{\bar{x}}^2$$

$$= 2\,(0.625)$$

$$\sigma^2 = 1.25$$

The second method is demonstrated in Table 9-5. In this method, the concept that the population variance is the average of all sample variances of a given sample size, n, will be employed. That is,

$$\sigma^2 = \frac{\Sigma s^2}{n'} \tag{9.20}$$

where n' = number of samples

Table 9-5 shows the calculation which produces $\sigma^2 = 1.25$. Thus, it is easy to see that the ratio $\dfrac{n\sigma_{\bar{x}}^2}{\dfrac{\Sigma s^2}{n'}} = \dfrac{\sigma^2}{\sigma^2} = 1.00$.

Now what happens in analysis of variance is that the numerator, $n\sigma_{\bar{x}}^2$, does not contain the *actual* value of $\sigma_{\bar{x}}^2$ because we don't select all possible samples of a given size. What is used in analysis of variance is an *estimate of* $\sigma_{\bar{x}}^2$ denoted $\hat{\sigma}_{\bar{x}}^2$, which is obtained from the actual samples collected. Similarly, in the denominator, the sum, Σs^2, contains only a few of the possible s^2's rather than all sixteen as in our demonstration. Still, however, the F ratio, which is formed by the two estimates of the population variance using a few samples, should vary from one only by

TABLE 9-5. *Determination of the Population Variance Using All the Sample Variances.*

SAMPLE NUMBER	OBSERVED VALUES	SAMPLE MEAN	SAMPLE VARIANCE $= \dfrac{\Sigma\,(X - \overline{X})^2}{n - 1}$
1	1, 1	1.0	0
2	1, 2	1.5	0.50
3	1, 3	2.0	2.00
4	1, 4	2.5	4.50
5	2, 1	1.5	0.50
6	2, 2	2.0	0
7	2, 3	2.5	0.50
8	2, 4	3.0	2.00
9	3, 1	2.0	2.00
10	3, 2	2.5	0.50
11	3, 3	3.0	0
12	3, 4	3.5	0.50
13	4, 1	2.5	4.50
14	4, 2	3.0	2.00
15	4, 3	3.5	0.50
16	4, 4	4.0	0
			20.00

$$\sigma^2 = \frac{\Sigma s^2}{n'}$$

$$\sigma^2 = \frac{20}{16}$$

$$\sigma^2 = 1.25$$

chance if there is just one population involved. Thus, the F ratio for analysis of variance when we do not use all possible samples of a given size is given in Equation 9.21.

$$F = \frac{n\left[\dfrac{\Sigma\,(\overline{X} - \hat{\mu})^2}{n' - 1}\right]}{\Sigma\dfrac{\left[\dfrac{\Sigma\,(X - \overline{X})^2}{n - 1}\right]}{n'}} \tag{9.21}$$

where $\quad \hat{\mu} = \Sigma \overline{X}/n'$
$n' =$ number of samples

In the numerator of this F ratio, the divisor is $n' - 1$ because $\hat{\mu}$ is an estimate of μ and not μ calculated from all possible sample means. This divisor, $n' - 1$, is the degrees of freedom for the numerator, and the degrees of freedom for the denominator is $n'\,(n - 1)$.

▶ **Example 9.10**   Suppose data on the daily productivity of three machine operators have been obtained. Operator A has produced in five days 100, 110, 92, 95, and 108 parts. Operator B has produced in five days 94, 97, 90, 101, and 98 parts. Operator C has produced 98, 104, 113, 97, and 103 parts in five days. Assuming normal populations with equal variances, determine, at the 0.05 level of significance, if the three operators are not producing at the same average daily rate.

$$H_0: \quad \mu_A = \mu_B = \mu_C$$

$$H_A: \quad \text{at least two means are different}$$

$$\alpha = 0.05$$

Reject H_0 if $F_{\text{test}} > 3.88$.

$$d.f. \text{ for numerator} = n' - 1$$
$$= 3 - 1 = 2$$

$$d.f. \text{ for denominator} = n' (n - 1)$$
$$(3 (5 - 1) = 12$$

Operator A		
X	$(X - \overline{X}_A)$	$(X - \overline{X}_A)^2$
100	−1	1
110	9	81
92	−9	81
95	−6	36
108	7	49
505		248

$$\overline{X}_A = \frac{\Sigma X_A}{n_A} = \frac{505}{5} = 101$$

$$s_A^2 = \frac{\Sigma (X - \overline{X}_A)^2}{n - 1} = \frac{248}{5 - 1} = 62$$

Operator B		
X	$(X - \overline{X}_B)$	$(X - \overline{X}_B)^2$
94	−2	4
97	1	1
90	−6	36
101	5	25
98	2	4
480		70

$$\overline{X}_B = \frac{\Sigma X_B}{n_B} = \frac{480}{5} = 96$$

$$s_B^2 = \frac{\Sigma (X - \overline{X}_B)^2}{n - 1} = \frac{70}{5 - 1} = 17.5$$

Operator C		
X	$(X - \overline{X}_C)$	$(X - \overline{X}_C)^2$
98	−5	25
104	1	1
113	10	100
97	−6	36
103	0	0
515		162

$$\overline{X}_C = \frac{\Sigma X_C}{n_C} = \frac{515}{5} = 103$$

$$s_C^2 = \frac{\Sigma (X - \overline{X}_C)^2}{n - 1} = \frac{162}{5 - 1} = 40.5$$

All Data

$$\hat{\mu} = \frac{\Sigma \overline{X}}{n'} = \frac{101 + 96 + 103}{3} = 100$$

$$F_{test} = \frac{n \left[\dfrac{\Sigma (\overline{X} - \hat{\mu})^2}{n' - 1} \right]}{\Sigma \dfrac{\left[\dfrac{\Sigma (X - \overline{X})^2}{n - 1} \right]}{n'}}$$

$$F_{test} = \frac{5 \left[\dfrac{(101 - 100)^2 + (96 - 100)^2 + (103 - 100)^2}{3 - 1} \right]}{\dfrac{62 + 17.5 + 40.5}{3}}$$

$$F_{test} = \frac{5 \left[\dfrac{1 + 16 + 9}{2} \right]}{\dfrac{62 + 17.5 + 40.5}{3}}$$

$$F_{test} = \frac{65}{40}$$

$$F_{test} = 1.625$$

Accept (do not reject) H_0; conclude that the operators are equally as good. ◀

The preceding discussion and example are designed to provide insight into how analysis of variance works. The computational effort associated with Equation 9.21, however, is laborious and cumbersome for routine use. Fortunately, there is a much more satisfactory solution approach, which makes use of the following equation:

$$\text{Sum of Squares} = \Sigma X^2 - \Sigma X\, (\overline{X}) \qquad\qquad (9.22)$$

To demonstrate how to use Equation 9.22, let us rework Example 9.10.

▶ **Example 9.11**

X_A	X_B	X_C	X_A^2	X_B^2	X_C^2
100	94	98	10000	8836	9604
110	97	104	12100	9409	10816
92	90	113	8464	8100	12769
95	101	97	9025	10201	9409
108	98	103	11664	9604	10609
505	480	515	51253	46150	53207

$$\overline{X}_A = \frac{\Sigma X_A}{n} = \frac{505}{5} = 101$$

$$\overline{X}_B = \frac{\Sigma X_B}{n} = \frac{480}{5} = 96$$

$$\overline{X}_C = \frac{\Sigma X_C}{n} = \frac{515}{5} = 103$$

$$\hat{\mu} = \frac{\overline{X}_A + \overline{X}_B + \overline{X}_C}{n'} = \frac{101 + 96 + 103}{3} = 100$$

Sum of Squares for A $= \Sigma X_A^2 - (\Sigma X_A)\,\overline{X}_A$
$$= 51253 - (505)\,(101)$$
$$= 51253 - 51005$$
Sum of Squares for A $= 248$

Sum of Squares for B $= \Sigma X_B^2 - (\Sigma X_B)\,\overline{X}_B$
$$= 46150 - (480)\,(96)$$
$$= 46150 - 46080$$
Sum of Squares for B $= 70$

Sum of Squares for C $= \Sigma X_C^2 - (\Sigma X_C)\,\overline{X}_C$
$$= 53207 - (515)\,(103)$$
$$= 53207 - 53045$$
Sum of Squares for C $= 162$

Sum of Squares for Total $= \Sigma X^2 - (\Sigma X)\,\hat{\mu}$
$$= (51253 + 46150 + 43207) - (1500)\,(100)$$
$$= 150610 - 150000$$
Sum of Squares for Total $= 610$

$$\text{Total variation} = \text{Within variation} + \text{Between variation}$$
$$610 = (248 + 70 + 162) + \text{Between variation}$$
$$\text{Between variation} = 610 - 480$$
$$= 130$$

Now, the F ratio is formed using the appropriate sum of squares (variation) and degrees of freedom as follows:

$$F_{\text{test}} = \frac{\text{Between Variation}/d.f. \text{ for Numerator}}{\text{Within Variation}/d.f. \text{ for Denominator}}$$

$$= \frac{130/2}{480/12}$$

$$F_{\text{test}} = 1.625 \blacktriangleleft$$

Notice in Example 9.11 that the result of the F ratio computation is the same as obtained in Example 9.10. However, the computations in Example 9.11 are much easier since the individual subtractions necessary to form the numerator and denominator of Equation 9.21 are eliminated.

And finally, the results of the analysis of variance computations are presented frequently in an analysis of variance summary table as shown in Table 9-6. Such a presentation is convenient and is commonly used for the presentation of results of regression analysis (presented in the Appendix to Chapter 12 and in Chapter 13).

TABLE 9-6. *Anova Summary Table.*

SOURCE OF VARIATION	SUM OF SQUARES	DEGREES OF FREEDOM	MEAN SQUARES	F
Between variation	130	2	65	1.625
Within variation	480	12	40	
Total	610	14		

General Comments

The use of the F ratio test is predicated on sampling from *normal populations* just like the t distribution. If the samples are large, the Central Limit Theorem can be used as the basis for normality.

The preceding section was a brief introduction to analysis of variance for samples of equal size. An example of what is called *one-way analysis* of variance was provided. More advanced texts should be consulted for expanded discussions or applications of analysis of variance.

9.1 $\quad \sigma_{\bar{x}_1-\bar{x}_2} = \sqrt{\dfrac{\sigma_1^2}{n_1} + \dfrac{\sigma_2^2}{n_2}}, \quad$ *page 193*

9.2 $\quad s_{\bar{x}_1-\bar{x}_2} = \sqrt{\dfrac{s_1^2}{n_1} + \dfrac{s_2^2}{n_2}}, \quad$ *page 194*

9.3 $\quad s_{\bar{x}_1-\bar{x}_2} = \sqrt{\dfrac{s_1^2\,(n_1-1) + s_2^2\,(n_2-1)}{n_1+n_2-2}}\,\sqrt{\dfrac{1}{n_1}+\dfrac{1}{n_2}}, \quad$ *page 194*

9.4 $\quad (1-\alpha)\%$ confidence interval for $\mu_1 - \mu_2 = (\overline{X}_1 - \overline{X}_2) \pm Z\sigma_{\bar{x}_1-\bar{x}_2}$, *page 194*

9.5 $\quad (1-\alpha)\%$ confidence interval for $\mu_1 - \mu_2 = (\overline{X}_1 - \overline{X}_2) \pm Zs_{\bar{x}_1-\bar{x}_2}$, *page 194*

9.6 $\quad (1-\alpha)\%$ confidence interval for $\mu_1 - \mu_2 = (\overline{X}_1 - \overline{X}_2) \pm ts_{\bar{x}_1-\bar{x}_2}$, *page 195*

9.7 $\quad Z_{\text{test}} = \dfrac{(\overline{X}_1 - \overline{X}_2) - (\mu_1 - \mu_2)}{\sigma_{\bar{x}_1-\bar{x}_2}}, \quad$ *page 198*

9.8 $\quad Z_{\text{test}} = \dfrac{(\overline{X}_1 - \overline{X}_2) - (\mu_1 - \mu_2)}{s_{\bar{x}_1-\bar{x}_2}}, \quad$ *page 198*

9.9 $\quad t_{\text{test}} = \dfrac{(\overline{X}_1 - \overline{X}_2) - (\mu_1 - \mu_2)}{s_{\bar{x}_1-\bar{x}_2}}, \quad$ *page 198*

9.10 $\quad \bar{d} = \dfrac{\Sigma d}{n}, \quad$ *page 200*

9.11 $\quad s_d = \sqrt{\dfrac{\Sigma\,(d-\bar{d})^2}{n-1}}, \quad$ *page 201*

9.12 $\quad \sigma_{\bar{d}} = \dfrac{\sigma_d}{\sqrt{n}}, \quad$ *page 201*

9.13 $\quad Z_{\text{test}} = \dfrac{\bar{d} - \overline{D}}{s_{\bar{d}}}, \quad$ *page 201*

9.14 $\quad t_{\text{test}} = \dfrac{\bar{d} - \overline{D}}{s_{\bar{d}}}, \quad$ *page 201*

9.15 $\quad \sigma_{p_1-p_2} = \sqrt{\dfrac{\pi_1\,(1-\pi_1)}{n_1} + \dfrac{\pi_2\,(1-\pi_2)}{n_2}}, \quad$ *page 203*

9.16 $\quad \hat{p} = \dfrac{n_1 p_1 + n_2 p_2}{n_1 + n_2}, \quad$ *page 203*

9.17 $\quad s_{p_1-p_2} = \sqrt{\dfrac{\hat{p}\,(1-\hat{p})}{n_1} + \dfrac{\hat{p}\,(1-\hat{p})}{n_2}}, \quad$ *page 203*

9.18 $\quad Z_{\text{test}} = \dfrac{(p_1 - p_2) - (\pi_1 - \pi_2)}{s_{p_1-p_2}}, \quad$ *page 204*

9.19 $F_{\text{test}} = \dfrac{s^2_{larger}}{s^2_{smaller}},$ *page 206*

9.20 $\sigma^2 = \dfrac{\Sigma s^2}{n'},$ *page 208*

9.21 $F = \dfrac{n\left[\dfrac{\Sigma\,(\overline{X} - \hat{\mu})^2}{n' - 1}\right]}{\dfrac{\Sigma\left[\dfrac{\Sigma\,(X - \overline{X})^2}{n - 1}\right]}{n'}},$ *page 209*

9.22 Sum of Squares $= \Sigma X^2 - \Sigma X\,(\overline{X})$ *page 212*

REVIEW QUESTIONS

1. Identify the use of each equation shown at the end of the chapter.
2. Under what conditions will the differences between sample means be normally distributed?
3. Under what conditions will the differences between sample means be t distributed?
4. What are independent samples?
5. What are dependent samples?
6. What is analysis of variance? (Refer to the Chapter 9 Appendix.)

PROBLEMS

1. From two populations, N_1 with numbers, 4, 5, and 6, and N_2 with numbers, 1, 2, 3, determine all possible pairs of sample means and their differences. Use a sample size of two from both populations.
2. Use the results of Problem 1 to
   a. plot a histogram of the differences of pairs of sample means.
   b. calculate the mean and standard deviation of the differences.
3. Use the data in Problem 1 to
   a. show that the average of the differences between all pairs of sample means equals $\mu_1 - \mu_2$.
   b. calculate the standard deviation of the difference between the two sample means using Equation 9.1. Is this the same answer as the standard deviation determined in Problem 2b?

4. Two random samples are selected from two normal populations with standard deviations equal to 4 and 6. If the first sample (from population with $\sigma_1 = 4$) is size 24 and yields a mean of 41 and the second sample (from population with $\sigma_2 = 6$) is size 36 and yields a mean of 40, construct a 95 percent confidence interval for the difference between the two population means.

5. Use the data in Problem 4 to test the null hypothesis $\mu_1 = \mu_2$ against the alternate hypothesis $\mu_1 \neq \mu_2$. Set $\alpha = 0.05$.

6. Use the data in Problem 4 to test the null hypothesis $\mu_1 = \mu_2$ against the alternate hypothesis $\mu_1 - \mu_2 < 0$. Set $\alpha = 0.05$.

7. The corporation that operates a chain of restaurants is considering the purchase of the Eatery chain. A random sample of 10 restaurants already in the chain showed an average volume of 1,500 meals per day with a standard deviation of 120, whereas a survey of 15 of the Eatery shops showed a mean volume of 1,675 meals with a standard deviation of 250. At the 0.01 level of significance, would the corporation be safe in acquiring the Eatery shops if it desires comparable operations?

8. Two different production runs on the same brand of golf ball produced the following results concerning the diameter of the balls.

	FIRST RUN	SECOND RUN
Mean	1.65 inches	1.70 inches
Standard deviation	0.04 inches	0.06 inches
Sample size	100	100

At the 0.10 level of significance, did the runs produce the same size golf ball?

9. A manufacturer of paper produces two types of stock. A random sample of 5 rolls from each type yielded the following tear strengths:

A	B
154	149
143	162
135	160
140	154
128	175

At the 0.05 level of significance, are the tear strengths different? State any assumptions necessary.

10. Two random samples are selected. The first sample of size 76 yields a mean of 135 and a standard deviation of 14. The second sample of size 61 yields a mean of 110 and a standard deviation of 10. Construct a 90% confidence interval for the difference between the two population means.

11. Use the data in Problem 10 to test the null hypothesis $\mu_1 = \mu_2$ against the alternate hypothesis $\mu_1 - \mu_2 < 0$. Set $\alpha = 0.10$.

12. Use the data in Problem 10 to test the null hypothesis $\mu_1 = \mu_2$ against the alternate hypothesis $\mu_1 - \mu_2 > 0$. Set $\alpha = 0.10$.

13. A random sample of 50 students who used the standard method to prepare for a national test revealed a mean score of 580 with a standard deviation of 45. In order to make a comparison, a random sample of 100 students who had been prepared for the test by Study Enterprises was obtained. If the mean and standard deviation of the 100 students were 620 and 55, respectively, at the 0.05 level of significance can Study Enterprises claim a superior preparation method?

14. Two random samples were taken of members of two different nationwide social clubs, The Dancers and The Rollers, to determine their annual incomes. The following results were determined from the samples.

THE DANCERS		THE ROLLERS	
Mean income	$7,100	Mean income	$7,300
Standard deviation	300	Standard deviation	400
Sample size	226	Sample size	226

At the 0.05 level of significance, does a significant difference in the average incomes of the two groups exist?

15. The management of a very large plant wished to investigate the effect of the 4-day work week on absenteeism. Two random samples of size 40 were selected; group I of employees worked 10-hour days (4-day week), and group II of employees worked 8-hour days (5-day week). If group I averaged 4 hours of absenteeism per week with a standard deviation of 1.2 and group II averaged 4.4 hours of absenteeism per week with a standard deviation of 1.5, should we conclude that the shorter work week reduces absenteeism? Set $\alpha = 0.05$.

16. Two organizations are meeting at the same convention hotel. A sample of 10 members of The Cranes revealed an average daily expenditure on food of $12.00 with a standard deviation of $1.00; and a sample of 15 members of The Penguins revealed an average daily expenditure on food of $14.00 with a standard deviation of $2.00.
    a. If $\alpha = 0.01$, are the average expenditures of the two organizations different?
    b. What assumptions were necessary to perform a test of hypothesis in part (a)?

17. Use the data in Problem 16 to see if the members of The Penguins are bigger spenders than the members of The Cranes.

18. A paint manufacturer is experimenting with a new, less expensive paint. A random sample of 144 gallons of the new paint produced an average coverage of 480 square feet with a standard deviation of 25 square feet; and a random sample of 138 gallons of the old paint produced an average coverage of 500 square feet with a standard deviation of 30 square feet. At the 0.05 level of significance, do you think the new paint provides greater coverage than the old paint?

19. Two production processes are designed to manufacture a particular bottle. If a random sample of 100 bottles produced by the first process yielded 8 defective bottles and a random sample of 200 bottles from the second process yielded 22 defective bottles, are the population proportions different? Set $\alpha = 0.10$.

20. If a random sample of 100 registered voters yielded 60 percent yes respondents, and another random sample of 150 voters yielded 56 percent yes respondents, can you conclude at the 0.05 level of significance that the samples came from the same population?

21. A sample of 150 automobiles in the southern part of a state revealed 30 cars that could not pass the safety inspection. In the northern part of the state, a sample of 250 automobiles revealed 55 cars that could not pass the safety inspection. At the 0.01 level of significance, determine if the population proportions are different.

22. Two switchboard operators met at a convention. After some discussion, it was determined that Operator A believed she received a larger proportion of long distance calls than Operator B. To decide the issue, each operator collected data on incoming calls. The results are recorded below. Using the results below, do you think Operator A is correct? Set $\alpha = 0.05$.

	OPERATOR A	OPERATOR B
Number of calls	500	400
Number of long distance calls	100	72

23. Determine if it is reasonable to assume that the variances in Problem 10 are equal. Set $\alpha = 0.05$.

24. Determine if it is reasonable to assume that the variances in Problem 16 are equal. Set $\alpha = 0.01$.

25. A bowling alley, in order to encourage participation, gave free instruction to all who desired it. From a sample of 20 of the participants, scores were obtained before and after the instruction period. Based on the results given below, would you regard the instruction as beneficial? Set $\alpha = 0.01$.

PARTICIPANT	BEFORE SCORE	AFTER SCORE
1	90	95
2	105	112
3	78	90
4	125	126
5	150	148
6	85	90
7	100	102
8	102	115
9	94	99

PARTICIPANT	BEFORE SCORE	AFTER SCORE
10	129	125
11	152	153
12	80	90
13	76	80
14	140	145
15	95	110
16	104	112
17	89	100
18	110	115
19	115	122
20	99	108

26. The following data were obtained from the members of three organizations at their last annual convention. Determine if the average daily food expenditures are different. Set $\alpha = 0.01$.

Daily Expenditure of Five Members of Each Organization		
THE CRANES	THE PENGUINS	THE ROBINS
$25	$17	$30
18	18	28
22	16	25
21	20	21
24	14	26

27. A paper manufacturer wishes to determine if there is any significant difference in the moisture prevention of four methods of storage. The results of testing the four methods of storage are given below.

Methods			
A	B	C	D
5.2	4.3	6.0	5.6
5.3	3.7	5.0	8.0
6.5	3.8	5.6	5.4
5.4	4.6	4.9	6.5
7.6	4.1	4.5	8.5

At the 0.01 level of significance, is there a difference in moisture content?

28. A manufacturer of portable calculators is considering three batteries to use in the new calculator. Data on the life between charges (in hours) have been collected on each battery as shown on p. 220.

Battery		
I	II	III
12	17	21
19	25	15
19	13	16
16	31	29
20	36	32
24	15	14
30	26	18
15	35	26

At the 0.05 level of significance, is there a difference in the average life between charges?

29. The following are hypothetical data for three random samples:

I	II	III
25	27	30
32	36	34
27	28	31
33	32	32
18	22	28

Under the assumption of normal populations and equal population variances,

a. Test the null hypothesis $\mu_I = \mu_{II}$ against the alternate hypothesis $\mu_I \neq \mu_{II}$. Set $\alpha = 0.05$.

b. Test the null hypothesis $\mu_{II} = \mu_{III}$ against the alternate hypothesis $\mu_{II} \neq \mu_{III}$. Set $\alpha = 0.01$.

c. Test the null hypothesis $\mu_I = \mu_{II} = \mu_{III}$. Set $\alpha = 0.01$.

Some Practical Aspects of Sampling

CHAPTER
10

The previous chapters on sampling focused more on the theoretical or conceptual aspects of sampling, though an attempt was made to show how the theories or concepts applied to real-world situations. This chapter, however, is more concerned with the practical aspects of sampling.

Why Sample?

Perhaps the most practical question one can ask about sampling is simply "Why take a sample?" Surely it is not to determine the flavor of jelly beans.

In a word, samples are useful in providing information about the characteristics of populations when *time* and *money* (or both) are important factors. Oftentimes in a real-world situation, a *complete enumeration* of all sample units in a population is too costly or too time consuming. Certainly in cases of infinite or extremely large populations a sample is required because a complete enumeration is impractical, if not impossible.

Also, consider destructive testing situations. In these cases, the sample element must be destroyed or consumed in order to obtain the necessary measurement. For example, to measure the length of time a certain brand of light bulb will last requires that bulbs be allowed to burn out. Or to test the flavor of wines requires that the wine be consumed. Automobile manufacturers often test the strength of various bumpers by crashing into some object and measuring the extent of the damages. Obviously, all these tests that destroy, damage, or consume the sample element require that the test be conducted by some method of sampling.

Why Not Sample?

Perhaps the question then should become "Why not sample?" The answer, of course, is because for the advantage of saving time, money, effort, and product, one gives up a margin of *accuracy* and *precision*. One might argue that a sample is not a complete count (enumeration) and only from a complete count or measure of every sample element in the population can one be 100 percent sure of the accuracy of the results. This answer is not entirely true. Let's look at precision and accuracy.

Precision and Accuracy

As you have seen in earlier chapters, for a simple random sample, we usually measure the *precision* of the sample by the *standard error*. For a particular universe, lower standard error means more precision (less variance in the sampling distribution). But using the standard error as a measure of precision relates only to the chance, or sampling error, involved. In other words, this recognizes that in a sampling situation there is always *internal error* due to the fact that a sample has been taken. This idea can be seen in the discussion of sampling distributions in Chapter 6 of this text. Because of this sampling error, we also recognize that the sample statistic is expected to deviate from the population parameter.

However, there are two other factors that may cause a sample estimator to deviate from the parameter that is to be estimated. Often these factors are overlooked in sample design, but they should not be for they are very important. These two factors are *bias* and *external error*. Therefore, there are really three factors that cause deviation between the sample statistic (the estimator) and its parameter value:

1. sampling (internal) error
2. external error
3. bias

All three of the factors may contribute to a difference between the estimator and its parameter.

Sampling or internal error is measured by the standard error and it has been thoroughly examined in previous chapters. It is caused by the fact that a sample is taken. Other things being equal, sampling error is a function of the sample size. As the sample size increases, the value of the standard error is decreased, thereby reducing the sampling error.

External error is error related to the practical considerations in taking a sample. These errors are usually caused by the people conducting the sample and processing the data. The main components of external error are *recording errors* (recording statistical data) and *processing errors* (editing and calculating data). This type of error usually increases as the sample size increases because individuals are dealing with larger quantities of data. For example, an interviewer who must call on 200 students in a sample survey is apt to make more mistakes in recording the information than an interviewer who calls on 100 students. Likewise, someone computing information from 100 items is likely to make fewer mistakes than when computing from 2,000 items.

External error can usually be reduced by reducing the number of items in the sample. Better trained personnel will also help reduce this type error.

While *external* and *internal* errors may be in either direction, *sample bias* is a factor that causes the sample estimator to deviate from the parameter in a single direction. If you have ever seen a tree where the wind blows constantly in one direction, and the tree is bent over, you have a good idea of sample bias.

Bias, viewed from a practical consideration, can be the most insidious of the three factors causing deviation between the estimator and the parameter because it may be difficult to detect. We expect internal error, and you can repeat calculations and otherwise detect external error. Yet we may base conclusions on sample results without recognizing that bias exists.

We know that there are times when we are using a *biased estimator*. For instance, we know that $s_{\bar{x}}$ is a biased (but consistent) estimator of $\sigma_{\bar{x}}$. These cases are not the ones that cause concern. The type of bias which makes trouble is the bias resulting from such things as:

error is smaller as n increases

1. a poorly defined universe
2. an inadequate sample design
3. improperly worded questions
4. any phenomenon providing the basis for respondents to distort the the actual answer. For instance, it is quite likely that data concerned with the ages (especially the over-30 crowd) are biased downward.

There are many classic examples of bias in sampling. Perhaps the best known one is the case of the survey taken by the *Literary Digest* in connection with the Roosevelt-Landon presidential election of 1936. Based on a sample using the telephone directories and automobile registrations as the sample frame, it was concluded, "Landon wins by a landslide." Of course, this result was not the case. The bias was due to a poor definition of the universe (names listed in the phone directories and automobile registers). Remember that in 1936, those who could afford to have telephones and automobiles were not the masses of unemployed, most of whom identified with the Democratic Party at that time.

If there is bias in a design or question or procedure, increasing the sample size from n to an N (a complete enumeration) will not eliminate the bias necessarily. External error increases with an increase in sample size. Thus, a complete enumeration, itself, will eliminate the *internal error,* but it will not eliminate the *bias,* and it will actually increase external error. Therefore, why not take a sample?

Random Versus Nonrandom Sampling

Judgmental

So far, all earlier chapters on sampling assumed random distributions of variables and estimators. In those chapters, estimations and statistical tests were based on theoretical distributions depending on random variables. In fact, the formulation of the standard errors was developed from simple random sample notions.

Sometimes nonrandom or *judgment samples* are used to determine characteristics of various populations. Perhaps the most common of these judgment samples is called the *quota sample.* Quota samples are so called because quotas of certain characteristics are established to complete the sample. For example, when the sample units are houses, the quotas may be established by the size of the house, the location of house, the structure of the house, the design of the house, etc. Therefore, a particular survey sheet may require an interview to sample a 1,700 to 2,000 square foot house, located in the northern quadrant of town, with a frame structure, single-story design. The quotas are established from known characteristics of the universe. The sample is "forced" to assume these characteristics. The concept underlying quota samples is that, if the sample is deterministically designed to assume known characteristics of the universe, then it should assume the unknown characteristics. This concept is questionable; certainly there is no statistical basis for it. However, quota samples have been known to estimate quite well.

This chapter (and the previous one) is concerned with random (probability) samples. The main advantages of random samples are

1. they depend on random outcomes (not the judgment of the sampler) and are more likely to be representative
2. the result is a probability distribution (normal or t) for which the magnitude of interval error can be estimated.

probability

Types of Random Samples

In a finite universe, random samples may be drawn *with* replacement or *without* replacement. Given N sample units in a universe, if we draw a sample of n units without replacement, the number of different possible samples is C_n^N. Therefore, if $N = 9$ and $n = 3$, the number of possible samples drawn is $C_3^9 = 84$.

If instead of sampling *without* replacement the sample was taken *with* replacement, the number of samples is N^n. In our example, 9^3 or 729 is the number of possible samples with replacement, a difference of 645 compared to a sample without replacement.

A simple random sample may be drawn *without* replacement. However, the *finite population correction factor* should be used when the population is quite finite. The *simple random sample* is the basic design for most probability samples; other probability sample designs may be considered modificators of sorts of this basic one.

Simple Random Sample

A simple random sample requires that the sample be taken in such a way that each sample in the universe has an equal probability of being selected. The technique for selecting a simple random sample is, in fact, reasonably simple. Suppose that you want to select a simple random sample of 100 customers from a list of 1,000 credit customers of a store. (The list of credit customers is called a *sample frame*). One way to select the sample is to place each credit customer's name on a slip of paper and then place the 1,000 pieces of paper in a container so that the slips can be mixed in a random fashion.

Then we select the names of 100 customers one at a time from the container without replacement. The number of different simple random samples of size n that can be taken from a universe of size N is equal to C_n^N. For example, if we want to take a sample of 100 units from a universe of 1,000 without replacement, the number of possible different samples that can be drawn is 1000!/(100! 900!).   If a simple random sample is selected, each of the possible samples has an equal probability of selection which in this case is $1/C_{100}^{1000}$.

It is quite possible that writing all the items on slips of paper and putting them in a container (a hat for instance) to select the samples will be cumbersome and unwieldy.   In this case, a table of random numbers (or computer random number generator) will facilitate the procedure in taking a simple random sample.   In using a table of random numbers (see Appendix M as an example), we take the sample frame of all units from which the selection will be made and assign a number to each item $(1, 2, 3, \ldots, N)$.   In the case of the illustration above, we would take the list (sample frame) of the 1,000 credit customers and assign a number to each one.   For example, we might start with the number 1 and assign consecutive digits to the 1,000 customers as shown below.

$$
\begin{array}{ll}
0001 & \text{Sue Smith} \\
0002 & \text{Frank Jones} \\
0003 & \text{Sam Owen} \\
& \quad . \\
& \quad . \\
& \quad . \\
0279 & \text{Pam Taylor} \\
& \quad . \\
& \quad . \\
& \quad . \\
1000 & \text{Vivian Lane}
\end{array}
$$

For this sample frame, we would need four single columns of random digits.   Once these columns are selected, we simply go down the columns and record the first 100 four-digit numbers (discarding duplications and numbers that are too large).   Then, we go to the sample frame and select the 100 names corresponding to the numbers selected.   For example, consider the following four column random digits:

$$
\begin{array}{l}
0988 \\
0210 \\
7772 \\
0279 \\
0899 \\
1464 \\
0279
\end{array}
$$

The item corresponding to number 988 would be in the sample. Units numbered 210, 279, and 899 would also be included. Since there are only 1,000 customers in the universe and we have numbered the units from 1 to 1,000, there is no unit numbered over 1,000 in the sample frame. Thus, there are no units corresponding to 7,772 or 1,464. We just discard these numbers. Notice the number 0279 occurs twice. On the second occasion we also discard this number since you do not want to include Pam Taylor (#279) twice in the same sample.

It may already have occurred to you that we will probably have to select substantially more than 100 numbers in order to complete the selection of our sample. Actually, the numbering system we used in the above illustration was not very efficient. We could have begun numbering with zero (0), instead of 1, and ended with 999 rather than 1,000. Therefore, we could have assigned numbers to the entire sample frame by using only three digits in each number. Had we done this, we would not have to discard any three-digit numbers because they were too large; and this could lead to the need to select considerably fewer random numbers in order to find the sample of 100.

Systematic Sampling

Systematic sampling may really be considered a method of selection when compared to simple random sampling. Thus, most random sample designs use one or both of these methods of selection.

A *systematic sample* is drawn by selecting every ith item in a universe. Referring back to our example with the credit customers, if we selected every 100th customer in our universe for the sample, we would be taking a systematic sample. Other examples are to select every 4th house in a series of blocks, every 50th pay voucher, or every 30th car passing an intersection.

The procedure requires that the sample units in the universe be ordered in some way. The ordering can be in explicit detail, or a rule may be invoked which results in an ordering. In the case of ordering in explicit detail, the ordering may be numerical, geographical, by time interval, alphabetical, or any other such ordering approach. For example, pay vouchers may be ordered by the date on which the liability occurs. Another example of explicit ordering is the alphabetical listing of names in the telephone directory. If the sample frame is the houses in a city, the houses could be ordered for a systematic sample by establishing a rule concerning the sequence in which the streets will be listed in the sample frame.

To maintain randomness in a systematic sample, the *start* of the sample is determined by chance. For instance, if every 50th item is to be selected, we start the sample selection between items 1 and 50. Then we select every 50th item from there on. The particular item between 1 and 50 to start the selection would be *probabilistically determined*. In this case, we might select a two-column list of random digits from a table of random numbers. The first number between 1 and 50 would be the start. Or else, we might simply draw from slips of paper numbered between 1 and 50; the number drawn would indicate the start. In any case, a random start is necessary to involve probability concepts in a systematic sample.

In addition to a random start, another consideration in conducting a systematic sample is *the interval at which selection is to be made*. That is, how do we determine whether every 10th, 50th, 100th, or *i*th item is to be chosen? This interval is important, because it will affect the size of the sample. On the other hand, the interval may be influenced by the sample size. For instance, if we want to select a sample of 100 items from a universe of 1,000 items, we would select every 10th item. That is,

$$i = \frac{N}{n}$$

(**10.1**)

where i = the interval
n = sample size
N = universe size

Let us consider again the universe with 9 units in it and a sample of size 3. That is, $N = 9$ and $n = 3$. How many different possible samples can we draw using the systematic sample selection procedure? It is also equal to i as shown in Equation 10.1. In this case, only three different samples are possible.

The major advantage of a systematic sample design is the ease of selection, especially in a situation where ordering of some sort is already available. Suppose the 1,000 customers in our example had their names and addresses listed alphabetically in a credit file. Using a random start, we can take every *i*th name listed in the file to make up the sample.

There is one real danger in using a systematic sample design. If there is a *pattern* in the ordering of the universe we may get a sample with statistics that do not represent the population parameters to be estimated. Suppose we were taking a systematic sample of items coming off a production line. Let us further assume that one of the important machines used to produce the items is faulty in that every 10th item produced is defective. If we were to select every 10th item (or multiple of 10, such as 20, 30, etc.), all the items in the sample will be defective if the first item selected is defective. Similarly, if the first item is not defective, the sample will contain 0 defectives.

Stratified Sample

Another modification of the simple random design is the *stratified sample*. This sample design gets its name from the division of sample elements in the population *strata* or layers. When a population is heterogeneous overall but within it are homogeneous subpopulations (or strata), it is usually desirable to divide the population into these strata. Items for the sample are selected from each of the strata rather than the universe as a whole, depending on the population to be sampled (age, GPA, height, weight, etc.). Let us consider an illustration of taking a sample of 200 students from a universe of 5,000 undergraduates in a small college. No doubt the student body consists of freshmen, sophomores, juniors, and seniors. Depending on the population to be measured, the four classes of students may form four homogeneous strata for the particular population and, under certain circumstances, it might be advantageous to recognize these strata. Rather than selecting 200 students from the entire universe of 5,000, we might divide the students into strata according to their classifications and select n_i from each stratum. Thus the sample size (n) would equal $n_1 + n_2 + n_3 + n_4$, where n_1 is the number of items selected from the first stratum, n_2 is the items from the second stratum, etc.

The basis for using a stratified sample design is the *homogeneity* of the population with respect to the particular variable under consideration. Where the population is homogeneous, little benefit can be obtained from stratification. In other cases we find (or have reason to suspect) that the population is fairly *heterogeneous,* but that the population forms into layers, or strata, and there is a wide spread between the strata. In our illustration, if we were interested in the level of courses taken by the students at our mythical college, it seems reasonable to expect great homogeneity within each class (freshman, sophomore, etc.), but there could be larger dispersions between classes on certain issues. That is, we would expect freshmen to take mainly lower-level courses, sophomores middle-level, juniors middle- to upper-level, and seniors primarily upper-level courses. Here is a classic case in which a stratified sample design is indicated. Because of the nature of the items in the universe, the population divides into four relatively homogeneous strata. In this instance, *greater precision* (smaller standard error) is expected from a stratified sample than from a simple random one for the same size sample.

There are various approaches for determining the number of items to select from each stratum. Perhaps the most common is to make the number of items selected from the stratum proportional to the size of the strata. Suppose in our college the student body is broken down as follows:

CLASS	NUMBER OF STUDENTS
Freshmen	2000
Sophomores	1250
Juniors	1000
Seniors	750
	5000

This is very similar to *quota sampling*, which has already been mentioned. Like quota samples, you "force" a certain proportion of the sample to be taken from each of the different strata.

If we were selecting items proportional to the size of the strata, we would select the sample as shown in Table 10-1. The last column of Table 10-1 was constructed using Equation 10.2.

$$n_i = \left(\frac{N_i}{N}\right) n \qquad (10.2)$$

where n_i = the stratum sample size for the ith stratum
n = the total sample size
N_i = the size of the ith stratum
N = the universe size

TABLE 10-1. *Sample Size for Each Stratum Using Proportional Parts.*

CLASS	NUMBER OF STUDENTS (N_i)	NUMBER SELECTED FOR SAMPLE (n_i)
Freshman	2000	80
Sophomores	1250	50
Juniors	1000	40
Seniors	750	30
	5000	200

$$n_1 = \frac{2000}{5000}(200) = 80 \qquad n_3 = \frac{1000}{5000}(200) = 40$$

$$n_2 = \frac{1250}{5000}(200) = 50 \qquad n_4 = \frac{750}{5000}(200) = 30$$

In selecting sample units proportional to the size of the strata, we assume that the standard deviation within each stratum is the same, or we simply choose to ignore the existence of any differences. We may wish to select the number of sample units proportional to the *size* of the stratum and the *standard deviation* within the stratum because this will allow us to increase the proportion of the sample that is selected from the strata that are larger *and* have more variability (larger standard deviations). This will provide greater precision for any given sample size. For example, in sampling the undergraduate students to obtain information on grade-point averages, most likely the GPA's for seniors are more homogeneous (lower standard deviation) than the GPA's for freshmen. So, even if the stratum for freshmen was the same size as the stratum for seniors, we would want to take a larger size sample from the freshmen than from seniors to adjust for the greater variability of GPA's for freshmen. Equation 10.3 is given to compute the sample size for each stratum (n_i) considering differences in strata size and variability.

$$n_i = \frac{N_i \sigma_i}{\Sigma [N_i \sigma_i]} \cdot n \qquad \textbf{(10.3)}$$

where n_i = the stratum sample size for the ith stratum
           n = the total sample size
           N_i = the size of the ith stratum
           σ_i = the standard deviation for the ith stratum

The procedure for applying this formula is quite similar to the one used for Equation 10.2. The difference here is to use the σ_i for each stratum. The problem with this procedure is that often σ_i is not known or ascertainable.

Another approach would be to select the number of items proportional to the *size* of the strata and the reciprocal of the *cost* of sampling from the strata. For example, it might be more (or less) expensive to select items from one stratum than another. In most situations, the number of items selected from each stratum would vary directly with the size of the stratum and inversely with the cost associated with selecting items from the stratum. If only cost and the relative size of each stratum are to be considered, Equation 10.4 may be used to determine n_i for each stratum.

$$n_i = \frac{N_i (1/c_i)}{\Sigma [N_i (1/c_i)]} \cdot n \qquad \textbf{(10.4)}$$

where n_i = the stratum sample size for the ith stratum
           n = the total sample size
           N_i = the size of the ith stratum
           c_i = the cost of sampling from the ith stratum

Referring to our mythical college student body again, we have assigned different costs for acquiring a sample unit from each stratum as shown in Table 10-2. Observe how this changes the sample size for each stratum.

TABLE 10-2. *Sample Size for Each Stratum Using Cost.*

CLASS	NUMBER OF STUDENTS (N_i)	COST OF SAMPLE UNIT (c_i)	$N_i \left(\dfrac{1}{c_i}\right)$	NUMBER SELECTED FOR SAMPLE (n_i)
Freshmen	2000	$1.00	2000	117
Sophomores	1250	2.00	625	36
Juniors	1000	2.00	500	29
Seniors	750	2.50	300	18
	5000		3425	200

$$n_1 = \frac{2000}{3425}(200) = 117 \qquad n_3 = \frac{500}{3425}(200) = 29$$

$$n_2 = \frac{625}{3425}(200) = 36 \qquad n_4 = \frac{300}{3425}(200) = 18$$

Finally, we may combine all three factors—strata size, variability, and cost—to select the number of sample units from each stratum. Equation 10.5 would be used in this case. Again, it is similar to the other formulas used for strata selection:

$$n_i = \frac{N_i \sigma_i \,(1/c_i)}{\Sigma \,[N_i \sigma_i \,(1/c_i)]} \cdot n \qquad\qquad (10.5)$$

where n_i = the stratum sample size for the *i*th stratum
n = the total sample size
N_i = the size of the *i*th stratum
σ_i = the standard deviation of the *i*th stratum
c_i = the cost of sampling from the *i*th stratum

The rule here is that in a given stratum n_i is larger if

1. the stratum is large
2. the stratum is more variable
3. sampling is less expensive in the stratum.

As a practical matter, the simplest approach one might use is simply to select an equal number of units from each stratum regardless of its size, cost, or variability. In our example of selecting 200 students in the sample, 50 would be selected from each of the four strata. In other words $n_i = n/k$ where k is the number of strata into which the universe is divided. However, we do not especially recommend this approach.

These are the most likely approaches one might employ in determining the number of items to be selected from each stratum. The actual selection of units may be conducted in the same manner as a simple random sample.

Cluster Sample

Another form of a modified random sample design is called the *cluster sample*. Like the stratified sample, the cluster sample design requires that the sample units be grouped. Unlike the stratified sample, the sample units are grouped by *clusters* in the *universe* not homogeneous strata in the population. The sample unit consists of more than one sample element. An entire cluster or group of elements is randomly selected, and all (or part) of the elements in the cluster are included in the sample. Thus, the sample unit is the cluster of elements.

A simple illustration of the use of a cluster sample design would be to take a sample of the 5,000 college students by selecting entire sections of courses. Rather than select individual students we would select, at random, courses and sections. As an example, Math 134, section 2, might be randomly selected. In this case all students enrolled in this section of the course are included in the sample.

When dividing a universe into clusters, care should be taken to *avoid any overlap* of the clusters. In our example, some student might be in the sections of two different courses that have been randomly selected. As in the case of the simple random sample, duplication of individual items, as well as clusters, should be avoided. Thus, if Gloria Ann appears in two randomly selected clusters, she should be omitted from the second cluster as well as any other clusters selected thereafter.

A more practical example of the use of a cluster sample will serve to illustrate an important advantage of this type of sample design. (See Example 10.1.)

▶ **Example 10-1**   Hy-take Supermarket, Inc., has one of its stores located near university student housing. Because it is felt that the average student purchase is lower than the average purchase among nonstudents, Hy-take has not used a trading stamp program in this store. Recently, however, the general manager questioned the assumption regarding the average purchase by students and has suggested that a sample of individual purchases be taken at the store to determine the average purchase.

The store keeps its used rolls of cash register tape for several months. Each of these rolls ordinarily contains the records of the purchases of from 60 to 90 customers.

If the rolls are treated as sample units, a simple random sample can be selected from the store's file of tapes. Each sample unit contains a *cluster* of sample elements (individual purchases in this case). That is, the roll of tape is a cluster of sample elements. In taking a random sample of tape rolls, the group at Hy-take is really taking a cluster sample. To have taken a simple random sample of purchases would have required that individual purchases (not rolls) be selected by a simple random method. ◄

The preceding example shows that there are circumstances when it is more expedient, or less expensive, to select a cluster of elements rather than select each individual element. However, we must be cautious about the effect the cluster sample will have on precision and the extent to which the cluster might distort the sample results. Consider a situation in which abnormally high (or low) purchases are made during certain times of the day. As a result, a cash register tape may have only the high (or low) purchases. In other words, *a segment or cluster may not contain anything like a representative distribution of sample elements.*

It may be interesting to note that while we originally compared the cluster sample to the stratified sample, it is much more similar to the systematic sample. Using a systematic selection with a random start really involves selecting a cluster of elements by selecting every ith element. As in the systematic design, in a cluster sample if there are k clusters, then there are only k possible samples regardless of how many elements are in each cluster.

Multistage Sampling

The sample designs that have been mentioned are considered the basic ones among the sample selection procedures. There are others, but these four, (1) simple random sample, (2) systematic sample, (3) stratified sample, and (4) cluster sample, provide the fundamental selection techniques for many other sample systems. One such sample system is called the *multistage sample* system. It is multi-stage because the selection procedure takes place in a hierarchy of stages. Yet these four fundamental selection techniques can be usefully applied in various combinations in a multistage sample system. Let us consider Example 10.2 to illustrate a multistage sample system.

▶ **Example 10-2**   The president of Hy-take Supermarket, Inc., was so pleased with the results from sampling purchases at the university store that he has decided a sample should be taken of the purchases at 150 stores in the country. The market researcher at the Hy-take home office has de-decided that a three-stage (multistage) sample should be taken. The first stage is to select, on the basis of a simple random sample, 15 of the 150 stores. She then recommends that cash register tape rolls be randomly selected at each of the 15 stores. The final stage of the sample is to select every 20th purchase on the tape rolls using a random start. ◀

The first stage is the selection of the *primary sample unit* (the 15 stores in the example); the second is the selection of the *secondary sample unit* (tapes at each store); the third is the selection of the *tertiary sample unit* (specific purchases). Each stage of the multistage procedure may involve any of the possible sample designs mentioned earlier.

One practical or real-world use of multistage sampling is a situation in which sampling units are dispersed over a wide geographical area. It is often less expensive to divide the geographical area into smaller geographical regions and let them be the primary sample unit, a number of which are selected. Then subregions become the secondary sample units and are selected (systematically or simple randomly). Then subregions may again be subdivided into tertiary sample units such as cities or even neighborhoods.

Final or *ultimate sample units* may be selected at any stage. If they had been selected after the primary units were selected, we would have a two-stage sample. If they were selected after the secondary units, we would have a three-stage sample; after the tertiary sample unit a four-stage sample, etc.

Multistage sampling has a wide range of applications, but in general it is useful in sampling large numbers of units, especially when cost saving is important.

Sampling Size and Cost Analysis

In statistical sampling, cost of the sample and sample size are so interrelated that it is difficult to consider one without the other. For, other things being equal, you would expect cost to vary with the sample size. Therefore, in developing a sample design for a given situation, cost and sample size are important considerations.

The two important factors affecting the size of a sample, other than costs, are

1. level of accuracy or precision desired, and
2. variability of the items in the population.

Notice that the size of the universe is not one of the two factors. If the universe is small in size, then the sample size necessary to achieve a given level of precision may be reduced. However, as a general rule, this is only considered when either the sample size is at least 10 percent of the population or the cost of each sample unit is very important.

As you know, the *level of accuracy* refers to the extent to which the estimator coincides with the parameter and considers the effects of internal error, external error, and bias; while the *level of precision* refers only to the internal error. For purposes of determining sample size, we are prone to consider only the level of precision because it is easily measured by the standard error. However, in designing a sample, we cannot overlook the importance of external error and bias because different sample designs may affect them. For example, it may be more accurate (less external error) to use a systematic selection of names from a list of registered voters, rather than a simple random one, if the list is compiled according to the date of registration. Similarly, using a telephone interview to gather information may evoke bias in the response. Finally data obtained from the use of a mail questionnaire may be biased if only those persons with relatively strong feelings on the subject return the questionnaire.

Nonetheless, in using formulas to determine sample size, we usually relate sample size to levels of precision, and the standard error is one component in these equations. Generally speaking, to lower the standard error and gain a higher level of precision the sample size must be larger.

The other important factor affecting sample size is the *variability* of items in the population. In short, more variability among elements usually requires larger samples. Yet, we have seen how to divide the population into strata and to sample from these strata. If the variance *between* strata is greater than the variance of items *within,* we can get more precision with a stratified sample than a simple random sample.

If the population divides into convenient groups or segments of sample elements, then we may select sample units consisting of a number of elements. Cluster sampling usually requires a smaller sample size because the sampling units determining the sample size for selection contain more than one sample element. If the variability of the elements within the cluster is greater than the variability between clusters, a cluster sample may give better accuracy or precision for the same sample size, or dollar of sample cost, than either the stratified sample or the simple random sample.

For purposes of illustration, let us see how sample size is computed for simple random samples, considering precision (as measured by the standard error) and variability (as measured by the standard deviation) for means and proportions.

Sample Size for Means

Using the standardized normal deviate

$$Z = \frac{\overline{X} - \mu}{\sigma_{\bar{x}}}$$

and remembering that $\sigma_{\bar{x}} = \sigma/\sqrt{n}$, then

$$Z = \frac{\overline{X} - \mu}{\sigma/\sqrt{n}}$$

$$Z = \left(\frac{\overline{X} - \mu}{\sigma}\right)\sqrt{n}$$

$$Z\sigma = (\overline{X} - \mu)\sqrt{n}$$

$$\sqrt{n} = \frac{Z\sigma}{\overline{X} - \mu}$$

$$n = \frac{Z^2\sigma^2}{(\overline{X} - \mu)^2} \tag{10.6}$$

Now, Equation 10.6 can be used to determine the sample size n necessary to provide an $\overline{X}$ *within a certain distance of* μ, $(\overline{X} - \mu)$, for a given level of confidence which will be provided by the normal distribution. The Z value will again be looked up in Appendix H using α. The difficult feature as far as using Equation 10.6 is that it assumes that the variance σ^2 is known. As we already know, this is not always the case; thus Equation 10.6 for application purposes is usually

$$n = \frac{Z^2 s^2}{(\overline{X} - \mu)^2} \tag{10.7}$$

▶ **Example 10.3**   Suppose the Board of Directors of a large corporation desires to know the average number of days vacation taken by its employees. Further, the corporation desires that the estimate be off by no more than 2 days from the true average, and that the level of confidence be 90 percent. This project has been delegated to the Statistics Department. The department knows that it is being asked to estimate μ such that the difference, $\overline{X} - \mu$, is no more than 2 days. The department also realizes that σ is unknown and will estimate σ from a preliminary sample of 100. Suppose this sample of 100 yielded a standard deviation of 15 days. Using Equation 10.7

$$n = \frac{Z^2 s^2}{(\overline{X} - \mu)^2} = \frac{(1.64)^2 (15)^2}{(2)^2} = \frac{(2.69)(225)}{4} = 151$$

Thus, to complete its assignment, the Statistics Department will take a sample of 151 observations to obtain $\overline{X}$ which will provide an estimate of μ within ± 2 days at the 90 percent confidence level. ◄

Sample Size for Proportions

It is possible to determine the sample size necessary to provide a sample proportion p, that is within a certain distance of the true population proportion, π.   Returning to the expression

$$Z = \frac{p - \pi}{\sigma_p}$$

where $\sigma_p = \sqrt{\dfrac{\pi(1 - \pi)}{n}}$

such that $Z = \dfrac{p - \pi}{\sqrt{\dfrac{\pi(1 - \pi)}{n}}}$

$$Z\sqrt{\frac{\pi(1 - \pi)}{n}} = p - \pi$$

$$Z^2\left[\frac{\pi(1 - \pi)}{n}\right] = (p - \pi)^2$$

$$n = \frac{\pi(1 - \pi)Z^2}{(p - \pi)^2} \tag{10.8}$$

► **Example 10.4**   Suppose a candidate has employed a consulting firm famous for political polls.   He has requested the firm to determine within 3 percentage points the proportion of people who currently will vote for him.   The level of confidence desired is 95 percent.   Thus, the firm will be using Equation 10.8 to decide on the necessary sample size such that

$$n = \frac{\pi(1 - \pi)Z^2}{(p - \pi)^2}$$

As no information is available concerning the candidate's previous popularity, π is assumed to be 0.50 which generates the largest value of n, and from Appendix H, $Z = 1.96$.

$$n = \frac{(0.50)(0.50)(1.96)^2}{(0.03)^2} = \frac{(0.2500)(3.84)}{0.0009} = 1067$$

The firm must use a sample of 1,067 voters to give the candidate the information he desires. ◄

Sampling Cost

Like most other situations, there are both *fixed* and *variable costs* associated with taking a sample. It is usually less expensive to take a sample than to do a complete enumeration (examine 100 percent of the universe). Yet the cost of taking a sample of *n* items may vary considerably from one sample design to another. In considering the cost of alternative designs, we must differentiate between the fixed cost and the variable cost. The fixed cost, of course, will be a fixed amount whether we take a sample size 1 or one of size 10,000.

The variable cost, on the other hand, varies with the sample size. Some designs may have a high fixed cost but the variable cost is low. Other designs may be just the opposite—the fixed cost is low but the variable cost is high. However, it is usually the variable cost that is more affected by different sample designs because of the effect on sample size and operating field expenses. In any event, if you are taking a sample for yourself, boss, client, instructor, or whomever, you must eventually address the question of cost.

▶ **Example 10.5** Suppose our political candidate is now considering running for the U.S. Senate from his state or for the U.S. Congress from his district, whichever he feels he'll have a better chance of winning. Much of his decision will depend upon the sample survey results.

He explains to you that he wants to take two political "polls" (sample surveys) of registered voters—one across the entire state and another across his congressional district. He wants the same level of accuracy for both. Now, he wants to know how much each poll will cost.

You examine the past political behavior of the state and district; you also compare certain characteristics of the registered voters for both geographical areas. Based on this analysis you determine that the variability of voting behavior is roughly the same when comparing the congressional district to the state as a whole.

Now, considering accuracy and variability, what sample size would you recommend, and consequently what cost estimates would you give the candidate?

If the variability is essentially the same between the state and the district and if the candidate wants the same level of accuracy or precision for each sample and if you use the same sample design, then the sample size would be approximately the same to sample either the district or the state. The state contains a much larger number of registered voters than the congressional district but this fact is of little consequence in determining the sample size. Finally, however, the state is the larger geographical area, so the cost to sample it may be higher if you expect to use face-to-face interviews or telephone calls to collect the information. ◀

Sample Efficiency

To summarize this chapter, let us say that, in designing a sample, we must be concerned with *sample efficiency*. Sample efficiency deals with achieving the greatest accuracy or precision of sample results for a given cost of the sample. Here, again, we are talking about cost, accuracy (or precision), variability, and sample size. Thus, in designing an efficient sample we should consider the following:

1. given a desired level of efficiency, a sample design should be selected which provides the *minimum cost for a certain size sample,* or
2. given a level of expenditure (total cost of the sample), the sample design selected should provide the *maximum precision.* In either case, we should try to optimize one or the other (cost or precision).

For example, we may find that by taking a stratified sample, rather than a simple random sample, we can reduce the sample size and retain the same standard error. Going further, if we take a systematic sample of the items in the stratum, perhaps we can also reduce the variable cost per sample unit. In another case, we might consider that by taking a smaller size sample the sampling (internal) error is increased, but the external error might be reduced proportionately more and thus the total cost lowered. Both the factors relating to total precision and the total cost of the sample should be considered in relation to the circumstances. Only then can the proper design be determined.

────────────── **SUMMARY OF EQUATIONS** ──────────────

10.1 $i = \dfrac{N}{n}$,   *page 228*

10.2 $n_i = \left(\dfrac{N_i}{N}\right) n$,   *page 230*

10.3 $n_i = \dfrac{N_i \sigma_i}{\Sigma\,(N_i \sigma_i)} \cdot n$,   *page 231*

10.4 $n_i = \dfrac{N_i\,(1/c_i)}{\Sigma\,[N_i\,(1/c_i)]} \cdot n$,   *page 231*

10.5 $n_i = \dfrac{N_i \sigma_i\,(1/c_i)}{\Sigma\,[N_i \sigma_i\,(1/c_i)]} \cdot n$,   *page 232*

10.6 $n = \dfrac{Z^2 \sigma^2}{(\overline{X} - \mu)^2}$,   *page 237*

10.7 $n = \dfrac{Z^2 s^2}{(\overline{X} - \mu)^2}$,   *page 237*

10.8 $n = \dfrac{\pi(1 - \pi)Z^2}{(p - \pi)^2}$,   *page 238*

—————————————— **REVIEW QUESTIONS** ——————————————

1. Explain the use of each equation shown at the end of the chapter.
2. Define
   a. simple random sample
   b. systematic sample
   c. stratified sample
   d. cluster sample
   e. multistage sampling
3. What is meant by sample precision?
4. What is the difference between internal and external error?
5. What is sample bias?

—————————————— **PROBLEMS** ——————————————

1. Assume you have 100 households numbered 00 to 99. Which house-holds would be in a random sample of size 10 if you use the table of ran-dom digits (Appendix M) and begin in row 25 and columns 10 and 11. (Proceed downward in the table to determine the households.)
2. Repeat the above procedure, but begin the selection of random digits in row 20 and columns 45 and 46.
3. Use the following population of numbers to select the specified samples. (8, 5, 3, 6, 2, 3, 7, 9, 3, 5, 6, 3)
   a. Estimate the population mean by selecting a systematic sample begin-ning with the third element and selecting every third element there-after.
   b. Repeat part (a), but begin with the second element.
   c. Compare the results of parts (a) and (b).
4. Suppose a retirement community has the following numbers in the given age groups:

AGE	f
55 and under 60	2000
60 and under 65	1500
65 and under 70	1000
70 and above	700

How many should be included in each stratum of a stratified sample of 300 if the method of proportional parts is used?

5. In Problem 4, if the cost of sampling is $1.00 per person for the youngest group and increases by 50¢ per person for each group, how will the sample size of 300 be distributed in each stratum?

6. Amicom specializes in "market surveys." Recently the firm received a contract from a county planning agency to conduct a sample to determine information regarding the income and spending habits of families in the area. In a conversation with an Amicom researcher, the agency head stated, "We don't know what sample technique you will use, but we would prefer a 20 percent sample rate."
   a. Discuss the problems of taking such a sample in your county.
   b. If you were with Amicom, what are some of the factors you should consider in designing this sample.
   c. What do you think of the statement made by the agency head?

7. The Washington Emergency Medical Service serves a geographical area that has 60,000 urban residents and 30,000 rural residents. From past experience, the Service knows that the costs of serving the residents is normally distributed and that the standard deviation of per customer costs is $5.00 for urban residents and $20.00 for rural residents. The WEMS is preparing to take a sample of its recent records to determine current costs. They have already decided that $n = 200$, but how should the sample be divided between the strata (urban and rural)?

8. The Washington Emergency Medical Service has just discovered that it can sample the records of its rural customers using its computer so that the cost of each sample unit will be 25¢. For its urban customers, however, it must use written records so that the cost per sample unit will be $1.00. Rework Problem 7 with this added information.

9. The mean average time for an aluminum fitting to be anodized is to be estimated from a simple random sample. Assume a standard deviation of 30 seconds, answer (a), (b), and (c).
   What size sample should you take
   a. If the desired probability of being wrong by more than 10 seconds (under or over) is 0.01?
   b. if the desired probability is 0.10 for a tolerable error of 30 seconds?
   c. if the desired probability is 0.10 for a tolerable error of 15 seconds?
   d. Assuming a standard deviation of 15 seconds, what sample size would you take under condition (a)?
   e. Compare the results of (a) and (d) and then (b) and (c).

10. Telephone, Inc., would like to estimate the proportion of families who prefer colored phones. If 60 percent of the families were known previously to prefer colored phones, what size simple random sample is needed to estimate the current proportion who prefer colored phones within 4 points of the actual proportion?   Set $\alpha = 0.10$.

11. New Cars, Inc., is interested in obtaining the average age of the customer who buys its model.   What size simple random sample would be necessary to make this age estimate within 2 years if a preliminary sample of 81 customers produced a sample mean of 45 years and a standard deviation of 8 years? .Set $\alpha = 0.05$.   *already have* $Z\sigma_{\bar{x}} = 1.74$

use 1yr

12. In a large city, a research analysis organization wants to estimate the average monthly rental expense for independent realtors for the Real Estate Association.   If an estimate of the standard deviation is $75.00, what size simple random sample would be necessary to assure that the difference between the sample mean and the population mean did not exceed $10.00 at the 95 percent level of confidence.

13. If a simple random sample of registered voters is to be taken to determine how they expect to vote on a proposal to increase the property tax millage, what sample size would be necessary to keep the sample proportion within 3 points of the population proportion?   Assume a 50–50 split and $\alpha = 0.05$.

14. A political candidate is interested in taking a poll of registered voters in his congressional district.   There are approximately 95,000 registered voters in the district consisting of one urban county of 45,000 voters and five rural counties with a total of 50,000 voters.   You have estimated that the fixed cost of the poll amounts to $500.   The average cost per interview for personal interviews comes to $5.00 in the urban county and $6.00 in the rural counties.   Suppose there are only two candidates involved in the election, and it is estimated that they are running very close to one another (roughly a 50–50 split).

Cost analysis

   a. Suppose your candidate wants the desired risk probability of being in error by more than 4 percentage points (in either direction) to be 0.05, what size simple random sample would you take.
   b. Suppose he wanted to reduce the error to 2 percentage points at the same risk probability.   What size simple random sample would you take?
   c. Compute the cost under (a) and (b) and compare.
   d. Do you think there would be any advantage in taking a stratified sample based on rural and urban classification of registered voters.
   e. In this type sample, what other bases for stratification might you consider?
   f. Would there be any basis for using a systematic sample in the case? A cluster sample?

15. A large television repair service is trying to improve the scheduling of its repair persons.   The manager feels that he needs an estimate of the mean length of travel time required for a service call that is accurate to

pilot study

within 2 minutes.   Moreover, he needs a 99 percent level of confidence for his answer.   A small random sample of the travel times came up with the following times (in minutes): 10, 15, 7, 6, 24.   Based upon this information, how large a sample would be necessary to meet the manager's requirements?   (Assume travel times are normally distributed.)

16. By comparing its total savings capital to that of its competitor, Fayetteville Federal Savings and Loan (FFS & L) estimates that it has 20 percent of the total savings deposits in the market it serves.   Currently, FFS & L is contemplating locating a new branch office in an area with 12,000 residents.   The FFS & L needs to know, with an error of no more than 0.5 percent and with $\alpha = 0.05$, the percent of the market it can expect in this new area.

   a. If past experience is used as an indicator of what to expect for the new branch, how large a sample should FFS & L take?

   b. Management has stated, "This is a new ball game and we don't know what to expect."   Now, how large a sample should FFS & L take?

First National Financial Services

First National Financial Services (FNF) of Atlanta, Georgia, has been operating as a financial company specializing in high-risk loans for the past 30 months. During the first three-quarters of its operation FNF showed an operating loss of $17,000, but its unearned interest seemed reasonable, and its loan volume was growing at its expected rate. During the last quarter of its first fiscal year, profits had turned with its increase volume, and FNF's Profit and Loss Statement showed a net operating profit of $6,000 for this quarter. However, with the increased volume, FNF began having operating problems. There were 5,500 loan accounts to be serviced by Mr. Elmar, a partner, aided by the manager and his four assistants. It was difficult to keep the kind of control required of high-risk loans with the increased number of individual loan accounts using a manual system. Mr. Elmar decided that the best approach would be to automate his system, especially his accounts receivable work.

Mr. Elmar called a representative from All Share, Inc., to see if it was feasible to convert his manual system to an on-line, real-time computer system. The representative from All Share assured Mr. Elmar that such a system could be developed and that the data conversion from his manual system using individual ledger cards was no problem at all. The software development and data conversion would take approximately two months. In the meantime, FNF employees would continue to operate on the manual system until the last week in October, at which time it was estimated by All Share personnel that the software development and data conversion would be complete.

All Share personnel began software development and data conversion for FNF during the first week of August. By the end of the second week in October the software development was not yet completed, and the data conversion had not started. At the end of October, when all procedures on the new system were to be fully implemented, an All Share representative contacted Mr. Elmar to notify him that complete conversion of FNF data to the new system was not yet operational. However, he was told that all new accounts and information would be put on the new system and stored until FNF was ready to go "on-line." During the next four months, All Share was never able to get the system completely operational or functional. In fact, many accounts which were fully paid and up-to-date were deemed delinquent, while others were considered active and were really delinquent and in arrears for four or more months. Still other accounts were simply lost and not recorded in the system.

Because of the difficulties he encountered with All Share, Mr. Elmar and his partner decided to terminate their contract with All Share and purchase a minicomputer for the office operations. At a contested but agreed on price of $12,500, All Share gave Mr. Elmar the FNF files and records on computer tape. In preparation for a transfer to a minicomputer system, FNF had hired a computer consultant, Dr. Marian Bund, to develop the computer software and direct the data conversion to the new minicomputer system. On each loan record, there are three bits of information that are very important for the operation of this type of finance company. They are the amount of the accounts receivable, the amount of unearned interest, and the principal balance. If all things balance, the principal balance equals the accounts receivable minus the unearned interest.

Although All Share representatives had declared that the information on the computer tape sold to Mr. Elmar was "penny perfect," Dr. Bund found many data er-

TABLE 1. *Dr. Bund's Enumeration of Accounts by Initial Year of Loan.*

	1977	1978	1979	TOTAL
Accounts Receivable	$102,950	$1,364,300	$464,890	$1,932,140
Interest Due	4,943	117,742	53,822	176,507
Principal Balance	99,044	1,271,491	411,132	1,781,667

rors on the tape. On many of the individual loans, the three balances did not agree, nor was there agreement in the totals of the loan accounts. Dr. Bund attempted to design several corrective programs to locate and adjust these errors, but since the number of accounts had grown to almost 8,500 by then, the problem of locating such errors became overwhelming. However, Dr. Bund suggested to Mr. Elmar that eventually many of these errors would surface and could then be corrected.

In the next several months the minicomputer system became operational and quite effective, but the overall totals in the loan accounts never seemed to balance with information in the general ledger. Dr. Bund explained that eventually most of these problems would "wash out."

In preparation for a formal audit, Dr. Bund was asked to write a computer program that would perform a total check of each individual loan account giving the balances and unearned interest of each account and a total of all accounts. There were now over 12,000 accounts on the system but some of them were paid out. Just two days before, Dr. Bund had delivered the computer output of this task and the totals to Mr. Elmar.

When Mr. Elmar and his partner carefully examined these totals, they declared that there was no way the totals could be correct. They called Dr. Bund to a meeting to discuss the matter. Mr. Elmar asked Dr. Bund if she was absolutely sure that her computer program for determining the totals was correct. Dr. Bund declared that her program was 100 percent correct, but Mr. Elmar showed at least one case where an account was omitted from the master file and the ac-

companying amounts from the total. The total amounts given to Mr. Elmar by Dr. Bund are presented in Table 1.

Dr. Bund was asked to run a series of samples to check the accuracy of the totals given by her computer program. At the end of the meeting Mr. Elmar's partner suggested that FNF employ a statistical consultant to run an independent sample of the loan accounts.

Given in Table 2 are the results of three "simple random samples" taken from all accounts by Dr. Bund, but there is reason to believe that her sample was not selected randomly. Given in Table 3 is the exact number of accounts on the total master file (computer tape) and the number broken down by year of loans.

The results of the stratified random sample (proportional to strata size) prepared by the statistical consultant are given in Table 4. If you were the statistical consultant, explain how you would approach the problem of analyzing these results. What advice would you give to Mr. Elmar? If it is absolutely necessary that the principal balance be no less than $1,780,000, would you say this criterion is met?

TABLE 2. *Dr. Bund's Sample Results.*

	SAMPLE		
	1	2	3
Mean Accounts Receivable	162.39	149.97	149.67
Mean Principal Balance	149.41	137.77	137.96
Sample Size	501	2000	2000

TABLE 3. *Number of Loans Made Each Year.*

1977	925
1978	10,434
1979	1,103

TABLE 4. *Consultant's Stratified Sample Results.*

	ACCOUNTS			
	1977	1978	1979	TOTAL
Mean Accounts Receivable	$124.11	$138.93	$469.66	$164.80
Standard Deviation	344.67	269.69	532.45	269.77
Mean Principal Balance	122.90	130.65	419.00	153.60
Standard Deviation	338.38	255.32	484.03	257.67
Sample	47	527	51	625

Nonparametric Statistical Tests

CHAPTER
11

In earlier chapters, we presented rather extensive discussions on hypothesis testing. In applying those testing procedures, you may recall, we either assumed some *population parameter was known* (or could be estimated), or we assumed that *we knew something about the form of the population distribution* from which the sample was drawn.

Nonparametric and Distribution-free Techniques

As you might suspect, situations arise when we can make *no assumption* about the value of a parameter, but some statistical test of a hypothesis would appear to be useful. In such cases, we would use a *nonparametric test*. In other instances, we may want to test a hypothesis, but we may not know or would not want to assume the precise form of the population distribution. To apply a statistical test in this situation, we would use some *distribution-free technique*.

Of the two terms, *nonparametric and distribution-free*, the term *nonparametric* seems to be the more widely accepted one.[1] So, rather than make a distinction between them, we will simply refer to the tests described in this chapter as *nonparametric tests*.

---

1. To a certain extent, however, distribution-free describes the quality of the test, which really makes it more interesting and useful to us, even though *nonparametric* is the more conventionally accepted term.

Chi Square

Probably the most widely used nonparametric test is the *chi square* (χ^2) *test.* While there are several applications of the chi square test, we will examine only three of them. But first, let us consider the nature of the *chi square distribution.*

In fact, there is not simply one chi square distribution. Like the Student *t* distribution, there is a different chi square (χ^2) distribution for every number of *degrees of freedom.* The distributions shown in Figure 11-1 are representative of three different numbers of degrees of freedom. Notice that with fewer degrees of freedom, the distributions are more positively skewed. With a greater number of degrees of freedom, the distribution becomes approximately normal.

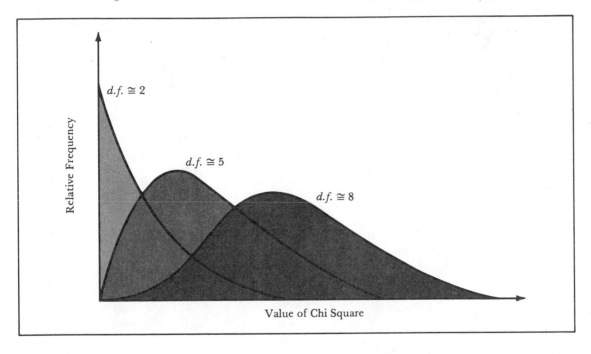

FIGURE 11-1. *Chi Square Distributions.*

The chi square distribution has several properties that make it easy to use. These properties are
1. The mean of the chi square distribution is equal to the number of degrees of freedom (*d.f.*)
2. The variance is two times the number of degrees of freedom [$2(d.f.)$]
3. The sum of two or more independent chi square variables yields a chi square variable.
4. The square of a standard normal variable is a chi square variable.

249

Degrees of Freedom

When you learned how to compute the sample standard deviation that can be used to estimate the population standard deviation, you encountered the term *degrees of freedom*. At that point in Chapter 3, it was explained that one degree of freedom is lost when you compute s as an estimate of σ. Similarly, when you are using chi square distributions to test hypotheses, degrees of freedom are lost. In other words, $d.f. <$ n. The value of $d.f.$ is determined in different ways, depending upon the way in which the chi square distribution is being used. Specifically, if a chi square technique is being applied to test the shape of a distribution:

$$d.f. = g - 1 - m$$

where g = the number of groups, or classes, in the frequency distribution

            m = the number of population parameters that must be estimated from sample statistics in order to test the hypothesis

If chi square is utilized to test for homogeneity or contingency, (these are discussed in detail in the following pages) then

$$d.f. = (r - 1)(c - 1)$$

where   r = the number of rows in the table
          c = the number of columns in the table

Recalling that there is a different χ^2 distribution for each number of degrees of freedom, it is important to determine the correct value of $d.f.$

Goodness-of-Fit Test

Chi square is frequently used to test *goodness-of-fit*. That is, it is used to ascertain whether the distribution of values in a sample supports a hypothesized population distribution. In order to apply the chi square distribution in this manner, the critical value of chi square is expressed as

$$\chi^2_{\text{test}} = \sum \frac{(f_o - f_e)^2}{f_e} \tag{11.1}$$

where f_o = observed frequency of the variable (sample)
          f_e = expected frequency (based on the hypothesized population distribution)

Example 11.1 shows how the chi square test is used in testing goodness-of-fit.

▶ **Example 11.1** Lawash Garments, Inc. produces ladies' sweaters. These are high-quality sweaters; so, each one is inspected by one of 10 company inspectors. Suppose we are asked by the Vice President for Marketing to test the hypothesis that the number of rejected sweaters is evenly distributed among the 10 quality control inspectors. Conversely, are some inspectors rejecting more sweaters than others?

H_0: Inspectors reject the same number of sweaters.
H_A: Inspectors do not reject the same number of sweaters.

In order to make this test, a random sample of 100 rejects was taken, and the initials of the inspector who rejected the sweater are recorded and classified as shown in Table 11-1.

TABLE 11-1. *Chi Square—Goodness-of-Fit Uniform Distribution.*

INITIAL OF INSPECTOR	COLUMN 1 OBSERVED FREQUENCY (NUMBER OF REJECTS) (f_o)	COLUMN 2 EXPECTED FREQUENCY OF REJECTS (f_e)	COLUMN 3 $(f_o - f_e)$	COLUMN 4 $(f_o - f_e)^2$	COLUMN 5 $\dfrac{(f_o - f_e)^2}{f_e}$
R. R.	15	10	+5	25	2.5
R. Y.	3	10	−7	49	4.9
G. G.	7	10	−3	9	0.9
S. B.	8	10	−2	4	0.4
P. S.	5	10	−5	25	2.5
S. G.	12	10	+2	4	0.4
B. B.	11	10	+1	1	0.1
S. W.	13	10	+3	9	0.9
J. B.	9	10	−1	1	0.1
P. P.	17	10	+7	49	4.9
	100	100			17.6

It is obvious that the number of rejects is not evenly distributed in the sample, but is there sufficient evidence to conclude that the population (all sweaters rejected) from which the sample was drawn is also not evenly distributed? To answer this question, look at Table 11-1 again and observe the following steps:

1. The observed frequencies f_o are listed in column 1.
2. With 10 inspectors, if there is an even distribution of rejects among them, then 10 percent of the sample rejects should allocate equally to each inspector. Thus, the expected frequencies f_e of each inspector are 100×0.10.

3. The remaining columns (3, 4, and 5) are simply further steps in the calculations as expressed in Equation 11.1
4. The sum of column 5 is the calculated chi square for the overall sample distribution.

Before we reach a decision concerning the hypothesis, we must determine the *critical chi square value*. In order to do this, the first step is to compute the number of degrees of freedom. Remember, degrees of freedom is defined as the number of observations that are free to vary. Since this is a test of a hypothesis about a distribution, then $d.f. = g - 1 - m$. But, in the example we are testing a uniform distribution and thus no population parameters have to be estimated. Consequently, $m = 0$ when working with distributions of this type. In this case, there are 10 inspectors, so $g = 10$ and $d.f. = 10 - 1 = 9$.

Now, we must decide what probability of Type I Error can be tolerated in testing the hypothesis. Assume an $\alpha = 0.05$ is reasonable. Now look in Appendix L, (Chi Square Distribution), and you will find that the value of chi square for $\alpha = 0.05$ at 9 *d.f.* is 16.919.

Referring back to our calculated chi square value, if $\chi^2_{test} > \chi^2_\alpha$, then we reject the null hypothesis that the numbers of rejected sweaters are evenly distributed among the 10 inspectors. Conversely, if $\chi^2_{test} < \chi^2_\alpha$, the null hypothesis is accepted. In this example, $(\chi^2_{test} = 17.6) > (\chi^2_\alpha = 16.919)$. Therefore, based on our sample, we reject the null hypothesis of an even distribution. It seems that some inspectors are rejecting more sweaters than others. ◄

The same procedure can be applied to test any hypothesized distribution: normal, Poisson, hypergeometric, etc. Like the uniform distribution in our example, you must be able to determine the "expected frequencies" and the number of degrees of freedom. As an illustration, suppose that you would like to test the data on per capita income given in Table 11-2 to determine if it is normally distributed. You can do so using Equation 11.1, but first we have to determine the expected frequencies, (f_e).

TABLE 11-2. *Per Capita Income in a Sample of 103 Counties.*

CLASS INTERVAL	CLASS MIDPOINT (M)	FREQUENCY (f_o)
$2500 and under $3100	$2800	12
3100 and under 3700	3400	15
3700 and under 4300	4000	31
4300 and under 4900	4600	20
4900 and under 5500	5200	15
5500 and under 6100	5800	10
		103

Back in Chapter 5, there is an equation (5.5) for determining the height of the normal curve at any value of X. Since the height of the curve represents the frequency, if we assume that the midpoint of each class in Table 11-2 represents all the values in the class, we can use Equation 5.5 to determine f_e for each class or group. However, a set of standardized ordinates is given in Appendix G and is somewhat easier to work with than Equation 5.5.

The arithmetic mean ($\overline{X}$) of the per capita incomes in Table 11-2 is \$4,239 and the standard deviation is \$875. Using these two values and Appendix G, we can determine the standardized ordinate for each class using the formula below

$$Z = \frac{M - \overline{X}}{s} \qquad (11.2)$$

where $\quad Z$ = the number of standard deviates
$\quad\quad\quad M$ = the midpoint of each class
$\quad\quad\quad \overline{X}$ = the arithmetic mean of the data
$\quad\quad\quad s$ = the standard deviation

Then, the following formula can be used to determine the expected frequency for each class:

$$f_e = O_z \left(\frac{n}{s}\right) i \qquad (11.3)$$

where $\quad f_e$ = expected frequency for each class (must be five or more for each class)
$\quad\quad\quad O_z$ = the ordinate at each class midpoint
$\quad\quad\quad n$ = the sample size
$\quad\quad\quad s$ = standard deviation
$\quad\quad\quad i$ = the class interval

TABLE 11-3. *The Expected Frequency for a Normal Distribution of Per Capita Incomes*

CLASS MIDPOINT (M)	$M - \overline{X}$	$\frac{M - \overline{X}}{s}$	O_z	EXPECTED FREQUENCY (f_e)
\$2800	−1439	−1.64	.1040	7
3400	− 839	−0.96	.2516	18
4000	− 239	−0.27	.3847	27
4600	+ 361	+0.41	.3668	26
5200	+ 961	+1.10	.2179	15
5800	+1561	+1.78	.0818	6
				99

Look at Table 11-3 and you can see how the expected frequency for each class was found. Using the first class as an example

$$M = \$2800, \overline{X} = \$4239, s = \$875, Z = \frac{2800 - 4239}{875} = -1.64$$

Referring to Appendix G, the ordinate is 0.1040 when $Z = -1.64$, so $O_Z = 0.1040$. Continuing with the example, $n = 103$ and $i = 600$

$$f_e = 0.1040 \left(\frac{103}{875}\right) 600 = 7$$

Actually, we should add classes at the beginning and end of the distribution until $\Sigma f_e = n$. (Notice $\Sigma f_e < n$ in Table 11.3.) However, $f_e < 5$ for each of these additional classes in our example so we must add their frequencies to the first and last classes, respectively. That is why, in Table 11.4, $f_e = 9$ for $M = \$2800$ and $f_e = 8$ when $M = \$5800$.

Now the hypothesis that the distribution of per capita income is normal can be tested by comparing the observed frequencies (f_o) to the expected frequencies.

    H_0: Per capita incomes are normally distributed
    H_A: Per capita incomes are not normally distributed
      $\alpha = 0.05$, $d.f. = g - 1 - m = 6 - 1 - 2 = 3$

Reject H_0 if $\chi^2_{test} > 7.81$.

The χ^2_{test} is determined using Equation 11.1, and in our example $\chi^2_{test} = 3.97$. Since $\chi^2_{test} < \chi^2_\alpha$, we do not reject the null hypothesis.

TABLE 11-4. *Chi Square Goodness-of-Fit Normal Distribution.*

CLASS MIDPOINT	f_o	f_e	$f_o - f_e$	$(f_o - f_e)^2$	$\dfrac{(f_o - f_e)^2}{f_e}$
$2800	12	9	+3	9	1.00
3400	15	18	−3	9	0.50
4000	31	27	+4	16	0.59
4600	20	26	−6	36	1.38
5200	15	15	0	0	0.00
5800	10	8	+2	4	0.50
	103	103			3.97

Tests of Contingency

At times you may wish to compare samples taken from two populations in order to determine if their attributes are statistically independent of the populations. In other words, does the population from which a sample was selected make a difference? Tests of statistical independence are known as contingency tests.

In Table 11-5 there is a tabulation of the number of responses received when a questionnaire was mailed using a postage meter and also the number returned when the same questionnaire was mailed

TABLE 11-5. *The Results of Two Mail Surveys Using Different Postage Types.*

| | POSTAGE | | |
	METER	COMMEMORATIVE	TOTAL
Response	100	150	250
Nonresponse	400	350	750
Total	500	500	1000

using a commemorative stamp. The question to be answered is whether the number of responses varies with the type of postage used. We can find an answer to this question by testing the following null hypothesis.

H_0:   The response rate is independent of the type of postage

H_A: The response rate is not independent of the type of postage

$\alpha = 0.01$ $d.f. = (r - 1)(c - 1) = (2 - 1)(2 - 1) = 1$

Reject H_0 if $\chi^2_{test} > 6.63490$

As you can see in Table 11-6, $\chi^2_\alpha < \chi^2_{test}$ so the null hypothesis is rejected. Apparently, the response rate may be contingent upon the type of postage used.

TABLE 11-6. *Computation of χ^2_{test} for Two Mail Surveys.*

| | EXPECTED RETURNS | |
	METER	COMMEMORATIVE
Response	$\dfrac{250}{1000}(500) = 125$	$\dfrac{250}{1000}(500) = 125$
Nonresponse	$\dfrac{750}{1000}(500) = 375$	$\dfrac{750}{1000}(500) = 375$

$$\chi^2_{test} = \sum \frac{(f_o - f_e)^2}{f_e}$$

$$= \frac{(100 - 125)^2}{125} + \frac{(400 - 375)^2}{375} + \frac{(150 - 125)^2}{125} + \frac{(350 - 375)^2}{375}$$

$$= 5.000 + 1.667 + 5.000 + 1.667$$

$$\chi^2_{test} = 13.334$$

Before you continue, two additional issues need to be considered. First, notice how the number of degrees of freedom was determined. As was pointed out earlier in this chapter, $d.f. = (r-1)(c-1)$ when testing for contingency. The second issue also deals with degrees of freedom. Theoretically, when $d.f. = 1$, a correction factor should be applied when computing χ^2_{test} so that the Equation 11.1 is changed to the following:

$$\chi^2_{test} = \sum \frac{(|f_o - f_e| - \frac{1}{2})^2}{f_e} \qquad (11.4)$$

In using Equation 11.4, if $(|f_o - f_e| - \frac{1}{2}) < 0$, treat it as if it is 0. Mindful of the need to use a correction factor when $d.f. = 1$, the example in Table 11-6 is reworked below using Equation 11.2.

$$\chi^2_{test} = \sum \frac{(|f_o - f_e| - \frac{1}{2})^2}{f_e}$$
$$= \frac{(|100-125| - \frac{1}{2})^2}{125} + \frac{(|400-375| - \frac{1}{2})^2}{375}$$
$$+ \frac{(|150-125| - \frac{1}{2})^2}{125} + \frac{(|350-375| - \frac{1}{2})^2}{375}$$
$$\chi^2_{test} = 4.802 + 1.601 + 4.802 + 1.601 = 12.806$$

We still reject H_0 because $12.806 > 6.63490$. Actually, when the sample is relatively large, say $n > 50$, using Equation 11.4 is not really necessary. Thus, our original work in Table 11-6 is acceptable but if $n < 50$, we should use Equation 11.4.

Tests of Homogeneity

You may also wish to test two or more samples to determine if they were drawn from the same population or populations with like distributions. When chi square is used for this purpose, it is called a test of homogeneity. Example 11.2 illustrates an application of a chi square test in this type situation.

▶ **Example 11.2**   Suppose that we wish to determine whether customers redeeming Lickum Trading Stamps select the same merchandise as customers turning in Stickum Stamps. A random sample of the selections of customers of each company yielded the distributions shown in the upper part of Table 11-7. The lower part of Table 11-7 shows the computations needed to test the following hypothesis:

$$H_0 \; : \quad \text{The redemptions are the same}$$
$$H_A \; : \quad \text{The redemptions are not the same}$$
$$d.f. = (\text{rows} - 1)\,(\text{columns} - 1)$$
$$d.f. = (5 - 1)\,(2 - 1) = 4$$
$$\alpha = 0.05$$

Reject H_0 if $\chi^2_{\text{test}} > 9.48773$.

TABLE 11-7. *Chi Square—Homogeneity Tests.*

MERCHANDISE SELECTED	OBSERVED		TOTAL
	LICKUM (f_L)	STICKUM (f_S)	
Hardware	12	20	32
Garden Tools	20	16	36
Clothes	18	34	52
Sporting Goods	10	10	20
Sundries	20	40	60
	80	120	200

MERCHANDISE SELECTED	EXPECTED		TOTAL
	LICKUM (f_e)	STICKUM (f_e)	
Hardware	$\dfrac{32}{200}(80) = 12.8$	$\dfrac{32}{200}(120) = 19.2$	32
Garden Tools	$\dfrac{36}{200}(80) = 14.4$	$\dfrac{36}{200}(120) = 21.6$	36
Clothes	$\dfrac{52}{200}(80) = 20.8$	$\dfrac{52}{200}(120) = 31.2$	52
Sporting Goods	$\dfrac{20}{200}(80) = 8$	$\dfrac{20}{200}(120) = 12$	20
Sundries	$\dfrac{60}{200}(80) = \dfrac{24}{80}$	$\dfrac{60}{200}(120) = \dfrac{36}{120}$	$\dfrac{60}{200}$

$$\chi^2_{\text{test}} = \sum \frac{(f_L - f_e)^2}{f_e} + \sum \frac{(f_S - f_e)^2}{f_e}$$

$$= \frac{(12 - 12.8)^2}{12.8} + \frac{(20 - 14.4)^2}{14.4} + \frac{(18 - 20.8)^2}{20.8} + \frac{(10 - 8)^2}{8} + \frac{(20 - 24)^2}{24}$$

$$+ \frac{(20 - 19.2)^2}{19.2} + \frac{(16 - 21.6)^2}{21.6} + \frac{(34 - 31.2)^2}{31.2} + \frac{(10 - 12)^2}{12} + \frac{(40 - 36)^2}{36}$$

$$= 0.050 + 2.178 + 0.377 + 0.500 + 0.667 + 0.033$$
$$+ 1.452 + 0.251 + 0.333 + 0.444$$

$$\chi^2_{\text{test}} = 6.285$$

Since $\chi_{test}^2 = 6.285 < \chi_\alpha^2 = 9.48773$, we accept (do not reject) the null hypothesis.

On the basis of our sample, it looks as if both Lickum and Stickum customers redeem their stamps for the same merchandise. ◄

Your introduction to chi square analysis has been brief. Even with this limited knowledge, however, you will find that the chi square test is a useful technique if you will follow these simple rules in using the test.

1. No classification should have an expected frequency of less than 5. This constraint may sometimes necessitate combining classifications in order to raise the number of expected frequencies, but be careful for in some cases, the combination may not be meaningful.
2. Chi square distributions are continuous. If a situation involves a discrete distribution, chi square analysis may still be used but a continuity correction factor may be necessary. If you desire further clarification or explanation of its use, consult an advanced text dealing with statistical methods.

Runs Test

We have frequently referred to *random* samples. In fact, you have learned that in order to use a sample statistic properly to make a probability statement concerning a population parameter, the sample must be random.

The importance of some assurance that the sample is random may seem rather trivial to you, for randomness may seem obvious. Yet experiments have shown that subjects' concepts of random differ greatly from what is actually random. For instance, have you ever noticed what is customarily called random tile flooring? Usually such flooring is not random, but has been laid purposely to look random.

In sampling, we try to maintain some control over the selection of the sample frame to assure randomness. There are occasions, however, where such control is not possible, practical, or desirable. The following example illustrates how a test called the *runs test* might be used to test whether or not a sequence is random.

► **Example 11.3**   A small town in West Texas recently added some females to its police force. The chief has been developing a work schedule, and one of his objectives is to arrange the working hours of his female officers so that they will be available to help with female prisoners.

In the past, the chief hasn't kept any records on the hour of the day in which arrests are made but he is inclined to believe that there is a definite pattern. To determine if he is correct, he orders that a record be kept of the next 64 arrests and that they be recorded by sex. The results were as follows:

F F MF MF MMMF MMF F MF MF F F MMF M
F F F F MMF MMF F MF F MF MF MF MMF MMM
F F F MF F MF MF MF F F

Question: Is the chief correct? That is, are the arrests random by sex or is there a pattern?

If the sex of the person arrested can be regarded as a random event, there will be an absence of pattern. Thus, if we can discover a pattern, we may reasonably conclude that the occurrence of the event is not random. If *F* denotes female and *M* denotes male, and if the recorded arrests were *M F M F M F*, we could be reasonably sure that the sequence is not random. Similarly, patterns of *M M M M M M M* or *F F F F F F* would be good evidence of nonrandomness.

The above patterns that cause us to doubt the randomness of the arrests consists of one or more runs. A run is one or more identical sequential outcomes (in this case, sex of the person arrested). The following series consists of five runs: $\overline{MM}\ \overline{FFF}\ \overline{M}\ \overline{F}\ \overline{MMM}$.

Moreover, if a sampling procedure is random, the distribution of runs *R* will be approximately normal when the sample is reasonably large. With this knowledge, we can use the number of runs to test for randomness by applying the following equations.

$$\overline{R} = \frac{2n_1 n_2}{n_1 + n_2} + 1 \tag{11.5}$$

$$s_R = \sqrt{\frac{2n_1 n_2 (2n_1 n_2 - n_1 - n_2)}{(n_1 + n_2)^2 (n_1 + n_2 - 1)}} \tag{11.6}$$

where $\overline{R}$ = the mean number of runs.

n_1 = the number of outcomes of one type.

n_2 = the number of outcomes of the other type.

s_R = the standard deviation of the distribution of the number of runs.

The application of these formulas in the runs test is illustrated in Table 11-8.

Given that $\overline{R}$ and s_R can be calculated, you are now in a position to test the chief's intuition concerning a pattern. As usual, you begin by establishing the null hypothesis. In this case, H_0 = the arrests are random. We will use $\alpha = 0.05$ as a reasonable criterion for the test. The number of runs (R) observed in the sample is then compared to the mean number of runs in the following way:

TABLE 11-8. *Runs Test.*

Record of 64 arrests	$\overline{FF}\ \overline{M}\ \overline{F}\ \overline{M}\ \overline{F}\ \overline{MMM}\ \overline{F}\ \overline{MM}\ \overline{FF}\ \overline{M}\ \overline{F}\ \overline{M}\ \overline{FFF}\ \overline{MM}\ \overline{F}\ \overline{M}\ \overline{FFFF}\ \overline{MM}$ $\overline{F}\ \overline{MM}\ \overline{FF}\ \overline{M}\ \overline{FF}\ \overline{M}\ \overline{F}\ \overline{M}\ \overline{F}\ \overline{M}\ \overline{FF}\ \overline{M}\ \overline{MMM}\ \overline{FFF}\ \overline{M}\ \overline{FF}\ \overline{M}\ \overline{F}$ $\overline{M}\ \overline{F}\ \overline{M}\ \overline{FFF}$

$$\text{Number of females} = n_1 = 35$$
$$\text{Number of males} \ \ = n_2 = 29$$
$$\text{Number of runs} \ \ \ = R \ = 41$$

$$\overline{R} = \frac{2n_1 n_2}{n_1 + n_2} + 1$$

$$\overline{R} = \frac{2\,(35)\,(29)}{35 + 29} + 1$$

$$\overline{R} = 31.72 + 1 = 32.72$$

$$s_R = \sqrt{\frac{2n_1 n_2\,(2n_1 n_2 - n_1 - n_2)}{(n_1 + n_2)^2\,(n_1 + n_2 - 1)}}$$

$$s_R = \sqrt{\frac{2\,(35)\,(29)\,[2\,(35)\,(29) - 35 - 29]}{(35 + 29)^2\,(35 + 29 - 1)}}$$

$$s_R = \sqrt{\frac{3,990,980}{258,048}} = \sqrt{15.466} = 3.93$$

$$Z_{\text{test}} = \frac{R \pm \frac{1}{2} - \overline{R}}{s_R} \tag{11.7}$$

We will reject H_0 if $Z_{\text{test}} < -1.96$ or $Z_{\text{test}} > 1.96$.   Now we can complete our test of the chief's intuition.[2]

$$Z_{\text{test}} = \frac{41 - \frac{1}{2} - 32.72}{3.93} = 1.98$$

According to this technique, it would appear that there is a pattern to the arrests, thus we reject H_0.   We could not conclude at the 0.05 level of significance that the arrests were random. ◄

Sign Test

You may encounter business problems where it is necessary to compare two sample series, consisting of matched pairs of observations in order to determine which series is the larger of the two, or at least if

2.   The $\pm 1/2$ is used in Equation 11.7 to convert the discrete data (number of runs) to a continuous distribution.   Use $+1/2$ if $R < \overline{R}$, and use $-1/2$ if $R > \overline{R}$.

there is any difference between them. If the magnitude of the differences is not important, then the two series may be compared on the basis of the signs (+ or −) of the differences. Thus, under the assumption that magnitudes are not to be considered, a hypothesis that the values in two series are not different in size can be tested using the *sign test*. Example 11.4 illustrates a situation for which the application of the sign test for matched pairs would be appropriate.

▶ **Example 11.4** The makers of GoNoMo gasoline wish to evaluate the influence of giveaway games on gasoline sales. They decide to run a test by observing the sales of 40 service stations for two months. During one month the stations would have a license-plate game in which customers could win valuable gifts based on the numbers on the license of the customer's car. During the other month the station would have no such game as a sales inducement. Let us denote sales in the month without the game as w and sales in the month with the game as w'. The results of this experiment are shown in Table 11-9. ◀

In comparing the sales in the two months, remember the magnitude of the differences is not important. Thus, when subtracting $(w' - w)$ only the sign of the difference—plus or minus—is retained. If the use of the games does not cause a difference in sales, then the number of pluses (k) will equal the number of minuses (m). Furthermore, in any given sample there would be a 0.5 probability that m would be greater than k. (This assumes a random sample and also that we ignore any sample units where there is no difference, i.e., where $w' - w = 0$.)

As usual, the test is begun by establishing the null hypothesis. In this case

$$H_0: \quad k - m = 0$$
$$H_A: \quad k > m$$

Now it is necessary to determine something about the expected number of pluses $E(k)$ and their sampling distribution. This can be readily accomplished if you recall early chapters dealing with sample proportions. If there is no difference in sales, then there should be equal numbers (or proportions) of pluses and minuses. In other words, if π is the proportion then π should equal 0.5 when the null hypothesis is correct. The expected number of pluses $E(k)$ in any given sample of size n is $n\pi$. (Since π is always equal to 0.5 in this type of problem, $E(m)$ is also $n\pi$.) Continuing to draw on our previous knowledge of sampling, the dispersion of the number of pluses in any sample can be measured by

$$s_k = \sqrt{n\pi(1-\pi)} \tag{11.8}$$

TABLE 11-9. *Sign Test.*

| STATION | SALES (in $1000) | | SIGN ($w' - w$) |
	WITHOUT GAME (w)	WITH GAME (w')	
1	9.0	10.0	+
2	9.5	10.1	+
3	10.2	10.1	−
4	13.0	12.8	−
5	8.1	8.0	−
6	11.9	12.2	+
7	9.6	9.8	+
8	11.2	10.5	−
9	6.3	6.3	(omit)
10	12.7	12.6	−
11	12.5	13.7	+
12	9.2	9.2	(omit)
13	11.0	11.9	+
14	11.3	12.1	+
15	6.8	6.7	−
16	7.0	7.5	+
17	7.0	6.9	−
18	14.2	14.3	+
19	10.0	9.8	−
20	13.1	13.0	−
21	14.0	13.8	−
22	12.7	12.8	+
23	11.9	11.5	−
24	12.0	12.0	(omit)
25	7.2	7.1	−
26	10.3	10.5	+
27	6.8	7.3	+
28	11.7	11.8	+
29	9.2	9.0	−
30	13.7	13.5	−
31	12.8	13.2	+
32	10.7	11.0	+
33	11.2	11.1	−
34	12.3	12.9	+
35	9.1	9.0	−
36	8.7	8.7	(omit)
37	9.9	9.2	−
38	13.1	12.5	−
39	12.4	12.3	−
40	11.7	12.0	+

$n = 36$

$E(k) = n\pi = 36\,(0.5) = 18$

$s_k = \sqrt{n\pi\,(1 - \pi)}$

$s_k = \sqrt{36\,(0.5)\,0.5} = \sqrt{9} = 3$

Proceeding with the example involving the game and GoNoMo sales, you can now test the null hypotheses that the introduction of the game made no difference in sales in the following manner:

$$Z_{test} = \frac{k \pm 1/2 - E\ (k)}{s_k} \tag{11.9}$$

$$Z_{test} = \frac{17 + 1/2 - 18}{3}$$

$$Z_{test} = -0.167$$

Assuming that we wished to test the null hypothesis with $\alpha = 0.01$, then reject H_0 if $Z_{test} > 2.33$. Therefore the null hypothesis is accepted. Apparently, the introduction of the game did not cause an increase in sales.

The sample distribution of pluses and minuses is actually a binomial distribution, but the sign test we have illustrated assumes that a normal distribution can be used to carry out the test with a reasonable degree of accuracy. As a general rule, this assumption can be made when the sample size n is at least 30. In the calculation of Z, the addition of $1/2$ to k is necessary because the binomial distribution represents discrete events, while the normal distribution is continuous. (From a practical standpoint, when n becomes large failing to subtract $1/2$ will not alter your solutions.) Carrying out a signs test using the binomial distribution (when n is less than 30) is a relatively easy, but sometimes lengthy, task. It simply requires that the binomial $[\pi + (1 - \pi)]$ be expanded to the n power so that the probability of obtaining a certain number or greater pluses can be determined. If this probability is less than the specified value of α for the test, the null hypothesis is rejected. Example 11.5 illustrates this type of test.

▶ **Example 11.5** Suppose that in our GoNoMo gasoline situation, we had obtained only every fourth observation in Table 11-9. This would have resulted in 7 usable data points (stations 4, 8, 16, 20, 28, 32, and 40), 4 pluses and 3 minuses. We then test the following null hypothesis at the 0.10 α level.

$$H_0: \quad \pi = 0.5 \text{ (probability of plus} = 0.5)$$

$$H_A: \quad \pi > 0.5 \text{ (probability of plus} > 0.5)$$

From the cumulative binomial distribution, Appendix E, we obtain:

$$P\ (X \geq 4) = 1.0 - P\ (X \leq 3)$$

$$P\ (X \geq 4) = 1.0 - 0.5000$$

$$P\ (X \geq 4) = 0.5000$$

Since $0.5000 > 0.10$, we conclude the outcome of 4 pluses from 7 occurrences could be due to chance, and we accept the null hypothesis. ◄

Mann-Whitney U-Test

In Chapter 9, we learned to use the t test to determine if two independent random samples were drawn from the same population or from different populations.   But the Student t distribution requires the assumptions that the populations from which the samples are drawn be normally distributed and have equal variances.   An alternative to using the t test is the *Mann-Whitney U-test*.   As a nonparametric test, it does not require that the preceding assumptions about the population be made.   It does require, however, the size of each of the two samples be 10 or more to use the normal approximation.[3]

The *U-test* is based upon a comparison of the rank of the values of one sample in an array of the combined values of the two samples. It is fairly obvious that if the two samples come from the same population, or two populations with equal means, the average ranks for each sample should be equal. U is the sample statistic for the distribution of the sum of the ranks (R) for one of the samples.

$$U_1 = n_1 n_2 + \frac{n_1 (n_1 + 1)}{2} - R_1 \qquad (\mathbf{11.10})$$

$$U_2 = n_1 n_2 + \frac{n_2 (n_2 + 1)}{2} - R_2 \qquad (\mathbf{11.11})$$

where $n_1 = $ the size of the first sample
$n_2 = $ the size of the second sample
$R_1 = $ the sum of the ranks of the first sample
$R_2 = $ the sum of the ranks of the second sample
$U = $ the lower value of U_1 or U_2.

Assuming that n_1 and n_2 are sufficiently large (≥ 10), the sampling distribution of U can be assumed to be normal with the following parameters:

$$\bar{U} = \frac{n_1 n_2}{2} \qquad (\mathbf{11.12})$$

$$s_u = \sqrt{\frac{n_1 n_2 (n_1 + n_2 + 1)}{12}} \qquad (\mathbf{11.13})$$

---

3.   Smaller samples may be tested using special tables of the Mann-Whitney U-statistic.

Now it is possible to establish the following null hypothesis and to test data such as in Table 11-10 using Equation 11.14.

H_0:   Both samples come from the same population

H_A: The samples do not come from the same population

$$Z_{\text{test}} = \frac{U - \overline{U}}{s_u} \tag{11.14}$$

TABLE 11-10. *Daily Sales.*

SAM SMITH	JOHN JONES
$100	$ 82
125	98
60	115
137	143
82	65
99	123
150	128
98	93
143	135
122	120
95	
110	

The two samples are random selections of the daily sales of two sales-men—Sam Smith (Sample 1) and John Jones (Sample 2). Using the data from Table 11-11, the computations for the test are

$$Z_{\text{test}} = \frac{58.5 - 60}{15.17}$$

$$= \frac{-1.5}{15.17}$$

$$Z_{\text{test}} = -0.099$$

Assume

$$\alpha = 0.05$$

Reject H_0 if $Z_{\text{test}} < -1.96$ or $Z_{\text{test}} > 1.96$.

Thus, based on the use of the U-test, Sam Smith and John Jones appear to sell equal amounts.

TABLE 11-11.   *Mann-Whitney U-test.*

ARRAY OF DAILY SALES	SALESMAN	RANK	SMITH RANK	JONES RANK
60	Smith	1	1	
65	Jones	2		2
82	Smith	3.5	3.5	
82	Jones	3.5		3.5
93	Jones	5		5
95	Smith	6	6	
98	Smith	7.5	7.5	
98	Jones	7.5		7.5
99	Smith	9	9	
100	Smith	10	10	
110	Smith	11	11	
115	Jones	12		12
120	Jones	13		13
122	Smith	14	14	
123	Jones	15		15
125	Smith	16	16	
128	Jones	17		17
135	Jones	18		18
137	Smith	19	19	
143	Smith	20.5	20.5	
143	Jones	20.5		20.5
150	Smith	22	22	
			139.5	113.5

$$U_1 = n_1 n_2 + \frac{n_1 (n_1 + 1)}{2} - R_1 \qquad U_2 = n_1 n_2 + \frac{n_2 (n_2 + 1)}{2} - R_2$$

$$U_1 = (12)(10) + \frac{12(12+1)}{2} - 139.5 \qquad U_2 = (12)(10) + \frac{10(10+1)}{2} - 113.5$$

$$U_1 = 58.5 \qquad\qquad\qquad U_2 = 61.5$$

Thus, $U = 58.5$

$$\overline{U} = \frac{n_1 n_2}{2} \qquad\qquad s_u = \sqrt{\frac{n_1 n_2 (n_1 + n_2 + 1)}{12}}$$

$$\overline{U} = 60 \qquad\qquad s_u = \sqrt{\frac{(12)(10)(12 + 10 + 1)}{12}}$$

$$s_u = \sqrt{230}$$

$$s_u = 15.17$$

The U-test not only has the advantage of not requiring assumptions regarding the population distribution, but it is also generally easier to apply. However, like most of the nonparametric techniques, it is a somewhat weaker test than comparable parametric tests as a general rule. Still, the nonparametric methods can be a useful addition to your bag of statistical tools.

Kruskal-Wallis Test

There may be instances when you have three or more *independent random samples* and you need a test to determine if they come from populations with identical distributions. This can be accomplished with the Kruskal-Wallis test. Using this method, a statistic called H is computed that is approximately chi square distributed with $d.f. = n' - 1$, where n' is the number of independent samples and no sample is smaller than 5. This statistic (H) is based upon the sum of the ranks of the values in each sample with the ranks being determined by first combining all samples and constructing an ordered array. The formula for computing H is given below.

$$H = \frac{12}{n\,(n+1)} \sum \frac{R_i^2}{n_i} - 3\,(n+1) \tag{11.15}$$

where n = the size of all samples combined

            R_i = the sum of the ranks of the ith sample

            n_i = the size of the ith sample.

▶ **Example 11.6**   Smokey Motors Corporation has developed three new diesel engines and needs to select one that might be suitable for use in a high-performance car it hopes to market. All three engines are equally suitable with respect to most performance and design criteria but the company has just completed testing the fuel consumption of each of the three engines. In order to determine the fuel consumption, they built a limited number of each of the engine types and tested them under "normal" road conditions. After looking at the results, shown in Table 11-12, the Vice President for Marketing remarked that it appeared that there might be differences in fuel consumption among the three engines. ◀

TABLE 11-12. *Fuel Consumption of Three Engine Types (kilometers per liter).*

	TYPE	
1	2	3
31	12	17
30	13	20
29	26	23
22	24	9
32	11	15
25	16	18
	27	19
	10	14
		8
		5

The situation described in Example 11.6 is one in which the Kruskal-Wallis test can be applied. The results are shown in Table 11-13. The degrees of freedom for this chi-square test are $d.f.$ = number of samples -1. In this instance, $d.f. = 3 - 1 = 2$. Thus, the following null hypothesis can be tested

H_0:   The distributions of the three populations are identical

H_A: The distributions are not identical

$$\alpha = 0.01$$

$$\chi_\alpha^2 = 9.21034$$

Since $H = 10.532 > \chi_\alpha^2 = 9.21034$, the null hypothesis is rejected. There may be a difference in the consistency of the fuel consumption of the three engines.

TABLE 11-13. *Computation of H for the Kruskal-Wallis Test.*

	ENGINE TYPE		
KPL	1	2	3
5			1
8			2
9			3
10		4	
11		5	
12		6	
13		7	
14			8
15			9
16		10	

(cont'd)

KPL	ENGINE TYPE		
	1	2	3
17			11
18			12
19			13
20			14
22	15		
23			16
24		17	
25	18		
26		19	
27		20	
29	21		
30	22		
31	23		
32	24		
R_i	123	88	89

$$\frac{R_1^2}{n_1} = \frac{(123)^2}{6} = 2521.5 \qquad \frac{R_2^2}{n_2} = \frac{(88)^2}{8} = 968 \qquad \frac{R_3^2}{n_3} = \frac{(89)^2}{10} = 792.1$$

$$H = \frac{12}{n\,(n+1)} \sum \frac{R_i^2}{n_i} - 3\,(n+1)$$

$$= \frac{12}{24\,(25)}\,(2521.5 + 968 + 792.1) - 3\,(25)$$

$$H = 0.02\,(4281.6) - 75 = 85.632 - 75 = 10.632$$

In our example, there were no "ties" in the ranks. If there had been, we would have assigned their mean rank to the tied values. To illustrate, had the 23rd and 24th items both been 32, we would have assigned each of them the rank 23.5. Then it would have been necessary to adjust the statistic H to one called H'.

$$H' = \frac{H}{C} \qquad\qquad \textbf{11.16}$$

$$C = 1 - \frac{\sum (t_i^3 - t_i)}{n^3 - n} \qquad\qquad \textbf{11.17}$$

where t_i = the number of values in each set of tied values
             n = the combined total of all samples

Unless there are several tied values in the samples, the values of H and H' will be very close to one another. In any case, H' is compared to χ_α^2 in testing of the hypothesis instead of H.

11.1 $\quad \chi^2_{\text{test}} = \sum \dfrac{(f_0 - f_e)^2}{f_e},$ *page 250*

11.2 $\quad Z = \dfrac{M - \overline{X}}{s},$ *page 253*

11.3 $\quad f_e = O_Z \left(\dfrac{n}{s}\right) i,$ *page 253*

11.4 $\quad \chi^2_{\text{test}} = \sum \dfrac{(\,|f_0 - f_e| - \frac{1}{2})^2}{f_e},$ *page 256*

11.5 $\quad \overline{R} = \dfrac{2n_1 n_2}{n_1 + n_2} + 1,$ *page 259*

11.6 $\quad s_R = \sqrt{\dfrac{2n_1 n_2 (2n_1 n_2 - n_1 - n_2)}{(n_1 + n_2)^2 (n_1 + n_2 - 1)}},$ *page 259*

11.7 $\quad Z_{\text{test}} = \dfrac{R \pm \frac{1}{2} - \overline{R}}{s_R},$ *page 260*

11.8 $\quad s_k = \sqrt{n \pi (1 - \pi)},$ *page 261*

11.9 $\quad Z_{\text{test}} = \dfrac{k \pm \frac{1}{2} - E(k)}{s_k},$ *page 263*

11.10 $\quad U_1 = n_1 n_2 + \dfrac{n_1 (n_1 + 1)}{2} - R_1,$ *page 264*

11.11 $\quad U_2 = n_1 n_2 + \dfrac{n_2 (n_2 + 1)}{2} - R_2,$ *page 264*

11.12 $\quad \overline{U} = \dfrac{n_1 n_2}{2},$ *page 264*

11.13 $\quad s_u = \sqrt{\dfrac{n_1 n_2 (n_1 + n_2 + 1)}{12}},$ *page 264*

11.14 $\quad Z_{\text{test}} = \dfrac{U - \overline{U}}{s_u},$ *page 265*

11.15 $\quad H = \dfrac{12}{n(n + 1)} \sum \dfrac{R_i^2}{n_i} - 3(n + 1),$ *page 267*

11.16 $\quad H' = \dfrac{H}{C},$ *page 269*

11.17 $\quad C = 1 - \dfrac{\sum (t_i^3 - t_i)}{n^3 - n},$ *page 269*

1. Identify the use of each equation given at the end of the chapter.
2. Why is there a need for nonparametric statistics?

PROBLEMS

1. A statistics instructor just completed preparing his grades for the semester and they are as follows:

GRADE	NUMBER
A	2
B	40
C	25
D	13
F	20

At the outset of the semester, he told his classes that, over time, his grades had been uniformly distributed. If the above grades are considered a random sample, test the instructor's statement. Set $\alpha = 0.01$.

 2. A single die was thrown 120 times with the following results. Based on the observed frequencies, is the die fair? Set $\alpha = 0.05$.

NUMBER OF SPOTS	OBSERVED FREQUENCY
1	18
2	21
3	17
4	22
5	23
6	19

 3. A student has maintained the following information on typing errors.

NUMBER OF ERRORS	NUMBER OF ASSIGNMENTS
0	11
1	14
2	15
3 or more*	12

*The total number of errors is 47.

Is this a uniform distribution? Set $\alpha = 0.01$.

4. The manager of Stop-and-Wait, a local fast-service food chain, has kept records of the number of people waiting in line to be served at his restaurant. The other day, a customer complained about the number of people waiting and the manager replied, "I know from experience that the number of people waiting at any one time is normally distributed." If the following information is considered to be a random sample, was the manager's statement correct?   Set $\alpha = 0.05$.

$\bar{X} = 8.12$

$S = 3.9618$

CUSTOMERS IN LINE	FREQUENCY
0 but less than 2	15
2 but less than 4	20
4 but less than 6	35
6 but less than 8	25
8 but less than 10	35
10 but less than 12	40
12 but less than 14	20
14 but less than 16	15

5. A random sample of the records of an automobile insurance company yielded the following information on the accident experience of its policy holders. After examining the distribution, the company statistician concluded the number of accidents is normally distributed. Using $\alpha = 0.01$, is he correct?

$\bar{X} = 4.169$

$S = 1.9059$

NUMBER OF ACCIDENTS	FREQUENCY
0	10
1	40
2	100
3	150
4	200
5	125
6	75
7	50
8	30
9	20

6. A local retail store has experimented with enclosing a return envelope in some of the monthly statements. The results are given below:

STATE OF PAYMENT	WITH ENVELOPE	WITHOUT ENVELOPE
Paid when due	150	120
Not paid when due	50	30

At the 0.01 level of significance is there any indication that the enclosure of an envelope promotes prompt payment?

7. The owner of the Specialty Shop has recorded information on 450 customers:

	FEMALE	MALE
Purchase	60	60
No Purchase	150	180

At the 0.05 level of significance, do you believe that whether a purchase is made or not is independent of the sex of the customer?

8. The Robb Insurance Agency has recently completed a study of 300 salesmen relating the price of automobiles driven and the salesmen's home office ratings.

	SALESMEN'S RATING		
AUTO PRICE	SUCCESSFUL	FAIR	POOR
High	100	40	15
Medium	35	30	15
Low	15	30	20

At the 0.01 level of significance, would you conclude that price of the car and salesmen's ratings are independent?

9. Premium Product Promotions, a market research firm, is trying to determine if there are differences in the brand of beer preferred by various customer groups. Formulate and test a hypothesis for them using $\alpha = 0.025$ if the following data are the preferences of a sample of 800.

	BRAND OF BEER		
CUSTOMER GROUP	PALE	GOLDEN	HEAVY
Housewives	75	20	5
Businessmen	50	130	20
Factory Worker	5	25	170
College Students	100	100	100

10. The following distribution represents even and odd digits from a table of numbers believed to be random. Test at the 0.05 level of significance to see if the digits are random.

E E O E O E E O E E E E O O O O E E O O O O O O
E E E O O E E O O O O E O E O O O E E O O

11. In the Student Union, the red and green tiles are supposedly laid on the floor as random tiles. Perform a runs test to see, at the 0.05 level of significance, if the red and green tiles are actually laid in a random manner. (We have used only one row of tiles in the sample.)

R R G R G R G G G R G R R G R R R R R R G G G
G R G G G R G R G G G R G R R R R R R G R G R
G R R G R G

12. Twelve students, randomly selected, participated in a memory course. After the course, 8 students scored higher on a reading comprehension test, 2 had the same score, and the second score of 2 was lower than the first. At the 0.05 level of significance, do you think the memory course improved the reading comprehension of the students?

13. The sales tickets of two clerks were compared at the end of a particularly busy day.

CLERK A	CLERK B
$10	$18
29	13
12	44
19	4
22	36
14	38
50	25
46	15
5	28
39	11
42	17
15	50
24	16

At the 0.05 level of significance, would you conclude that the sales are from the same population?

14. The Specialty specializes in two types of dinners, a steak dinner or a seafood dinner. Two chefs are employed, one for each type of dinner. The management, in order to determine if one chef is better than the other, has maintained a record of complaints for the past 10 weeks. Determine, at the 0.05 level, if the chefs are equally popular.

NUMBER OF COMPLAINTS	
STEAK CHEF	SEAFOOD CHEF
5	3
9	0
12	13
2	20
11	7
6	8
1	5
15	10
19	14
4	1

15. The Dean has developed some data on the academic performance of students from three high schools. The following is a sample of that data.

SLEEPY HIGH	BIG TIME HIGH	GRUBBY HIGH
1.6	2.5	0.5
2.0	2.8	1.4
1.8	3.0	3.2
2.7	2.9	1.3
1.0	3.5	2.6
2.1		2.4
2.3		

Test the Dean's hypothesis that there is no difference in the distribution of the academic performance between students from the various high schools. Set $\alpha = 0.05$.

16. Mr. S. Lagree is the office manager for a company that provides part-time office help. He insists that the four secretaries he employs be highly accurate typists but lately he has begun to suspect that the distribution of number of errors they make is different. To test this, he has taken a sample of their work for several days and recorded the errors per page. Test Mr. Lagree's notion. Set $\alpha = 0.01$.

ERRORS PER PAGE			
ANN	BILL	CINDY	DONNIE
2.1	0.0	1.2	4.1
3.3	1.9	1.3	2.8
3.8	1.5	0.1	2.4
2.7	2.2	0.7	4.2
5.1	2.3	2.6	3.5
4.0	0.8	1.1	3.4

Correlation and Regression Analysis: The Simple Linear Case

CHAPTER
12

Until now we have focused our attention on describing and drawing inferences relating to a single variable. This chapter addresses itself to describing, in a linear fashion, the relationship between two variables. The next chapter will be devoted to describing the relationship between two variables in a nonlinear fashion and to the relationship between more than two variables.

Nature of Correlation and Regression Analysis

Correlation and regression analysis involves determining and measuring the relationship (covariation or going-together) between two or more variables. Regression deals with determining a quantitative expression to describe the relationship, while the purpose of correlation is to measure the degree of the relationship.

Frequently, statistical analysts attempt to differentiate further between correlation and regression analysis. We have made a distinction between them only for the purpose of defining how each contributes to the general analysis and interpretation of the data. But further distinctions will not be attempted because we feel that the two go together like ham and eggs. In an analytical sense, they are intertwined, and it serves no useful purpose here to make further distinctions.

This chapter deals with simple linear correlation and regression analysis. The word *simple* in this case does not mean *easy*. Rather, it refers to the notion that we are concerned with the relationship between only two variables—a *dependent* variable and an *independent* variable. The idea of *linear* correlation and regression analysis relates to the assumption that the best representation of perfect correlation is a *straight* (linear) regression line fitted to the observed data. That is, in a linear analysis we describe the mathematical relationship by a straight line using the general form given in Equation 12.1.

$$Y = \alpha + \beta X \tag{12.1}$$

So, simple means only *two variables,* and linear means a *straight line (arithmetic) relationship* between them.

Before proceeding with the chapter, let us now take a bird's-eye view of the steps involved in conducting simple linear correlation and regression analysis. These steps are

1. Plot the scatter diagram
2. Compute the regression equation
3. Compute the standard error of estimate
4. Test hypothesis concerning β
5. Perform variation analysis
6. Compute correlation coefficient

Each of these steps will be developed in the sections that follow in this chapter.

The Scatter Diagram

A scatter diagram consists of coordinate points of the two variables plotted on a graph. The points are not connected together. Rather, they look like a scattering of dots, not too unlike a pattern of bird shot. In simple regression analysis there are only two variables—the independent variable is plotted on the X axis (the abscissa) and is sometimes referred to as the X *variable,* while the dependent, or Y *variable,* is plotted on the Y axis (the ordinate). Figure 12-1 illustrates a hypothetical scatter diagram.

The Variables

Before we get too involved with the scatter diagram per se, let us consider the variables which are the subject of the analysis. First, what is the nature of the data that comprise the two variables? Second, how do we determine which is the dependent and which is the independent variable? The answer to the first question may seem very simple. Yet, there are some serious implications related to your response. Specifically, will the data be treated as an *enumeration* of a population, or shall they be treated as a *sample*? If treated as an enumeration, then measures computed from the data (α and β) will be treated as *parameters.* But if the data represent a sample, the measures (a and b) com-

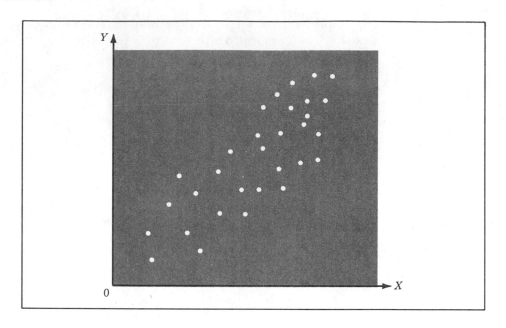

FIGURE 12-1. *Illustration of a Two-variable Scatter Diagram.*

puted will be treated as *estimators*. In most cases (perhaps in all cases), data used for correlation and regression analysis should be considered sample data. Most, if not all, of the time, data represent a slice (sample), either out of time or space; and the correlation and regression analysis projects to a much broader base—a population. Thus, using sample data we will determine estimates of α and β, denoted a and b, and form the *regression equation*.

$$Y_c = a + bX \qquad (12.2)$$

▶ **Example 12.1** The faculty at State U. has been concerned about the number of hours its students spend in the library. In fact, the Dean of Women has decided to do a study relating the number of hours spent in the library with the student grade point average. There are 1,000 students at State. While it might be expensive, a census of all 1,000 students could be taken. An estimate of the average number of hours spent per week in the library and the student's cumulative grade point averages could be obtained for all 1,000 students, and a correlation and regression analysis could then be made. In this case, can the Dean assume that she is dealing with a population and not a sample? After all, she has 1,000 values for average hours in the library and 1,000 values for grade point averages. In other words, she has information on all 1,000 students. Is she really dealing with a complete universe? Maybe not! Let us see what she might do with correlation and regression analysis and these data. ◀

278

If the Dean's analysis deals only with the present 1,000 pairs of values and nothing more, she is concerned with a population and any measures computed may be considered parameter values.   But, if she plans to make any projection or any inferences regarding the State U. students during other semesters, the Dean is dealing with a sample. Thus, any measures computed are sample estimators, not parameters. For instance, if the Dean generalizes (e.g., students who spend more time in the library make higher GPA's) in any way, she is treating the 1,000 observations as a sample.   Generalizations might extend across time—next semester or next year—or to other students or other campuses.   If the extension is to other semesters or other years (except for the time the actual study was performed), the data represent a sample out of time (a temporal sample).   If the results are extended to other campuses or other student bodies, the sample is a spatial one.   Moreover any time a projection of the relationship is extended beyond the limits of the observed data, the data must be considered as a sample and treated accordingly.

The second point to consider regarding the variables is how to determine which one is the independent variable and which one is the dependent variable.   In correlation and regression analysis, one variable is expressed as a function of the other in mathematical terms. But in terms of actual data, the analysis itself cannot determine which variable is a function of the other.   This determination is made on the basis of knowledge of the subject matter itself.   Consider Example 12.1.   Is the grade point average a function of the number of hours spent in the library, or is the number of hours spent in the library a function of the grade point average?   Which variable is the independent one and which is the dependent one?   Much depends on the relationship that is under consideration.   In this illustration, the faculty of State U. is probably trying to show that the more hours spent in the library, the higher the grade point average.   Thus, the number of hours spent in the library is treated as the independent variable, and the GPA is treated as the dependent variable.   Thus, hours are plotted on the X axis and grade points are plotted on the Y axis.

Purpose of the Scatter Diagram

The purpose of the scatter diagram is to provide a picture of any relationship between the independent and dependent variables.   To this extent, the scatter diagram may assist the analyst in three ways:

1. It indicates generally whether or not there is an apparent relationship between the two variables.
2. If there is a relationship, it may suggest whether it is *linear* or *nonlinear*.

3. If the relationship is linear, the scatter diagram will show whether the relationship is *positive* or *negative*.

Figure 12-2 shows four scatter diagrams. The first one (Figure 12-2a) shows two variables in which there is no apparent correlation. Figures 12-2b and 12-2c show linear correlation between the two variables plotted. In Figure 12-2b, the linear correlation is positive— shows a positive relationship, but the scatter diagram of Figure 12-2c indicates a negative relationship. The fourth illustration, Figure 12-2d, does not suggest a linear relationship at all. In fact, the scatter strongly indicates that there is a nonlinear relationship (curvilinear) between the X and Y variables. Since this chapter deals exclusively with simple linear correlation and regression analysis, we will wait until Chapter 13 for further discussion of this type of relationship.

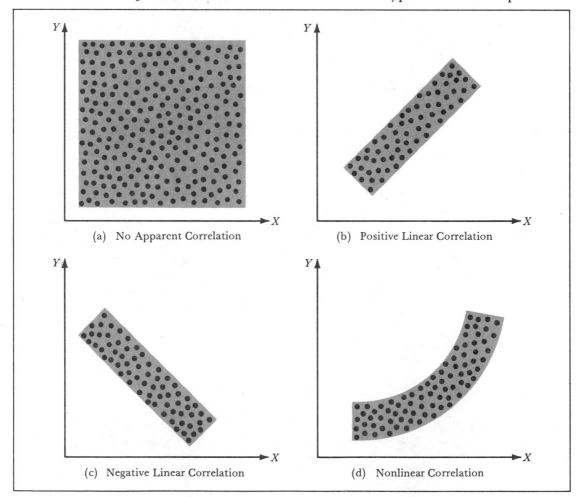

FIGURE 12-2. *Four Scatter Diagrams.*

The Regression Equation

The *regression equation* expresses the relationship between the X and Y variables more precisely. It is one thing to say, "Y is a function of X." It is another thing (more explicit) to express this relationship in a mathematical equation. Given the assumption of linearity, the equation showing the relationship between the X and Y variables is just a mathematical function for a straight line.

$$Y_c = a + bX \qquad (12.2)$$

In this equation we must distinguish between the actual dependent variable represented by Y, and the computed or regression value of the dependent variable, represented by Y_c. When the actual dependent variable Y is plotted with X, the result is the scatter diagram. However, when the Y_c value is plotted with X and the points are connected together, the result is called the *regression line*.

Let us look at the equation again:

$$Y_c = a + bX \qquad (12.2)$$

where Y_c = the estimate of the average value of Y for a specified value of X

X = the actual value of the independent variable

a = the estimate of the Y intercept (α) of the regression line (a constant)

b = the estimate of the slope (β) of the regression line (a constant)

The regression equation serves three important purposes:

1. It provides a precise mathematical definition of the relationship.
2. It represents a standard of perfect correlation.
3. It provides the basis for making predictions (Y_c) of the values of the dependent variable Y.

Mathematical Definition

One of the important functions of the regression equation, or the resulting regression line, is to provide a mathematical expression defining the relationship between the X and Y variables. The accuracy of this definition depends on the extent to which the equation or line actually represents the data. If the scatter diagram falls in a linear pattern, as in Figures 12-2b and 12-2c, the line may be reasonably representative of the actual data. On the other hand, a straight line fitted to the data in Figure 12-2d is not very representative, because the coordinated points of the scatter diagram do not form a linear pattern.

Thus, the more closely the scatter diagram conforms to a straight line (positive or negative) the more accurately the line describes the relationship.

Standard of Perfect Correlation

Another way to view the relationship between the dots in the scatter diagram and the regression line is that the regression line is a standard of perfect correlation. Thus the closer the dots conform to the line, the higher the correlation between X and Y. The regression line provides the base to which we compare the actual observations (the coordinate points) to determine the extent of the correlation. If the regression line has any slope to it (positive or negative) and all dots fall directly on it, there is perfect correlation. Of course, we must remember that the occurrence of perfect correlation is highly improbable. Rather, it is a defined state of perfection. As such, then, the regression line is a standard of perfect correlation.

Determination of the Regression Equation

There are several ways to determine the regression equation. One can simply "free-hand" a line that appears to represent the relationship between the two variables as displayed in a scatter diagram. The regression equation can then be written for the free-hand line. There are, however, some rather obvious problems with determining the regression line in this fashion. Specifically, the placement of the line depends upon individual judgment and no two people could be expected to draw the line in exactly the same place.

A frequently used technique for determining the regression equation is the *least-squares method*. Since the values of a, the ordinate intercept, and b, the slope, are mathematically determined, the results are reproducible. That is, two or more people using the same data set and the least-squares method will arrive at identical regression equations, barring arithmetic errors. The line determined using this technique is unique in that *the sum of the squared deviations* of the values of Y from the regression line are a *minimum*. Thus, the least-squares regression line and the arithmetic mean are similar in this respect. Moreover, the sampling distribution associated with a least-squares regression line has similar characteristics to the sampling distribution of the arithmetic mean.

Before you apply the least-squares method of regression to a set of data, you should consider the following assumptions associated with its use.

1. There is a linear relationship between X and Y.
2. Y is a random variable and the values of X are known constants.
3. For every value of X, there is a subset of values of the dependent variable Y, and each of these subsets of Y values has the same standard deviation σ. The condition described by this assumption, equal standard deviations, is called *homoscedasticity*.
4. The deviations $Y - Y_c$ between the actual values of Y and the estimated average value using regression are independent.
5. The subsets of Y values for each X value are normally distributed. (This assumption is necessary only if the researcher desires to make inferences about the population parameters.)

We sometimes apply the least-squares method to a data set that does not have all the characteristics described by these assumptions. When we do this, we are simply adding another assumption. That is, there is not enough difference to be important. However, the assumptions still exist and you should recognize that your results, and any decisions based upon them, may be incorrect if the data set does not meet the assumptions.

To compute the values of a and b in the linear regression equation, we use two *normal equations*. These basic normal equations are

$$\Sigma Y = na + b\Sigma X \tag{12.3}$$

$$\Sigma XY = a\Sigma X + b\Sigma X^2 \tag{12.4}$$

where X and Y refer to the actual values of the independent and dependent variables respectively, and n is the number of observations.

To use the normal equations in their present forms, with two unknowns in each equation, we must employ some method (such as the elimination procedure) to solve for a and b. Consider the following normal equations. By the elimination process, we compute the values of a and b as follows:

$$
\begin{aligned}
\text{I} \quad & 40 = 5a + 20b \\
\text{II} \quad & 100 = 20a + 140b
\end{aligned}
$$

1. Multiply equation I by -4 so the coefficients of a are the same in both equations.

$$-4\,(40) = -4\,(5a + 20b)$$

Thus,

$$
\begin{aligned}
\text{I} \quad & -160 = -20a - 80b \\
\text{II} \quad & 100 = 20a + 140b
\end{aligned}
$$

2. Sum the two equations. That is, fuse them into one equation with one unknown.

$$-60 = 60b$$

3. Solve for the remaining unknown.

$$b = -1$$

4. Substitute and solve for the other unknown.

$$-160 = -20a - 80\,(-1)$$
$$a = 12$$

5. Check results:

$$100 = 20\,(12) + 140\,(-1)$$
$$100 = 240 - 140$$

Another way to solve for two unknowns in two equations is to solve for one unknown in terms of the other in one equation and then substitute in the second equation. The results of such a procedure would be the following equations.

$$b = \frac{n\Sigma XY - (\Sigma X)\,(\Sigma Y)}{n\,(\Sigma X^2) - (\Sigma X)^2} \qquad (12.5)$$

$$a = \frac{\Sigma Y}{n} - b\left(\frac{\Sigma X}{n}\right) = \overline{Y} - b\overline{X} \qquad (12.6)$$

▶ **Example 12.2** Dime's Department Store has recently hired a consultant to help boost its sales and thereby the morale of its management (and hopefully its employees). The consultant, a marketing specialist, has reasoned that sales are a function of advertising expenditures. Thus he has decided to correlate sales (the dependent variable) and advertising expenditure (the independent variable) for the department store. The sales and advertising expenditure data from the company's records are given in Table 12-1. Looking at the sales and advertising data given in Table 12-1, we can now compute the regression equation for the consultant.

TABLE 12-1. *Sales and Advertising Expenditures Dime's Department Store.*

YEAR	ADVERTISING EXPENDITURES	SALES
1961	$100,000	$9,000,000
1962	105,000	8,000,000
1963	90,000	5,000,000

YEAR	ADVERTISING EXPENDITURES	SALES
1964	80,000	2,000,000
1965	80,000	4,000,000
1966	85,000	6,000,000
1967	87,000	4,000,000
1968	92,000	7,000,000
1969	90,000	6,000,000
1970	95,000	7,000,000
1971	93,000	5,000,000
1972	85,000	5,000,000
1973	85,000	4,000,000
1974	70,000	3,000,000
1975	85,000	3,000,000

X ($1,000)	Y ($1,000,000)	XY	X²	Y²
100	9	900	10,000	81
105	8	840	11,025	64
90	5	450	8,100	25
80	2	160	6,400	4
80	4	320	6,400	16
85	6	510	7,225	36
87	4	348	7,569	16
92	7	644	8,464	49
90	6	540	8,100	36
95	7	665	9,025	49
93	5	465	8,649	25
85	5	425	7,225	25
85	4	340	7,225	16
70	3	210	4,900	9
85	3	255	7,225	9
1,322	78	7,072	117,532	460

$$b = \frac{(15)\,(7072) - (1322)\,(78)}{(15)\,(117532) - (1322)^2} = 0.194$$

$$a = \frac{(78)}{(15)} - (0.194)\,\frac{(1322)}{(15)} = -11.9$$

$$Y_c = -11.8802 + 0.1938X \blacktriangleleft$$

Predictions Based on the Regression Equation

One of the purposes of regression analysis is to make predictions of the values of the dependent variable based on certain values of the independent variable. If we look at the equation

$$Y_c = -11.8802 + 0.1938X$$

we see that once we have computed values for a and b, we can assign values to X and solve for Y_c. Through this procedure Y_c becomes an estimate of the average value of Y given a value of X. Using the equation developed for the data in Example 12.2, the Y_c values may be computed as

X ($1,000)	Y_c ($1,000,000)
100	7.4998
105	8.4688
90	5.5618
80	3.6238
80	3.6238
85	4.5928
87	4.9804
92	5.9494
90	5.5618
95	6.5308
93	6.1432
85	4.5928
85	4.5928
70	1.6858
85	4.5928

By plotting Y_c and X, the regression line is determined. We need only plot any two points here, because the regression line is a straight line. However, it may be a good idea to plot at least three points to avoid mistakes. If there has been no error in computing Y_c, given the assigned values of X, the regression line should pass through all three points. Once the regression line is plotted (see Figure 12-3) it becomes fairly obvious that, in this example, there is reasonably high correlation.

Standard Error of Estimate

The *standard error of estimate is the standard deviation for the observed values of Y from the regression line.* It's a computed value that shows the average amount by which the individual Y values vary from the regression

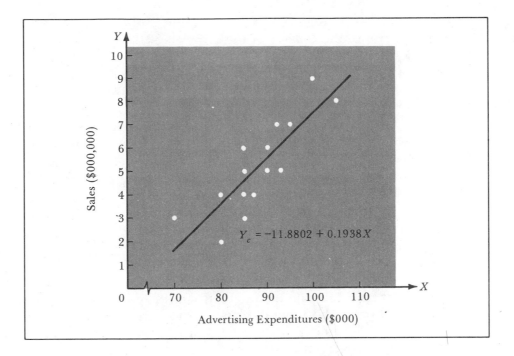

FIGURE 12-3. *Scatter Diagram and Regression Line.*

line. Graphically, the standard error of estimate shows, on the average, how much the individual dots (observations) in a scatter diagram vary from the regression line. Since the regression line represents a standard of perfect correlation, the larger the standard error of estimate relative to the magnitude of the Y_c values, the lower the coefficient of correlation. If the standard error of estimate were 0, all the dots in the scatter diagram would fall precisely on the regression line, and therefore there would be perfect correlation.

In this section we will present two formulas for computing the standard error of estimate. The first formula, Equation 12.7, is sometimes referred to as the long method and it is the one that most graphically represents the relationship between the individual Y values and the regression line values, Y_c.

$$s_{yx} = \sqrt{\frac{\Sigma\,(Y - Y_c)^2}{n - 2}} \qquad (12.7)$$

where s_{yx} = the standard error of estimate
Y = actual (observed) values of the dependent variable
Y_c = regression values computed from the regression equation
n = number of observations

As is shown in this equation, the standard error represents the sum of the squares of the differences between the actual values of the dependent variable Y and the estimates of values of the dependent variable Y_c computed from the regression equation. Thus, the standard error of estimate relates the actual dependent variable Y to the computed dependent variable Y_c. *The degrees of freedom in this case are represented by $n - 2$ because two parameters (α and β) have been estimated.* Using them in this equation recognizes the idea that we are computing an estimator of a parameter for the standard error of estimate. It assumes that we are involved in a sampling situation.

The short method for computing the standard error of estimate requires a longer formula, Equation 12.8, than Equation 12.7.

$$s_{yx} = \sqrt{\frac{\Sigma Y^2 - a\Sigma Y - b\Sigma XY}{n - 2}} \qquad (12.8)$$

Equation 12.7 requires that each value of Y_c be computed from the regression equation and then subtracted from Y and then squared; whereas the formula for the short method requires only one additional element, ΣY^2, to be computed. The other information in the formula has already been obtained in computing the regression equation. That is, we have already determined ΣY and ΣXY. Obviously we know the constants a and b from the regression equation. The value of ΣY^2 can be easily computed by squaring each value of Y and summing the squares. Both the long and short calculation methods are illustrated in Example 12.3.

▶ **Example 12.3**   The consultant for Dime's Department Store has decided he needs to measure the dispersion about the regression line.

Y	Y_c	$(Y - Y_c)$	$(Y - Y_c)^2$
9	7.4998	1.5002	2.2506
8	8.4688	−0.4688	0.2198
5	5.5618	−0.5618	0.3156
2	3.6238	−1.6238	2.6367
4	3.6238	0.3762	0.1415
6	4.5928	1.4072	1.9802
4	4.9804	−0.9804	0.9612
7	5.9494	1.0506	1.1038
6	5.5618	0.4382	0.1920
7	6.5308	0.4692	0.2201
5	6.1432	−1.1432	1.3069
5	4.5928	0.4072	0.1658
4	4.5928	−0.5928	0.3514
3	1.6858	1.3142	1.7271
3	4.5928	−1.5928	2.5370
			16.1097

Long Method

$$s_{yx} = \sqrt{\frac{\Sigma (Y - Y_c)^2}{n - 2}}$$

$$s_{yx} = \sqrt{\frac{16.1097}{15 - 2}}$$

$$s_{yx} = \sqrt{1.2392} = 1.113$$

Short Method

$$s_{yx} = \sqrt{\frac{\Sigma Y^2 - a\Sigma Y - b\Sigma XY}{n - 2}}$$

$$= \sqrt{\frac{460 - (-11.8802)\,(78) - (0.1938)\,(7072)}{15 - 2}}$$

$$s_{yx} = \sqrt{1.2386} = 1.113 \blacktriangleleft$$

Inference about Regression

Now that we have determined a, b, and s_{yx}, it is necessary to investigate whether the regression really exists.   That may seem to be an unnecessary step since you might be reasonably sure that your calculations of a and b are accurate.   However, if the data with which you are working are sample data, the values of a and b are only estimates of α and β, respectively.   Since the values in your data set are assumed to be randomly selected, the values you calculate for a and b may not accurately reflect α and β.   Put another way, we must ask whether b, the estimate of β, is really different from 0; *because if $\beta = 0$ there is no slope to the regression line and thus no relationship between X and Y.*

In order to determine if the relationship really exists, it is necessary to test the null hypothesis that $\beta = 0$ using the test statistic shown in Equation 12.9.   Also either Equation 12.10 or 12.11 has to be used

$$t_{\text{test}} = \frac{b - \beta_H}{s_b} \tag{12.9}$$

where b = estimate of β from normal equations
               β_H = hypothesized value of β
               s_b = standard error (standard deviation) of b

to calculate s_b, the standard error of the sample b's.   The degrees of freedom for the test are $n - 2$.   And finally, to legitimately use Equa-

$$s_b = \frac{S_{yx}}{\sqrt{\Sigma \, (X - \overline{X})^2}} \qquad (12.10)$$

$$s_b = \frac{S_{yx}}{\sqrt{\Sigma X^2 - \frac{(\Sigma X)^2}{n}}} \qquad (12.11)$$

tion 12.9, we must assume that the Y values will be normally distributed for any Y_c or have a large n in order to rely on the Central Limit Theorem.

▶ **Example 12.4**   The consultant for Dime's Department Store now must determine if the regression line he has developed really has a significant slope.

$$H_0: \quad \beta = 0$$

$$H_A: \quad \beta \neq 0$$

$$\alpha = .05$$

Reject H_0 if $t_{\text{test}} < -2.160$ or $t_{\text{test}} > +2.160$

$$t_{\text{test}} = \frac{b - \beta_H}{s_b}$$

$$t_{\text{test}} = \frac{b - \beta_H}{\dfrac{S_{yx}}{\sqrt{\Sigma X^2 - \dfrac{(\Sigma X)^2}{n}}}}$$

$$= \frac{0.1938 - 0}{\dfrac{1.113}{\sqrt{117532 - \dfrac{(1322)^2}{15}}}}$$

$$= \frac{0.1938 - 0}{\dfrac{1.113}{\sqrt{117532 - \dfrac{1747684}{15}}}}$$

$$t_{\text{test}} = \frac{0.1938 - 0}{\dfrac{1.113}{\sqrt{117532 - 116521.26}}}$$

$$= \frac{0.1938 - 0}{\dfrac{1.113}{\sqrt{1019.74}}}$$

$$= \frac{0.1938 - 0}{\dfrac{1.113}{31.933}}$$

$$= \frac{0.1938 - 0}{0.0349}$$

$$t_{\text{test}} = 5.559$$

Reject H_0, conclude that $\beta \neq 0$ and that the regression relationship does exist. ◄

Confidence Intervals

Earlier in the chapter we utilized the expression, $Y_c = a + bX$, to estimate values of Y for given values of X. Actually, these estimated values of Y, designated Y_c, are estimates of the *average value of Y for a given* X. If we are using sample data, we can use these averages to construct confidence intervals for the Y values for a given X. In particular, we will be able to construct two sets of confidence intervals. The first interval is for the average value of Y (denoted Y_c) that might have been obtained from the sample information, and the second interval is for an individual value of Y, which is sometimes referred to as the forecasted value of Y.

Confidence Interval for Y_c

Construction of a confidence interval for Y_c is actually a construction of a range for the possible average value for Y for a given X. Such an interval exists because from regression analysis, only *one* of many possible regression lines is obtained from a single sample. Since a and b are subject to sampling variation, any estimates that use a and b are subject to sampling error. Y_c reflects this variation, and so we need a measure of dispersion to account for this sampling error. This measure will be called the *standard error of* Y_c, and we will use Equation 12.12 to calculate this measure of dispersion. Then, to construct confidence intervals for Y_c, use Equation 12.13.

$$s_{yc} = s_{yx} \sqrt{\frac{1}{n} + \frac{(X - \overline{X})^2}{\Sigma X^2 - \frac{(\Sigma X)^2}{n}}} \qquad (12.12)$$

$(1 - \alpha)$ confidence interval for $Y_c = (a + bX) \pm ts_{yc}$ **(12.13)**

▶ **Example 12.5**   Continuing with Dime's Department Store, let us construct a 95 percent confidence interval for Y_c if X is 80 thousand dollars.

$$95\% \text{ C.I. for } Y_c = (a + bX) \pm ts_{yx} \sqrt{\frac{1}{n} + \frac{(X - \bar{X})^2}{\Sigma X^2 - \frac{(\Sigma X)^2}{n}}}$$

$$= [-11.8802 + (0.1938)(80)] \pm (2.16) \, 1.113 \sqrt{\frac{1}{15} + \frac{(80 - 88.13)^2}{117532 - 116512.27}}$$

$$= (-11.8802 + 15.504) \pm 2.4041 \sqrt{\frac{1}{15} + \frac{66.0969}{1019.73}}$$

$$= 3.6238 \pm 2.4041 \sqrt{0.0667 + 0.0648}$$

$$= 3.6238 \pm 2.4041 \, (0.3626)$$

$$= 3.6238 \pm 0.8717$$

$$2.7521 \le Y_c \le 4.4955$$

Thus, the average value of sales when the advertising expenditure is 80 thousand dollars has a lower limit of 2.7521 million dollars and an upper limit of 4.4955 million dollars. ◀

Confidence Interval for Y

The confidence interval previously constructed is for the average value of sales (Y_c) when X is a particular value. At times, however, it is more desirable to construct a confidence interval for the single *predicted* value (Y) rather than an average value. In order to do this it is necessary to have the proper measure of dispersion. Such a measure includes the sampling error present in s_{yc}, but it also must include the variation due to the individual observations being scattered about the regression line. We will call this measure of dispersion the *standard error of a predicted value of Y* and use Equation 12.14 to calculate s_{yp}.

$$s_{yp} = s_{yx} \sqrt{1 + \frac{1}{n} + \frac{(X - \bar{X})^2}{\Sigma X^2 - \frac{(\Sigma X)^2}{n}}} \qquad (12.14)$$

Then, to construct confidence intervals for a predicted value of Y for a given X, use Equation 12.15.

$$(1 - \alpha) \text{ confidence interval for } Y = (a + bX) \pm ts_{yp} \qquad (12.15)$$

▶ **Example 12.6** Let us now construct a 95 percent confidence interval for the predicted sales of Dime's Department Store when the advertising expenditure is 80 thousand dollars.

$$95\% \text{ C.I. for } Y = (a + bX) \pm ts_{yx} \sqrt{1 + \frac{1}{n} + \frac{(X - \overline{X})^2}{\Sigma X^2 - \frac{(\Sigma X)^2}{n}}}$$

$$= [-11.8802 + 0.1938 \,(80)] \pm (2.16)\, 1.113 \sqrt{1 + 0.0667 + 0.0648}$$

$$= 3.6238 \pm 2.4041 \sqrt{1.1315}$$

$$= 3.6238 \pm 2.4041 \,(1.0637)$$

$$= 3.6238 \pm 2.5572$$

$$1.0666 \leq Y \leq 6.1810 \blacktriangleleft$$

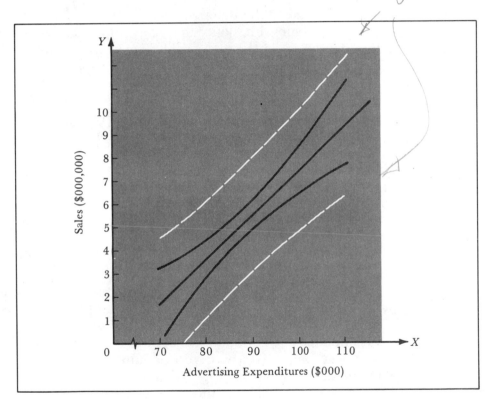

FIGURE 12-4. *Regression Line and Confidence Intervals.*

You should notice that the interval for the predicted value of an individual Y is larger than the interval for the average value of Y. This is true because predicting individual values is subject to more sampling variation, thus more variability, than predicting average values.

Figure 12-4 shows the situation. The dotted limits are for the predicted values of Y for the possible X's. Also the interval becomes larger as the value of the independent variable X moves away from $\overline{X}$. Thus, the predictors become more vulnerable or, suspect, as we let X vary more and more from $\overline{X}$.

Variation Analysis

The main objective of correlation analysis is to try to relate the changes in the values of the dependent variable Y with the movements or changes in the values of the independent variable X. That is, the dependent variable is varying (otherwise it would not be considered a variable), and we are concerned with the extent of the variation which can be related to the independent variable.

To begin with, notice that in Example 12.3 the standard error is the square root of the mean of the squared variations of Y and Y_c, adjusted for lost degrees of freedom. *The unexplained variation is that portion of the total variation of Y which is not associated with, or which cannot be related to, X.* Thus the more unexplained variation there is in analyzing a two-variable situation, the less correlation there is between the two variables X and Y under consideration. The concepts of unexplained variation, total variation, and explained variation will be presented within the next few pages.

Figure 12-5 shows a graphic analysis representing the three variations involved in a simple linear correlation and regression analysis. This figure depicts a scatter diagram with a horizontal line drawn through it at the value of the mean of Y (Figure 12-5a). Of course the mean line for the Y variable is a horizontal line, because the mean of Y is the same for every value of X. On the other hand, the regression line, which is shown in the Figure 12-5b diagram, would be a straight line with either a negative or positive slope. Since the total variation of Y represents the variation around the mean of Y, the total variation may be graphically depicted by the scatter about the horizontal (mean) line. We can depict the unexplained variation graphically by the scatter about the regression line. So, total variation is depicted by the scatter about the mean line, and unexplained variation is depicted by the scatter about the regression line.

Now let us consider the graphic representation of the explained variation. Perhaps it is not quite so obvious as the other two, but it is the variation in Y that is related to changes in the independent variable. This portion of the variation of the dependent variable can be viewed as the variation of Y_c about the mean and represented by the difference between the regression line and the mean line. In fact, the explained variation is graphically depicted by the distance between the mean line and the regression line at each value of X as shown in Figure 12-5c.

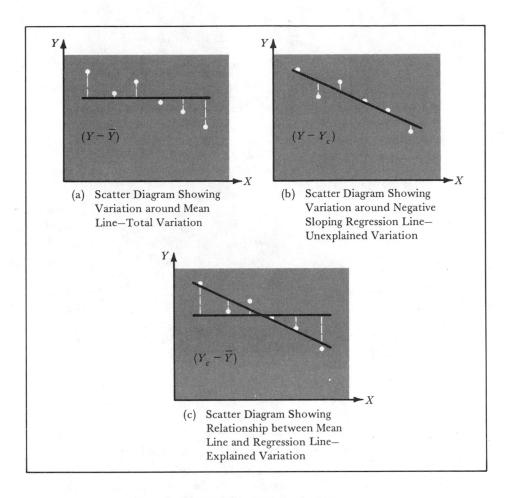

FIGURE 12-5. *Variation concepts.*

Another way to view the explained, unexplained, and total variation is in terms of *squared variation,* commonly referred to as *sum of squares.*   This is shown in Equations 12.16 through 12.18.

Total Sum of Squares

$$SST = \Sigma\,(Y - \overline{Y})^2 \qquad\qquad \textbf{(12.16)}$$

Regression (Explained) Sum of Squares

$$SSR = \Sigma\,(Y_c - \overline{Y})^2 \qquad\qquad \textbf{(12.17)}$$

Error (Unexplained) Sum of Squares

$$SSE = \Sigma\,(Y - Y_c)^2 \qquad\qquad \textbf{(12.18)}$$

As a matter of fact, the total sum of squares can be partitioned into two components: regression sum of squares and error sum of squares.

$$SST = SSR + SSE \tag{12.19}$$

or $$\Sigma \, (Y - \overline{Y})^2 = \Sigma \, (Y_c - \overline{Y})^2 + \Sigma \, (Y - Y_c)^2 \tag{12.20}$$

The calculations of the sum of squares for Dime's Department store are shown in Table 12-2.

TABLE 12-2. *Total, Regression, and Error Sum of Squares.*

Y	Y_c	$(Y - \overline{Y})$	$(Y - \overline{Y})^2$	$(Y - Y_c)$	$(Y - Y_c)^2$	$(Y_c - \overline{Y})$	$(Y_c - \overline{Y})^2$
9	7.4998	3.8	14.44	1.5002	2.2506	2.2998	5.2891
8	8.4688	2.8	7.84	−0.4688	0.2198	3.2688	10.6851
5	5.5618	−0.2	0.04	−0.5618	0.3156	0.3618	0.1309
2	3.6238	−3.2	10.24	−1.6238	2.6367	−1.5762	2.4844
4	3.6238	−1.2	1.44	0.3762	0.1415	−1.5762	2.4844
6	4.5928	0.8	0.64	1.4072	1.9802	−0.6072	0.3687
4	4.9804	−1.2	1.44	−0.9804	0.9612	−0.2196	0.0482
7	5.9494	1.8	3.24	1.0506	1.1038	0.7494	0.5616
6	5.5618	0.8	0.64	0.4382	0.1920	0.3618	0.1309
7	6.5308	1.8	3.24	0.4692	0.2201	1.3308	1.7710
5	6.1432	−0.2	0.04	−1.1432	1.3069	0.9432	0.8896
5	4.5928	−0.2	0.04	0.4072	0.1658	−0.6072	0.3687
4	4.5928	−1.2	1.44	−0.5928	0.3514	−0.6072	0.3687
3	1.6858	−2.2	4.84	1.3142	1.7271	−3.5142	12.3496
3	4.5928	−2.2	4.84	−1.5928	2.5370	−0.6072	0.3687
			54.40 (SST)		*16.1097 (SSE)		*38.2996 (SSR)

$$SST = SSR + SSE$$

*Rounding errors cause the sum of *SSR* and *SSE* to differ slightly from *SST*.

The Coefficients—Measuring the Relationship

Earlier in this chapter we said that correlation analysis was primarily concerned with measuring the relationship between the variables. For instance, we may ask, "How will income levels and education levels correlate?" In correlation analysis such relationships between variables are measured by computing certain coefficients. There are four such coefficients associated with simple linear correlation and regression analysis:[1]

---

1.   In this text, r, r^2, k and k^2 are all sample statistics. To be more precise estimates of the population parameters, the sample statistics should be adjusted for degrees of freedom.

1. the coefficient of correlation, r
2. the coefficient of determination, r^2
3. the coefficient of alienation, k
4. the coefficient of nondetermination, k^2

While the *coefficient of correlation, r,* is perhaps the best known of the four, we shall soon learn that the *coefficient of determination* is more meaningful. Also note that the coefficient of correlation, r, is the square root of the coefficient of determination, r^2. The same relationship exists between the *coefficient of alienation* and the *coefficient of nondetermination.*

Computation and Meaning of the Coefficients

The coefficient of determination, r^2, is defined as the proportion (or percent) of the variation in the values of the dependent variable that is "explained" by variations in the values of the independent variable. Thus, if the total variation squared (SST) is 54.40 and the explained variation squared (SSR) is 38.2996, then the coefficient of determination equals 0.704 or 70.4 percent. That is:

$$r^2 = \frac{\Sigma (Y_c - \overline{Y})^2}{\Sigma (Y - \overline{Y})^2} \qquad (12.21)$$

$$r^2 = \frac{38.2996}{54.40}$$

$$r^2 = 0.704 \text{ or } 70.4\%$$

Or using computational formulas for determining the numerator and denominator, Equation 12.21 becomes

$$r^2 = \frac{SSR}{SST} = \frac{SST - SSE}{SST} = \frac{\left[\Sigma Y^2 - \frac{(\Sigma Y)^2}{n}\right] - \left[\Sigma Y^2 - a\Sigma Y - b\Sigma XY\right]}{\Sigma Y^2 - \frac{(\Sigma Y)^2}{n}} \qquad (12.22)$$

$$= \frac{\left[460 - \frac{(78)^2}{15}\right] - \left[460 - (-11.8802)(78) - (0.1938)(7072)\right]}{460 - \frac{(78)^2}{15}}$$

$$= \frac{[460 - 405.6] - [460 + 926.6556 - 1370.5536]}{460 - 405.60}$$

$$r^2 = \frac{54.40 - 16.102}{54.40} = \frac{38.298}{54.40} = 0.704$$

And if the actual sums of squares are not needed, the following equation (called the product-moment formula) is a more direct method for determining r and r^2. This equation, however, does not distinguish between the dependent and independent variables.

$$r = \frac{n\Sigma XY - \Sigma X \Sigma Y}{\sqrt{[n\Sigma X^2 - (\Sigma X)^2][n\Sigma Y^2 - (\Sigma Y)^2]}} \qquad (12.23)$$

$$r^2 = (r)^2$$

$$r = \frac{15(7072) - (1322)(78)}{\sqrt{[15(117532) - (1322)^2][15(460) - (78)^2]}}$$

$$= \frac{106080 - 103116}{\sqrt{(1762980 - 1747684)(6900 - 6084)}}$$

$$= \frac{2964}{\sqrt{(15296)(816)}} = \frac{2964}{\sqrt{12481536}} = \frac{2964}{3533}$$

$$r = 0.839 \qquad r^2 = 0.704$$

The coefficient of correlation is an abstract measure of the degree of relationship between variables. The measure is based on a scale value between 0 and 1 in the case of positive correlation and 0 and −1 in the case of negative correlation. A value of 0 means there is no correlation, and a value of 1 or −1 means perfect correlation. The sign of r (+ or −) depends on the sign of "b" in the regression equation. If "b" is positive, the correlation is positive; if "b" is negative, the correlation is inverse or negative.

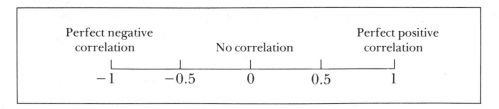

Like all scale values, the coefficient of correlation is difficult to interpret. Unlike the coefficient of determination, it cannot be given a direct interpretation. Yet, we can say that, in general, the higher the absolute value (to allow for negative correlation) of the coefficient, the higher the correlation.

The coefficient of nondetermination is the proportion of the variation in the dependent variable that is not explained by changes in the values of the independent variable. It is equal to the unexplained variation divided by the total variation. Using Example 12.3,

$$k^2 = \frac{\Sigma (Y - Y_c)^2}{\Sigma (Y - \overline{Y})^2.} \qquad (12.24)$$

$$k^2 = \frac{16.09}{54.40}$$

$$k^2 = 0.296 \text{ or } 29.6\%$$

From our previous discussion of variations, it can be shown that there is another way to calculate the coefficient of nondetermination. The two coefficients (determination and nondetermination) must sum to one (1) if all variation is either explained or unexplained. Thus, if r^2 is already known, then Equation 12.25 may be used.

$$k^2 = 1 - r^2 \qquad (12.25)$$

The coefficient of alienation is an abstract measure of the lack of correlation, and is equal to the square root of the coefficient of non-determination. That is, $k = \sqrt{k^2}$. Like the coefficient of correlation, k is a rating on a scale. But unlike r, k cannot be positive or negative. It is an unsigned (absolute value) rating. The sign attached to r means direction of a relationship. The coefficient of alienation k means the absence of a relationship, no direction; so it can take on no sign.

Earlier we discussed the relationship between correlation and regression analysis and sampling. If we consider that we are dealing with a sample situation in the analysis, then the coefficients that we computed are really not parameter values, but they are estimators based on the observed (sample) data. However, r is a biased estimator of the population coefficient of correlation in that the sample coefficient of correlation tends to be greater than the population value. This is particularly true when you are working with small samples of population values. While we will not present the method here, in more advanced texts you will find a technique for adjusting r so that it becomes an unbiased estimator of the population coefficient of correlation.

Types of Relationships

One frequent error we are prone to make in correlation and regression analysis is to interpret a high correlation between the variables as a cause-and-effect relationship. Correlation analysis measures a relationship; it does not define the basis for it. Thus, a high coefficient of determination is interpreted to mean a high proportion of explained variation, but the term "explained" as used in this context does not mean "causal." It means that the variables are closely related in some undefined way. The relationship between the variables may be purely coincidental (a casual relationship), or it may be due to seriality where some third phenomenon or variable is providing the basis for the relationship.

Suppose it can be shown that there is high correlation between the salaries of professional marriage counselors and the divorce rate in America. It is highly unlikely that one variable here is causing the other. They may show high, positive correlation because each variable is being affected by a changing cultural pattern in the country—a serial effect. Yet, we might find that the high negative correlation between popcorn sales and the incidence of television viewing is causal since popcorn sales occur at movie theaters.

In no case does correlation analysis itself, then, define the type of relationship. At best, one might say there must be correlation for a meaningful relationship to exist. Thus, if two variables show little or no correlation, it would be difficult to conclude that change in one is possibly causing change in the other.

A Complete Example

A student organization at State U. is interested in the relationship between student grade point averages (GPA) and the score made on the school's entrance examination. Moreover, the organization leaders are also wondering whether any real meaningful predictions of students' GPA's can be made from these entrance test scores.

To answer the questions, an investigation of the relationship between GPA and entrance test scores was undertaken. A sample of 30 records has been randomly selected from the entire senior class (all fourth-year students). The GPA's and entrance test scores for the 30 students are shown below:

STUDENT NUMBER	ADMISSION TEST SCORE (200 PTS. POSSIBLE) (X)	GPA (4.0 POSSIBLE) (Y)
128	92	1.75
934	145	2.81
156	180	3.50
408	142	2.75
389	102	1.95
515	77	1.65
423	155	3.20
056	135	2.80
109	70	1.40
310	170	3.49
915	72	1.54
405	105	2.00
985	168	3.46

(cont'd)

STUDENT NUMBER	ADMISSION TEST SCORE (200 PTS. POSSIBLE) (X)	GPA (4.0 POSSIBLE) (Y)
361	87	1.65
020	112	2.35
490	108	2.05
581	96	2.02
095	100	1.91
430	175	3.60
756	132	2.75
615	165	2.41
324	148	3.05
049	170	3.29
519	101	1.93
804	102	3.15
439	119	2.28
650	130	2.50
522	168	3.45
888	185	3.75
158	180	2.75

1. Figure 12-6 shows a scatter diagram between test scores (X) and the dependent variable, GPA (Y).

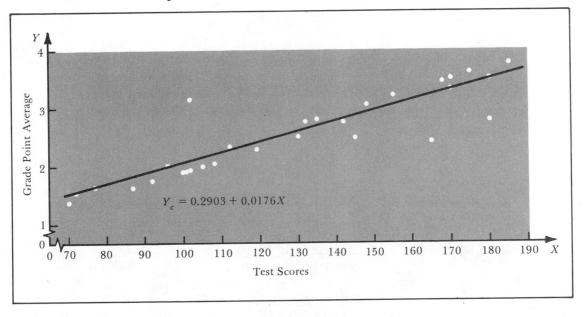

FIGURE 12-6. *Scatter Diagram and Regression Line Showing Pattern of Relationship Between Grade Point Average and Entrance Test Scores at State U for 30 Randomly Selected Seniors.*

Also the following values are determined from the sample data.

$$n = 30 \qquad\qquad \Sigma Y^2 = 213.12$$
$$\overline{X} = 129.7 \qquad\qquad \Sigma X = 3891$$
$$\overline{Y} = 2.573 \qquad\qquad \Sigma XY = 10677.68$$
$$\Sigma Y = 77.19 \qquad\qquad \Sigma X^2 = 542455$$

2. The regression equation is computed as follows:

$$b = \frac{30\,(10677.68) - (3891)\,(77.19)}{30\,(542455) - (3891)^2}$$

$$b = \frac{19984.11}{1133769} = 0.0176$$

Since $\dfrac{\Sigma Y}{n} = \overline{Y}$ and $\dfrac{\Sigma X}{n} = \overline{X}$

$$a = 2.573 - 129.7\,(0.0176)$$

$$a = 0.2903$$

$$Y_c = 0.2903 + 0.0176X$$

The regression line is shown in Figure 12-6.

3. The standard error of estimate is:

$$s_{yx} = \sqrt{\frac{213.12 - (77.19)\,(0.2903) - (10677.68)\,(0.0176)}{30 - 2}}$$

$$s_{yx} = \sqrt{\frac{2.78}{28}} = \sqrt{0.0993} = 0.3151$$

4. The correlation coefficients are computed as follows:

$$r = \frac{n\Sigma XY - \Sigma X \Sigma Y}{\sqrt{[n\Sigma X^2 - (\Sigma X)^2]\,[n\Sigma Y^2 - (\Sigma Y)^2]}}$$

$$r = \frac{30\,(10677.68) - (3891)\,(77.19)}{\sqrt{[30\,(542455) - (3891)^2]\,[30\,(213.12) - (77.19)^2]}}$$

$$= \frac{320330.4 - 300346.29}{\sqrt{(16273650 - 15139881)\,(6393.6 - 5958.3)}}$$

$$= \frac{19984.11}{\sqrt{(1133769)\,(435.3)}} = \frac{19984.11}{\sqrt{493529645.7}}$$

$$r = \frac{19984.11}{22215.52} = 0.90$$

$$r^2 = (r)^2$$

$$r^2 = 0.81$$

$$k^2 = 1 - r^2$$

$$= 1 - 0.81$$

$$k^2 = 0.19$$

$$k = \sqrt{k^2}$$

$$k = 0.4359$$

5. If we want to test the value of b, then:

$$H_0: \quad \beta = 0$$

$$H_A: \quad \beta \neq 0$$

$$\alpha = 0.05$$

Reject H_0 if $t_{\text{test}} < -2.048$ or $t_{\text{test}} > 2.048$.

$$t_{\text{test}} = \frac{b - \beta_H}{\dfrac{s_{yx}}{\sqrt{\Sigma X^2 - \dfrac{(\Sigma X)^2}{n}}}}$$

$$t_{\text{test}} = \frac{0.0176 - 0}{\dfrac{0.3151}{\sqrt{542455 - \dfrac{(3891)^2}{30}}}}$$

$$= \frac{0.0176}{\dfrac{0.3151}{\sqrt{37792.3}}}$$

$$t_{\text{test}} = \frac{0.0176}{0.0016} = 11.0$$

Reject H_0 and conclude the regression relationship does exist.

6. To make a 95 percent confidence interval estimate of the mean of Y_c at $X = 150$:

$$(1 - \alpha) \text{ C.I. for } Y_c = [a + bX] \pm ts_{yx} \sqrt{\frac{1}{n} + \frac{(X - \overline{X})^2}{\Sigma X^2 - \dfrac{(\Sigma X)^2}{n}}}$$

$$= [0.2903 + 0.0176\,(150)]$$

$$\pm\,2.048\,(0.3151)\,\sqrt{\frac{1}{30} + \frac{(150 - 129.7)^2}{542455 - \frac{(3891)^2}{30}}}$$

$$= [0.2903 + 2.64] \pm 0.645\,\sqrt{\frac{1}{30} + \frac{412.09}{37792.3}}$$

$$= 2.93 \pm 0.645\,\sqrt{0.033 + 0.011}$$

$$= 2.93 \pm 0.645\,\sqrt{0.044}$$

$$= 2.93 \pm 0.645\,(0.210)$$

$$= 2.93 \pm 0.135$$

$$2.795 \le Y_c \le 3.065$$

7. To determine the 95 percent confidence interval estimate for a single value of Y when $X = 150$:

$$(1 - \alpha)\ \text{C.I. for } Y = [a + bX] \pm ts_{yx}\,\sqrt{1 + \frac{1}{n} + \frac{(X - \overline{X})^2}{\Sigma X^2 - \frac{(\Sigma X)^2}{n}}}$$

$$= [0.2903 + 0.0176\,(150)]$$

$$\pm\,2.048\,(0.3151)\,\sqrt{1 + \frac{1}{30} + \frac{(150 - 129.7)^2}{542455 - \frac{(3891)^2}{30}}}$$

$$= [0.2903 + 2.64] \pm 0.645\,\sqrt{\frac{31}{30} + \frac{412.09}{37792.3}}$$

$$= 2.93 \pm 0.645\,\sqrt{1.033 + 0.011}$$

$$= 2.93 \pm 0.645\,\sqrt{1.044}$$

$$= 2.93 \pm 0.645\,(1.022)$$

$$= 2.93 \pm 0.659$$

$$2.271 \le Y \le 3.589$$

     This example can also serve to illustrate another point. The highest possible score on the entrance test is 200 points, and the highest grade-point average is 4.0 Thus X cannot exceed 200 and Y cannot be greater than 4.0. In other words, there are limits to this particular regression equation.

APPENDIX TO CHAPTER 12

Testing β with Analysis of Variance

In the Appendix at the end of Chapter 9, there is a discussion of analysis of variance and the F distribution. This procedure provides an alternative for testing the null hypothesis of $\beta = 0$. Analysis of variance yields the same results as the t-test in Equation 12.9 when you are testing the slope of a simple linear regression equation. However, analysis of variance has broader possibilities for use in that it can be applied to multiple regression equations to test several independent variables at one time, while the t-test can be applied only to each independent variable separately.

When testing regression equations, analysis of variance utilizes the regression sum of squares and the error sum of squares. Each sum of squares is divided by the appropriate degrees of freedom to calculate what are known as mean squares. The two mean squares are called *regression mean square (MSR)* and *error mean square (MSE)* and are shown in Equations 12.26 and 12.27.

$$MSR = \frac{SSR}{m} \tag{12.26}$$

where $\quad SSR = \Sigma\,(Y_c - \overline{Y})^2 \tag{12.17}$

$$MSE = \frac{SSE}{n - m - 1} \tag{12.27}$$

where $\quad SSE = \Sigma\,(Y - Y_c)^2 \tag{12.18}$
and $\quad\quad m$ = number of independent variables
$\quad\quad\quad\; n$ = sample size

These mean squares are sample variances and can be used in the F ratio to test for regression. The test statistic is:

$$F_{\text{test}} = \frac{MSR}{MSE} \tag{12.28}$$

and the results are frequently presented in an ANOVA summary table as shown in Table 12-3.

For Table 12-3, the following computational equations can be used to determine the sum of squares.

$$SST = \Sigma Y^2 - \frac{(\Sigma Y)^2}{n} \tag{12.29}$$

$$SSE = \Sigma Y^2 - a\Sigma Y - b\Sigma XY \tag{12.30}$$

$$SSR = SST - SSE \tag{12.31}$$

TABLE 12-3. *Anova Summary Table: General Presentation.*

SOURCE OF VARIATION	SUM OF SQUARES	DEGREES OF FREEDOM	MEAN SQUARES	F_{test}
Regression	SSR	m	$MSR = \dfrac{SSR}{m}$	$\dfrac{MSR}{MSE}$
Error	SSE	$n - m - 1$	$MSE = \dfrac{SSE}{n - m - 1}$	
Total	SST	$n - 1$		

TABLE 12-4. *Anova Summary Table.*

SOURCE OF VARIATION	SUM OF SQUARES	DEGREES OF FREEDOM	MEAN SQUARES	F_{test}
Regression	38.298	1	38.298	30.92
Error	16.102	13	1.2386	
Total	54.40	14		

Thus, using the data in Table 12-2, we can test β in the following manner.

$$H_0: \quad \beta = 0$$
$$H_A: \quad \beta \neq 0$$
$$\alpha = 0.05$$

Reject H_0 if $F_{\text{test}} > 4.67$.

$$SST = \Sigma Y^2 - \frac{(\Sigma Y)^2}{n} = 460 - \frac{(78)^2}{15}$$

$$SST = 54.40$$

$$SSE = \Sigma Y^2 - a\Sigma Y - b\Sigma XY$$

$$= 460 - (-11.8802)(78) - 0.1938(7072)$$

$$= 460 + 926.6556 - 1370.5536$$

$$SSE = 16.102$$

$$SSR = 54.40 - 16.102 = 38.298$$

$$F_{\text{test}} = \frac{\dfrac{38.298}{1}}{\dfrac{16.102}{13}} = 30.92$$

Since $F_{\text{test}} > 4.67$, reject H_0.

In the above example, the critical value of F (4.67) was found by using the table in Appendix K. The degrees of freedom are m and $(n - m - 1)$, with $d.f. = m$ across the top and $d.f. = (n - m - 1)$ down the side.

If you look back at Example 12.4, you can see that the results using analysis of variance are consistent with that achieved using the t-test.

SUMMARY OF EQUATIONS

12.1 $\quad Y = \alpha + \beta X,$ *page 277*

12.2 $\quad Y_c = a + bX,$ *page 278*

12.3 $\quad \Sigma Y = na + b\Sigma X,$ *page 283*

12.4 $\quad \Sigma XY = a\Sigma X + b\Sigma X^2,$ *page 283*

12.5 $\quad b = \dfrac{n\Sigma XY - (\Sigma X)(\Sigma Y)}{n(\Sigma X^2) - (\Sigma X)^2},$ *page 284*

12.6 $\quad a = \dfrac{\Sigma Y}{n} - b\left(\dfrac{\Sigma X}{n}\right) = \overline{Y} - b\overline{X},$ *page 284*

12.7 $\quad s_{yx} = \sqrt{\dfrac{\Sigma(Y - Y_c)^2}{n - 2}},$ *page 287*

12.8 $\quad s_{yx} = \sqrt{\dfrac{\Sigma Y^2 - (a\Sigma Y) - (b\Sigma XY)}{n - 2}},$ *page 288*

12.9 $\quad t_{\text{test}} = \dfrac{b - \beta_H}{s_b},$ *page 289*

12.10 $\quad s_b = \dfrac{s_{yx}}{\sqrt{\Sigma(X - \overline{X})^2}},$ *page 290*

12.11 $\quad s_b = \dfrac{s_{yx}}{\sqrt{\Sigma X^2 - \dfrac{(\Sigma X)^2}{n}}},$ *page 290*

12.12 $\quad s_{yc} = s_{yx}\sqrt{\dfrac{1}{n} + \dfrac{(X - \overline{X})^2}{\Sigma X^2 - \dfrac{(\Sigma X)^2}{n}}},$ *page 291*

12.13 $\quad (1 - \alpha)$ confidence interval for $Y_c = (a + bX) \pm ts_{yc},$ *page 291*

12.14 $\quad s_{yp} = s_{yx}\sqrt{1 + \dfrac{1}{n} + \dfrac{(X - \overline{X})^2}{\Sigma X^2 - \dfrac{(\Sigma X)^2}{n}}},$ *page 292*

12.15 $\quad (1 - \alpha)$ confidence interval for $Y = (a + bX) \pm ts_{yp},$ *page 292*

12.16 $SST = \Sigma (Y - \overline{Y})^2,$ *page 295*

12.17 $SSR = \Sigma (Y_c - \overline{Y})^2,$ *page 295*

12.18 $SSE = \Sigma (Y - Y_c)^2,$ *page 295*

12.19 $SST = SSR + SSE,$ *page 296*

12.20 $\Sigma (Y - \overline{Y})^2 = \Sigma (Y_c - \overline{Y})^2 + \Sigma (Y - Y_c)^2,$ *page 296*

12.21 $r^2 = \dfrac{\Sigma (Y_c - \overline{Y})^2}{\Sigma (Y - \overline{Y})^2},$ *page 297*

12.22 $r^2 = \dfrac{SSR}{SST} = \dfrac{SST - SSE}{SST} = \dfrac{\left[\Sigma Y^2 - \dfrac{(\Sigma Y)^2}{n}\right] - [\Sigma Y^2 - a\Sigma Y - b\Sigma XY]}{\Sigma Y^2 - \dfrac{(\Sigma Y)^2}{n}},$

page 297

12.23 $r = \dfrac{n\Sigma XY - \Sigma X\Sigma Y}{\sqrt{[n\Sigma X^2 - (\Sigma X)^2][n\Sigma Y^2 - (\Sigma Y)^2]}},$ *page 298*

12.24 $k^2 = \dfrac{\Sigma (Y - Y_c)^2}{\Sigma (Y - \overline{Y})^2},$ *page 299*

12.25 $k^2 = 1 - r^2,$ *page 299*

12.26 $MSR = \dfrac{SSR}{m},$ *page 305*

12.27 $MSE = \dfrac{SSE}{n - m - 1},$ *page 305*

12.28 $F_{\text{test}} = \dfrac{MSR}{MSE},$ *page 305*

12.29 $SST = \Sigma Y^2 - \dfrac{(\Sigma Y)}{n},$ *page 305*

12.30 $SSE = \Sigma Y^2 - a\Sigma Y - b\Sigma XY,$ *page 305*

12.31 $SSR = SST - SSE,$ *page 305*

REVIEW QUESTIONS

1. Explain the use of each equation at the end of the chapter.

2. Define
   a. regression
   b. correlation
   c. scatter diagram
   d. standard error of the estimate
   e. coefficient of determination
   f. coefficient of nondetermination

3. What is the explained variation?
4. What is the unexplained variation?
5. What is the purpose of testing the null hypothesis $\beta = 0$?
6. In the regression equation $Y_c = a + bX$, what are the a and b?

------------------------------ **PROBLEMS** ------------------------------

1. A company has recorded the following data on performance rating and employment test scores. (Save your results for use in Problem 2.)

PERFORMANCE (Y)	TEST SCORES (X)
10	3
15	5
25	6
40	8
30	11
50	15

a. Plot a scatter diagram for the data.
b. Determine the regression equation.
c. Is the regression significant at $\alpha = 0.05$?
d. Estimate the performance rating for an employee with a test score of 12.
e. Construct a 95 percent confidence interval for Y_c when $X = 12$.
f. Construct a 95 percent confidence interval for a predicted value of Y when $X = 12$.

2. For the above data determine the coefficient of correlation and the coefficient of determination.

3. A series of tests to investigate the relationship between highway speeds and gasoline mileage yielded the following results. (Save your results for use in Problems 4 and 5.)

X SPEED (MPH)	Y GASOLINE MILEAGE (MPG)
50	16
90	8
70	13
60	15
80	10
55	15
65	14
45	18
75	12
85	10

a. Plot a scatter diagram for the data.
b. Determine the regression equation.
c. Test the null hypothesis $\beta = 0$ at the 0.05 level of significance.

4. Use the data and results of Problem 3 to
   a. Construct a 99 percent confidence interval for Y_c when $X = 60$.
   b. Construct a 99 percent confidence interval for an individual value of gasoline mileage when speed = 60.

5. Determine the coefficients of correlation and determination for the data in Problem 3.

6. Assume a sample was taken to collect data on family income and food expenditures for home consumption. The results are given below. (Save your work to use in Problem 7.)

X ($1000) FAMILY INCOME	Y ($1000) EXPENDITURE
24.0	3.6
15.0	2.9
18.2	2.9
12.6	2.6
8.0	2.0
9.5	2.4
21.0	3.0
11.4	2.5
6.4	1.8
13.2	2.4

a. Plot a scatter diagram for the data.
b. Determine the regression equation.
c. Test for regression with $\alpha = 0.05$.
d. Determine the coefficient of determination.

7. Use the data and results of Problem 6 to
   a. Estimate the average expenditure for home consumption if the family income is $20,000.
   b. Construct a 95 percent confidence interval for the value estimated in part (a).
   c. Construct a 95 percent confidence interval for the forecasted home consumption expenditure when family income is $20,000.

8. Data have been collected on family income and away-from-home food expenditures. (Save your results for use in Problem 9.)
   a. Plot a scatter diagram for the data.
   b. Determine the regression equation.
   c. Test the null hypothesis, $\beta = 0$, at the 0.10 level of significance.
   d. Determine the coefficient of determination.

Y ($1000) EXPENDITURES	X ($1000) FAMILY INCOME
1.8	24.0
0.9	15.0
1.1	18.2
0.6	12.6
0.3	8.0
0.4	9.5
1.2	21.0
0.8	11.4
0.1	6.4
0.7	13.2

9. Use the data and results of Problem 8 to:
   a. estimate the average away-from-home expenditure for a family with annual income of $20,000.
   b. construct a 90 percent confidence interval for the value estimated in part (a).
   c. construct a 90 percent interval for the predicted away-from-home expenditure when family income is $20,000.

10. Delivery, Ltd. has recorded data on delivery time and weight load on deliveries that are about the same distance.

Y DELIVERY TIME (HOURS)	X WEIGHT (HUNDREDS OF POUNDS)
2	20
3	22
1.5	15
6	45
4	32
2.5	29
3	25
5	40

   a. Determine the regression equation.
   b. Calculate the standard error of the estimate.
   c. Test for regression at the 0.05 level of significance.
   d. Determine the coefficient of determination.

11. $n = 12$ $\Sigma XY = 2000$
    $\overline{Y} = 19$ $\Sigma Y^2 = 4612$
    $\overline{X} = 8$ $\Sigma X^2 = 1502$
    a. Determine the regression equation.
    b. Test for regression at the 0.05 level of significance.

12. If $n = 25$, $\Sigma (Y - \overline{Y})^2 = 500$, $\Sigma (Y - Y_c)^2 = 125$, and $\Sigma (Y_c - \overline{Y})^2 = 375$:

    a. Determine the standard error of the estimate.

    b. Determine the coefficient of determination.

    c. Determine the coefficient of nondetermination two ways.

13. If $n = 100$, $\Sigma (Y - \overline{Y})^2 = 8000$, and $\Sigma (Y_c - \overline{Y})^2 = 7000$:

    a. Determine the coefficient of correlation.

    b. Determine the standard error of the estimate.

14. The owner of the Freshaire Hotel has been concerned with the number of complaints received from guests about poor room service. The hotel has 400 rooms on floors 2 through 7. For the past two months records have been maintained according to the number of room service complaints by floor. These data are shown below. Is there a relationship between floor number and number of complaints? Set $\alpha = 0.05$.

X FLOOR NUMBER	Y NUMBER OF COMPLAINTS
2	50
3	40
4	45
5	36
6	32
7	25

15. Management of the Postal Inspection Service is interested in predicting the salary level of employees based on years of education. A random sample of 10 employees yielded the following results:

YEARS OF EDUCATION	SALARY ($1,000)
12.5	20.2
14.0	23.7
12.0	19.8
16.0	21.5
11.0	21.8
16.0	35.7
18.0	39.2
11.2	22.6
15.0	25.9
12.7	17.3

    a. Plot a scatter diagram for the data.

    b. Determine the regression equation.

    c. Test the null hypothesis, $\beta = 0$, at the 0.05 level of significance.

    d. Determine the coefficient of determination.

    e. Based on the results from parts (c) and (d), what can you tell management about using years of education to predict salaries.

16. Henning and Kazmier recently conducted a study to compare I.Q. scores made by learning disabled students on both the Peabody Picture Vocabulary Test Form and the Wechsler Intelligence Scale for Children-Revised Full-Scale I.Q. test.   Nineteen verified learning disabled students, ages 7 through 12 from three Hamilton County, Tennessee, public schools, were included in the study.   The following data were obtained:

PVVT	WISC-R FULL SCALE
125	89
107	71
121	90
105	80
108	88
107	88
104	87
88	76
93	82
90	81
93	84
93	88
86	84
109	110
100	102
78	82
78	84
77	85
85	96

a. Plot a scatter diagram for the data.
b. Determine the regression equation using WISC-R Full-Scale as the dependent variable.
c. Test for regression at the 0.05 level of significance.
d. Determine the coefficient of correlation.
e. Based on the results of the previous four parts of this problem, do you think it is necessary to give students both examinations?   Why?

17. A local law enforcement officer has noticed what he believes to be a direct relationship between the maximum temperature on a given day and the incidence of major crimes in his city.   He has selected a random sample of 10 days and determined both the high temperature and the number of major crimes.

Use linear least-squares regression.
a. At $\alpha = 0.05$, is the officer correct?
b. As a predictor of major crimes, what is the quality of daily high temperatures?
c. With 95 percent confidence, what is the average number of crimes the officer should expect when the high temperature is 80 degrees?

d. If the officer wants to have enough manpower on hand to handle the number of major crimes 95 percent of the time, how many crimes should he have officers for when he expects the temperature to be 60 degrees?

HIGH TEMPERATURES (DEGREES)	NUMBER OF MAJOR CRIMES
12	2
89	15
40	6
52	8
75	14
60	12
50	7
20	3
32	5
90	16

Correlation and Regression Analysis: The Nonlinear and Multivariate Cases

CHAPTER
13

In the previous chapter, the subject of correlation and regression analysis was discussed under the following assumptions:

1. The regression equation was assumed to be linear—a straight line.
2. There were two variables—a dependent variable and an independent one.

Relationships in the real world are frequently neither linear nor confined to two variables. Thus, in this chapter we relax the two assumptions made in the preceding chapter and look at correlation and regression analysis where

1. The relationship is *nonlinear* but there are *only two variables*.
2. The relationship is *linear* but there are more *than two variables*—one dependent variable and several independent ones.

Nonlinear Two-Variable Cases

There may be times when the two variables which we are attempting to correlate should not be described by a linear regression equation simply because the relationship is not linear. In such cases, it would not be appropriate to use a straight line as the standard of perfect correlation. For instance, we might find that, while there is a definite relationship between X and Y, Y is changing geometrically (by a given rate) while X is changing arithmetically (by a given amount).   Or, we might find that X and Y are related, but their relationship is simply nonlinear.   Thus, using a regression equation, which depicts the dependent variable as changing by varying amounts as the independent variable varies, would be more appropriate.   Put another way, the regression line should not have a constant slope.

Exponential Regression and Correlation Analysis

Let us consider the first of the two nonlinear cases mentioned above, where Y is changing by a certain rate while X is changing by a certain amount.   If we were to plot a scatter diagram of such a situation on regular arithmetic graph paper, the scatter might appear something like the pattern shown in Figure 13-1a.   But if the data were plotted on semilog graph paper, then the scatter would appear as shown in Figure 13-1b.   The linear relationship shown in this figure, 13-1b, indicates that the regression equation should be an exponential one.   That is, the dependent variable is changing by a constant rate rather than by a constant amount.

The Exponential Regression Equation

There are two ways in which this type of relationship can be expressed as a regression equation.   It can be expressed in terms of Y_c itself, or it can be expressed in terms of the log Y_c.   In the first case, the equation reads as follows:

$$Y_c = ab^X \qquad (13.1)$$

In terms of log Y_c, the equation is

$$\log Y_c = \log a + X \log b \qquad (13.2)$$

A close examination of the second equation reveals that it is a linear equation in its logarithmic form.[1]   This bears out what the scatter diagram in Figure 13-1b indicates.   In the log form of Y, the relationship is linear.

Given this linear relationship for log Y and X, the procedure for finding Y_c requires that you first convert all Y values to log Y.   Then, you compute log b and log a using Equations 13.3 and 13.4.   After finding log b and log a, you can use Equation 13.2 to compute log Y_c.   Next, you find Y_c by determining the antilog of log Y_c.

---

1.   If you do not understand logarithms or do not remember how to determine antilogs from logs, look at the section at the beginning of Appendix A.

$$\log b = \frac{[n\Sigma\,(X\log Y)] - [(\Sigma X)\,(\Sigma\log Y)]}{n\,(\Sigma X^2) - (\Sigma X)^2} \qquad \textbf{(13.3)}$$

$$\log a = \frac{\Sigma\log Y}{n} - \log b\,\left(\frac{\Sigma X}{n}\right) \qquad \textbf{(13.4)}$$

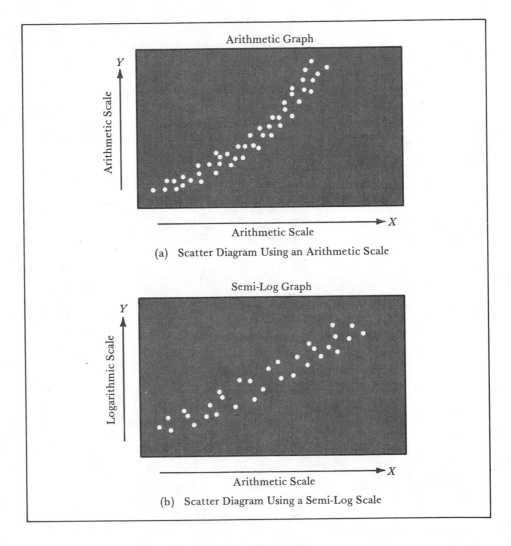

FIGURE 13-1. *Scatter Diagrams.*

▶ **Example 13.1**   The Vice President for Sales of Our Favorite Publisher, Inc., has been collecting data from the expense reports of the company's book salesmen.   The salesmen frequently entertain prospective customers by taking them to dinner.   In casually looking at the data, the Vice President has decided that the entertainment cost appears to increase faster than the size of the party being entertained.   The Vice President decided to pursue his idea a little further and came up with the data and calculations in Table 13-1.

TABLE 13-1.   *Entertainment Cost of Our Favorite Publisher Salesmen (Logarithmic regression equation).*

SIZE OF PARTY X	ENTERTAINMENT COST Y	$\log Y$	$X \log Y$	X^2	$(\log Y)^2$
1	$ 5.00	0.69897	0.69897	1	0.488559
2	10.20	1.00860	2.01720	4	1.017274
3	15.35	1.18611	3.55833	9	1.406857
4	20.50	1.31175	5.24700	16	1.720688
5	25.95	1.41414	7.07070	25	1.999792
6	32.20	1.50786	9.04716	36	2.273642
7	38.50	1.58546	11.09822	49	2.513683
8	46.00	1.66276	13.30208	64	2.764771
9	53.80	1.73078	15.57702	81	2.995599
10	62.00	1.79239	17.92390	100	3.212662
55	$309.50	13.89882	85.54058	385	20.393527

$$\log b = \frac{n \Sigma (X \log Y) - (\Sigma X)(\Sigma \log Y)}{n (\Sigma X^2) - (\Sigma X)^2}$$

$$= \frac{(10)(85.54058) - (55)(13.89882)}{(10)(385) - (55)^2}$$

$$= \frac{855.4058 - 764.4351}{3850 - 3025}$$

$$= \frac{90.9707}{825}$$

$$\log b = 0.11027$$

$$\log a = \frac{\Sigma \log Y}{n} - \log b \left(\frac{\Sigma X}{n}\right)$$

$$= \frac{13.89882}{10} - (0.11027)\left(\frac{55}{10}\right)$$

$$= 1.389882 - 0.606485$$

$$\log a = 0.783397$$

$$\log Y_c = 0.78340 + 0.11027 X \blacktriangleleft$$

The equation shown in Table 13-1 can also be written in antilog form.

$$Y_c = 6.073 \ (1.289)^X$$

Looking at the equation written in this fashion, we can interpret the b coefficient. Specifically, b is one plus the *rate* of increase in the dependent variable as the independent variable increases one unit. In this case, the value of b tells us that entertainment costs increase 28.90 percent each time the party size increases by 1.

Just as with other regression equations, it is possible to substitute a value for X and calculate Y_c. Suppose we use the equation from Table 13-1 to calculate the value of Y_c when the size of the party (X) is 20.

$$\log Y_c = 0.78340 + 0.11027 \ X$$

$$\log Y_c = 0.78340 + 0.11027 \ (20)$$

$$= 0.78340 + 2.20540$$

$$\log Y_c = 2.98880$$

$$Y_c = \text{antilog} \ (2.98880)$$

$$Y_c = \$974.50$$

Perhaps you are surprised at the magnitude of Y_c in this example. You should remember that in using exponential equations we are assuming a *constant rate of change*, and this can cause our estimate of the dependent variables (Y_c) to *explode* with relatively small increases in the magnitude of the independent variable.

Coefficients of Correlation and Determination

The quality of the relationship between the two variables can also be measured using the exponential equation as the standard of perfect correlation. In the same manner as shown in Chapter 12, we can calculate the coefficient of correlation using Equation 13.5. This calculation using the data from Example 13.1 is shown below.

$$r = \frac{n\Sigma \ (X \ \log Y) - \Sigma X \ (\Sigma \ \log Y)}{\sqrt{[n\Sigma X^2 - (\Sigma X)^2] \ [n\Sigma \ (\log Y)^2 - (\Sigma \ \log Y)^2]}} \quad \textbf{(13.5)}$$

$$r = \frac{10 \ (85.54058) - 55 \ (13.89882)}{\sqrt{[10 \ (385) - (55)^2] \ [10 \ (20.393527) - (13.89882)^2]}}$$

$$= \frac{855.4058 - 764.4351}{\sqrt{(3850 - 3025) \ [10 \ (20.393527) - 193.17719]}}$$

$$r = \frac{90.9707}{\sqrt{(825)\,(203.93527 - 193.17719)}}$$

$$= \frac{90.9707}{\sqrt{(825)\,(10.75808)}}$$

$$= \frac{90.9707}{\sqrt{8875.416}}$$

$$r = \frac{90.9707}{94.2094} = 0.9656$$

Thus, with an $r = 0.9656$, the coefficient of determination (r^2) is 0.9324. In other words, using the exponential regression equation, variations in the independent variable explain 93.2 percent of the variation in the dependent variable.

Second-Degree (Arithmetic) Regression and Correlation Analysis

There may be times when neither a linear regression equation nor an exponential equation fits the data displayed in a scatter diagram. In such a case, one might consider a *second-degree* (or second order) *arithmetic regression equation.* If the data in a scatter diagram appear curve-shaped in one of the four ways shown in Figure 13-2 (a, b, c, and d), the use of a second-degree regression equation may be justified.

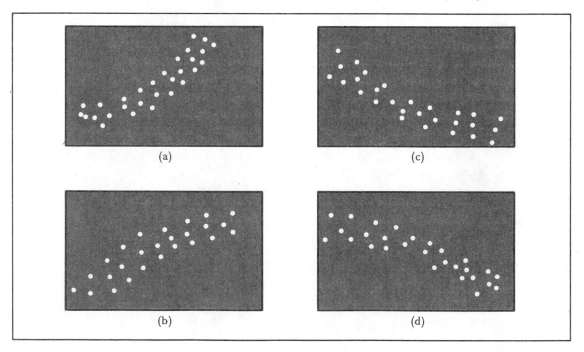

(a)

(c)

(b)

(d)

FIGURE 13-2. *Examples of Curvilinear Scatter Diagrams.*

The Second-Degree Regression Equation

The general form of the regression equation for a second-degree case is

$$Y_c = a + bX + cX^2 \qquad (13.6)$$

Unlike the linear equation and the log form of the exponential equation, *the slope of the second-degree curve is not constant.* Rather, the slope varies as X varies. However, a is still the intercept (the value of Y_c when $X = 0$).

If we use the least squares method to solve for the values of a, b, and c in the equation, we utilize the following normal equations:

$$\Sigma Y = na + b\Sigma X + c\Sigma X^2 \qquad (13.7)$$

$$\Sigma XY = a\Sigma X + b\Sigma X^2 + c\Sigma X^3 \qquad (13.8)$$

$$\Sigma X^2Y = a\Sigma X^2 + b\Sigma X^3 + c\Sigma X^4 \qquad (13.9)$$

▶ **Example 13.2**

After seeing the results of the exponential regression, the Our Favorite Vice President thinks that some other form of regression equation might be more appropriate. His work computing a second-degree regression equation is shown in Table 13-2.

TABLE 13-2. *Entertainment Cost of Our Favorite Publisher Salesmen (second-degree regression equation).*

SIZE OF PARTY X	ENTERTAINMENT COST Y	XY	X^2	X^2Y	X^3	X^4	Y^2
1	$ 5.00	5.00	1	5.00	1	1	25.00
2	10.20	20.40	4	40.80	8	16	104.04
3	15.35	46.05	9	138.15	27	81	235.6225
4	20.50	82.00	16	328.00	64	256	420.25
5	25.95	129.75	25	648.75	125	625	673.4025
6	32.20	193.20	36	1159.20	216	1296	1036.84
7	38.50	269.50	49	1886.80	343	2401	1482.25
8	46.00	368.00	64	2944.00	512	4096	2116.00
9	53.80	484.20	81	4357.80	729	6561	2894.44
10	62.00	620.00	100	6200.00	1000	10000	3844.00
55	$309.50	2218.10	385	17708.20	3025	25333	12831.845

I $\Sigma Y = na + b\Sigma X + c\Sigma X^2$
   $309.50 = 10a + 55b + 385c$

II $\Sigma XY = a\Sigma X + b\Sigma X^2 + c\Sigma X^3$
   $2218.10 = 55a + 385b + 3025c$

III $\Sigma X^2Y = a\Sigma X^2 + b\Sigma X^3 + c\Sigma X^4$
   $17708.20 = 385a + 3025b + 25333c$

$$\frac{55}{10} = 5.5$$

I (−5.5)	−1702.25 =	−55a − 302.5b − 2117.5c
II	2218.10 =	55a + 385b + 3025c
(I + II)	515.85 =	82.5b + 907.5c

$$\frac{385}{55} = 7$$

II (−7)	−15,526.70 =	−385a − 2695b − 21175c
III	17,708.20 =	385a + 3025b + 25333c
(II + III)	2,181.50 =	330b + 4158c

$$\frac{330}{82.5} = 4$$

(I + II) (−4)	−2063.40 =	−330b − 3630c
	2181.50 =	330b + 4158c
	118.1 =	528c

$$c = 0.223674$$

(I + II)

$$515.85 = 82.5b + 907.5c$$
$$515.85 = 82.5b + 907.5 (0.223674)$$
$$515.85 = 82.5b + 202.98415$$
$$312.86585 = 82.5b$$
$$b = 3.792313$$

I $\quad 309.50 = 10a + 55b + 385c$
$$309.50 = 10a + 55 (3.792313) + 385 (0.223674)$$
$$309.50 = 10a + 208.577215 + 86.114490$$
$$309.50 = 10a + 294.691705$$
$$10a = 14.808295$$
$$a = 1.480830$$

$$Y_c = 1.480830 + 3.792313X + 0.223674X^2$$

The second degree equation for the expenses of Our Favorite salesmen provides a significantly different Y_c value when X is equal to 20 than does the exponential equation.

$$Y_c = 1.480830 + 3.792313X + 0.223674X^2$$

$$Y_c = 1.480830 + 3.792313 (20) + 0.223674 (20)^2$$

$$= 1.480830 + 75.84626 + 89.4696$$

$$Y_c = \$166.80 \blacktriangleleft$$

Intuitively, at least, this appears to be a more reasonable estimate of what might be expected as an estimate of expenses for the salesmen in our examples. Comparing the two equations in this instance serves to emphasize a point made earlier. The value of b in the exponential equation will tend to *explode* as X increases in size. Obviously, as b becomes larger, this happens at smaller values of X. Generally, a second-degree equation has a similar tendency to explode but at higher values of X than exponential equations.

Second-Degree Correlation Analysis

The quality of the relationship between the two variables, as expressed by a second-degree regression equation, can be measured using the following equation for the coefficient of correlation.

$$r = \sqrt{\frac{a\Sigma Y + b\Sigma XY + c\Sigma X^2 Y - n\overline{Y}^2}{\Sigma Y^2 - n\overline{Y}^2}} \qquad (13.10)$$

$$r = \sqrt{\frac{1.4808\,(309.5) - 3.7923\,(2218.1) - 0.22367\,(17708.2) - 10\,(30.95)^2}{12831.845 - 10\,(30.95)^2}}$$

$$= \sqrt{\frac{458.3076 + 8411.7006 + 3960.7931 - 9579.025}{12831.845 - 9579.025}}$$

$$= \sqrt{\frac{3251.7763}{3252.82}}$$

$$= \sqrt{0.9997}$$

$$r = 0.99984$$

The coefficient of determination is still r^2, so $r^2 = 0.9997$. In other words, using the second-degree equation form, changes in the independent variable explain 99.97 percent of the changes in the dependent variable in Example 13.2.

Other Comments on Nonlinear Regression and Correlation Analysis

The previous discussion has been only a brief introduction to nonlinear analysis. Basically, our intent is simply to introduce you to nonlinear analysis. However, you should recognize that the various statistical measures you studied in linear regression (Chapter 12) can also be applied here.

Multivariate Regression and Correlation Analysis

Multivariate linear correlation and regression analysis is just an extension of the simple linear case, so that there is more than one independent variable. Thus, the assumption regarding only two variables (one dependent and one independent) is relaxed. There may be 2, 3, 4, or more independent variables. Each independent variable, however, is linearly related to the single dependent variable.[2]

The basis for multivariate correlation and regression analysis may be explained in terms of the variation of the dependent variable. In the previous chapter, we discussed the total variation of Y (the dependent variable) as consisting of two components—the explained variation and the unexplained variation. The larger the proportion of the total variation that can be associated (or explained) with the independent variable, the higher the coefficient of correlation.

In simple regression analysis, we attempt to associate the variation of the dependent variable (Y) with a single independent variable (X) and measure the proportion of explained (associated) variation. Yet, we often may find that a fairly high proportion of the total variation is still unexplained. For instance, a retail gasoline service station may find that in correlating monthly gross gasoline sales with the number of automobiles passing the service station during the month, the coefficient of correlation is 0.60. If r equals 0.60, then r^2 equals 0.36 which means 36 percent of the variation in monthly gross sales Y is associated with the number of automobiles passing the service station during the month X. Even so, there is still 64 percent of the variation of Y left unexplained. Another variable (for example, number of repair jobs during the month) may explain 25 percent of the total variation. And together the two variables might explain 49 percent of the total variation. Thus $R^2 = 0.49$ and $R = 0.70$. We use a capital letter, R and R^2, to indicate the coefficient of multiple correlation and the coefficient of multiple determination.

Note that r^2 from the first independent variable and r^2 from the second independent variable need not (indeed would not) total R^2. In the above example, the r^2 for the number of cars passing the service station was 36 percent and r^2 for the number of repair jobs was 25 percent. Yet, R^2 for both variables was 49 percent, not 61 percent $(25 + 36)$. The explanation for this is simple. Part of the 25 percent explained by the second r^2 overlaps with the 36 percent explained by the first r^2. Figure 13-3 shows diagrammatically how this

---

2.   Note: This statement is technically correct but nonlinearity can be introduced to multivariate regression analysis by transforming the values of one or more of the variables. (The values may be transformed by squaring them, converting to logarithms, etc.) The transformed values are then used in the regression equation.

relationship works. Let the entire rectangle represent the total variation of Y. Smaller rectangles show the part of the total variation that is
associated with X_1 and X_2. Note that the two smaller rectangles overlap one another. This is important, because too much overlap among
the independent variables suggests that they are related. When the
independent variables are related to one another, multicolinearity is
said to exist, and this can distort our analysis. In fact, in conducting
multiple correlation and regression analysis we assume that the independent variables are also independent of one another.

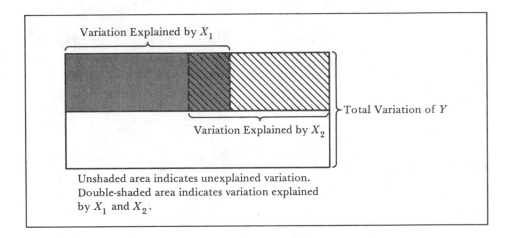

FIGURE 13.3. *Diagram Showing Contribution to Explained Variation by
Variables X_1 and X_2 and Their Relationships.*

Scatter Diagram

Figure 13-4 is a facsimile of a scatter diagram with two independent
variables. The points in the scatter diagram represent the coordinate
points of Y, X_1, and X_2. For instance, point A in Figure 13-4 shows the
point at which $Y = 7$, $X_1 = 4$, and $X_2 = 8$ intersect. Obviously, it
would not be possible to show the same sort of relationship when there
were four such variables—a dependent and three independent variables. However, a series of two-variable scatter diagrams might be
constructed in which each independent variable in the multivariate
analysis would be related to the dependent variable. Thus, there
would be a separate scatter diagram for each independent variable.
In each case, the scatter diagram would indicate the linear relationship
between the dependent variable and the particular variable being
used.

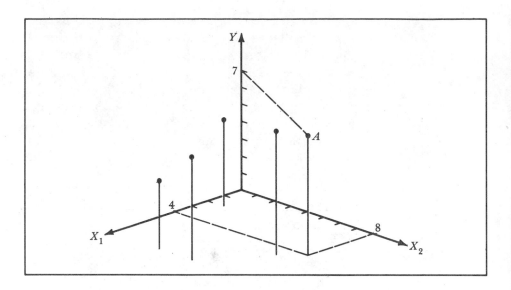

FIGURE 13-4. *Facsimile of a Three-Dimensional Scatter Diagram.*

Multivariate Regression Equation

The multivariate regression equation takes the following form:

$$Y_c = a + b_1X_1 + b_2X_2 + \cdots + b_mX_m \qquad (13.11)$$

Here again, Y_c is the computed regression value for the dependent variable. The value a is still the Y intercept, but it is the value of Y_c when all independent variables (X's) equal zero. There is a separate b value (or slope) for each of the independent variables. These b values represent the slope of the relationship for each respective independent variable if the remaining independent variables are held constant. For example, b_1 is the slope of X_1 where all remaining independent variables are held constant. In other words, the value of b_1 is the average change in the dependent variable, assuming other independent variables are held constant, when X_1 varies by one unit.

The b values in a multiple regression equation are called *coefficients of partial regression,* while the b value in a simple linear regression equation is called the *coefficient of simple regression.*

The constants $a, b_1, b_2, \ldots b_m$ in the equation are computed from normal equations. Thus, the multivariate equation assumes the characteristics of a least-squares equation. The general forms of the normal equations are

$$\Sigma Y = na + b_1\Sigma X_1 + b_2\Sigma X_2 + \cdots + b_m\Sigma X_m \qquad (13.12)$$

$$\Sigma X_1 Y = a\Sigma X_1 + b_1\Sigma X_1^2 + b_2\Sigma X_1 X_2 + \cdots + b_m\Sigma X_1 X_m \quad (13.13)$$

$$\Sigma X_2 Y = a\Sigma X_2 + b_1\Sigma X_1 X_2 + b_2\Sigma X_2^2 + \cdots + b_m\Sigma X_2 X_m \quad \textbf{(13.14)}$$

$$\Sigma X_m Y = a\Sigma X_m + b_1\Sigma X_1 X_m + b_2\Sigma X_2 X_m + \cdots + b_m\Sigma X_m^2 \quad \textbf{(13.15)}$$

Consider a regression equation with three variables—a dependent variable and two independent ones. The equation would appear as

$$Y_c = a + b_1 X_1 + b_2 X_2 \quad \textbf{(13.16)}$$

and the normal equations would be

$$\Sigma Y = na + b_1\Sigma X_1 + b_2\Sigma X_2 \quad \textbf{(13.17)}$$

$$\Sigma X_1 Y = a\Sigma X_1 + b_1\Sigma X_1^2 + b_2\Sigma X_1 X_2 \quad \textbf{(13.18)}$$

$$\Sigma X_2 Y = a\Sigma X_2 + b_1\Sigma X_1 X_2 + b_2\Sigma X_2^2 \quad \textbf{(13.19)}$$

When we presented simple linear regression in Chapter 12, we discussed testing the b coefficient of the equation to determine if it was significantly different from zero. Multiple regression requires that we do the same thing except now we must test the b value for each of the variables in the equation. Just as in simple linear regression, if the value of b for a particular variable is not significantly different from zero, we cannot assume that there is a relationship between the dependent variable and that particular independent variable.

In order to test each b value in the multiple regression equation, we must first calculate an estimate of the standard error of multiple regression (this is sometimes called *error variance*). This measure is similar to the standard error of estimate of a simple linear regression equation in that it is a measure of the average amount that the Y values differ from the estimated (Y_c) values. Thus, the standard error of multiple regression (s_m) is a measure of closeness-of-fit for the regression equation.

Now using the estimate of the standard error of multiple regression, it is possible to estimate the standard error of each coefficient of partial regression (s_{b_i}). Then, the null hypothesis $\beta_i = 0$ can be tested, using a t-test, for each independent variable to determine if, in the multiple regression equation, there is any relationship between that independent variable X_i and the dependent variable. We will look at a detailed example later, but you should be aware that an independent variable may be accepted as having a significant regression relationship with a given dependent variable in a simple linear regression equation, but may be rejected, using the same significance level, when one or more additional independent variables are added to the equation. If this happens, it is because *multicolinearity* exists between the independent variables. That is, two or more of the independent variables have relatively high correlation among themselves.

After testing each b_i value to determine if it is significant, it is now time to consider testing the total regression effect of the independent variables on the dependent variable. In other words, there is a way to test the entire equation. This can be accomplished by the use of analysis of variance, in the form initially presented in the Appendix to Chapter 12, to test the null hypothesis $\beta_1 = \beta_2 = \cdots \beta_m = 0$, where m is the number of independent variables. Just as in other applications of analysis of variance, the test statistic, a value of F, is found using the following formula:

$$F_{\text{test}} = \frac{MSR}{MSE} \qquad\qquad (13.20)$$

where $MSR = \dfrac{SSR}{m}$ $\qquad\qquad (13.21)$

$MSE = \dfrac{SSE}{n - m - 1}$ $\qquad\qquad (13.22)$

Multiple Correlation

Like simple linear correlation, we can compute a measure of the quality of the relationship between the dependent variable and the independent variables. The measure is called the *coefficient of multiple correlation* (R) *and* R^2 is the *coefficient of multiple determination*. R^2 can be interpreted to mean the proportion of the total variation in the dependent variable which is explained by changes in all the independent variables. Equation 13.23 is used to compute R^2.

$$R^2 = \frac{SSR}{SST} \qquad\qquad (13.23)$$

where SSR = regression sum of squares
SST = total sum of squares

Multiple Regression and the Computer

Our discussion of multiple correlation and regression has been in general terms up until now. In addition, you may have noticed that you were given very few equations. As a matter of fact, the computations required for multiple regression analysis are rather time consuming. Consequently, these computations are usually carried out by using either computers or, for problems with relatively few independent variables, programmable electronic calculators, for multiple regression programs are readily available for both of them.

TABLE 13-3. *Data for a Multiple Regression Example.*

YEAR	NUMBER OF BACHELOR'S DEGREES GRANTED	MEDIAN FAMILY INCOME (1974 DOLLARS)	UNEMPLOYMENT RATE	POPULATION AGED 18 TO 24 (1,000's)
1960	368,323	$ 9,358	5.5	16, 128
1961	368,857	9,457	6.7	17,004
1962	387,830	9,710	5.5	17,688
1963	416.421	10,065	5.7	18,268
1964	466,486	10,444	5.2	18,783
1965	501,248	10,874	4.5	20,293
1966	520,248	11,445	3.8	21,376
1967	558,075	11,717	3.8	22,327
1968	633,923	12,236	3.6	22,883
1969	728,167	12,689	3.5	23,723
1970	791,510	12,531	4.9	24,683
1971	839,730	12,523	5.9	25,776
1972	887,273	13,103	5.6	25,901
1973	922,130	13,373	4.9	26,381
1974	945,870	12,836	5.6	26,908
1975	975,000	12,570	8.5	27,597

Source: Social Indicators 1976, U.S. Department of Commerce, December 1977.

```
SAMPLE COMPUTER OUTPUT FROM A MULTIPLE REGRESSION PROGRAM WITH THREE INDEPENDENT VARIABLES


                                                      PROBABILITY
                        REGRESSION    STANDARD         OF A LARGER
VARIABLE                COEFFICIENT   ERROR    T STATISTIC  ABSOLUTE T   MEAN

POPULATION                 16.978     16.888     1.0053     0.334577    22232.438

INCOME                    111.060     46.978     2.3641     0.035785    11558.188

UNEMPLOYMENT RATE      47187.503  13357.437     3.5237      0.004126      5.2


MEAN FOR DEPENDENT VARIABLE (BACHELOR DEGREES) IS 644443.188

CONSTANT FOR SIMPLIED MODEL = -1262041.195

ERROR MEAN SQUARE = 1065953620.578 WITH 12 D.F.

ERROR SUM OF SQUARES = 12791443446.919

REGRESSION MEAN SQUARE = 245129596268.506 WITH 3 D.F.

REGRESSION SUM OF SQUARES = 73538878805.518

TOTAL SUM OF SQUARES = 748180232252.438


MULTIPLE R**2 = 0.98290324

F STATISTIC = 229.9625 WITH 3 AND 12 D.F.

THE PROBABILITY OF OBTAINING F THIS LARGE OR LARGER BY CHANCE
WHEN THE HYPOTHESIS OF NO CORRELATION IS TRUE = 0.00000006
```

FIGURE 13-5. *Sample Computer Output From a Multiple Regression Program With Three Independent Variables.*

In Table 13-3 there is a set of data which has been used to develop an illustration of the output that might be provided from a multiple regression computer program. The example of this computer output is given in Figure 13-5.

In looking at the illustration, you should note that the number of bachelor's degrees granted per year is the dependent variable and there are three independent variables. The form of the equation in the illustration is

$$Y_c = -1262041.195 + 16.978X_1 + 111.060X_2 + 47187.503X_3$$

If you wish to test the various values of b_i, the output also contains information useful for that. Given the following three sets of hypotheses they can be tested at any level of α with the data in the col-

$$H_0: \quad \beta_1 = 0 \qquad H_0: \quad \beta_2 = 0 \qquad H_0: \quad \beta_3 = 0$$
$$H_A: \quad \beta_1 = 0 \qquad H_A: \quad \beta_2 = 0 \qquad H_A: \quad \beta_3 = 0$$

umn headed "Probability of a Larger Absolute T." If $\alpha = 0.05$, you would reject $H_0: \quad \beta_2 = 0$ and $H_0: \quad \beta_3 = 0$, but you would not reject $H_0: \quad \beta_1 = 0$ because the probability of a larger t value is greater than 0.05. In other words, there does not appear to be a regression rela-

SAMPLE COMPUTER OUTPUT FROM A MULTIPLE REGRESSION PROGRAM WITH TWO INDEPENDENT VARIABLES

VARIABLE	REGRESSION COEFFICIENT	STANDARD ERROR	T STATISTIC	PROBABILITY OF A LARGER ABSOLUTE T	MEAN
INCOME	157.881	6.159	25.6361	0.000000	11558.188
UNEMPLOYMENT RATE	58826.729	6664.907	8.8263	0.000001	5.2

MEAN FOR DEPENDENT VARIABLE (BACHELOR DEGREES) IS 644443.188

CONSTANT FOR SIMPLIED MODEL = -1486276.182

ERROR MEAN SQUARE = 1066831560.605 WITH 13 D.F.

ERROR SUM OF SQUARES = 13868810287.860

REGRESSION MEAN SQUARE = 367155710982.289 WITH 2 D.F.

REGRESSION SUM OF SQUARES = 734311421964.578

TOTAL SUM OF SQUARES = 748180232252.438

MULTIPLE R**2 = 0.98146325

F STATISTIC = 344.1548 WITH 13 D.F.

THE PROBABILITY OF OBTAINING F THIS LARGE OR LARGER BY CHANCE WHEN THE HYPOTHESIS OF NO CORRELATION IS TRUE = 0.00000006

FIGURE 13-6. *Sample Computer Output From a Multiple Regression Program With Two Independent Variables.*

tionship between the amount of population between the ages of 18 and 24 and the number of bachelor's degrees granted, but there does appear to be one between the unemployment rate and the number of degrees granted as well as between constant dollar median family income and number of degrees granted.

Now that you have discovered this, you may wish to recompute the multiple regression equation using only two independent variables —family income and unemployment rate. Figure 13-6 contains an illustration of the output your computer might yield. First of all, notice that the partial regression coefficients are not the same as those shown in Figure 13-5 but they are still significant at a level of $\alpha = 0.05$. The values of the coefficients of partial regression become changed because we no longer have the effects of the third independent variable. The printout also contains the Coefficient of Multiple Determination (R^2) as well as an F statistic. This latter is the same as F_{test} for either the regression equation or R^2. Finally, at the end of the sample output is a statement concerning the probability of an F value this large or larger. In this instance, it can be interpreted to mean that the null hypothesis $\beta_1 = \beta_2 = 0$ would be rejected at any $\alpha > 0.00000006$ because $F_{\text{test}} = 344.1548$ with $d.f. = 2$ and 13.

Actually, this is a very simplified example of multiple regression analysis using a computer. The output resulting from many, if not most, computer programs contains somewhat more information than that presented here. Moreover, the equations presented in Figures 13-5 and 13-6 are not all the ones that should be considered if your objective is to analyze the data thoroughly. For example, we have not shown you an equation using just unemployment rates and population as independent variables. As a matter of fact, in some of the problems at the end of this chapter, we ask you to do that regression, along with some others. Try them, and you may be surprised by some of your answers.

--------------------------- **APPENDIX TO CHAPTER 13** ---------------------------

Adjusted Sums of Squares Method of Determining Multiple Regression Equations

Solving the normal equations required to carry out multiple regression analysis can be rather laborious. Using the adjusted sums of squares method reduces the number of normal equations by one and, therefore, simplifies the task somewhat. Using this procedure for a problem with two independent variables, we can rewrite Equations 13.17, 13.18, and 13.19 as Equations 13.24 and 13.25.

Note that instead of X_1 and X_2 we now have x_1 and x_2.

$$\Sigma x_1 y = b_1 \Sigma x_1^2 + b_2 \Sigma x_1 x_2 \qquad (13.24)$$

$$\Sigma x_2 y = b_1 \Sigma x_1 x_2 + b_2 \Sigma x_2^2 \qquad (13.25)$$

where

$$\Sigma y^2 = \Sigma Y^2 - n\overline{Y}^2 \qquad (13.26)$$

$$\Sigma x_1^2 = \Sigma X_1^2 - n\overline{X}_1^2 \qquad (13.27)$$

$$\Sigma x_2^2 = \Sigma X_2^2 - n\overline{X}_2^2 \qquad (13.28)$$

$$\Sigma x_1 y = \Sigma X_1 Y - n\overline{X}_1\overline{Y} \qquad (13.29)$$

$$\Sigma x_2 y = \Sigma X_2 Y - n\overline{X}_2\overline{Y} \qquad (13.30)$$

$$\Sigma x_1 x_2 = \Sigma X_1 X_2 - n\overline{X}_1\overline{X}_2 \qquad (13.31)$$

If the adjusted sums of squares method is used, the value of a in the multiple regression equation is found using Equation 13.32.

$$a = \overline{Y} - b_1\overline{X}_1 - b_2\overline{X}_2 \qquad (13.32)$$

Similarly, in order to test each b value, we can write the equations for the standard error of b_1 and b_2 using adjusted sums of squares. In order to do so, though, we first calculate the standard error of multiple regression using Equation 13.33.

$$s_m = \sqrt{\frac{\Sigma y^2 - b_1 \Sigma x_1 y - b_2 \Sigma x_2 y}{n - 3}} \qquad (13.33)$$

Then we can use Equations 13.34 and 13.35 to compute the *standard error of b_1 and b_2.*

$$s_{b_1} = \frac{s_m}{\sqrt{\Sigma x_1^2 - \dfrac{(\Sigma x_1 x_2)^2}{\Sigma x_2^2}}} \qquad (13.34)$$

$$s_{b_2} = \frac{s_m}{\sqrt{\Sigma x_2^2 - \dfrac{(\Sigma x_1 x_2)^2}{\Sigma x_1^2}}} \qquad (13.35)$$

Now we are ready to establish the hypotheses and test the b values using Equation 13.36.

$$t_{\text{test}} = \frac{b_i - \beta_H}{s_{b_i}} \qquad (13.36)$$

It is also possible to write an equation for R^2 using this method.

$$R^2 = \frac{b_1 \Sigma x_1 y + b_2 \Sigma x_2 y}{\Sigma y^2} \qquad (13.37)$$

Now that you have at your disposal a set of equations that can be used to do multiple regression analysis, let's look at an example that illustrates their application.

▶ **Example 13.3**   Dr. Mac Michael, a professor of statistics, has been a regular Friday afternoon customer of the Library (an establishment dedicated to quenching the thirst of the university community) for many years. During his visits to the Library, Dr. Mac has noticed a great variation in the number of customers. Consequently, he has also observed that the more customers the Library has, the longer it takes for him to be served. Having stimulated his intellectual curiosity, as well as his thirst, Dr. Mac has suggested to the Library management that they employ him as a consultant, and it did.

Dr. Mac decided that his first task should be to find some way to make a reasonably accurate estimate, using multiple regression analysis, of how many customers the Library might expect so that it could improve planning and hope to provide faster service. Since most of the establishment's customers are university students, it seemed logical to the professor that the number might be related to the size of the student body. He also recognized that students came to the Library only if they had money to spend and, while he does not know the income of students, he does know how much they contribute to the Student Mascot Fund. Dr. Mac decided that contributions to the fund might be a good indicator of student income and, therefore, might also be related to the number of customers at the Library.

The data in Table 13-4 are what Dr. Mac had to work with. The tables also include his calculations for a multiple regression equation.

Now Dr. Mac is ready to establish the hypothesis and test the b values. First, however, he must determine an appropriate value for the level of significance. While he frequently uses relatively small alpha values, he finally decided that $\alpha = 0.50$ would be good enough for the problem at hand. After all, that would mean that Dr. Mac would be rejecting a correct H_0 only 50 percent of the time.

$$H_0: \quad \beta_1 = 0 \qquad H_0: \quad \beta_2 = 0$$

$$H_A: \quad \beta_1 \neq 0 \qquad H_A: \quad \beta_2 \neq 0$$

TABLE 13-4. *Dr. Mac's Multiple Regression Problem.*

CUSTOMERS Y	STUDENT ENROLLMENT X_1	STUDENT CONTRIBUTION TO MASCOT FUND X_2	Y_2	X_1^2	X_2^2	YX_1	YX_2	X_1X_2
10	700	$ 50	100	490000	2500	7000	500	35000
15	750	65	225	562500	4225	11250	975	48750
20	760	80	400	577600	6400	15200	1600	60800
15	800	100	225	640000	10000	12000	1500	80000
20	842	105	400	708964	11025	16840	2100	88410
16	910	85	256	828100	7225	14560	1360	77350
18	965	90	324	931225	8100	17370	1620	86850
22	1010	100	484	1020100	10000	22220	2200	101000
24	1070	110	576	1144900	12100	25680	2640	117700
30	1100	100	900	1210000	10000	33000	3000	110000
190	8907	$885	3890	8113389	81575	175120	17495	805860

$$\Sigma y^2 = \Sigma Y^2 - n\bar{Y}^2$$
$$= 3890 - 10\,(19)^2$$
$$= 3890 - 3610$$
$$\Sigma y^2 = 280$$

$$\Sigma x_1^2 = \Sigma X_1^2 - n\bar{X}_1^2$$
$$= 8113389 - 10\,(890.7)^2$$
$$= 8113389 - 7933464.9$$
$$\Sigma x_1^2 = 179924.1$$

$$\Sigma x_2^2 = \Sigma X_2^2 - n\bar{X}_2^2$$
$$= 81575 - 10\,(88.5)^2$$
$$= 81575 - 78322.5$$
$$\Sigma x_2^2 = 3252.5$$

$$\Sigma x_1 y = \Sigma X_1 Y - n\bar{X}_1\bar{Y}$$
$$= 175120 - 10\,(890.7)\,(19)$$
$$= 175120 - 169233$$
$$\Sigma x_1 y = 5887$$

$$\Sigma x_2 y = \Sigma X_2 Y - n\bar{X}_2\bar{Y}$$
$$= 17495 - 10\,(88.5)\,(19)$$
$$= 175120 - 169233$$
$$\Sigma x_2 y = 680$$

$$\Sigma x_1 x_2 = \Sigma X_1 X_2 - n\bar{X}_1\bar{X}_2$$
$$= 805860 - 10\,(890.7)\,(88.5)$$
$$= 805860 - 788269.5$$
$$\Sigma x_1 x_2 = 17590.5$$

I $\quad \Sigma x_1 y = b_1\Sigma x_1^2 + b_2\Sigma x_1 x_2$

II $\quad \Sigma x_2 y = b_1\Sigma x_1 x_2 + b_2\Sigma x_2^2$

I $\quad 5887 = 179924.1\,b_1 + 17590.5\,b_2$

II $\quad 680 = 17590.5\,b_1 + 3252.5\,b_2$

I $\quad 5887 = 179924.1\,b_1 + 17590.5\,b_2$

II $(-10.2285) - 6955.4 = -179924.1\,b_1 - 33268.2\,b_2$

$\quad -1068.4 = -15677.7\,b_2$

$\quad 0.06815 = b_2$

$$680 = 17590.5\,b_1 + 3252.5\,(0.06815)$$
$$680 = 17590.5\,b_1 + 221.65788$$
$$458.34212 = 17590.5\,b_1$$
$$0.02606 = b_1$$

$$a = \bar{Y} - b_1\bar{X}_1 - b_2\bar{X}_2$$
$$a = 19.0 - 0.02606\,(890.7) - (0.06815)\,(88.5)$$
$$= 19.0 - 23.21164 - 6.03128$$
$$a = -10.24292$$
$$Y_c = -10.24292 + 0.02606 X_1 + 0.06815 X_2$$

Using the values from Table 13-4, the calculations for the standard error of the estimate of each of b values are shown below

$$s_m = \sqrt{\frac{\Sigma y^2 - b_1 \Sigma x_1 y - b_2 \Sigma x_2 y}{n - 3}}$$

$$s_m = \sqrt{\frac{280 - (0.02606)\,(5887) - (0.06815)\,(680)}{10 - 3}}$$

$$= \sqrt{\frac{280 - 153.41522 - 46.34200}{7}} = \sqrt{\frac{80.24278}{7}} = \sqrt{11.46325}$$

$$s_m = 3.38574$$

$$s_{b_1} = \frac{s_m}{\sqrt{\Sigma x_1^2 - \dfrac{(\Sigma x_1 x_2)^2}{\Sigma x_2^2}}} = \frac{3.38574}{\sqrt{179924.1 - \dfrac{(17590.5)^2}{3252.5}}}$$

$$= \frac{3.38574}{\sqrt{179924.1 - \dfrac{309425690.25}{3252.5}}} = \frac{3.38574}{\sqrt{179924.1 - 95134.7}}$$

$$= \frac{3.38574}{\sqrt{84789.4}} = \frac{3.38574}{291.19}$$

$$s_{b_1} = 0.01163$$

$$s_{b_2} = \frac{s_m}{\sqrt{\Sigma x_2^2 - \dfrac{(\Sigma x_1 x_2)^2}{\Sigma x_1^2}}} = \frac{3.38574}{\sqrt{3252.5 - \dfrac{(17590.5)^2}{179924.1}}}$$

$$= \frac{3.38574}{\sqrt{3252.5 - \dfrac{309425690.25}{179924.1}}} = \frac{3.38574}{\sqrt{3252.5 - 1719.76}}$$

$$= \frac{3.38574}{\sqrt{1532.74}} = \frac{3.38574}{39.15}$$

$$s_{b_2} = 0.08648$$

Continuing with the calculations on Dr. Mac's regression equation:

$$H_0: \quad \beta_1 = 0 \qquad H_0: \quad \beta_2 = 0$$
$$H_A: \quad \beta_1 \neq 0 \qquad H_A: \quad \beta_2 \neq 0$$
$$\alpha = 0.50 \qquad\qquad \alpha = 0.50$$

Reject H_0 if $t_{\text{test}} < -0.711$ or $t_{\text{test}} > 0.711$

$$t_{\text{test}} = \frac{b_1 - \beta_H}{s_{b_1}} \qquad\qquad t_{\text{test}} = \frac{b_2 - \beta_H}{s_{b_2}}$$

$$t_{test} = \frac{0.02606 - 0}{0.01163} \qquad t_{test} = \frac{0.06815 - 0}{0.08648}$$

$$t_{test} = 2.241 \qquad\qquad t_{test} = 0.788$$

Reject H_0 Reject H_0

Thus, at the 0.50 level of significance the professor can accept both independent variables as having a regression relationship with the number of customers. However, it should be noted that b_1 is significant at a much lower α level than is b_2.

Once again using the values in Table 13-4, we can compute R and R^2 to check the quality of the multiple regression equation.

$$R = \sqrt{\frac{b_1 \Sigma x_1 y + b_2 \Sigma x_2 y}{\Sigma y^2}}$$

$$R = \sqrt{\frac{(0.02606)\, 5887 + (0.06815)\, (680)}{280}}$$

$$= \sqrt{\frac{153.41522 + 46.34200}{280}}$$

$$R = \sqrt{\frac{199.75722}{280}} = \sqrt{0.71342} = 0.84464$$

$$R^2 = 0.71342$$

Changes in the two independent variables explain 71.34 percent of the changes in the dependent variable. Remember that we said earlier that R^2 would not equal the sum of the r^2's for the independent variables because of overlap. The two independent variables do have some overlap in this example. In other words, they are not fully independent of one another. If the R^2 is not large enough, then we may decide to try another independent variable instead of the present X_2 or we can add a third, or fourth, or more independent variables. We can't have more than eight in our example because we would exhaust our degrees of freedom since $d.f. = n - (1 + $ number of independent variables).

Finally, Dr. Mac has just learned that students have contributed $100.00 to the Student Mascot Fund. Because he already knows that enrollment is 1,200 students, he can now make an estimate as to how many customers the Library should expect.

$$Y_c = -10.24292 + 0.02606X_1 + 0.06815X_2$$
$$Y_c = -10.24292 + 0.02602\, (1200) + 0.06815\, (100)$$
$$= -10.24292 + 31.27200 + 6.81500$$
$$Y_c = 27.84408$$

If Dr. Mac is correct, the Library can make plans to serve approximately 28 customers. ◄

13.1 $Y_c = ab^X$,   *page 316*

13.2 $\log Y_c = \log a + X \log b$,   *page 316*

13.3 $\log b = \dfrac{[n\Sigma\,(X \log Y)] - [(\Sigma X)\,(\Sigma \log Y)]}{n\,(\Sigma X^2) - (\Sigma X)^2}$,   *page 317*

13.4 $\log a = \dfrac{\Sigma \log Y}{n} - \log b \left(\dfrac{\Sigma X}{n}\right)$,   *page 317*

13.5 $r = \dfrac{n\Sigma\,(X \log Y) - \Sigma X\,(\Sigma \log Y)}{\sqrt{[n\Sigma X^2 - (\Sigma X)^2]\,[n\Sigma\,(\log Y)^2 - (\Sigma \log Y)^2]}}$,   *page 319*

13.6 $Y_c = a + bX + cX^2$,   *page 321*

13.7 $\Sigma Y = na + b\Sigma X + c\Sigma X^2$,   *page 321*

13.8 $\Sigma XY = a\Sigma X + b\Sigma X^2 + c\Sigma X^3$,   *page 321*

13.9 $\Sigma X^2Y = a\Sigma X^2 + b\Sigma X^3 + c\Sigma X^4$,   *page 321*

13.10 $r = \sqrt{\dfrac{a\Sigma Y + b\Sigma XY + c\Sigma X^2Y - n\overline{Y}^2}{\Sigma Y^2 - n\overline{Y}^2}}$,   *page 323*

13.11 $Y_c = a + b_1X_1 + b_2X_2 + \cdots + b_mX_m$,   *page 326*

13.12 $\Sigma Y = na + b_1\Sigma X_1 + b_2\Sigma X_2 + \cdots + b_m\Sigma X_m$,   *page 326*

13.13 $\Sigma X_1Y = a\Sigma X_1 + b_1\Sigma X_1^2 + b_2\Sigma X_1X_2 + \cdots + b_m\Sigma X_1X_m$,
page 326

13.14 $\Sigma X_2Y = a\Sigma X_2 + b_1\Sigma X_1X_2 + b_2\Sigma X_2^2 + \cdots + b_m\Sigma X_2X_m$,
page 327

13.15 $\Sigma X_mY = a\Sigma X_m + b_1\Sigma X_1X_m + b_2\Sigma X_2X_m + \cdots + b_m\Sigma X_m^2$,
page 327

13.16 $Y_c = a + b_1X_1 + b_2X_2$,   *page 327*

13.17 $\Sigma Y = na + b_1\Sigma X_1 + b_2\Sigma X_2$,   *page 327*

13.18 $\Sigma X_1Y = a\Sigma X_1 + b_1\Sigma X_1^2 + b_2\Sigma X_1X_2$,   *page 327*

13.19 $\Sigma X_2Y = a\Sigma X_2 + b_1\Sigma X_1X_2 + b_2\Sigma X_2^2$,   *page 327*

13.20 $F_{\text{test}} = \dfrac{MSR}{MSE}$,   *page 328*

13.21 $MSR = \dfrac{SSR}{m}$,   *page 328*

13.22 $MSE = \dfrac{SSE}{n - m - 1}$,   *page 328*

13.23 $R^2 = \dfrac{SSR}{SST}$,   *page 328*

13.24 $\Sigma x_1 y = b_1 \Sigma x_1^2 + b_2 \Sigma x_1 x_2$,   *page 332*

13.25 $\Sigma x_2 y = b_1 \Sigma x_1 x_2 + b_2 \Sigma x_2^2$,   *page 332*

13.26 $\Sigma y^2 = \Sigma Y^2 - n\overline{Y}^2$,   *page 332*

13.27 $\Sigma x_1^2 = \Sigma X_1^2 - n\overline{X}_1^2$,   *page 332*

13.28 $\Sigma x_2^2 = \Sigma X_2^2 - n\overline{X}_2^2$,   *page 332*

13.29 $\Sigma x_1 y = \Sigma X_1 Y - n\overline{X}_1 \overline{Y}$,   *page 332*

13.30 $\Sigma x_2 y = \Sigma X_2 Y - n\overline{X}_2 \overline{Y}$,   *page 332*

13.31 $\Sigma x_1 x_2 = \Sigma X_1 X_2 - n\overline{X}_1 \overline{X}_2$,   *page 332*

13.32 $a = \overline{Y} - b_1 \overline{X}_1 - b_2 \overline{X}_2$,   *page 332*

13.33 $s_m = \sqrt{\dfrac{\Sigma y^2 - b_1 \Sigma x_1 y - b_2 \Sigma x_2 y}{n-3}}$,   *page 332*

13.34 $s_{b_1} = \dfrac{s_m}{\sqrt{\Sigma x_1^2 - \dfrac{(\Sigma x_1 x_2)^2}{\Sigma x_2^2}}}$,   *page 332*

13.35 $s_{b_2} = \dfrac{s_m}{\sqrt{\Sigma x_2^2 - \dfrac{(\Sigma x_1 x_2)^2}{\Sigma x_1^2}}}$,   *page 332*

13.36 $t_{\text{test}} = \dfrac{b_i - \beta_H}{s_{b_i}}$,   *page 332*

13.37 $R = \sqrt{\dfrac{b_1 \Sigma x_1 y + b_2 \Sigma x_2 y}{\Sigma y^2}}$,   *page 333*

REVIEW QUESTIONS

1. Explain the use of the equations at the end of the chapter.
2. Differentiate between nonlinear and multiple linear regression analysis.
3. Explain why R^2 is not the sum of the simple r^2s in multiple correlation analysis.

1. A service company has kept records on the number of employees it needs for various levels of business activity. A portion of those records is given below.

NUMBER OF SERVICE CALLS	NUMBER OF EMPLOYEES REQUIRED
10	1
15	2
20	3
25	5
30	8
35	12
40	15

   a. Fit an exponential regression equation to the data.
   b. Using the regression equation, estimate the number of employees required if the level of business reaches 45 calls.
   c. Compute the coefficient of correlation.
   d. What is the percentage increase in personnel required for each additional service call?

2. A certain stretch of highway has become infamous for the number of traffic accidents. Records kept by the highway department for several years are presented in the following table:

YEAR	JULY 4 TRAFFIC COUNT	NUMBER OF ACCIDENTS
1969	452	2
1970	425	1
1971	500	6
1972	517	5
1973	350	1
1974	480	4
1975	603	7
1976	611	10
1977	650	15
1978	718	18

   a. Fit an exponential regression equation to the data using number of accidents as the dependent variable and traffic count as the independent variable.
   b. In 1979 the traffic count on July 4 was 700. Estimate the number of accidents in that year.
   c. Compute the coefficient of correlation.

3. In one of the rural counties of Texas, agricultural employment has been declining for several years. The county chamber of commerce has collected the data below (see hint):

YEAR	AGRICULTURAL EMPLOYMENT
1968	1050
1969	1000
1970	900
1971	800
1972	750
1973	675
1974	500
1975	450
1976	425
1977	350
1978	300

a. Fit a simple linear regression equation to the data.
b. Fit an exponential regression equation to the data.
c. Fit a second-degree regression equation to the data.
d. Estimate agricultural employment in the county in 1988. Looking at your estimates, which equation seems to be the most logical one, intuitively, to use? Why?
e. Compute the coefficient of determination for the exponential equation.
f. Compute the coefficient of determination for the second-degree equation:
(Hint: The arithmetic in the problem will be simplified if you subtract 1973 from each year.)

4. The owners of the Pizza Place have been keeping records on their average cost per pizza and their volume of business.

NUMBER OF PIZZAS SOLD	AVERAGE COST
100	$0.90
110	0.88
120	0.85
130	0.80
140	0.72
150	0.65
160	0.70
170	0.73
180	0.79

a. Make a scatter diagram with the data.
b. Compute a second-degree regression equation for the data set.

5. Economic Analysis Associates has been retained to study the market for dingbats. They have experimented with the production process and also carried out an extensive test-marketing program. The data in the table below were developed from their work.

If the client has enough capital to produce only 350,000 dingbats, estimate his profit potential using exponential regression equations.

DINGBAT DATA		
QUANTITY (1,000)	PRICE (PER UNIT)	COST (PER UNIT)
100	$1.00	$0.50
200	0.90	0.55
300	0.80	0.62
400	0.72	0.70
500	0.65	0.80
600	0.60	0.88
700	0.55	0.95

6. After having completed the Dingbat Study described in Problem 5, EAA was employed to develop some additional information on the market for dingbats. Based on previous research, they feel that dingbat consumption may be related to median family income. To test this notion, they gathered the data below. Construct a scatter diagram, determine an appropriate regression equation, and compute R^2. Finally, based on your work, what conclusions can you draw about income levels and dingbat consumption.

DINGBAT CONSUMPTION IN VARIOUS MARKET AREAS		
MARKET	MEDIAN FAMILY INCOME	DINGBAT CONSUMPTION
1	$10,500	1,500
2	8,200	1,000
3	20,100	2,000
4	15,600	2,500
5	18,000	2,200
6	13,000	2,100
7	12,700	2,000
8	17,300	2,400

7. The Maynerd Company, a manufacturer of dog food, sells its product through supermarkets. The sales manager is seeking a way of predicting sales when the company enters a new market and decides to investigate the feasibility of doing this on the basis of a regression equation derived from the information on page 342:

AREA	COMPANY SALES ($1,000's)	NUMBER OF DOG LICENSES (100's)	NUMBER OF SUPERMARKETS CARRYING PRODUCT
1	76	15	25
2	38	10	29
3	25	7	15
4	98	21	25
5	93	14	11
6	54	8	13
7	78	14	22
8	85	24	14
9	65	9	13
10	88	23	11

a. Calculate the regression equation, test the partial regression coefficients and compute R^2.

b. Should the sales manager use this equation to predict sales in new markets?

8. The matrix below contains the correlation coefficients for the variables in Table 13-3. Based on these data, what conclusions would you reach concerning the use of variables 3, 4, and 5 as independent variables in the same multiple regression equation?

VARIABLE	1	2	3	4	5
1	1.0000	0.9869	0.9473	0.1591	0.9941
2	0.9869	1.0000	0.9329	0.2106	0.9810
3	0.9473	0.9329	1.0000	−0.1286	0.9652
4	0.1591	0.2106	−0.1286	1.0000	0.1006
5	0.9941	0.9810	0.9652	0.1006	1.0000

9. Figures 13-5 and 13-6 contain regression equations for the data in Table 13-3 with bachelors degrees as Y and median income, the unemployment rate, and population as independent variables. Use these same data and compute a regression equation with unemployment and the population as the independent variables. Next, compute a simple linear regression equation to estimate bachelor degrees using population as the only independent variable.

Compare your results to those in Figures 13-5 and 13-6. Which equation among the four should you use?

10. Sailboat sales at the Waterway Marina have varied a great deal over time but recently the boats have been gaining in popularity. Use a second-degree regression equation and project sales in 1981.

YEAR	NUMBER OF SAILBOATS SOLD
1970	100
1971	90
1972	85
1973	90
1974	75
1975	80
1976	86
1977	110
1978	115

(Hint: If you subtract 1974 from each year, you will simplify your work.)

11. In the table below are some data for beer sales at a convenience store. The sales for a random selection of days and the high temperature, as well as the traffic count on the street, are given for each day.

TEMPERATURE	TRAFFIC COUNT	NUMBER OF 6-PAKS OF BEER SOLD
78	720	61
80	623	83
40	301	15
70	850	90
68	352	45
45	424	38
92	605	103
42	457	58
75	783	79
50	311	27

a. Compute a multiple regression equation to predict beer sales using temperature and traffic count as the independent variables.
b. Test the b coefficients at $\alpha = 0.2$.
c. Compute the coefficient of multiple correlation for the equation in part a above.

12. The strawberry crop seems to vary with the amount of rainfall and the amount of fertilizer used. Records of one farmer are given in the table on page 348.
a. Construct a multiple regression equation using rainfall as X_1 and fertilizer as X_2.
b. Test the b coefficients with $\alpha = 0.01$.
c. Compute the coefficient of multiple determination.

AMOUNT OF RAINFALL (INCHES)	FERTILIZER USED (TONS)	STRAWBERRY CROP (CRATES)
16	510	1000
22	450	950
23	500	1200
13	425	700
17	450	800
25	475	1100
18	515	1050
20	500	1150
21	490	1000
19	510	950
22	525	1300

Index Numbers

**CHAPTER
14**

"In the good old days, things were a lot cheaper and a dollar was worth something." Does that sound familiar? Probably so, for it is a common expression, particularly when the bills arrive each month. The phrase obviously implies that prices are higher today than in the "good old days." Is that really true, and if it is, how can we measure the change?

In order to interpret a measurement meaningfully, we often use a *standard* or *yardstick*. You use a great many such standards—miles to measure distance, pounds to measure weight, quarts to measure volume, etc. These standards remain fixed. That is, a mile is a mile yesterday, today, and tomorrow.

Change, however, is one of those things for which no stable standard of measurement exists. Thus, to handle this situation, the statistician uses *index numbers*. An index number is a ratio multiplied by 100 that can be used as a specialized form of a *yardstick*.

The purpose of this chapter is to show you how to construct and use index numbers. We will investigate simple and composite index numbers for both price and quantity changes. The primary effort, however, will concern price indexes.

Simple Price Index

A *simple index* is merely the ratio of each value in a series to one particular value in the series, which serves as the base or base period value. Table 14-1 shows prices and quantities sold for three items available at the Pizza Parlor. A simple price index, PI, for each item can be constructed using Equation 14.1.

$$(PI)_i = (p_i/p_b)(100) \qquad\qquad (14.1)$$

where $(PI)_i$ = index number for time period i

                 p_i = price for time period i

                 p_b = price for time period b (base period)

TABLE 14-1. *Historical Records of Pizza Parlor.*

	UNIT	PRICE PER UNIT			QUANTITY SOLD		
ITEM	SIZE	1976	1977	1978	1976	1977	1978
Cola	8 ounce	$0.20	$0.25	$0.30	2000	2200	2500
Salad	4 ounce	0.50	0.60	0.75	1000	1200	1500
Pizza	12 inch	2.00	2.10	2.30	1500	1750	2000

Table 14-2 shows the results of using Equation 14.1 to construct a simple price index for each item in Table 14-1. In Table 14-2, a different base period year was used for each item. This is noted by the year that is set equal to 100. For cola, 1976 = 100 is the base period; for salad, 1977 = 100 is the base period; and for pizza, 1978 = 100 is the base period.

TABLE 14-2. *Construction of Simple Price Index.*

	INDEX NUMBER		
YEAR	COLA: 1976 = 100	SALAD: 1977 = 100	PIZZA: 1978 = 100
1976	(0.20/0.20)(100) = 100	(0.50/0.60)(100) = 83.3	(2.00/2.30)(100) = 87.0
1977	(0.25/0.20)(100) = 125	(0.60/0.60)(100) = 100	(2.10/2.30)(100) = 91.3
1978	(0.30/0.20)(100) = 150	(0.75/0.60)(100) = 125	(2.30/2.30)(100) = 100

Looking at the first column of Table 14-2, we can see that the price of cola increased 25 percent in 1977 over the base period 1976, and the price of cola increased 50 percent in 1978 over the base period 1976. Notice that these increases can be read directly by subtracting 100 from the appropriate index. This method of determining the percentage increase is valid, however, only if the base period year is compared to some other year. That is, it would be *incorrect* to say that the percentage increase in the cola price from 1977 to 1978 is 25 (150 − 125). This is actually the percentage point increase. The percentage increase is determined as (150 − 125)/125 = 20 percent.

All of the indexes in Table 14-2 have as their base period a single year. At times, however, it is more convenient to use more than one year to form the base period. When this is the case, p_b in Equation 14.1 becomes the *arithmetic mean price of the years chosen for the base period.* Table 14-3 shows a price index for the salad prices with the base period as 1976 − 1977 = 100.

TABLE 14-3. *Simple Price Index for Salad: 1976-77 = 100.*

YEAR	INDEX NUMBERS FOR SALAD: 1976-77 = 100
1976	$\dfrac{0.50}{\dfrac{0.50 + 0.60}{2}} (100) = 90.9$
1977	$\dfrac{0.60}{\dfrac{0.50 + 0.60}{2}} (100) = 109.1$
1978	$\dfrac{0.75}{\dfrac{0.50 + 0.60}{2}} (100) = 136.4$

Notice in Table 14-3 there is not an index number exactly equal to 100. This is so because the average price for the base period was $0.55, and no individual yearly price of salad was exactly $0.55.

Composite Price Index

There are times when a simple price index is not sufficient. That is, suppose the Pizza Parlor really wants to know about the overall price change in the three items rather than the individual price changes. In this situation, we must resort to the construction of a *composite index*. Four methods of constructing a composite price index will be presented.

Aggregate Method

An aggregate price index is constructed by summing the individual prices of a chosen year and dividing by the sum of prices in the base year as shown in Equation 14.2. Table 14-4 shows the calculation of the aggregate price index using Equation 14.2 for the three items of the Pizza Parlor.

$$(\text{PI})_i = \frac{\Sigma p_i}{\Sigma p_b} (100) \tag{14.2}$$

where
$(\text{PI})_i$ = index number for time period i
p_i = prices in time period i
p_b = prices in time period b (base period)

TABLE 14-4.   *Calculation of Composite Price Index by the Aggregate Method.*

| | PRICE PER UNIT | | | | PRICE INDEX |
YEAR	COLA	SALAD	PIZZA	TOTAL	1976 = 100
1976	$0.20	$0.50	$2.00	$2.70	100
1977	0.25	0.60	2.10	2.95	109.3
1978	0.30	0.75	2.30	3.35	124.1

In this method each price is treated without regard to units when calculating the index.   Yet the unit of the product for which the price is given will make a difference in the numerical outcome of the calculated index.   To overcome this deficiency, the weighted aggregate method is frequently used.

Weighted Aggregate Method

The *weighted aggregate method* utilizes some measure to *assign importance* to the individual prices.   For our local Pizza Parlor, it would be appropriate to weight our items according to the quantities sold given in Table 14-1.

When *quantities* are used as the weights, the quantities chosen for weights are *usually those in the base period or current period*.   When base period weights are used, the index is referred to as a *Laspeyres index;* and when current period weights are used, the index is referred to as a *Paasche index.*[1]   In practice, the Laspeyres index is probably more prevalent because it requires only one period of quantity information.   Equations 14.3 and 14.4 are used to calculate weighted aggregate price indexes.

$$(PI)_{Li} = \frac{\Sigma\,(p_i q_b)}{\Sigma\,(p_b q_b)}(100) \quad \text{(Laspeyres)} \tag{14.3}$$

where $(PI)_{Li}$ = price index for the time period i
p_i = prices in time period i
p_b = prices in time period b (base period)
q_b = quantities in time period b (base period)

$$(PI)_{Pi} = \frac{\Sigma\,(p_i q_i)}{\Sigma\,(p_b q_i)}(100) \quad \text{(Paasche)} \tag{14.4}$$

where $(PI)_{Pi}$ = price index for time period i
p_i = prices in time period i

---

1.   The indexes are named after the men who developed each concept.

$$p_b = \text{prices in time period } b \text{ (base period)}$$
$$q_i = \text{quantities in time period } i$$

Table 14-5 shows the calculation of the price index for the items of the Pizza Parlor using the weighted aggregate method and base period weights. That is, we have used Equation 14.3 to construct Table 14-5. Notice that only one year of quantity data is used. The index provides a measure of the price change focused on the base year quantities. Thus, in 1977 at the Pizza Parlor it would take $1.09 to buy the same quantity that one dollar purchased in 1976.

TABLE 14-5. *Calculation of Composite Price Index by the Weighted Aggregate Method: Laspeyres Weighting.*

	PRICE PER UNIT			1976 BASE YEAR	$p_i q_b$		
ITEM	1976	1977	1978	QUANTITY	1976	1977	1978
Cola	$0.20	$0.25	$0.30	2000	$ 400	$ 500	$ 600
Salad	0.50	0.60	0.75	1000	500	600	750
Pizza	2.00	2.10	2.30	1500	3000	3150	3450
					$3900	$4250	$4800

$$\text{Price Index} = \frac{\Sigma (p_i q_b)}{\Sigma (p_b q_b)} (100) \qquad \frac{3900}{3900}(100) \qquad \frac{4250}{3900}(100) \qquad \frac{4800}{3900}(100)$$

$$\text{Price Index: } (1976 = 100) \qquad\qquad 100 \qquad\qquad 109.0 \qquad\qquad 123.1$$

TABLE 14-6. *Calculation of Composite Price Index by the Weighted Aggregate Method: Paasche Weighting, 1976 = 100.*

	1976		1977		1978						
	PRICE	QUANTITY	PRICE	QUANTITY	PRICE	QUANTITY					
ITEM	p_b	q_b	p_1	q_1	p_2	q_2	$p_b q_b$	$p_b q_1$	$p_b q_2$	$p_1 q_1$	$p_2 q_2$
Cola	$0.20	2000	$0.25	2200	$0.30	2500	$ 400	$ 440	$ 500	$ 550	$ 750
Salad	0.50	1000	0.60	1200	0.75	1500	500	600	750	720	1125
Pizza	2.00	1500	2.10	1750	2.30	2000	3000	3500	4000	3675	4600
							$3900	$4540	$5250	$4945	$6475

$$(PI)_i = \frac{\Sigma (p_i q_i)}{\Sigma (p_b q_i)} (100)$$

$$(PI)_{1976} = \frac{(3900)}{(3900)} (100) = 100$$

$$(PI)_{1977} = \frac{(4945)}{(4540)} (100) = 108.9$$

$$(PI)_{1978} = \frac{(6475)}{(5250)} (100) = 123.3$$

Table 14-6 shows the calculation of the price index for the items of the Pizza Parlor using the weighted aggregate method and current period weights. That is, we have used Equation 14.4 to construct Table 14-6. This index provides a measure of the price change focused on current quantities. That is, the quantities purchased in 1977 cost 8.9 percent more than if they had been purchased in 1976. Also, it should be noted that even though the indexes in Tables 14-5 and 14-6 are very nearly the same they do not have the same interpretation.

Average of Relatives Method

Another rather simple composite index which does not weight the individual prices is the *average of relatives method*. This method sums the simple price indexes and averages them by dividing by the number of items as shown in Equation 14.5

$$(PI)_i = \frac{\Sigma\left[(p_i/p_b)(100)\right]}{N} \tag{14.5}$$

where p_i = prices in period i

p_b = prices in period b (base period)

N = number of items

TABLE 14-7. *Calculation of Composite Price Index by the Average of Relatives Method.*

	SIMPLE PRICE INDEX			AVERAGE OF RELATIVES $PI = \frac{\Sigma\left[(p_i/p_b)\,100\right]}{N}$
YEAR	COLA: 1976 = 100	SALAD: 1976 = 100	PIZZA: 1976 = 100	
1976	(.20/.20) (100) = 100	(.50/.50) (100) = 100	(2.00/2.00) (100) = 100	300/3 = 100.0
1977	(.25/.20) (100) = 125	(.60/.50) (100) = 120	(2.10/2.00) (100) = 105	350/3 = 116.7
1978	(.30/.20) (100) = 150	(.75/.50) (100) = 150	(2.30/2.00) (100) = 115	415/3 = 138.3

Table 14-7 shows how to calculate this composite index for the Pizza Parlor data. This method, however, has a weakness in that equal dollar changes in prices of different products may not have the same effect on the value of the index. To overcome this weakness, let us look at the weighted average of relatives method.

Weighted Average of Relatives Method

Equations 14.6 and 14.7 are used to calculate weighted average of relative price indexes according to the Laspeyres and Paasche methods, respectively. If you look closely, you will see that Equation 14.6 is equivalent to Equation 14.3 (the weighted aggregate Laspeyres method) and Equation 14.7 is equivalent to Equation 14.4 (the weighted aggregate Paasche method). Thus, when you apply either set of equations, you should get the same answers.

$$(PI)_{Li} = \frac{\Sigma\left[(p_i/p_b)\,(100)\,(p_bq_b)\right]}{\Sigma\,(p_bq_b)} \qquad (14.6)$$

$$(PI)_{Pi} = \frac{\Sigma\left[(p_i/p_b)\,(100)\,(p_bq_i)\right]}{\Sigma\,(p_bq_i)} \qquad (14.7)$$

The need for the weighted average of relatives method might then be questioned. It has an important place, however, because it can be applied to a set of index numbers already constructed. That is, in Equations 14.6 and 14.7, the expression, $(p_i/p_b)\,(100)$, could be simple index numbers already calculated. Tables 14-8 and 14-9 show the index numbers resulting from using Equations 14.6 and 14.7. As you can see the results from the weighted average of relatives method are identical to the results from the weighted aggregate method for both the Laspeyres and Paasche weighting. And again, you should be reminded that even though the Laspeyres and Paasche approaches may sometimes give very similar results, the interpretation of the two sets of index numbers is not the same.

TABLE 14-8. *Calculation of Composite Price Index by the Weighted Average of Relatives Method: Laspeyres Weighting.*

	SIMPLE PRICE INDEX p_i/p_b (100): 1976 = 100			p_bq_b	WEIGHTED PRICE INDEX (p_i/p_b) (100) (p_bq_b)		
ITEM	1976	1977	1978	1976 = 100	1976	1977	1978
Cola	100	125	150	400	40,000	50,000	60,000
Salad	100	120	150	500	50,000	60,000	75,000
Pizza	100	105	115	3000	300,000	315,000	345,000
				3900	390,000	425,000	480,000

$$\text{Price Index} = \frac{\Sigma\left[(p_i/p_b)\,(100)\,(p_bq_b)\right]}{\Sigma\,(p_bq_b)}$$

				$\dfrac{390,000}{3900}$	$\dfrac{425,000}{3900}$	$\dfrac{480,000}{3900}$
Price Index: (1976 = 100)				100	109.0	123.1

TABLE 14-9. *Calculation of Composite Price Index by Weighted Average of Relatives Method: Paasche Weighting.*

	SIMPLE PRICE INDEX p_i/p_b (100): 1976 = 100						WEIGHTED PRICE INDEX (p_i/p_b) (100) $(p_b q_i)$		
ITEM	1976	1977	1978	$p_b q_b$	$p_b q_1$	$p_b q_2$	1976	1977	1978
Cola	100	125	150	400	440	500	40000	55000	75000
Salad	100	120	150	500	600	750	50000	72000	112500
Pizza	100	105	115	3000	3500	4000	300000	367500	460000
				3900	4540	5250	390000	494500	647500

$$\text{Price Index} = \frac{\Sigma\,[(p_i/p_b)\,(100)\,(p_b q_i)]}{\Sigma\,(p_b q_i)} \qquad \frac{390000}{3900} \quad \frac{494500}{4540} \quad \frac{647500}{5250}$$

Price Index: (1976 = 100)   100   108.9   123.3

Specific Price Indexes

Two of the most commonly quoted price indexes are the Consumer Price Index and the Producer Price Index (formerly the Wholesale Price Index). Since these are indexes you have heard of, or probably will hear about, a brief description of each is presented below.

Consumer Price Index

Beginning in January 1978, the Bureau of Labor Statistics altered substantially the Consumer Price Index, CPI, that it publishes each month. As a matter of fact, the Bureau now publishes two CPI's. One is the traditional Consumer Price Index that is designed to measure the price change in a fixed market basket of goods and services purchased by urban wage earners and clerical workers, including their families. The second CPI published each month is for all urban consumers. This new index covers about 80 percent of the population, whereas the urban wage and clerical worker CPI encompasses approximately 50 percent of the nation's population.

The 1978 revision of the CPI also included several other changes. Prior to the revision, the weights assigned to the items included in the index were based on a survey of consumer expenditures conducted in the period 1960 to 1961. A new consumer expenditure survey was carried out during 1972 and 1973, and the current CPI uses weights derived from that survey. Prior to 1978, the index was based on a fixed market basket of 400 specific items, both goods and services. With the implementation of the revised index, items included in the market basket are described in broader terms so that now the index is comprised of 250 items that are more general.

Essentially, the CPI is a weighted average of relatives index. The base year for the index is 1967 = 100. While the CPI is not exactly what we think of as a Laspeyres index, it is calculated in a very similar manner.

The CPI has many uses. It is a measure of changes in prices; it can be used as a deflator for other economic series; and a third major use is to escalate income payments. However, it is not a cost-of-living index although it is frequently called that. There are several reasons why the CPI is not a true cost-of-living index. For example, the market basket is relatively fixed so that it does not reflect changes in consumer spending patterns caused by changes in the relative prices of items. Moreover, the CPI includes only those taxes directly associated with retail prices, such as sales taxes, and excludes income and social security taxes, among others.

One further point should be remembered when you use the CPI. Many prices vary on a seasonal basis. That is, the price of a hotel room in Miami increases each winter and then declines in the summer. Consequently, the index is released in two forms—one in which the index has not been adjusted for seasonal price changes and the other where an adjustment has been made. While both reflect changes in the price level from month to month, you need to be sure you use the one that meets your needs.

Producer Price Index

The Producer Price Index (formerly known as the Wholesale Price Index) is also published monthly by the Bureau of Labor Statistics. The index is designed to measure changes in prices of commodities in primary markets in all stages of processing (essentially three main groups: crude materials, intermediate goods, and finished goods). The Producer Price Index is a weighted average of relatives index that includes about 2,800 commodities. Prices are obtained from representative producers and manufacturers and weighted by the values of shipments obtained from a 1972 study. Thus, the Producer Price Index differs from the Consumer Price Index in that the Producer Price Index attempts to measure changes in prices for primary transactions that occur in other than retail markets whereas the Consumer Price Index attempts to measure changes in consumer (retail) prices.

Some Other Common Indexes

No doubt you have seen, or heard reference to, common stock price indexes. The Standard and Poor's composite index is computed daily using a group of 500 stocks, with a base period of 1940 to 1943 = 10. There are five New York Stock Exchange price indexes each with a base of December 31, 1965 = 50. These five indexes (composite, industrials, transportation, utilities, and finance) include the prices of all stocks listed on the New York Exchange and are developed from closing prices each day.

Other frequently used price indexes include those for prices received and paid by farmers. There are also price indexes, called *implicit price deflators,* for the gross national product, gross private domestic investment, and exports, among others.

All the price index series mentioned here are available in several different publications. However, one publication that includes all of them, as well as many other frequently used economic series, is *Economic Indicators.* This is a U.S. Government publication that is released each month.

Working with Index Numbers

When working with index numbers that have already been constructed, it is possible that you may have to modify the index before you use it. Two useful modifications are shifting the base and splicing two index series.

Shifting the Base

You may encounter series of index numbers with *different base periods* or a series with a *rather old base year.* In either case, whether it is the need to put several series on the same base year or merely to update the base period, it is necessary to *shift the base.* Shifting the base is accomplished by taking the index number for the year desired as the base and dividing this index number into every other index number in the series. Table 14-10 shows a shift in the base of the hypothetical data for Product A. As you can see the base year was shifted from 1973 = 100 to 1975 = 100. This shift was accomplished by taking the 1975 index number of 108 and dividing it into the original price index producing the new price index with 1975 = 100.

TABLE 14-10. *Calculations for Shifting the Base.*

YEAR	PRODUCT A PRICE INDEX 1973 = 100	PRODUCT A PRICE INDEX 1975 = 100
1973	100	(100/108) (100 = 92.6
1974	105	(105/108) (100) = 97.2
1975	108	(108/108) (100) = 100
1976	110	(110/108) (100) = 101.9

Splicing Two Index Series

Sometimes, when using published index numbers, the period of interest will include the point in time where the base was changed. The result is that it becomes necessary to *splice* the two series, forming *one continuous series* of index numbers with a common base. Table 14-11 shows such a situation and the resulting spliced series.

TABLE 14-11. *Spliced Series for Hypothetical Data.*

YEAR	PRODUCT A PRICE INDEX 1964 = 100	PRODUCT A PRICE INDEX 1972 = 100	SPLICED SERIES 1972 = 100
1973	130		(105)
1974	134		(108)
1975	139	112	112
1976		117	117
1977		122	122
1978		128	128

The circled index numbers in Table 14-11 were obtained by using the following approach. If you look closely, you will see that new values for the spliced index are being constructed in such a manner as to maintain the same percentage change that existed in the 1964 = 100 index values.

$$I_{1974} = \frac{134}{139}(112) = 108$$

$$I_{1973} = \frac{130}{134}(108) = 105$$

spliced index are being constructed in such a manner as to maintain the same percentage change that existed in the 1964 = 100 index values.

Use of Price Index Numbers

In addition to showing the change in prices over time, price indexes are useful in the analysis of many sets of data expressed in money terms. Two common usages are (1) the determination of *real income* and (2) the *deflating of* sales data presented in the following two examples.

▶ **Example 14.1**   Suppose a particular wage earner has had salary figures of $8,000, $8,500, $9,200, $10,000, and $11,000 for the years 1969-73. Let us also suppose that the Consumer Price Index (CPI) is representative of goods and services he must purchase each year. Now let us see what his purchasing power has actually been each year. Table 14-12 shows our wage earner's real wage in terms of 1967 dollars. As you can see, our wage earner's salary increased by $3,000 or 37.5 percent, but his buying power in terms of what could be purchased with 1967 dollars increased only $978 or 13.4 percent. Thus, our wage earner is better off, but not by as much as you first might think.

TABLE 14-12.   *Demonstration of Determining Real Wage.*

YEAR	MONEY WAGE	CPI* 1967 = 100	1967 DOLLARS REAL WAGE
1969	$ 8,000	109.8	($ 8,000/109.8) (100) = $7286
1970	8,500	116.3	($ 8,500/116.3) (100) = 7309
1971	9,200	121.3	($ 9,200/121.3) (100) = 7585
1972	10,000	125.3	($10,000/125.3) (100) = 7981
1973	11,000	133.1	($11,000/133.1) (100) = 8264

*Source: *Monthly Labor Review*, U.S. Department of Labor, April, 1974.   ◀

▶ **Example 14.2**   A manufacturer has a record of sales and of his produce price index series for the years 1974 to 1978. In order to obtain a clearer picture of the growth in sales, the manufacturer deflates the sales figures as shown in Table 14-13. Now in terms of 1974 dollars, the increase in sales was $2,364 rather than $4,000, or 16.9 percent rather than 28.6 percent.

TABLE 14-13.   *Deflating Sales.*

YEAR	PRODUCT SALES	PRODUCT PRICE INDEX 1974 = 100	DEFLATED SALES
1974	$14,000	100	$14,000
1975	15,000	103	($15,000/103) (100) = 14,563
1976	15,200	105	($15,200/105) (100) = 14,476
1977	17,000	108	($17,000/108) (100) = 15,740
1978	18,000	110	($18,000/110) (100) = 16,364

The deflated sales figures reflected only the change in sales that are the result of changes in the quantity of the product sold. Thus, in terms of current dollars (column 2) the period 1975-76 looks like a period where sales quantity increased, but the deflated sales column (constant dollars) shows the true picture. That is, the year 1976 actually had a small decline from the quantity sold in 1975. ◀

Quantity Index

Virtually all that has been said about price indexes in the preceding pages could be repeated with respect to quantity indexes. However, to expedite discussion, only the equations necessary for calculating quantity indexes are shown in Table 14-14. The only difference in the equations involves a transposition of the p's and q's as can be seen in Table 14-14. Each method of calculating an index has the same limitations or advantages whether applied to prices or quantities. And the quantity index is used to measure change in quantity over a period of time.

TABLE 14-14. *Equations for Price and Quantity Indexes.*

METHOD	PRICE INDEX	QUANTITY INDEX
Simple	$(p_i/p_b)\,100$	$(q_i/q_b)\,100$
Simple Aggregate	$\dfrac{\Sigma p_i}{\Sigma p_b}(100)$	$\dfrac{\Sigma q_i}{\Sigma q_b}(100)*$
Weighted Aggregate		
Laspeyres	$\dfrac{\Sigma\,(p_i q_b)}{\Sigma\,(p_b q_b)}(100)$	$\dfrac{\Sigma\,(q_i p_b)}{\Sigma\,(q_b p_b)}(100)$
Paasche	$\dfrac{\Sigma\,(p_i q_i)}{\Sigma\,(p_b q_i)}(100)$	$\dfrac{\Sigma\,(q_i p_i)}{\Sigma\,(q_b p_i)}(100)$
Simple Average of Relatives	$\dfrac{\Sigma\,[\,(p_i/p_b)\,(100)\,]}{N}$	$\dfrac{\Sigma\,[\,(q_i/q_b)\,(100)\,]}{N}$
Weighted Average of Relatives		
Laspeyres	$\dfrac{\Sigma\,[\,(p_i/p_b)\,(100)\,(p_b q_b)\,]}{\Sigma\,(p_b q_b)}$	$\dfrac{\Sigma\,[\,(q_i/q_b)\,(100)\,(q_b p_b)\,]}{\Sigma\,(q_b p_b)}$
Paasche	$\dfrac{\Sigma\,[\,(p_i/p_b)\,(100)\,(p_b q_i)\,]}{\Sigma\,(p_b q_i)}$	$\dfrac{\Sigma\,[\,(q_i/q_b)\,(100)\,(q_b p_i)\,]}{\Sigma\,(q_b p_i)}$

*Use of this equation requires homogeneous units.

--------------------- SUMMARY OF EQUATIONS ---------------------

See Table 14-14.

1. Explain the use of each equation shown in Table 14-14.
2. Define
   a. simple index
   b. Consumer Price Index (CPI)
   c. Producer Price Index (PPI)
3. Explain the difference between Laspeyres and Paasche indexes.
4. What is meant by "shifting the base" of a series of index numbers?
5. What is meant by "splicing" two index series?

PROBLEMS

1. The Chattanooga Food Price Index (CFPI) is a statistical measure of changes in the price of food bought for home consumption by urban wage earners and clerical workers and their families. The index is based on actual prices of 93 food items in a sample of retail outlets representing both chain and independent grocers in the Chattanooga Metropolitan area. The study of CFPI was conducted by The University of Tennessee Center for Business and Economic Research (CBER). The prices of five items in Chattanooga from November 1975 through March 1976 are given in the table below

Price Per Unit

ITEM	UNIT	NOV. 1975	DEC. 1975	JAN. 1976	FEB. 1976	MARCH 1976
White, all-purpose flour	5 lb.	0.966	0.984	0.976	0.945	0.961
Frying chicken	2 lb.	1.232	1.196	1.261	1.210	1.192
American process cheese	½ lb.	0.842	0.841	0.847	0.858	0.856
Carrots	1 lb.	0.238	0.252	0.240	0.231	0.220
Canned coffee	1 lb.	1.454	1.471	1.480	1.478	1.493

   a. Construct a simple price index for a 5-lb bag of white, all-purpose flour with November, 1975 = 100.
   b. Shift the base of the index constructed in part (a) to January, 1976.
   c. Construct a simple price index for ½ lb. American process cheese with November, 1975 = 100.
   d. Construct a simple price index for canned coffee (1 lb) with December, 1975 = 100.

2. Use the data in Problem 1 and calculate a composite price index for each month using the simple aggregate method with November, 1975 = 100.

3. On May 19, 1976, the office of the Governor of Tennessee released the results of a study indicating the April, 1976, Chattanooga Food Price Index:

Chattanooga Food Price Index, Components Index, April, 1976. (November 1975 = 100)

Foods at home	99.3
Cereals and baking products	102.7
Meats, poultry, fish	90.6
Dairy products	105.1
Fruits and vegetables	105.2
Other foods at home	100.7

a. What is the percentage increase in fruits and vegetables from November, 1975, to April, 1976?

b. What has happened to the price of food at home in Chattanooga from November, 1975, to April, 1976?

4. Mrs. Moneymuncher has kept a record on grocery costs.

		PRICE PER UNIT			QUANTITY PURCHASED		
ITEM	UNIT SIZE	1973	1974	1975	1973	1974	1975
Coffee	lb.	$1.00	$1.10	$1.20	30	32	35
Cookies	lb.	0.30	0.35	0.40	6	7	8
Fruit	lb.	0.40	0.42	0.44	12	14	16
Sugar	5 lbs.	2.10	2.40	1.80	20	22	25

a. Construct a simple price index for coffee with 1973 = 100.

b. How many percentage points did the price of coffee increase from 1974 to 1975?

c. What was the percentage increase in the price of coffee from 1974 to 1975?

d. Construct a simple price index for sugar with 1975 = 100.

e. Construct a simple quantity index for fruit with 1973 = 100.

5. Use the data in Problem 4 and calculate a composite price index for each year using the simple aggregate method with 1973 = 100. Explain the confusion associated with using this index.

6. Use the data in Problem 4 and calculate a composite quantity index for each year using the simple aggregate method with 1973 = 100.

7. Use the data in Problem 4 and calculate a composite price index for each year using the average of relatives method with 1973 = 100.

8. Use the data in Problem 4 and calculate a composite quantity index for each year using the average of relatives method with 1973 = 100.

9. Use the data in Problem 4 and calculate a composite price index for each year using the weighted aggregate method with 1973 = 100.
   a. Use Laspeyres weights.
   b. Use Paasche weights.
   c. Interpret the results of parts (a) and (b).

10. Use the data in Problem 4 and calculate a composite price index for each year using the weighted average of relatives method with 1973 = 100.
    a. Use Laspeyres weights.
    b. Use Paasche weights.
    c. Compare the answers in Problems 9 and 10.

11. Use the data in Problem 4 and calculate a composite quantity index for each year using the weighted aggregate method with 1973 = 100.
    a. Use Laspeyres weights.
    b. Use Paasche weights.

12. Use the data in Problem 4 and calculate a composite quantity index for each year using the weighted average of relative method with 1973 = 100.
    a. Use Laspeyres weights.
    b. Use Paasche weights.
    c. Compare the answers in Problems 11 and 12.

13.

		PRICE PER UNIT			QUANTITY PURCHASED		
ITEM	UNIT SIZE	1973	1974	1975	1973	1974	1975
Fish	1 lb.	$0.25	$0.40	$0.50	200	250	300
Beef	1 lb.	1.50	1.60	1.65	500	600	650
Liver	1 lb.	0.75	0.90	1.10	100	200	250

   a. Construct a composite price index for each year using the weighted aggregate approach and Laspeyres weights with 1973 = 100.
   b. What is the percentage increase in prices from 1973 to 1975 based on your index from part (a)?
   c. What is the percentage increase in prices from 1974 to 1975 based on your index from part (a)?

14. Repeat Problem 13 for quantities rather than prices.

15. A small manufacturer has maintained records on certain basic expenses for two recent years.

	PRICE/UNIT		QUANTITY USED (UNITS)	
EXPENSE	1977	1978	1977	1978
Electricity	$ 0.42	$ 0.63	1000	1200
Insurance	10.00	12.00	250	280
Wages	2.00	2.25	3000	4000

a. Determine the composite price index for 1978 using Laspeyres weights and 1977 = 100. Use the weighted average of relatives method.

b. Interpret the answer in part (a).

c. Determine the composite price index for 1978 using Paasche weights and 1977 = 100. Use the weighted average of relatives method.

d. Interpret the answer in part (c).

16. Over several years, a manufacturer charged the following prices for Product R.

YEAR	PRICE/UNIT
1973	$1.10
1974	1.06
1975	1.12
1976	1.15
1977	1.24
1978	1.29

a. Construct a simple price index for Product R with 1973 = 100.

b. Shift the base of the index constructed in part (a) to 1975.

17. Splice the following two series of price indexes together.

YEAR	1960 = 100	1967 = 100
1971	129	
1972	132	
1973	134	115
1974		118
1975		124
1976		138

a. Maintain 1967 = 100.

b. Maintain 1960 = 100.

18. From the data below, determine if the average real wage of the employees is keeping pace with increased living costs. If wages are falling behind price increases, what wage is necessary in 1973 to be equivalent to $105 in 1969?

YEAR	AVERAGE WEEKLY WAGE	CPI* (1967 = 100)
1969	$105	109.8
1970	110	116.3
1971	112	121.3
1972	115	125.3
1973	120	133.1

*Source: Monthly Labor Review, U.S. Department of Labor, April, 1974.

19. Mr. Job Hopper is about to move again. He has been offered a new job with a salary of $13,500 per year in a city with a "cost of living" index of 132. If he presently earns $12,500 per year in a town with a "cost of living" index of 117, will he be better off financially in the new job?

20. New Products, Inc., has reported the following data for one product.

YEAR	SALES	PRODUCT PRICE INDEX 1970 = 100
1970	$17,200	100
1971	19,500	102
1972	19,900	106
1973	21,000	109
1974	22,500	112

Evaluate the growth recorded for this product.

21. If the consumer price index increased from 133.1 in 1973 to 147.7 in 1974, a clerical worker earning $2.90 per hour in 1973 must earn how much in 1974 in order to maintain a constant purchasing power?

22. Mr. Homeowner purchased his new home in 1972 at a cost of $31,000. It is now the year 1975, and Mr. Homeowner knows that residential construction costs have risen. In fact, he has discovered that a residential construction cost index (1970 = 100), which was 106 in 1972, is now 130. His home is currently insured for $32,500. Do you recommend he consider increasing his insurance? Why?

23. Assume that your real income, stated in 1967 dollars, increased from $8,000 in 1967 to $11,000 in 1973. If prices have increased 25 percent during the same period, what is your actual dollar income in 1973?

Time Series Analysis: Trend

A time series is a group of numbers arranged in chronological order. *Time series analysis* is the analysis of variables that are measured through time—days, weeks, months, years, etc. These variables might be sales, rainfall, unemployment, or income, to mention a few.

In an earlier chapter, you learned about regression analysis. The values of the variables that you worked with may have been taken at different points in time, but we were concerned only with variable quantities and not when (the time) those quantities occurred. Chapter 14, in the discussion of index numbers, implied that values of a variable may change with time. There was no attempt, however, to measure the relationship between time and index numbers. *Time series analysis deals with bivariate (two variables) data, where time is one of the variables.* The primary assumption of time series analysis is that the values of some variables are a function of time. In this analysis, time is always the independent variable.

Time Series Models

In time series analysis, the values of the variable may be expressed for any interval of time. Since time is infinitely divisible—a continuous variable—we could conceivably measure the dependent variable at very brief intervals—minutes, seconds, or microseconds. In most instances, however, the time periods used for measurement are days, weeks, months, quarters (3 months), and years. For example, we might analyze a time series in which weekly sales was the dependent variable.

The relationship between the two variables can be evaluated over different time intervals. That is, we can analyze the behavior of a variable when it is measured monthly, and we can analyze the same variable measured annually. The significance of the choice of time period rests with the sources, or causes, of the behavior of the dependent variable. In general, these sources may be categorized as (1) *seasonal,* (2) *cyclical,* (3) *irregular,* and (4) *trend factors.*

Seasonal variation in a dependent variable is a type of variation that exists within a calendar year and repeats itself from year to year. Factors that contribute to seasonal variation are weather, customs, holidays, etc. June weddings, swimming in the summer, snow skiing in the winter, and buying new clothes before returning to school each year are all seasonal events. Monthly or quarterly data are ordinarily required to examine seasonal variation. Thus, you would expect seasonal variations in the rental of tuxedos, sales of swimwear, tourists in Colorado, and the sale of school shoes. Figure 15-1 is a graph example of seasonal variation.

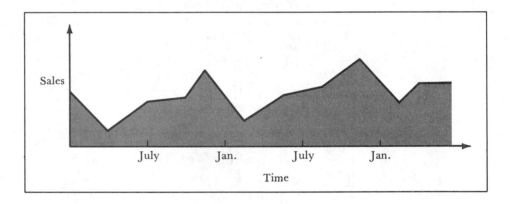

FIGURE 15-1. *Seasonal Variation.*

Cyclical variation in a dependent variable occurs over a longer time period than seasonal variation. Cyclical variation is characterized by a period of general expansion (growth) followed by a period of contraction. The periods of expansion and contraction vary in length and usually it requires several years to complete a cycle. The wavelike nature of cyclical variation is illustrated in Figure 15-2.

Irregular variation might best be called random variation. Unlike the other three types of variation, there is no pattern to its behavior. Random variations are unpredictable, both as to when they will occur and as to their duration. Periods of extreme fashions (fads), wars, and unseasonable variations in the weather are examples of sporadic events that contribute to this form of variation in economic data.   The graph in Figure 15-3 illustrates the irregular variation in the sales of Our Favorite Textbook Company.

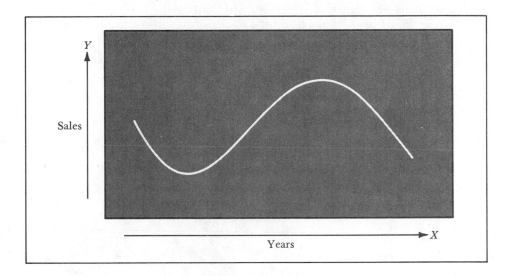

FIGURE 15-2.   *Cyclical Variation.*

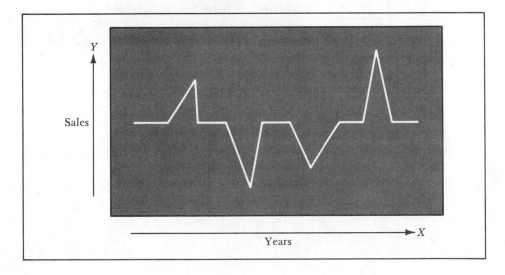

FIGURE 15-3.   *Irregular Variation.*

Trend is the long-term behavior of a variable. It is a generalized expression of the growth, positive (increasing) or negative (decreasing), in the size of the variable. Trend may be the result of the acceptance of a product by consumers, the development of a company, population growth, or any other factor that influences the size of an economic variable over a long period. In Figure 15-4, trend is represented by the straight line superimposed over the dashed lines. The dashed lines would represent the actual sales over time, and the trend is a representation of the long-term directional movement of the data. In this case, trend is presented as a straight line because it seems appropriate; but as you will learn later in this chapter, there are other shapes to describe the general behavior of the variable.

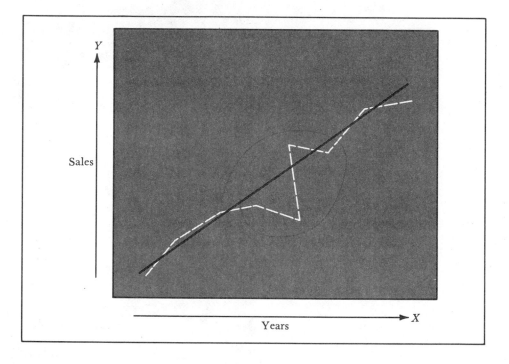

FIGURE 15-4. *Trend.*

At any given point in time, then, the value of a variable (economic data) is determined by the four types of variations (seasonal, cyclical, irregular, and trend). A specific value of the variable may be considered to be the product of these types of variation or their sum. In equation form, the relationships are expressed as either a product model (Equation 15.1) or an additive model (Equation 15.2).

$$Y = T \times S \times C \times I \qquad\qquad (15.1)$$

$$Y = T + S + C + I \qquad\qquad (15.2)$$

where Y = the value of dependent variable at some point in time

T = that part of the value of the dependent variable that is attributable to trend—stated as an absolute value in both models

S = that part of the value of the dependent variable that is attributable to seasonal variation—stated as an index in the product model and as an absolute value in the additive model

C = that part of the value of the dependent variable that is attributable to cyclical variation—stated as an index in the product model and as an absolute value in the additive model

I = that part of the value of the dependent variable that is attributable to random variation—stated as an index in the product model and as an absolute value in the additive model

Annual data exhibit no seasonal variation; so the models for annual data are

$$Y = T \times C \times I \tag{15.3}$$

$$Y = T + C + I \tag{15.4}$$

The additive model assumes that each component of the model is independent of the other components. For example, such an assumption implies that the seasonal variation for each´ month is the same amount every year regardless of the general trend or the cyclical position of the data. The product model, however, assumes that the components are related even though their causes are different. Thus, in the product model, one of the components (usually trend) is expressed in absolute units, and the other components are expressed as a percentage of the first component. The product model is more widely used and will receive most of the discussion in this text.

Justification for time series analysis, and the models, stems from two principal activities. One is the study of the effect on data of each of the types of variation. The other is the utilization of the knowledge of the behavior of the data to forecast future performance. If the latter use can be carried out with some degree of accuracy, and if you understand its limitations, you will have an extremely useful statistical tool at your disposal.

Editing Data for Time Series Analysis

Since the data to be analyzed have been gathered over a broad span of time, there are several problems that must be dealt with before we can begin a time series analysis. The following is not intended as an all-inclusive list, but rather, to introduce you to the type of potential difficulties to be aware of when you are analyzing time series data.

1. Data gathered over a long time span, or from many sources, may have a particular problem of consistency or comparability. Definitions may change, the geographic area covered may vary, or the method of calculation may be altered. As some examples of this, unemployment has not always been defined as it now is; accountants determine costs in several ways; and the United States has not always included 50 states. An investigation of the data should be made so that, if necessary, alternative sources of data can be sought or adjustments in the data can be made. Even when no alternative data sources exist that will improve the consistency of the data, the investigation will produce an awareness by the analyst of any weaknesses in the data.

2. Months, except February, always contain the same number of days each year but they do not always have the same number of "working days." There are also other calendar variations, such as the precise location of holidays and the beginning of the college year. Data may reflect this calendar variation so that, for example, data for June of one year may not be strictly comparable to that for June of another year. The possibility of calendar variations is often ignored in time series analysis, but in some types of monthly data, or data for shorter time periods, it may be very significant.

3. Time series analysis is bivariate analysis. That is, there is only one independent and one dependent variable. In measuring the relationship between two variables, you should exclude, as far as possible, the influence of other independent variables on the dependent variable. One such independent variable, when working with data measured in dollars, that will influence the dependent variable is price changes. For example, if you do time series analysis on gross national product data expressed in given year dollars, both real output changes and price changes will be included in the dependent variable. Generally, then, data should be converted to constant dollars (you learned how to do this in Chapter 14) before the analysis begins.

　　　Population changes can present a problem similar to that created by price changes. Consider the situation where consumption of a product increases by 20 percent but, at the same time, population increases by 50 percent. An analysis of the data without considering the population change would certainly lead to conclusions quite different from an analysis of the data adjusted for population changes. If the dependent variable is one that population changes can influence, converting the data to a per capita basis will remove the potential distortion.

Trend

In the beginning of trend analysis, two decisions must be made: (1) what shape line, or curve, will be used to represent the relationship between the two variables, and (2) what procedure will be used to determine the coefficients for an equation, corresponding to the shape selected, that will describe the relationship. Both decisions are critical, but at your present level in statistics, the first is the most significant. If you go on to study time series analysis in greater depth, many questions will be raised concerning methods of determining coefficients, or as it is sometimes called, "fitting curves to data." The procedures presented in this text are relatively easy to understand, and they are probably the most commonly used techniques when time series analysis is applied to business and economic data. They are, however, only a limited introduction to the techniques available for your use at more sophisticated levels of analysis.

Selecting the Shape of the Trend

Early in the text (Chapter 1) you learned about the use of graphs in statistical analysis. Scatter diagrams (a graph type) are basic to selecting the shape of the curve to describe the trend. Figure 15-5 illustrates some of the results you might obtain when you make a scatter diagram using your own data.

Notice in the graphs (Figure 15-5) that we have plotted only *annual data*. For trend analysis, the use of data for shorter time periods (monthly data, quarterly data, etc.) is unnecessary. Furthermore, the use of something less than annual data would involve a greater amount of work (arithmetic) and would probably lead to a greater number of errors. If you find that you need to know how to express trend in some form other than annual data, look in the appendix to this chapter, but only after you have learned what is presented in the pages between here and there.

Graph (a) in Figure 15-5 shows a set of data in which the dependent variable appears to have a positive, *linear* relationship with the independent variable. The relationship is positive because the dependent variable is generally increasing as time elapses. *Linear* describes the relationship because the dependent variable is changing by a reasonably uniform amount as time elapses. A relationship of this nature can be represented by a straight line, for which the general equation is $Y_c = a + bX$. In this and subsequent equations, Y_c is the estimated value of the dependent variable, and X represents time, the independent variable. The values (coefficients) a and b are constants that are determined for each set of data.

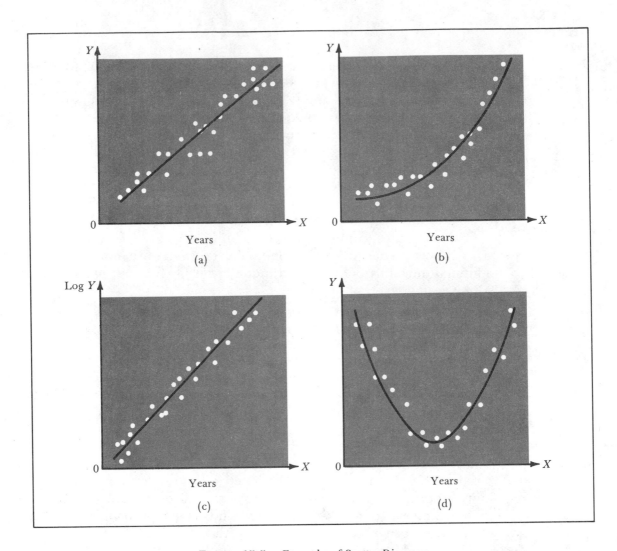

FIGURE 15-5. *Examples of Scatter Diagrams.*

The data scatter in graph (b) requires that the relationship be expressed in a different manner. The trend is positive, but a straight line does not accurately express the general behavior of the dependent variable. Rather than increasing by a constant amount (as a straight line implies), it appears to be increasing at a constant rate. This type of behavior can be represented by a *power curve*.

Fortunately there is another method of displaying the type of relationship shown in graph (b). Graph (c) shows data plotted with a *logarithmic* rather than arithmetic scale on the ordinate. Look at how the pattern of the scatter has changed. Data that are described by a power curve on an arithmetic graph appear as a straight line on a semi-log graph. Thus, $Y_c = ab^X =$ antilog $(\log a + X \log b)$.

Graph (d) in Figure 15-5 presents the scatter of still another possible set of data. This relationship between the variables is distinctly different from that presented in the first two graphs. In the earlier years the relationship appears negative, but in the most recent time periods it appears positive. Neither a straight line nor a power curve will adequately describe such behavior. If the trend changes its direction, a second-degree curve is needed to portray the relationship between the variables. The general equation for a second de-

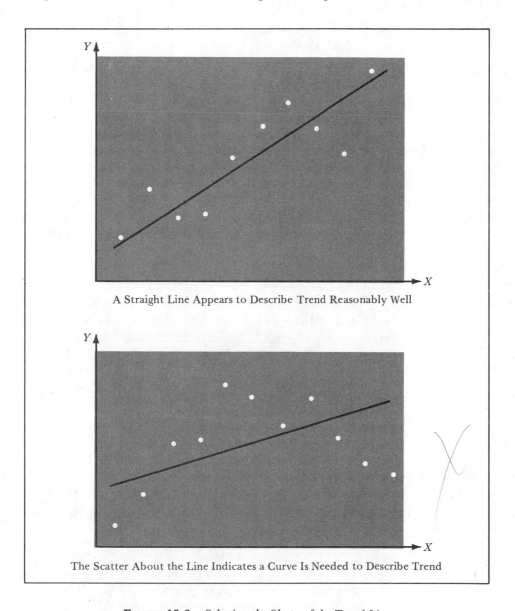

A Straight Line Appears to Describe Trend Reasonably Well

The Scatter About the Line Indicates a Curve Is Needed to Describe Trend

FIGURE 15-6. *Selecting the Shape of the Trend Line.*

gree curve is $Y_c = a + bX + cX^2$. The additional coefficient c has been added to the straight-line equation. At another point, we will discuss the limitations of using this type of curve in time series analysis.

Selecting the appropriate equation to describe the relationships in a set of data is not always so obvious as our illustrations might lead you to believe. If you encounter data that appear to have no pattern in their scatter, or you cannot visualize how well a given shape will describe the pattern, do what we have done in Figure 15-6. Simply "rough in" a straight line on the scatter diagram. If the scatter alternates both above and below the line, but there is no substantial interval where all the points are either above or below, then a straight line is probably the best choice to generalize on the behavior. Should you find that there is a pattern to the arrangement of the points about the line you have drawn, a curve may be needed. The existence of change in direction in the relationship can justify selecting a second-degree curve. Otherwise, the power curve would probably be the most suitable. From a practical standpoint, as your experience with trend analysis increases, you will generally be able to select the appropriate shape from a visual inspection of the scatter diagram.

Determining the Coefficients for a Linear Trend Equation

When studying Chapters 12 and 13, you learned how to "fit a line (or curve) to the data." The same concepts you learned in those chapters are applicable now, since trend analysis is simply regression analysis with time as the independent variable. In the general equation for a straight line, $Y_c = a + bX$, it is necessary to determine a and b. In this chapter, two methods of determining a and b will be examined: (1) freehand and (2) least-squares.

FREEHAND METHOD. The freehand method can be used to *approximate* a linear trend equation. The procedure is to plot the data, and then draw a straight line through the data, trying to keep the sums of the areas above and below the trend line about the same. Figure 15-7 illustrates the procedure.

Notice in Figure 15-7, the values of a and b are determined by observation. Thus, from the figure, a is observed to be 82. To determine b, the value of Y is observed for the first and last point on the regression line. That is, $Y = 82$ in 1972 and $Y = 152$ in 1978. Then,

$$b = \frac{\text{last } Y \text{ value} - \text{first } Y \text{ value}}{\text{number of periods}}$$

$$b = \frac{152 - 82}{6} = \frac{70}{6} = 11.67$$

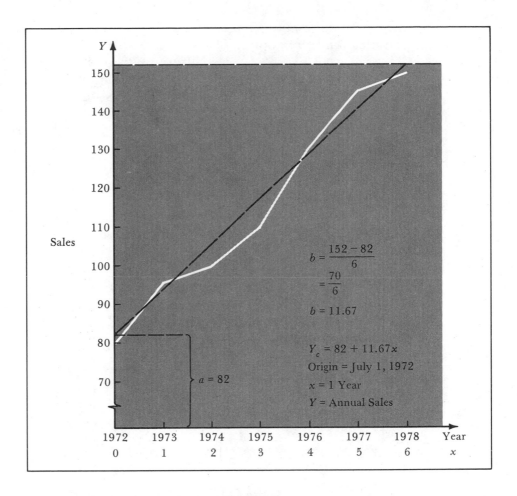

FIGURE 15-7. *Trend: Freehand Method.*

And, the regression equation is $Y_c = 82 + 11.67X$, which is shown on Figure 15-7. Also given on Figure 15-7 are the origin and the units of the variables. Notice that the origin is July 1, 1972. The middle of the year is used because the total sales for a year are accumulated over the year, and thus the annual sales value does not represent any point in time. Consequently, convention has been to plot data of this nature in the middle of the period from which they were collected.

Notice also that X and Y are defined so that anyone can identify easily the dependent variable and the period for the independent variable. In Figure 15-7, the time period (X) is one year, and it is shown in *coded* form as the x value under each year. The coding is necessary because the value of x has to be 0 at the origin. Thus, with the origin set as 1972, then $x = 0$ in 1972, $x = 1$ in 1973, and so forth. This coding of the X variable will prove very useful throughout this chapter.

The advantage of the freehand method is that it is very easy to understand and apply. The disadvantage is that it is a method whereby different people can get different answers for the same problem. Thus, it is not a consistently reproducible approach.

THE METHOD OF LEAST-SQUARES.   The least-squares method has been discussed previously in Chapter 12. Tables 15-1 and 15-2 illustrate the least-squares method for determining the trend (regression) equation for an odd number of years and an even number of years, respectively. In both tables, the coded form of the independent variable is used to reduce the arithmetic and the level of algebra required to solve for the constants, a and b.

In Table 15-1, the origin is established as the middle year, and all of the other values of x are obtained by subtracting 1975 from each of the other years. That is, $x = +1 = (1976-1975)$, and $x = -1 = (1974-1975)$, and so on. Actually, there is a quicker way to code the data than by subtracting the middle year from the other values of the

TABLE 15-1. *Fitting a Straight Line, least-squares method (odd number of years).*

YEAR X	x	x^2	SALES Y	xY
1972	-3	9	$ 80	-240
1973	-2	4	95	-190
1974	-1	1	100	-100
1975	0	0	110	0
1976	$+1$	1	130	130
1977	$+2$	4	145	290
1978	$+3$	9	150	450
	0	28	810	$+340$

$$Y_c = a + bx$$

$$a = \frac{\Sigma Y}{n} \qquad\qquad b = \frac{\Sigma xY}{\Sigma x^2}$$

$$a = \frac{810}{7} = 115.71 \qquad\qquad b = \frac{340}{28} = 12.14$$

$$Y_c = 115.71 + 12.14x$$

$$x = 1 \text{ year}, Y = \text{annual sales}$$

$$\text{origin} = \text{July 1, 1975}$$

$$Y_{c\ 1980} = 115.71 + 12.14\ (+5)$$

$$= 115.71 + 60.70$$

$$Y_{c\ 1980} = 176.41$$

independent variable. Simply designate $x = 0$ as the middle year, $x = +1$ for the following year, $x = -1$ for the preceding year, and so forth, until all years have been assigned a coded x value. Now study Table 15-1 observing the coding, the solution procedure, and the designation of the origin and the x and Y units.

In Table 15-2 the computations are the same as in Table 15-1. However, there are an even number of years, and the coding has to be handled differently to assure that $\Sigma x = 0$. Since it is an even number of years, there is no middle time period. That is, the middle of the data does not occur at the middle of a year; it occurs at the end of one year or at the beginning of another year. Thus, in Table 15-2, $x = 0$ occurs at the end of 1974 or, if you prefer, at the beginning of 1975. As discussed earlier, however, data that are associated with a period of time are usually plotted in the center of the period. Thus, to get the data centered opposite the year they came from, it is necessary to

TABLE 15-2. *Fitting a Straight Line, least-squares method (even number of years).*

YEAR X		x	x^2	OUTPUT (1000) Y	xY
1971	7	−3.5	12.25	12	−42.0
1972	5	−2.5	6.25	13	−32.5
1973	3	−1.5	2.25	15	−22.5
1974	1	− .5	.25	17	− 8.5
1975		+ .5	.25	18	9.0
1976		+1.5	2.25	19	28.5
1977		+2.5	6.25	21	52.5
1978		+3.5	12.25	23	80.5
		0	42.00	138	+65.0

$$Y_c = a + bx$$

$$a = \frac{\Sigma Y}{n} \qquad\qquad b = \frac{\Sigma xY}{\Sigma x^2}$$

$$a = \frac{138}{8} = 17.25 \qquad\qquad b = \frac{65}{42} = 1.5476$$

$$Y_c = 17.25 + 1.55x$$

$$x = 1 \text{ year}, Y = \text{annual output}$$

$$\text{origin} = \text{December } 31, 1974$$

$$Y_{c\ 1980} = 17.25 + 1.55\ (5.5)$$

$$Y_{c\ 1980} = 25{,}775$$

move 6 months ($x = 0.5$) forward or six months ($x = -0.5$) backward. So, for 1974, $x = -0.5$, and for 1975, $x = 0.5$. From then on, the movement is 12 months such that $x = 1.5$ for 1976, $x = 2.5$ for 1977, and so on.

As previously mentioned, the computations in Table 15-2 are the same as in Table 15-1. Notice, however, that because of the difference in coding the origins are not at the same part of the year. When there is an odd number of years, the origin is at the middle of the year (July 1), whereas when there is an even number of years, the origin is at the end of the year (December 31 or January 1).

The solutions shown in the preceding tables use the least squares method based on the following set of normal equations:

$$\Sigma Y = na + b\Sigma X \qquad (15.5)$$

$$\Sigma XY = a\Sigma X + b\Sigma X^2 \qquad (15.6)$$

where Y = the values of the dependent variable
n = the number of values of the independent (or dependent) variable
X = the values of the independent variable

The values of a and b can always be found by simultaneously solving the normal equations, but if you code the dependent variable so that $\Sigma x = 0$ as was done in our tables, the difficulty is greatly reduced. In this case, the equations are reduced.

$$a = \frac{\Sigma Y}{n} \qquad (15.7)$$

$$b = \frac{\Sigma xY}{\Sigma x^2} \qquad (15.8)$$

We have used x rather than X in the equations so that you will not forget that they are to be used only when the data are properly coded. Using the coded value (x), the straight-line trend equation is

$$Y_c = a + bx \qquad (15.9)$$

where Y_c = the trend value
x = the coded value of X
a = the intercept of the trend line. The value of Y_c when $x = 0$ (the origin)
b = the slope of the trend line. The change in Y_c when x is changed one time period

Finally look at the last calculation in Tables 15-1 and 15-2. These last entries are both forecasts for the year 1980. Notice how the forecast is done. The appropriate value of x is determined for the year of the forecast, and the derived trend equation is used to make the forecast.   Forecasting is one of the primary uses of trend analysis. The results, however, have to be used carefully, since we are basing the forecast for the future on past history.   If the future holds numerous uncertainties due to changing economic conditions, product development, or technology changes, or other causes of variation, the forecast should be used with caution and/or altered to reflect the changing situation.

Determining the Coefficients for a Logarithmic Straight-Line Trend Equation

In nearly the same way as fitting a straight line, you can fit a logarithmic straight line to the data.   Remember, the logarithmic straight line is an exponential, or power, curve.   The general equation is as follows:

$$Y_c = ab^X \tag{15.10}$$

If the dependent variable data are converted to logarithms, you can see the similarity to the arithmetic straight line.

$$\log Y_c = \log a + X \log b \tag{15.11}$$

where $\log Y_c$ = the logarithm of the trend value for any given value of the independent variable

$\log a$ = the logarithm of the intercept of the trend line ($\log Y_c$).   The value of $\log Y_c$ when $X = 0$.

$\log b$ = the logarithm of the slope of the trend line.   The change in $\log Y_c$ when X is changed one unit.   The antilog of $\log b$ is the average annual rate of growth when $X = 1$ year

X = any given value of the independent variable

The equations used to solve for a and b obtained using the method of least squares are

$$\Sigma (\log Y) = n (\log a) + \log b (\Sigma X) \tag{15.12}$$

$$\Sigma (X \log Y) = \log a (\Sigma X) + \log b (\Sigma X^2) \tag{15.13}$$

If we assume that the independent variable is coded so that $\Sigma x = 0$, the above equations are reduced to the following:

$$\log a = \frac{\Sigma (\log Y)}{n} \tag{15.14}$$

$$\log b = \frac{\Sigma (x \log Y)}{\Sigma x^2} \tag{15.15}$$

An example of the type of data that fit the exponential curve is given in Table 15-3. Notice that the change in the dependent variable is not constant from year to year, but that the increases get larger as time goes on. The data presented in Table 15-3 are plotted in Figures 15-8 and 15-9.

TABLE 15-3. *Data for the Exponential Trend Equation.*

YEAR (X)	SALES (Y)
1970	150
1971	165
1972	185
1973	200
1974	228
1975	252
1976	280
1977	315
1978	356

In Figure 15-8, the data are plotted on arithmetic paper so that you can see the continual increasing rate of Y. In Figure 15-9, the data are plotted on semilogarithmic paper. Notice now that the graph approximates a straight line. Hence, the non-

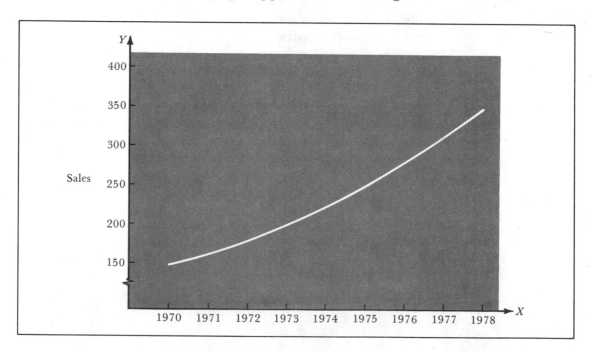

FIGURE 15-8. *Arithmetic Graph of the Data for the Exponential Trend Equation.*

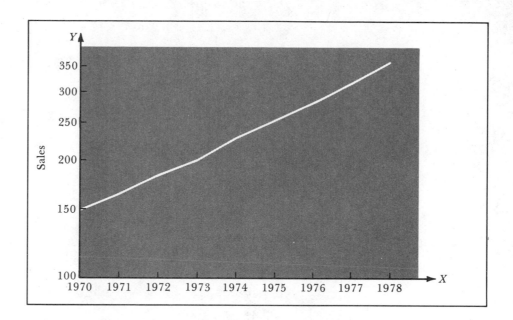

FIGURE 15-9. *Semilogarithmic Graph of the Data for the Exponential Trend Equation.*

linear data (Figure 15-8) are transformed to linear data (Figure 15-9) and are easier to work with than the nonlinear data.

Now that you have seen graphs of the data, look at Table 15-4 to see how the logarithmic trend equation is developed.

TABLE 15-4. *Fitting a Logarithmic Straight Line, least-squares method (exponential curve).*

YEAR (X)	x	x^2	SALES (Y)	log Y	x log Y
1970	−4	16	150	2.17609	−8.70436
1971	−3	9	165	2.21748	−6.65244
1972	−2	4	185	2.26717	−4.53434
1973	−1	1	200	2.30103	−2.30103
1974	0	0	228	2.35793	0
1975	1	1	252	2.40140	2.40140
1976	2	4	280	2.44716	4.89432
1977	3	9	315	2.49831	7.49493
1978	4	16	356	2.55145	10.20580
	0	60		21.21802	2.80428

$$\log a = \frac{\Sigma \log Y}{n} = \frac{21.21802}{9} = 2.35756$$

$$\log b = \frac{\Sigma (x \log Y)}{\Sigma x^2} = \frac{2.80428}{60} = 0.04674$$

$$\log Y_c = \log a + x \log b$$

$$\log Y_c = 2.35756 + 0.04674x$$

$$x = 1 \text{ year}; Y = \text{annual sales}$$

$$\text{origin} = \text{July 1, 1974}$$

$$\log Y_{c\ 1978} = 2.35756 + 0.04674\ (4)$$

$$\log Y_{c\ 1978} = 2.35756 + 0.18696 = 2.54452$$

$$Y_{c\ 1978} = 350.4$$

Notice, at the end of Table 15-4, sales for 1978 were calculated to be \$350.40 as compared to the actual value of \$356.00. The difference represents an error of only 1.57 percent. Now, let us use the logarithmic trend equation to forecast sales for 1980.

$$\log Y_{c\ 1980} = 2.35756 + 0.04674\ (6)$$

$$= 2.35756 + 0.28044 = 2.63800$$

$$\log Y_{c\ 1980} = \$434.50$$

Determining the Coefficients for a Second-Degree Trend Equation

Sometimes the trend in time series data appears to change direction. That is, it may have a positive slope over part of its length and a negative slope over the other part. When this happens, the trend might be appropriately expressed by a parabola, or second degree curve. This can be done by the least-squares method but the procedure is slightly more complicated than the other two you have learned. A second-degree curve is described by the equation:

$$Y_c = a + bX + cX^2 \tag{15.16}$$

Since you have three unknowns (a, b, and c), you need three equations to solve for them. Using the least squares method of fitting the line, these three equations are

$$\Sigma Y = na + b\Sigma X + c\Sigma X^2 \tag{15.17}$$

$$\Sigma XY = a\Sigma X + b\Sigma X^2 + c\Sigma X^3 \tag{15.18}$$

$$\Sigma X^2 Y = a\Sigma X^2 + b\Sigma X^3 + c\Sigma X^4 \tag{15.19}$$

Fortunately, coding the independent variable so that $\Sigma x = 0$ reduces the complexity of solving for a, b, and c. With this form of coding, the formulas are as follows:

$$\Sigma Y = na + c\Sigma x^2 \tag{15.20}$$

$$\Sigma xY = b\Sigma x^2 \tag{15.21}$$

$$\Sigma x^2 Y = a\Sigma x^2 + c\Sigma x^4 \tag{15.22}$$

(If $\Sigma x = 0$, Σx^3 is also equal to zero.)

Now the value of b can be solved for directly and a and c values can be found by using the two remaining equations.

$$b = \frac{\Sigma xY}{\Sigma x^2} \tag{15.23}$$

On the other hand, you may prefer to use the following approach for determining the values of a and c. Either way, you will get the same results.

$$c = \frac{n\Sigma x^2 Y - \Sigma Y \Sigma x^2}{n\Sigma x^4 - (\Sigma x^2)^2} \tag{15.24}$$

$$a = \frac{\Sigma Y - c\Sigma x^2}{n} \tag{15.25}$$

TABLE 15-5. *Fitting a Second-Degree Curve, least-squares method.*

YEAR X	x	x^2	x^4	SALES Y	xY	x^2Y
1974	−2	4	16	$100	−200	400
1975	−1	1	1	80	− 80	80
1976	0	0	0	105	0	0
1977	+1	1	1	120	+120	120
1978	+2	4	16	145	+290	580
	0	10	34	$550	+130	1180

$$Y_c = a + bx + cx^2$$

$$b = \frac{\Sigma xY}{\Sigma x^2} = \frac{130}{10} = 13.0$$

$$c = \frac{n\Sigma x^2 Y - \Sigma Y \Sigma x^2}{n\Sigma x^4 - (\Sigma x^2)^2} = \frac{5\,(1180) - (550)\,10}{5\,(34) - (10)^2} = \frac{400}{70} = 5.71$$

$$a = \frac{\Sigma Y - c\Sigma x^2}{n} = \frac{550 - (5.71)\,10}{5} = 98.58$$

$$Y_c = 98.58 + 13.0x + 5.71x^2$$

$$x = 1 \text{ year}, Y = \text{annual sales}$$

$$\text{origin} = \text{July 1, 1976}$$

$$Y_{c\,1978} = 98.58 + 13\,(+2) + 5.71\,(4)$$

$$= 98.58 + 26 + 22.84 = 147.43$$

Using the coded values of X, (x), the second-degree trend equation is

$$Y_c = a + bx + cx^2 \qquad \textbf{(15.26)}$$

Now, look at Table 15-5 for an example of fitting a second-degree curve.

Adjusting Data for Trend

The presence of trend in business and economic data may impair the data for evaluation purposes. For example, if there is a long-term (secular) growth trend of a 5 percent increase per year in a company's sales, extracting the trend from the data may facilitate the evaluation of the short-term performance of sales. In addition, measures of other types of economic fluctuation, seasonal, cyclical, and random, must be free of "trend bias." Therefore, either the trend must be removed from the data before these measures are determined, or the method of measurement must compensate for the presence of trend in the data.

Assuming the product model, trend is removed by dividing the actual data, Y, by the trend estimates, Y_c. The logic for this method is as follows:

$$\text{IF} \qquad Y = T \times S \times C \times I$$

$$\text{AND} \qquad T \cong Y_c$$

$$\text{THEN} \qquad Y = Y_c \times S \times C \times I$$

$$\text{AND} \qquad \frac{Y}{Y_c} = S \times C \times I$$

Trend is extracted most typically from T·C·I

When trend is removed, what remains of the data is the influence of seasonal, cyclical, and irregular variations or cyclical and irregular if the data are annual data. Extracting trend in this manner leaves the residual $S \times C \times I$ in the form of a ratio rather than units (dollars, pounds, etc.) Study Table 15-6 to see an example of how trend is removed in this manner.

TABLE 15-6. *Removing Trend, product model.*

YEAR X	SALES Y	Y_c	Y/Y_c
1972	$ 80	$ 80.55	0.993
1973	95	91.80	1.035
1974	100	103.05	0.970
1975	110	114.30	0.962
1976	130	125.55	1.035
1977	140	136.80	1.023
1978	145	148.05	0.979

Trend equations can be manipulated to make them more useful. Two particular adjustments that we can do are (1) change the units and (2) shift the origin.

Changing the Units

Assume that the trend equation for the number of customers stopping at the Roadside Inn is

$$Y_c = 1800 + 60x$$

$$x = 1 \text{ year}$$

$$Y = \text{hundreds of customers}$$

$$\text{origin} = \text{July 1, 1972}$$

Both *a* and *b* represent annual change, *so to convert them to monthly values we divide our trend equation by 12 as shown below.* This conversion

$$Y_c = \frac{1800}{12} + \frac{60}{12}x$$

$$Y_c = 150 + 5x$$

has not, however, affected the *x* variable, which is still in years. Thus, this trend equation reflects the monthly change for an annual period. That is, it gives the change for a given month in one year to the same month in the next year such as from April, 1974, to April, 1975. Consequently, one more step is necessary to get our original trend equation into the form that reflects month-to-month changes. Now we must also *divide x by 12* as shown below.

$$Y_c = \frac{1800}{12} + \left(\frac{60}{12}\right)\left(\frac{x}{12}\right)$$

$$Y_c = 150 + 0.417x$$

$$x = 1 \text{ month}$$

$$Y = \text{hundreds of customers}$$

$$\text{origin} = \text{July 1, 1972}$$

In general, the conversion of an annual trend equation to a monthly trend equation is as shown on p. 384.

$$Y_c = a + bx$$

$$x = 1 \text{ year}$$

$$Y = \text{annual totals}$$

$$Y_c = \frac{a}{12} + \left(\frac{b}{12}\right)\left(\frac{x}{12}\right)$$

$$x = 1 \text{ month}$$

$$Y = \text{monthly totals}$$

Conversion of an annual trend equation to a quarterly one follows the same procedure with a, b, and x divided by 4.

Shifting the Origin

Another adjustment may be necessary before two trend equations can be compared. Specifically, the origins of the equations must be equal. Thus, if for any reason the origins of two equations are not the same point in time, then you may wish to change one, or both, of the equations to reflect a different origin. This is done in the following manner. When the trend is an arithmetic straight line:

$$Y_c = a + b \left(x + (x_n)\right)$$

where x_n = the new origin.

Using the equation from Table 15-1 the origin can be changed to 1977 by substituting in the formula.

$$Y_c = 115.71 + 12.14x \qquad (\text{origin} = 1975)$$

$$Y_c = 115.71 + 12.14 (x + 2) = 115.71 + 24.28 + 12.14x$$

$$Y_c = 139.99 + 12.14x \qquad (\text{origin} = 1977)$$

If trend is expressed by a second-degree line, the procedure for changing the origin is much the same.

$$Y_c = a + bx + cx^2 \qquad (\text{original origin})$$

$$Y_c = a + b (x + x_n) + c (x + x_n)^2 \qquad (\text{new origin})$$

Perhaps a more useful way to express the procedure is in its expanded form.

$$Y_c = a + b (x + x_n) + c \left[x^2 + 2x (x_n) + x_n^2\right]$$

The origin of the equation in Table 15-5 is changed to 1974 in the following example:

$$Y_c = 98.58 + 13x + 5.71x^2 \qquad (\text{origin} = 1976)$$

$$Y_c = 98.58 + 13\,(x - 2) + 5.71\,(x - 2)^2$$

$$= 98.58 + 13\,(x - 2) + 5.71\,(x^2 - 4x + 4)$$

$$= 98.58 + 13x - 26 + 5.71x^2 - 22.84x + 22.84$$

$$Y_c = 95.42 - 9.84x + 5.71x^2 \qquad (\text{origin} = 1974)$$

────────────── **SUMMARY OF EQUATIONS** ──────────────

15.1 $Y = T \times S \times C \times I,$ *page 366*

15.2 $Y = T + S + C + I,$ *page 366*

15.3 $Y = T \times C \times I,$ *page 367*

15.4 $Y = T + C + I,$ *page 367*

15.5 $\Sigma Y = na + b\Sigma X,$ *page 376*

15.6 $\Sigma XY = a\Sigma X + b\Sigma X^2,$ *page 376*

15.7 $a = \dfrac{\Sigma Y}{n},$ *page 376*

15.8 $b = \dfrac{\Sigma xY}{\Sigma x^2},$ *page 376*

15.9 $Y_c = a + bx,$ *page 376*

15.10 $Y_c = ab^X,$ *page 377*

15.11 $\log Y_c = \log a + X \log b,$ *page 377*

15.12 $\Sigma\,(\log Y) = n\,(\log a) + \log b\,(\Sigma X),$ *page 377*

15.13 $\Sigma\,(X \log Y) = \log a\,(\Sigma X) + \log b\,(\Sigma X^2),$ *page 377*

15.14 $\log a = \dfrac{\Sigma\,(\log Y)}{n},$ *page 377*

15.15 $\log b = \dfrac{\Sigma\,(x \log Y)}{\Sigma x^2},$ *page 377*

15.16 $Y_c = a + bX + cX^2,$ *page 380*

15.17 $\Sigma Y = na + b\Sigma X + c\Sigma X^2,$ *page 380*

15.18 $\Sigma XY = a\Sigma X + b\Sigma X^2 + c\Sigma X^3,$ *page 380*

15.19 $\Sigma X^2Y = a\Sigma X^2 + b\Sigma X^3 + c\Sigma X^4,$ *page 380*

15.20 $\Sigma Y = na + c\Sigma x^2,$ *page 380*

15.21 $\Sigma xY = b\Sigma x^2,$ *page 380*

15.22 $\Sigma x^2 Y = a\Sigma x^2 + c\Sigma x^4$,   *page 381*

15.23 $b = \dfrac{\Sigma xY}{\Sigma x^2}$,   *page 381*

15.24 $c = \dfrac{n\Sigma x^2 Y - \Sigma Y\Sigma x^2}{n\Sigma x^4 - (\Sigma x^2)^2}$,   *page 381*

15.25 $a = \dfrac{\Sigma Y - c\Sigma x^2}{n}$,   *page 381*

15.26 $Y_c = a + bx + cx^2$,   *page 382*

----------------------------- **REVIEW QUESTIONS** -----------------------------

1. Explain the use of each equation given at the end of the chapter.
2. Name the four components of time series models.
3. Name the two types of time series models.
4. How do the representations of components of the two time series models differ?

----------------------------- **PROBLEMS** -----------------------------

1. Assume the following data represent the corn production for a particular farm.   (Keep this work for later use.)

YEAR	BUSHELS
1972	200
1973	205
1974	215
1975	210
1976	230
1977	240
1978	245

Save for
73-78

    a. Plot the data on graph paper.
    b. Fit a straight-line trend equation to the data using the least-squares method.
    c. Estimate the 1979 production.

2. The data on p. 387 are population data for the United States:
    a. Plot the data on graph paper.
    b. Fit a straight-line trend equation to the data using the least-squares method.
    c. Estimate the 1979 population.

YEAR	TOTAL POPULATION (MILLIONS)
1967	198.7
1968	200.7
1969	202.7
1970	204.9
1971	207.1
1972	208.8
1973	210.4
1974	211.9
1975	213.6
1976	215.1

Source: *Statistical Abstract of the United States*, 1977, p. 5..

3. The sales data for two organizations are given below.

YEAR	I SALES ($1000)	II SALES ($1000)
1972	356	420
1973	373	430
1974	388	438
1975	398	447
1976	417	453
1977	430	462
1978	445	470

    a. Use the method of least-squares, and fit a linear trend equation to each set of data.

    b. Assuming that current trends continue, determine the year that I sales will first exceed II sales.

4. The following data are recent income security payments by the Federal government.

FISCAL YEAR	INCOME SECURITY PAYMENTS (BILLIONS)
1969	37.3
1970	43.1
1971	55.4
1972	63.9
1973	73.0
1974	84.4
1975	108.6
1976	126.6
1977	137.0

Source: *The Budget of the United States Government, Fiscal Year 1979.* U.S. Government Printing Office, p. 481.

    a. Fit a linear trend equation to the data using the least-squares method.

    b. Estimate the 1978 expenditure.

    c. Fit a linear trend equation to the data using the freehand method.

5. The following data are the debt of the U.S. Government.

FISCAL YEAR	FEDERAL DEBT (BILLIONS)
1967	$341.3
1968	369.8
1969	367.1
1970	382.6
1971	409.5
1972	437.3
1973	468.4
1974	486.2
1975	544.1
1976	631.9
1977	709.1

Source: The Budget of the United States Government, Fiscal Year 1979. U.S. Government Printing Office, p. 486.

   a. Fit a linear trend equation to the data using the least-squares method.
   b. Estimate the 1978 debt, and comment on the estimate.

6. Consider the following data to be sales of New Store, Inc.

YEAR	SALES (THOUSANDS)
1972	$10
1973	12
1974	18
1975	22
1976	30
1977	40
1978	52

   a. Plot the data on graph paper.
   b. Fit a linear trend equation to the data using the least-squares method.
   c. Fit a logarithmic trend equation to the data using the least-squares method.
   d. Which of the two curves appears to have the better fit?
   e. Estimate the sales for 1979 using the equation developed in (c) above.

7. The following data represent a situation where the potential market is growing linearly but the sales are increasing exponentially.

YEAR	POTENTIAL MARKET ($1,000)	COMPANY SALES ($1,000)
1972	88	33
1973	94	38
1974	101	45
1975	106	53
1976	111	63
1977	117	74
1978	125	87

a. Determine the linear trend equation for the potential market using the least-squares method.
b. Determine the exponential trend equation for the sales data using the least-squares method.
c. Project the market and sales for 1982.

8. Our Favorite Publisher has published the following number of new books during the period 1970 to 1978.

YEAR	NEW BOOKS
1970	8
1971	10
1972	13
1973	17
1974	18
1975	21
1976	22
1977	26
1978	28

a. Fit a linear regression equation to the data.
b. Estimate the number of new books in 1982.

9. Use the following trend equation to answer the questions below:

$$Y_c = 284 + 14.4x$$
$$\text{origin} = \text{July 1, 1970}$$
$$x = 1 \text{ year}$$
$$Y = \text{annual sales in thousands of dollars}$$

a. If 1978 was the last year of the data, how many years of data were used to determine the trend equation assuming the origin has not been moved?
b. Project sales for 1985.
c. What is the annual dollar increase in sales?

10. Use the following trend equation to answer the questions below:

$$Y_c = 320 + 36x$$
$$\text{origin} = \text{January 1, 1971}$$
$$x = 1$$
$$Y = \text{annual sales in thousands of dollars}$$

a. If 1978 was the last year of data, how many years of data were used to determine the trend equation if the origin has not been changed?
b. Project sales for 1980.
c. What is the annual dollar increase in sales?

11. Use the following trend equation to answer the questions below:

$$Y_c = 50 + 12x$$
$$\text{origin} = \text{December } 31, 1972$$
$$x = 1 \text{ year}$$
$$Y = \text{annual sales in thousands of dollars}$$

a. If 1977 was the last year of data, how many years of data were used to determine the trend equation if the origin has not been changed?
b. What is the annual dollar increase in sales?
c. Project sales for 1980.

12. Remove the trend from the data of Problem 1 assuming the multiplicative model.

13. Assume the following data represent employment in a large plant. Fit a second-degree least-squares equation to these data.

YEAR	EMPLOYMENT
1972	3000
1973	3500
1974	3700
1975	4000
1976	3800
1977	3600
1978	3200

14. Fit a second-degree equation to the following data using the method of least-squares.

FARM INCOME STABILIZATION PAYMENTS	
FISCAL YEAR	BILLIONS OF DOLLARS
1969	5.3
1970	4.6
1971	3.7
1972	4.6
1973	4.1
1974	1.5
1975	0.8
1976	1.6
1977	5.5

Source: The Budget of the United States Government, Fiscal Year 1979. U.S. Government Printing Office, p. 479.

15. Fit a logarithmic trend equation, using the method of least squares, to the following data letting $x = 1$ year.

	SALES OF PRODUCT
YEAR	BILLIONS OF DOLLARS
1971	4.6
1972	7.2
1973	8.9
1974	9.9
1975	11.2
1976	13.4
1977	14.1
1978	17.2

16. Use the hypothetical data to answer the question below.

	GASOLINE SALES (MILLIONS OF GALLONS)				
YEAR	QUARTER				TOTAL
	I	II	III	IV	
1974	4.3	4.9	5.3	5.0	19.5
1975	4.4	4.9	5.4	5.1	19.8
1976	4.5	5.3	5.5	5.3	20.6
1977	4.6	5.3	5.6	5.4	20.9
1978	4.6	5.3	5.8	5.7	21.4

a. Determine the straight-line trend equation for the yearly totals using the method of least squares. (Keep this work for Chapter 16.)

b. Convert the equation developed in part (a) to a quarterly equation with Y units expressed as quarterly sales.

c. Shift the origin of the quarterly equation to the middle of the third quarter of 1976.

17. Estimate 1979 production using the following equation:

$$\log Y_c = 6.43921 + 0.23510x$$
$$\text{origin} = \text{July 1, 1973}$$
$$x = 1 \text{ year}$$
$$Y = \text{annual production in pounds}$$

18. Use the equation

$$Y_c = 100 + 10x$$
$$\text{origin} = \text{December 31, 1973}$$
$$x = 1 \text{ year}$$
$$Y = \text{annual sales in millions}$$

a. to estimate sales for 1982.

b. and shift the origin to December 31, 1976.

19. Use the trend equation in Problem 9 of this chapter.
   a. Shift the origin to July 1, 1973.
   b. Convert the trend equation to monthly sales with $x = 1$ month and set the origin at July 15, 1970.
   c. Estimate sales for September, 1982.

20. Use the equation

$$Y_c = 120 + 3.6x$$
$$\text{origin} = \text{December 31, 1972}$$
$$x = 1 \text{ year}$$
$$Y = \text{annual production in millions of pounds}$$

   a. and shift the origin to July 1, 1978.
   b. and convert the given trend equation to monthly data with $x = 1$ month.

Illindoh Petroleum Company

The Illindoh Petroleum Company is an independent gasoline distributor operating exclusively in the states of Illinois, Indiana, and Ohio. The company's headquarters is located in Akron, Ohio.

In 1962 the company's market share of gasoline sales in its three-state market area was 1.91 percent, representing total sales 160,892,000 gallons of gasoline. From 1962 to 1967, its market share showed a slight but steady increase; however, since 1967 Illindoh's market share has been declining. Its market share in 1978, for instance, was 1.77 percent. For this reason and in the interests of sound planning, the company's management has asked its planning and research division to make a thorough analysis and forecast of gasoline consumption for each of the states in its market area.

You have been assigned the task of making this analysis. Specifically, you have been asked to provide the company the following information:

1. a forecast of gasoline consumption by state for 1979,
2. a five-year forecast of gasoline consumption by state for 1980 to 1984, and
3. a five-year forecast of market share for the three-state market area.

You have been provided the following information: Motor gasoline consumption for the three states and the United States, motor gasoline sales for the Illindoh market area, motor vehicle registration for the three states and the United States, highway travel for the three states and the United States, and the percent of market share for the Illindoh Petroleum Company for its market area.

In analyzing these data and making your analysis and forecast, you should prepare a formal report to the management of Illindoh outlining your procedure and method of analysis, the basis for your forecast, and the forecast itself. You should also present all necessary data in the form of tables, graphs, and charts. If you feel that additional data are necessary, you should go to your own sources of information and provide these data.

Market Share, Illindoh Petroleum Company, 1962–78.

YEAR	PERCENT SALES
1962	1.91
1963	1.98
1964	1.96
1965	2.02
1966	2.03
1967	2.04
1968	1.96
1969	1.89
1970	1.87
1971	1.89
1972	1.87
1973	1.85
1974	1.84
1975	1.86
1976	1.81
1977	1.82
1978	1.77

Motor Gasoline Consumption for Selected States and the United States, 1962–78 (thousands of gallons).

YEAR	ILLINOIS	INDIANA	OHIO	UNITED STATES
1962	3,277,932	1,840,861	3,305,508	62,531,373
1963	3,382,046	1,890,093	3,359,222	64,779,104
1964	3,506,222	1,956,911	3,523,939	67,663,848
1965	3,638,663	2,037,715	3,657,408	70,337,126
1966	3,819,091	2,125,426	3,830,041	73,638,872
1967	3,879,185	2,169,585	3,934,925	76,139,326
1968	4,160,075	2,311,357	4,168,373	80,867,012
1969	4,312,652	2,410,523	4,383,205	85,415,920
1970	4,408,829	2,508,880	4,574,876	89,405,266
1971	4,596,788	2,590,540	4,695,627	93,907,352
1972	4,836,883	2,740,984	4,960,046	100,068,575
1973	5,048,307	2,846,485	5,194,284	104,579,479
1974	5,024,751	2,739,043	5,006,097	101,855,749
1975	5,087,328	2,762,822	5,095,243	104,271,488
1976	5,354,304	2,784,248	5,273,960	109,655,206
1977	5,537,366	2,806,522	5,469,096	112,286,931
1978	6,180,798	2,826,729	5,517,224	113,387,734

Motor Gasoline Sales for Illindoh Market Area, 1962–78 (thousands of gallons).

YEAR	ILLINOIS	INDIANA	OHIO	TOTAL
1962	65,558	23,010	72,324	160,892
1963	69,622	23,190	77,893	170,705
1964	71,224	24,234	80,931	176,389
1965	76,636	24,403	88,053	189,092
1966	77,279	26,697	94,305	198,281
1967	78,284	28,539	98,360	204,183
1968	80,476	28,793	98,950	208,219
1969	81,532	29,052	99,247	209,831
1970	83,091	31,057	101,033	215,181
1971	84,180	34,070	105,984	224,234
1972	86,580	34,342	113,297	234,219
1973	88,129	36,025	118,622	242,776
1974	87,533	34,702	112,691	234,926
1975	88,031	34,944	117,762	240,737
1976	89,917	35,573	118,350	243,840
1977	92,552	39,095	119,475	251,122
1978	94,187	41,949	120,645	256,781

Motor Vehicle Registration for Selected States, 1962–78.

YEAR	ILLINOIS	INDIANA	OHIO	UNITED STATES
1962	3,976,709	2,173,920	4,301,914	79,150,336
1963	4,105,402	2,253,222	4,468,985	82,696,732
1964	4,259,433	2,321,717	4,670,976	86,313,262
1965	4,437,191	2,427,044	4,935,295	90,357,667
1966	4,704,624	2,550,539	5,238,458	93,949,852
1967	4,818,259	2,631,944	5,305,391	96,905,876
1968	4,990,073	2,739,206	5,441,963	100,898,074
1969	5,162,398	2,806,206	5,739,576	105,096,369
1970	5,237,876	2,817,991	5,973,901	108,418,197
1971	5,417,021	2,868,016	6,043,109	112,986,342
1972	5,643,853	3,064,327	6,414,345	118,796,671
1973	5,951,948	3,232,321	6,644,610	125,653,934
1974	6,174,102	3,268,516	6,965,481	129,933,556
1975	6,343,875	3,315,371	7,178,933	132,959,950
1976	6,502,472	3,339,573	7,566,596	133,757,710
1977	6,561,644	3,365,622	7,622,589	135,068,535
1978	6,604,951	3,554,096	7,759,795	136,014,015

Highway Travel 1967–78 (millions of vehicle miles).

YEAR	ILLINOIS	INDIANA	OHIO	UNITED STATES
1967	49,858	27,399	49,722	965,132
1968	52,139	29,550	52,834	1,019,726
1969	53,873	30,617	55,821	1,066,108
1970	55,313	32,578	56,014	1,114,098
1971	57,390	34,292	61,051	1,183,524
1972	59,383	36,785	63,505	1,264,614
1973	60,772	38,727	65,171	1,316,207
1974	59,210	36,993	63,084	1,282,790
1975	60,943	37,359	64,134	1,330,074
1976	61,418	37,545	64,769	1,355,345
1977	64,857	40,361	65,287	1,377,030
1978	65,117	40,684	68,160	1,430,735

Time Series Analysis: Seasonality

CHAPTER

16

Whenever the units of the data are less than annual periods of time (ordinarily, monthly or quarterly data), the time series data may contain seasonal variation. This may be attributable to such factors as holidays, the seasonal weather pattern, or established conventions (such as weddings in June and county fairs in the fall). The subject of this chapter is the measurement of this seasonal variation and the applications of its measures.

The Nature and Importance of Seasonal Analysis

Recalling the time series models discussed in the preceding chapter, seasonal variation may be expressed in absolute dollars or as an index. The latter, an index, is ordinarily calculated even when the additive model is to be used, for it is with an index that seasonality is commonly measured. Thus, a *seasonal index* is a measure of the variation in time series data that is attributable to annually recurring factors.

In order to measure seasonality over a year, there must be a series of seasonal index numbers. Monthly data require twelve indexes to express the seasonality—one for each month of the year. Similarly, quarterly data require a series of four indexes.

Like all indexes, a seasonal index must have a base. Generally speaking, the base used is either the *average year* in the time series or an *average period* (less than a year) in the time series. When the base is an average period, the period may be a month, a quarter, or any other time segment of less than a year in length. The base (average year or average period) used in computing an index will cause minor differences in the methods of computation and, more important, differences in the interpretation and application of the index. The example in Table 16-1 illustrates the use of alternative bases to measure seasonality.

TABLE 16-1. *Seasonal Index, alternative bases.*

| | INDEX | |
QUARTER	BASE-AVERAGE YEAR	BASE-AVERAGE QUARTER
I	20	80
II	30	120
III	35	140
IV	15	60
	100	400

When the base is an average period (quarters in Table 16-1), the index expresses seasonal variation for each period in the year as a percent of the average period. (Notice, however, that the percent is *not stated.*) Referring to Table 16-1, the index of 80 for Quarter I means that sales in that quarter are typically 80 percent of average quarterly sales for a year. Since the base is assumed to be equal to 100, the difference between the index for any quarter and 100 (Index minus 100) is the amount of variation attributable to seasonality. Applying this to our example, sales in Quarter II are typically 20 percent (120 − 100) above the average quarterly sales for a year.

If the average year is used as a base for the seasonal index, then the index for Quarter I (shown in Table 16-1) is 20. Stated in this manner, the index number means that typically 20 percent of a given year's sales occur in Quarter I.

The choice of which base to use in measuring seasonality depends on the circumstances in which the seasonal index is being used. You may decide to use a particular base because it facilitates the interpretation, or communication, of the measures of seasonality. For example, if you are using monthly sales data in a report, the use of an average month base for a seasonal index of sales would retain the consistency of the data in your report. If you used an average year as the base, and your sales figures are in months, the reader might have a little trouble making the transition from months, to years, and back to months. However, it is relatively simple to convert from one base to the other. Index numbers with the average quarter as a base are *four* times the size of quarterly index numbers computed using an average year as the base. Monthly indexes with the average month as the base are *twelve* times the size of monthly indexes with the average year as the base. Using this relationship, it should be relatively easy for you to determine how to convert a monthly seasonal index, a quarterly seasonal index, or one for other time periods, from one base to another.

While we are considering alternative bases for a seasonal index, it may be worthwhile to take notice of a small technicality. The *sum* of the index numbers for time periods in a year will vary as the selection of a base varies. If an average year is used as the base, then the series of index numbers should have a sum of 100; if the average quarter is used, the sum should be 400; and when using the monthly average, the sum is 1200. The logic of this is evident if you consider that the base is an arithmetic mean (average) equal to 100. All this may seem obvious, and perhaps trivial, to you, but it provides a ready method for checking the arithmetic accuracy of your calculations when computing a series of seasonal index numbers.

Adjusting for Seasonality

Given that you have a seasonal index, what can you do with it? In analyzing business and economic data, short-term fluctuations (seasonal variations) may camouflage, at least temporarily, more basic changes in the data. Deseasonalizing the actual data removes the distortion caused by seasonal variation. In addition, if a forecast of annual data (sales, output, etc.) is available, the seasonal index offers a capability to estimate how the series will behave during the year.

▶ **Example 16.1** Assume that a toy shop has a November seasonal index of 150 and a December index equal to 200. If, in 1974, sales in November were $15,000 and in December they were $18,000, one might assume that the toy business was better in the month of December than it was in November. That is correct, for sales in December were actually 20 percent above November's sales. However, according to the index numbers, sales in December are normally 33.3 percent greater than sales in November! Perhaps business was not so good in December after all. ◀

Removing Seasonality

"Deseasonalizing data" is just another way of saying "removing the seasonal influence from the data." The procedure for deseasonalizing requires that the actual data be *divided* by the seasonal index, with the result multiplied by 100.

$$\frac{\text{actual data}}{\text{seasonal index}} \times 100 = \text{deseasonalized data}$$

The origin of this procedure is evident if you recall the product model you studied in Chapter 15.

$$Y = T \times S \times C \times I$$

$$Y/S = T \times C \times I = \text{deseasonalized data}$$

Table 16-2 provides an example of applying a seasonal index to remove the seasonal influence from data. Theoretically, the deseasonalized data reflect only the influence of trend, cyclical, and random influences. Of course, this is true only to the extent that the seasonal index is an accurate measure of seasonal influence.

TABLE 16-2. *Deseasonalizing Data.*

QUARTER	ACTUAL SALES	SEASONAL INDEX	DESEASONALIZED SALES
I	$400	90	$444
II	540	110	491
III	630	120	525
IV	430	80	538

Sample Calculation:

$$\frac{\$400 \text{ (actual sales)}}{90 \text{ (index)}} \times 100 = \$444 \text{ (deseasonalized sales)}$$

It is difficult to decide, from observing actual sales, whether sales are increasing or decreasing because of seasonal variation that occurs within the year. However, deseasonalized sales data, as shown in the last column of Table 16-2, allow the analyst to observe the trend of sales after the effect of seasonal variation has been removed. Actually, it should be remembered that the last column (sometimes called seasonally adjusted sales) of Table 16-2 contains the TCI components of time series data. However, the direction (increasing or decreasing) of the deseasonalized data is a reasonably good indicator of the trend.

Adding Seasonality

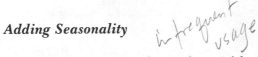

Frequently you may have data (either actual or a forecast) available only in an annual form. For purposes of anticipating short-term (seasonal) fluctuations, you can apply a seasonal index to the data. If the index has as its base the average year, this requires only that you multiply the annual data by the seasonal index.

$$\text{annual data} \times \frac{\text{seasonal index}}{100} = \text{seasonalized data}$$

If the base is an average time period, an intermediate step is necessary. In this case,

$$\frac{\text{annual data}}{\substack{\text{number of base time} \\ \text{periods in a year}}} \times \frac{\text{seasonal index}}{100} = \text{seasonalized data}$$

Both of these procedures are illustrated in Table 16-3.

TABLE 16-3. *Seasonalizing data.*

AVERAGE YEAR = BASE

QUARTER	INDEX	ANNUAL SALES	SEASONALIZED DATA
I	20	$20,000	$ 4,000
II	30	20,000	6,000
III	35	20,000	7,000
IV	15	20,000	3,000
			$20,000

$$\$20,000 \text{ (annual sales)} \times \frac{20 \text{ (index)}}{100} = \$4,000 \text{ (seasonalized data)}$$

AVERAGE QUARTER = BASE

QUARTER	INDEX	ANNUAL SALES ÷ 4	SEASONALIZED DATA
I	80	$5,000	$ 4,000
II	120	5,000	6,000
III	140	5,000	7,000
IV	60	5,000	3,000
			$20,000

$$\frac{\$20,000 \text{ (annual sales)}}{4 \text{ (number of periods)}} \times \frac{80 \text{ (index)}}{100} = \$4,000 \text{ (seasonalized data)}$$

The preceding discussion is intended to cover the utilization of seasonal indexes in a very general sense. Perhaps you can already visualize how these techniques may prove useful as you approach problems in the functional areas of business administration. Knowledge of seasonal variations can assist the businessman in planning his personnel needs from month-to-month, in determining his inventory requirements, and in anticipating changes in his cash flow. You will undoubtedly find that seasonal analysis is one of the most frequently used of all the statistical tools.

Determining the Existence of Seasonality

When we said that "data may contain seasonal variation" we implied that not all variables are subject to seasonal influences. This is correct. Furthermore, on occasion you may encounter a series of data where seasonality exists but its effect is so heterogeneous, or dispersed, that it is impossible, or at best misleading, to measure seasonality. An example of this is the monthly cash income of farmers. Variations in weather can cause "harvest time" to fluctuate between months. A farmer's cash this year may be greater in August than in September. Next year, September's income may be the larger. Another possibility is that the seasonal pattern is changing through time. That is, there is a trend in the measures of seasonal influences.

How do you determine if seasonality exists in a set of data? The best way for you to find the answer to this question is to construct a set of scatter diagrams using the ratio of the actual value to the average for the year. A set is needed since you must prepare a separate one for each seasonal time period. That is, if you are working with monthly data, scatter diagrams are prepared for January, February, and so forth, for each of the twelve months. Figure 16-1 provides some examples of how these scatter diagrams might appear.

Graph (a) in the figure represents a situation where the values for a hypothetical month are said to be heterogeneous, or dissimilar. Put another way, there appears to be little *central tendency* in the data. The measure of dispersion would be large regardless of the average you selected to represent the data. Should you encounter data where this pattern exists for one or more months, it would be misleading to seek to compute, and use, a seasonal index.

In some cases, the problem may be alleviated by lengthening the index time period, perhaps from monthly to quarterly. Failing to observe the requirement that there actually be some noticeable central tendency to the cluster of values for every time period will not prohibit you from continuing on to calculate a series of index numbers. However, since there would be no "typical" or "normal" seasonal influence, the resulting index numbers could hardly be construed as measuring the typical seasonal pattern.

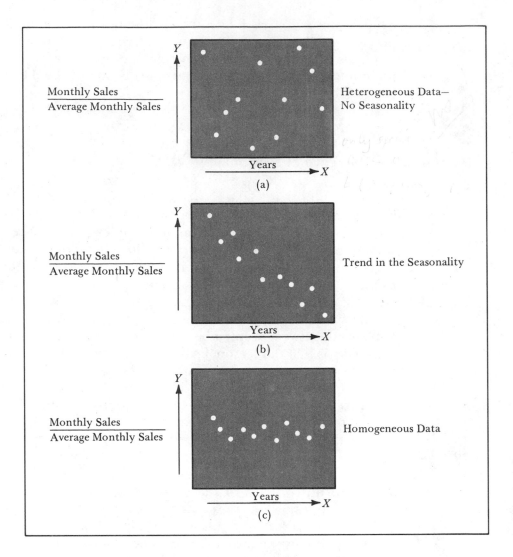

FIGURE 16.1 *Possible Variations for a Particular Month.*

     Graph (b) represents another situation in which caution must be exercised when a seasonal index is computed. When you first examine the graph you may conclude that it is very much like Graph (a). This is not the case, however. There is a distinct central tendency but it is *moving*. When a scatter of this nature occurs, there is said to be a *trend in the seasonality of the data*. Computation of an index requires that a trend equation be determined for each month or quarter. Thus, the seasonal index for a time period would change each year. While this condition may exist quite often in business and economic data, in practice it is seldom recognized since (1) it generally requires a large number of years of data to detect the existence of trend, and

(2) the amount of computation required to determine the trend is so extensive as to be generally prohibitive in any except computerized procedures.

Graph (c) in Figure 16-1 depicts the distribution of data most frequently assumed to exist when a seasonal index is computed. There is a definite central tendency to the data that can be adequately represented by an arithmetic mean. If there exist a few extreme deviations from the cluster at the midpoints, the influence of these values can be omitted by utilizing a modified arithmetic mean—assuming enough values are available to provide a reasonable estimate of the central tendency.

The methods presented in the following pages will carry an implicit assumption that there is a central tendency to the data and that this central tendency is reasonably fixed. In other words, the data are assumed to take on the approximate appearance of that in Graph (c) of Figure 16-1.

Measuring Seasonality

There are large numbers of variations to the general procedure of measuring seasonality. We have chosen to present only two methods that are fairly typical of the techniques that are generally utilized.

In measuring seasonality, a principal distinction in methodology rests upon the manner in which the influence of trend, cyclical, and random variations are removed from the data. This can be a very real problem since (1) ideally, the seasonal index must be devoid of any influence from other types of fluctuations; (2) trend may, or may not, have been measured prior to the computation of the seasonal index; and (3) cyclical and irregular fluctuations are generally measured, chronologically, after seasonal analysis has taken place.

Moving Average Method

Look at Tables 16-4 and 16-5. They illustrate the moving average method of computing a seasonal index. Using this technique the index is determined from raw data. That is, the data in Column 2 of Table 16-4 represent actual sales data just as they appear in the accounting reports of the firm.

Procedurally, there are seven basic steps in the calculation of the seasonal index by this method. The first five steps are shown in Table 16-4.

1. List the data in chronological order. This is shown in Column 2 of Table 16-4.
2. Take a 4-quarter moving total of the data. (If you are working with monthly data, it is a 12 month moving total.) The 4-quarter moving totals are taken for the entire time span of the data by dropping the sales figure for the earliest quarter and adding the next quarterly sales figure on the list. Put another way, when you drop Quarter I for one year from the total you add Quarter I of the next year to it. By looking at Column 3 in Table 16-4 this step in the procedure should become evident. The purpose of this step is to average out the seasonal and irregular components of the multiplicative model.
3. An 8-quarter (or 24-month) moving total is the next step. Actually, this is a moving total of the 4-quarter moving totals. This step is necessary because the 4-quarter totals are not centered on time periods—the middle of a 4-quarter period occurs between two quarters. A total of two successive 4-quarter totals is, therefore, represented as centered on a quarter.
4. Column 5 shows the fourth step. It is a centered quarterly moving average and is computed by dividing the 8-quarter moving totals by 8. When using monthly data, the monthly moving average is found by dividing the 24-month moving totals by 24. (Steps 2, 3, and 4 isolate the trend and cyclical components of the multiplicative model.)
5. The specific seasonals are shown in column 6. These are found by dividing the actual quarterly (monthly) data (column 2) by the quarterly (monthly) moving averages (column 5 and multiplying by 100. (Dividing TCSI by TC leaves the seasonal and random components of the multiplicative model.)

TABLE 16-4. *Moving Average Method, Quarterly Data.*

(1)		(2)	(3)	(4)	(5)	(6)
				CENTERED	CENTERED	
			4 QUARTER	8 QUARTER	MOVING AVERAGE	SPECIFIC SEASONAL
QUARTER		SALES	MOVING TOTAL	MOVING TOTAL	(4)/8	[(2)/(5)][100]
1970	I	8				
	II	12				
			42			
	III	12		86	10.750	111.63
			44			
	IV	10		91	11.375	87.91
			47			
1971	I	10		96	12.000	83.33
			49			
	II	15		98	12.250	122.45
			49			
	III	14		97	12.125	115.46
			48			
	IV	10		95	11.875	84.21
			47			

(cont'd)

(1)	(2)	(3)	(4)	(5)	(6)
			CENTERED	CENTERED	
		4 QUARTER	8 QUARTER	MOVING AVERAGE	SPECIFIC SEASONAL
QUARTER	SALES	MOVING TOTAL	MOVING TOTAL	(4)/8	[(2)/(5)][100]
1972 I	9	47	94	11.750	76.60
II	14	49	96	12.000	116.67
III	14	50	99	12.375	113.13
IV	12	52	102	12.750	94.12
1973 I	10	55	107	13.375	74.77
II	16	56	111	13.875	115.32
III	17	58	114	14.250	119.30
IV	13	60	118	14.750	88.14
1974 I	12	61	121	15.125	79.34
II	18	60	121	15.125	119.01
III	18	59	119	14.875	121.01
IV	12	59	118	14.750	81.36
1975 I	11	60	119	14.875	73.95
II	18	61	121	15.125	119.01
III	19	62	123	15.375	123.58
IV	13	63	125	15.625	83.20
1976 I	12	65	128	16.000	75.00
II	19	67	132	16.500	115.15
III	21	69	136	17.000	123.53
IV	15	74	143	17.875	83.92
1977 I	14	79	153	19.125	73.20
II	24	82	161	20.125	119.25
III	26	83	165	20.625	126.06
IV	18	84	167	20.875	86.23
1978 I	15	84	168	21.000	71.43
II	25	83	167	20.875	119.76
III	26				
IV	17				

The specific seasonal indexes shown in Column 6 of Table 16-4 identify whether sales in a given quarter were above or below the quarterly average for the year. For example, the third quarter sales of 1977 were 26.06 percent (126.06 − 100) above the quarterly average for the year. These specific seasonals, however, contain seasonal and irregular variations. Table 16-5 presents the final two steps necessary to eliminate the irregular variation and determine a seasonal index for each quarter.

6. This step involves the computation of an average specific seasonal for each time period. In other words, when working with quarterly data, it is necessary to compute an average specific seasonal for Quarter I, Quarter II, etc. In Table 16-5, the "modified mean" approach is used, whereby an average is obtained from the modified total (the sum, after elimination of the high and low values, of the specific seasonals). The average specific seasonals are considered

to be unadjusted indexes since they do not necessarily sum to 400 (1,200 for monthly indexes). Step 7 explains the adjustment process.

7. The final step involves computing the *adjusted seasonal index*. If you are working with quarterly data, the seasonal indexes should sum to 400. The difference between the unadjusted and adjusted indexes involves *forcing* the series to add to 400, or 1,200 in the case of monthly data. To accomplish, this simply sum the unadjusted seasonal indexes and make the sum the denominator of a fraction whose numerator is 400 (1,200 for monthly indexes). Next, multiply the unadjusted index for each quarter by the fraction. In Table 16-5, the multiplier is $400 \div 398.59 = 1.00354$. The result is the *seasonal index* (adjusted).

TABLE 16-5. *Calculating Seasonal Indexes.*

YEAR	I	II	III	IV	TOTAL
1970			~~111.63~~	87.91	
1971	~~83.33~~	~~122.45~~	115.46	84.21	
1972	76.60	116.67	113.13	~~94.12~~	
1973	74.77	115.32	119.30	88.14	
1974	79.34	119.01	121.01	~~81.36~~	
1975	73.95	119.01	123.58	83.20	
1976	75.00	~~115.15~~	123.53	83.92	
1977	73.20	119.25	~~126.06~~	86.23	
1978	~~71.43~~	119.76			
Modified Total	452.86	709.02	716.01	513.61	
Average Specific Seasonal (Unadjusted)	75.48	118.17	119.34	85.60	398.59
Seasonal Index (Adjusted)	75.75	118.59	119.76	85.90	400.00

The procedure described above yields an index with the average time period as a base. You might ask how the procedure removes the influence of trend, cyclical, and irregular variations from the data. This is assumed to be accomplished by using the 8-quarter moving average to determine the specific seasonals and averaging the specific seasonals. While this may not guarantee that the influence of the other types of fluctuations will be completely eliminated, it will almost certainly reduce their influence and it is a widely accepted technique for computing seasonal indexes. Now look at Tables 16-6 and 16-7 to see how monthly seasonal indexes are computed using the moving average method.

TABLE 16-6. *Moving Average Method, monthly data.*

YEAR	MONTH	(1) SALES	(2) 12 MONTH MOVING TOTAL	(3) CENTERED 24 MONTH MOVING TOTAL	(4) = (3) ÷ 24 CENTERED 24 MONTH MOVING. AVERAGE	(5) = [(1) ÷ (4)] [100] SPECIFIC SEASONAL
1972	Jan.	30				
	Feb.	26				
	March	29				
	April	36				
	May	40				
	June	40	451			
	July	41	456	907	37.79	108.49
	Aug.	44	460	916	38.17	115.27
	Sept.	42	468	928	38.67	108.61
	Oct.	45	472	940	39.17	114.88
	Nov.	40	476	948	39.50	101.27
	Dec.	38	480	956	39.83	95.41
1973	Jan.	35	484	964	40.17	87.13
	Feb.	30	485	969	40.38	74.29
	March	37	488	973	40.54	91.27
	April	40	489	977	40.71	98.26
	May	44	490	979	40.79	107.87
	June	44	492	982	40.92	107.53
	July	45	491	983	40.96	109.86
	Aug.	45	494	985	41.04	109.65
	Sept.	45	493	987	41.12	109.44
	Oct.	46	493	986	41.08	111.98
	Nov.	41	493	986	41.08	99.81
	Dec.	40	491	984	41.00	97.56
1974	Jan.	34	488	979	40.79	83.35
	Feb.	33	490	978	40.75	80.98
	March	36	491	981	40.88	88.06
	April	40	493	984	41.00	97.56
	May	44	496	989	41.21	106.77
	June	42	498	994	41.42	101.40
	July	42	502	1000	41.67	100.79
	Aug.	47	504	1006	41.92	112.12
	Sept.	46	508	1012	42.17	109.08
	Oct.	48	512	1020	42.50	112.94
	Nov.	44	515	1027	42.79	102.83
	Dec.	42	518	1033	43.04	97.58
1975	Jan.	38	524	1042	43.42	87.52
	Feb.	35	526	1050	43.75	80.00
	March	40	528	1054	43.92	91.07
	April	44	529	1057	44.04	99.91
	May	47	530	1059	44.12	106.53
	June	45	528	1058	44.08	102.09

YEAR	MONTH	(1) SALES	(2) 12 MONTH MOVING TOTAL	(3) CENTERED 24 MONTH MOVING TOTAL	(4) = (3) ÷ 24 CENTERED 24 MONTH MOVING AVERAGE	(5) = [(1) ÷ (4)] [100] SPECIFIC SEASONAL
	July	48		1054	43.92	109.29
	Aug.	49	526	1051	43.79	111.90
	Sept.	48	525	1050	43.75	109.71
	Oct.	49	525	1049	43.71	112.10
	Nov.	45	524	1048	43.67	103.05
	Dec.	40	524	1049	43.71	91.51
1976	Jan.	36	525	1053	43.88	82.04
	Feb.	34	528	1058	44.08	77.13
	March	40	530	1063	44.29	90.31
	April	43	533	1069	44.54	96.54
	May	47	536	1074	44.75	105.03
	June	46	538	1081	45.04	102.13
	July	51	543	1090	45.42	112.29
	Aug.	51	547	1095	45.62	111.79
	Sept.	51	548	1101	45.88	111.16
	Oct.	52	553	1113	46.38	112.12
	Nov.	47	560	1127	46.96	100.09
	Dec.	45	567	1142	47.58	94.58
1977	Jan.	40	575	1154	48.08	83.19
	Feb.	35	579	1163	48.46	72.22
	March	45	584	1171	48.79	92.23
	April	50	587	1178	49.08	101.87
	May	54	591	1185	49.38	109.36
	June	54	594	1189	49.54	109.00
	July	55	595	1195	49.79	110.46
	Aug.	56	600	1205	50.21	111.53
	Sept.	54	605	1212	50.50	106.93
	Oct.	56	607	1216	50.67	110.52
	Nov.	50	609	1224	51.00	98.04
	Dec.	46	615	1234	51.42	89.46
1978	Jan.	45	619	1243	51.79	86.89
	Feb.	40	624	1252	52.17	76.67
	March	47	628	1259	52.46	89.59
	April	52	631	1264	52.67	98.73
	May	60	633	1268	52.83	113.57
	June	58	635	1273	53.04	109.35
	July	60	638			
	Aug.	60				
	Sept.	57				
	Oct.	58				
	Nov.	52				
	Dec.	49				

TABLE 16-7. *Calculating Seasonal Indexes.*

	YEAR	JAN.	FEB.	MARCH	APRIL	MAY	JUNE	JULY	AUG.	SEPT.	OCT.	NOV.	DEC.	TOTAL
	1972							108.49	~~115.27~~	108.61	~~114.88~~	101.27	95.41	
	1973	87.13	74.29	91.27	98.26	107.87	107.53	109.86	~~109.65~~	109.44	111.98	99.81	97.56	
	1974	83.35	~~80.98~~	~~88.06~~	97.56	106.77	~~101.40~~	~~100.79~~	112.12	109.08	112.94	102.83	~~97.58~~	
	1975	~~87.52~~	80.00	91.07	99.91	106.53	102.09	109.29	111.90	109.71	112.10	~~103.05~~	91.51	
	1976	~~82.04~~	77.13	90.31	~~96.54~~	~~105.03~~	102.13	~~112.29~~	111.79	~~111.16~~	112.12	100.09	94.58	
	1977	83.19	~~72.22~~	~~92.23~~	~~101.87~~	109.36	109.00	110.46	111.53	~~106.93~~	~~110.52~~	~~98.04~~	89.46	
	1978	86.89	76.67	89.59	98.73	~~113.57~~	~~109.35~~							
Modified Total		340.56	308.09	362.24	394.46	430.53	420.75	438.10	447.34	436.84	449.14	404.0	379.06	
Average Specific Seasonal (Unadjusted)		85.14	77.02	90.56	98.62	107.63	105.19	109.53	111.84	109.21	112.29	101.0	94.77	1202.80
Seasonal Index (Adjusted)		84.94	76.84	90.35	98.39	107.38	104.95	109.28	111.58	108.96	112.03	100.76	94.55	1200.01

$$\text{Multiplier for Adjustment} = \frac{1200}{1202.80} = 0.9976721$$

Detrended Method

It is possible to isolate the influence of seasonal variation other than by the moving average method. In Chapter 15 you learned how to estimate trend. If a trend estimate is available, the trend can be removed from the data by using the method described near the end of chapter 15. One technique for doing this is illustrated in Table 16-8.

TABLE 16-8. *Detrending the Data.*

YEAR	QUARTER	Y SALES	$Y_c = 17.86 + 0.419x$ TREND	Y/Y_c
1974	I	12	13.670	0.878
	II	18	14.089	1.278
	III	18	14.508	1.241
	IV	12	14.927	0.804
1975	I	11	15.346	0.717
	II	18	15.765	1.142
	III	19	16.184	1.174
	IV	13	16.603	0.783
1976	I	12	17.022	0.705
	II	19	17.441	1.089
	III	21	17.860	1.176
	IV	15	18.279	0.821
1977	I	14	18.698	0.749
	II	24	19.117	1.255
	III	26	19.536	1.331
	IV	18	19.955	0.902
1978	I	15	20.374	0.736
	II	25	20.793	1.202
	III	26	21.212	1.226
	IV	17	21.631	0.786

After the data have been detrended as shown in Table 16-8, then a modified mean *of the detrended values* for each time period can be determined in order to average out the random component of the multiplicative model. Each mean is adjusted to force the sum of the indexes to 400. This is the final step and yields the seasonal index for each period (Table 16-9).

TABLE 16-9. *Calculating the Seasonal Indexes.*

	YEAR	I	II	III	IV	TOTAL
				QUARTER		
	1974	~~0.878~~	~~1.278~~	1.241	0.804	
	1975	0.717	1.142	~~1.174~~	~~0.783~~	
	1976	~~0.705~~	~~1.089~~	1.176	0.821	
	1977	0.749	1.255	~~1.331~~	~~0.902~~	
	1978	0.736	1.202	1.226	0.786	
Modified Total		2.202	3.599	3.643	2.411	
Unadjusted Index		73.40	119.97	121.43	80.37	395.17
Adjusted Index		74.30	121.44	122.91	81.35	400.00

$$\text{Multiplier for Adjustment} = \frac{400}{395.17} = 1.0122226$$

Detrending the data using the trend estimate offers a less time consuming approach to time series analysis than the moving average method. Furthermore, the moving average method causes a loss of the first one-half year and the last one-half year of data from the index calculations. (This can be significant if you have data for only a few years.) On the other hand, the detrending method does not remove the cyclical component so effectively as the moving average method. Thus, unless you are sure that the cyclical effect is slight, the moving average method is probably the better choice.

Adjusting Data for Calendar Variation

There are many instances where calendar variations can influence the size of a variable. The monthly payroll of a company will vary if employees are paid weekly simply because some months will contain more "paydays" than other months. Furthermore, the number of "paydays" in a given month will vary from year to year. Monthly sales will vary for much the same reason—months contain varying numbers of "business days." Weekly sales will even vary for the same reasons. These calendar variations can cause significant distortions in the economic and business data you may encounter. Removing the influence of calendar variations is best illustrated by Example 16.2.

▶ **Example 16.2** A given store is normally open 5 days a week, 52 weeks a year. However, the store takes eight holidays during the year. Thus, it is open 252 days a year, for an average of approximately 4.81 days a week or 21 days a month. In February the store is open 20 days and has sales of $4,000. In the next month (March) it has sales of $5,000 and is open 24 days. The following adjustments should be made to the monthly sales before a comparison of sales in the two months is made.

$$\text{February} \quad \$4,000 \times \frac{21}{20} = \$4,200 \text{ (adjusted sales)}$$

$$\text{March} \quad \$5,000 \times \frac{21}{24} = \$4,375 \text{ (adjusted sales)} \blacktriangleleft$$

In many cases, variations of this type have a negligible influence on data. However, you should assure yourself that this is the case before undertaking extensive analysis of time series data. If the influence proves to be significant, the data should be adjusted before you attempt to measure seasonal, cyclical, or irregular variations.

─────────── **REVIEW QUESTIONS** ───────────

1. What is meant by deseasonalizing the data?
2. Why do we deseasonalize data?
3. How can seasonality be identified?

─────────── **PROBLEMS** ───────────

1. The following data represent the monthly electric bill (in dollars) for a residential customer. Calculate a monthly seasonal index for this customer using a centered 12-month moving average.

YEAR	JAN	FEB	MAR	APR	MAY	JUNE	JULY	AUG	SEP	OCT	NOV	DEC
1972	28	29	25	21	29	35	40	50	45	53	40	32
1973	30	30	26	23	32	40	52	60	47	57	41	37
1974	34	34	31	29	28	44	54	57	53	60	45	34
1975	37	40	35	27	30	48	57	65	62	60	52	43
1976	39	41	40	35	40	52	58	71	65	65	60	49
1977	43	50	52	38	48	57	60	74	66	70	63	57
1978	56	58	55	46	59	62	65	78	72	75	70	64

2. Calculate monthly seasonal indexes using a centered 12-month moving average.

MONTHLY SALES
(in $1,000)

YEAR	JAN	FEB	MAR	APR	MAY	JUNE	JULY	AUG	SEP	OCT	NOV	DEC
1971	3	2	3	4	4	5	6	7	6	6	5	4
1972	4	3	5	7	8	8	9	10	10	8	8	6
1973	5	5	7	10	12	11	11	11	10	9	8	7
1974	6	5	6	6	7	8	9	10	10	10	9	9
1975	9	8	10	11	13	14	15	16	17	16	16	15
1976	13	13	14	15	16	17	17	17	17	16	14	12
1977	10	10	12	12	13	15	15	16	15	15	15	14
1978	14	13	15	15	14	15	16	18	19	19	18	18

3. Calculate monthly seasonal indexes using a centered 12-month moving average.

MONTH	SALES ($1,000)							
	1971	1972	1973	1974	1975	1976	1977	1978
Jan.	26.4	28.2	32.4	38.4	42.4	50.3	48.0	46.2
Feb.	32.8	38.0	42.3	48.0	50.1	56.4	56.0	56.2
Mar.	34.5	42.3	48.0	50.3	62.4	70.1	64.3	70.0
Apr.	32.4	34.4	40.1	46.0	54.0	56.2	52.4	54.4
May	30.2	34.3	38.1	42.2	50.0	54.2	50.3	52.1
June	30.1	32.1	36.4	42.1	48.2	52.0	50.0	50.3
July	26.4	26.0	32.3	36.1	40.3	48.1	44.4	48.2
Aug.	28.0	26.3	32.2	38.3	42.1	50.4	48.1	44.3
Sep.	28.1	30.1	32.4	40.0	44.3	50.1	48.0	44.4
Oct.	28.4	30.3	36.0	40.3	40.4	50.2	50.0	46.1
Nov.	26.2	24.4	30.3	34.4	40.3	48.2	42.2	44.4
Dec.	26.3	26.0	30.4	34.4	38.4	44.2	42.3	44.0

4. Calculate quarterly seasonal indexes using a centered 4-quarter moving average.

Composting! quarters and values are identical

		SALES FOR THE DEPARTMENT STORE		
		QUARTER		
YEAR	I	II	III	IV
1974	1	2	3	4
1975	2	3	4	5
1976	3	4	5	6
1977	4	5	6	7
1978	5	6	7	8

5. Calculate seasonal indexes for the data given below using a centered 4-quarter moving average.

		GASOLINE SALES (MILLIONS OF GALLONS)			
		QUARTER			
YEAR	I	II	III	IV	TOTAL
1974	4.3	4.9	5.3	5.0	19.5
1975	4.4	4.9	5.4	5.1	19.8
1976	4.5	5.3	5.5	5.3	20.6
1977	4.6	5.3	5.6	5.4	20.9
1978	4.6	5.3	5.8	5.7	21.4

6. Calculate quarterly seasonal indexes using a centered four-quarter moving average.

		SALES		
		QUARTER		
YEAR	I	II	III	IV
1974	8	12	15	10
1975	10	20	25	15
1976	15	25	30	20
1977	20	30	35	25
1978	25	35	40	30

7. Quarterly sales and seasonal indexes for Big Enterprise are given in the table that follows.

Omit wrong answer in back

	I	II	III	IV
Quarterly Sales	$110,560	$350,340	$422,220	$150,000
Seasonal Index	49.7	115.4	125.9	109.0

Determine the seasonally adjusted sales.

8. Sales and seasonal indexes for the July Corporation are given below.

	JAN	FEB	MAR	APR
Sales	$6000	$4200	$5600	$7500
Seasonal Index	70	80	100	110

a. What are the seasonally adjusted sales?
b. The corporation forecaster believes that the average monthly sales will be $7,000. If the May seasonal index is 105, what will he forecast for May sales?

9. The sales of Cloudy Plastic, Ltd., were $72,000 for the first quarter of 1978. If the seasonal indexes by quarter are 75, 90, 125, 110:
   a. What are the deseasonalized sales for the first quarter?
   b. Assuming average quarterly sales of $100,000, estimate the sales for the third quarter.

10. The trend equation for the sales of Lotta Manufacturing is $Y_c = 50 + 1.2x$ where Y is $1,000$, $x = 1$ quarter, and the origin is the middle of the third quarter, 1975. If the seasonal indexes are 70, 80, 140, and 110 for each quarter, estimate sales for each quarter of 1979, ignoring the cyclical and irregular fluctuations, using the multiplicative model.

11. The bank has determined the following seasonal indexes for cash demands.

	JUNE	JULY	AUG.
Seasonal Index	128	140	150

If the average monthly demand for cash is $2,000,000, how much cash should be available in July?

12. Determine the seasonal indexes for the data in Problem 5 using the detrending method. Use the trend equation developed in Problem 16 (c) of Chapter 15.

13. The following data represent the average of the specific seasonals for the months shown. Determine the adjusted seasonal index for each month.

JAN	89.50	MAY	99.85	SEP	114.75
FEB	72.46	JUNE	105.36	OCT	106.84
MAR	81.28	JULY	125.43	NOV	93.96
APR	95.63	AUG	115.60	DEC	88.88

14. The following data represent the average of the specific seasonals for the quarters shown. Determine the adjusted seasonal index for each quarter.

	QUARTER		
I	II	III	IV
114.86	94.63	98.48	92.80

Expected Values and Decision Making

CHAPTER
17

In this chapter we explain how to combine statistics with economic information in order to develop a framework for decision making. Situations under which decisions are made may be divided into three categories: (1) *certainty*, (2) *risk*, and (3) *uncertainty*.

When there is perfect knowledge about which events will occur, we call this condition one of *certainty*. *Uncertainty* exists when there is no knowledge available concerning which of several events will occur. A third category known as *risk*, is a condition between certainty and uncertainty. *Risk* is the condition where the occurrence of the possible events can be assigned probabilities or described with some probability distribution. That is, we know what events are possible, and the probability of each event can be specified. This chapter deals with decisions under the situation of risk.

Expected Values

One way to make decisions under risk conditions is to use *expected values*. The expected value of a random variable is an average value of the variable. The expected value of a discrete random variable can be calculated using Equation 17.1.

$$E(X) = \Sigma[x \cdot P(X = x)] \qquad (17.1)$$

where $E(X)$ = expected value of the random variable, X
x = the possible values (outcomes) of the random variable, X
$P(X = x)$ = the probability that the random variable, X, is equal to x.

Expected value is sometimes called *mathematical expectation.* The concept is probably easiest understood through examples about games of chance.

▶ **Example 17.1**   Suppose you have been invited to toss a coin with the understanding that if a head occurs you win $6.00 but that if a tail occurs you lose $6.00. If the probability of a head is 1/2 and that of a tail is 1/2, what is the expected value of this game to you?

$$E\ (X) = \Sigma\ [x \cdot P\ (X = x)]$$

$$E\ (X) = \$6\ (1/2) + (-\$6)\ 1/2$$

$$= \$3 - \$3$$

$$= 0$$

Conclusion: If you play long enough, the expected value of the game to you is zero (0). A game or situation with an expected value of 0 is said to be a "fair game." ◀

▶ **Example 17.2**   Suppose we shift the above example from the flip of a coin to the roll of a single fair die. Assume you win $6.00 if you roll a one or a two, but you lose $6.00 if you roll three, four, five, or six. The probability of each of the six possible outcomes is 1/6. Now, what is the expected value of this game to you?

$$E\ (X) = \Sigma\ [x \cdot P\ (X = x)]$$

$$= \$6 \cdot P\ (1\ \text{or}\ 2) + (-\$6)\ P\ (3\ \text{or}\ 4\ \text{or}\ 5\ \text{or}\ 6)$$

$$= \$6\ (1/6 + 1/6) + (-\$6)\ (1/6 + 1/6 + 1/6 + 1/6)$$

$$= \$6\ (2/6) + -\$6\ (4/6)$$

$$= \$2 - \$4$$

$$= -\$2$$

Conclusion: This game is a long run losing proposition to you, and your average loss will be $2 per play. ◀

▶ **Example 17.3**   A magazine distributor has been keeping records on the frequency of sales of BUM magazine as shown.

NUMBER OF MAGAZINES SOLD WEEKLY	FREQUENCY	PROBABILITY OF SALES
14	40	40/100 = 0.40
15	30	30/100 = 0.30
16	30	30/100 = 0.30
	100	

Based on this record of 100 sales, what have been the average weekly sales?

$$E (X) = \Sigma [x \cdot P (X = x)]$$

$$= 14 (0.40) + 15 (0.30) + 16 (0.30)$$

$$= 5.60 + 4.50 + 4.80$$

$$E (X) = 14.90$$

Conclusion: The average weekly sales have been 14.90 magazines. ◀

Payoff Tables

If we are trying to decide on which of several alternative choices to make, one way to analyze the result or outcome of the choice or decision is to construct a *payoff table*.

A payoff table shows the outcome for every action–event combination. *An action is a decision, an event is a state of nature, and an outcome is a result of the two,* usually a profit or loss. For instance, if you order two units of a product (the decision or choice) the payoff table will show how much profit (the outcome) you will make if you sell both, only one, or none of the units (the possible events). The table should also show your profit for other alternatives, such as ordering three units, one unit, or any other of several quantities.

▶ **Example 17.4**   Mr. Robert Mackey is a small farmer in southern California. Each planting season he must decide whether or not he will plant soybeans, sugar cane, or sorghum on a 10-acre tract reserved for this purpose. The amount of contribution to profit for each commodity planted depends on the season—dry, wet, or mixed. His alternative choices are which of three crops to plant. The events are the states of weather (dry, wet, or mixed), and the outcomes are the contribution to profit for each decision–event combination. Table 17-1 shows the payoffs for this example.

TABLE 17-1. *Payoff Table of Contribution to Profit Per Acre of Farm Crop Planted.*

PROBABILITY OF STATE OF NATURE	EVENTS (i) STATES OF NATURE	ACTIONS (j) DECISION ALTERNATIVE		
		PLANT SOYBEANS	PLANT SUGAR CANE	PLANT SORGHUM
0.6	Dry	$600	$100	$400
0.1	Wet	200	800	100
0.3	Mixed	700	600	800
	Expected value	590	320	490

◀

As another illustration, suppose an organization known as Tastree Partner has the sweets concession at the local Big Theater. A single item, known as Tastrees, is sold at the theater. Demand for the sweets has ranged from 30 to 50 dozen, so that the manager of Tastree Partner has to decide how many dozen Tastrees to make for a particular performance. She knows that a dozen Tastrees sell for $10.00 and cost $8.00. She also knows that any unsold Tastrees can be sold the next day (salvaged) for $4.00 a dozen. With this information she has decided to construct a payoff table to help with her decision.

TABLE 17-2. *A Payoff ($) Table for Tastree Partner.*

EVENTS (i) DOZENS OF TASTREES DEMANDED	ACTIONS (j) DOZENS OF TASTREES PREPARED				
	30	35	40	45	50
30	60	40	20	0	−20
35	60	70	50	30	10
40	60	70	80	60	40
45	60	70	80	90	70
50	60	70	80	90	100

Table 17-2 shows the payoff table the manager made. The entries in the table show the profits that *could* result from each decision she might make. If you examine the table closely, you will see that the first column represents the profit from preparing 30 dozen Tastrees. Since demand is always at least 30, there may be some unsatisfied demand (hungry customers), but there will never be any leftover Tastrees. The profit, $60.00, was found by using Equation 17.2.

$$\text{payoff}_{ij} = (P - C) \cdot j \qquad (17.2)$$

where P = price per unit
C = cost per unit
j = number units prepared or produced
i = number units demanded

In the illustration, the price is $10.00 per dozen and the cost is $8.00 per dozen. Thus, *when production is equal to or less than demand,* the potential profits are calculated as follows:

(for $i \geq 30$)
$\text{payoff}_{i30} = (\$10 - \$8)\, 30 = \$60$

(for $i \geq 35$)
$\text{payoff}_{i35} = (\$10 - \$8)\, 35 = \$70$

(for $i \geq 40$)
$\text{payoff}_{i40} = (\$10 - \$8)\, 40 = \$80$

$$\text{(for } i \geq 45)$$
$$\text{payoff}_{i45} = (\$10 - \$8)\ 45 = \$90$$

$$\text{(for } i \geq 50)$$
$$\text{payoff}_{i50} = (\$10 - \$8)\ 50 = \$100$$

Table 17-2 shows the above payoffs in the columns *below and on* the diagonal line. All the payoffs on and below this line are for situations where demand is equal to or greater than production. Thus, there are never any Tastrees left when these combinations of actions and events occur.

Now consider those decisions that the manager can make that will cause some Tastrees to be left over. This will happen anytime *the number prepared is greater than the number demanded.* The payoffs for these action–event combinations can be found by using Equation 17.3.

$$\text{payoff}_{ij} = (P - C)\,i + (S - C)\,(j - i) \qquad (\textbf{17.3})$$

where P = price per unit
C = cost per unit
i = number of units demanded
S = salvage value per unit
j = number of units prepared

In the payoff table, all the conditions where demand is less than supply are above the diagonal line as positioned in Table 17-2. The payoffs in the table for each possible action *when demand is 30* are calculated below.

$$\text{(for } j = 35)$$
$$\text{payoff}_{30,\,35} = (\$10 - \$8)\ 30 + (\$4 - \$8)\ (35 - 30) = \quad \$40$$

$$\text{(for } j = 40)$$
$$\text{payoff}_{30,\,40} = (\$10 - \$8)\ 30 + (\$4 - \$8)\ (40 - 30) = \quad \$20$$

$$\text{(for } j = 45)$$
$$\text{payoff}_{30,\,45} = (\$10 - \$8)\ 30 + (\$4 - \$8)\ (45 - 30) = \quad \$0$$

$$\text{(for } j = 50)$$
$$\text{payoff}_{30,\,50} = (\$10 - \$8)\ 30 + (\$4 - \$8)\ (50 - 30) = -\$20$$

The remainder of the payoff table is completed by considering each level of demand and calculating the payoffs when supply is greater than demand using Equation 17.3.

In looking at the payoff table in Table 17-2, you can see that the most conservative decision would be to prepare (stock) 30 dozen Tastrees for every performance of the Big Theater and be assured of a $60.00 payoff (profit). On the other hand, the riskiest decision would be to prepare 50 dozen Tastrees and, depending on attendance, experience a profit as high as $100.00 or a loss of $20.00. The maximum gain or loss from each action–event combination is of interest to decision makers because of the marginal utility of money. Perhaps the easiest way to visualize the significance of this information is to assume that the business in our example has only $10.00 in assets. Thus, a loss of more than $10.00 would bankrupt the firm. Under this condition, the manager will probably not decide to produce 50 dozen Tastrees unless she is certain of selling 35 dozen or more. (See the last column of Table 17-2.)

Decision Making Using Expected Money Values

Decision makers may find that information on the probability of the occurrence of each possible event would be helpful. That is, so far we have recognized only that different quantities may be demanded but we have said nothing about the *likelihood* that each event (level of demand) will occur.

Table 17-3 shows the probability of each level of demand for Tastrees. Let's assume that the information in the table was derived from the sales records of the company. Thus, 10 percent of the time 30 Tastrees are sold (demanded), 35 percent of the time 35 are demanded, etc. With this information, we can now calculate the *expected money value* (EMV) of each production decision.

TABLE 17-3. *Historical Demand for Tastrees.*

DOZENS OF TASTREES	PROBABILITY OF SALES
30	0.10
35	0.35
40	0.30
45	0.15
50	0.10

To see how the probabilities and payoffs are combined, look at Table 17-4. This table illustrates the calculation of the expected money value of producing 50 dozen Tastrees. As you can see, the payoff for each level of demand, given that 50 dozen are produced, is multiplied by the probability of that many Tastrees being demanded. The resulting expected values are then summed to give the expected money value of the action.

TABLE 17-4. *Determination of Expected Money Value.*

PROBABILITY OF DEMAND	PROFITS ASSOCIATED WITH ACTION: PREPARE 50 DOZEN	EXPECTED VALUE
0.10	$-20	$-2.00
0.35	10	3.50
0.30	40	12.00
0.15	70	10.50
0.10	100	10.00
		$34.00

Expected Money Value (Prepare 50 Dozen) = $34.00

The preceding procedure is repeated for each of the actions (possible production levels).   The expected money values are shown in the bottom row of Table 17-5.

TABLE 17-5. *A Payoff Table Including Probabilities of Demand and Expected Money Values for Tastree Partner.*

PROBABILITY OF DEMAND	EVENTS (i) DOZENS OF TASTREES DEMANDED	ACTIONS (j) DOZENS OF TASTREES PREPARED				
		30	35	40	45	50
0.10	30	60	40	20	0	-20
0.35	35	60	70	50	30	10
0.30	40	60	70	80	60	40
0.15	45	60	70	80	90	70
0.10	50	60	70	80	90	100
Expected Money Value		$60	$67	$63.50	$51	$34

The manager now knows that the decision to produce 35 dozen Tastrees each night will yield the highest average profit.   Thus, if the objective is to maximize profits, the company will always produce this quantity.

Expected Profit Under Certainty

While producing the quantity that yields the highest expected money value will maximize average profits under conditions of risk, it may be possible to have a higher average profit.   If decision makers had perfect information about how many units would be demanded, then they could always produce the exact quantity that was needed. This condition is a zero risk situation and, as you learned earlier, is called *certainty*.

The expected profit under certainty (EPUC) is the average payoff in a zero risk situation. Returning to our Tastree example, 10 percent of the time 30 dozen would be produced; 35 percent of the time the decision would be to produce 35 dozen, etc. The circled numbers in Table 17-6 are the profits that would be earned for each of the action–event combinations when exactly the quantity prepared would be demanded. To calculate the expected profit under certainty, it is necessary to multiply each payoff by the probability that the particular level of demand will occur. In our example, the long-run average profit for each night's operation, if the management has perfect information, is $78.00, as shown at the bottom of Table 17-6.

TABLE 17-6. *A Payoff Table Used to Calculate the Expected Profit Under Certainty.*

	EVENTS (i)	ACTIONS (j)				
PROBABILITY	DOZENS OF	DOZENS OF TASTREES PREPARED				
OF DEMAND	TASTREES DEMANDED	30	35	40	45	50
0.10	30	⦵60	40	20	0	−20
0.35	35	60	⦵70	50	30	10
0.30	40	60	70	⦵80	60	40
0.15	45	60	70	80	⦵90	70
0.10	50	60	70	80	90	⦵100

Expected Profit under Certainty = (0.10) (60) + (0.35) (70) + (0.30) (80) +
$$\qquad (0.15) (90) + (0.10) 100$$
$$= 6.00 + 24.50 + 24.00 + 13.50 + 10.00$$
$$= \$78.00$$

Expected Value of Perfect Information

Now that you know how to determine expected money value under risk and the expected profit under certainty, you have the knowledge required to calculate the *expected value of perfect information* (EVPI). That is, the average amount of additional profit that could be earned by moving from a risk state to one of certainty. To calculate EVPI, you must subtract the maximum expected money value under risk from the expected profit under certainty. In our example, it is $78 − $67 = $11. This represents the maximum amount of money that should be spent to acquire additional information that would reduce risk. Thus, if the manager is going to purchase additional information, she should not spend more than the EVPI, which is $11.00.

Revised Probabilities

In the example in the preceding part of this chapter, the probabilities were assumed to be historical in nature. That is, the probabilities reflected the likelihood of certain quantities being demanded and were determined from past sales records. These probabilities represent, however, a consolidation of sales for many performances. Thus, for any one particular performance, this historical probability distribution may be somewhat inadequate. It is the purpose of this section of the chapter to demonstrate how sampling can be used to revise the historical probabilities so that more useful information will be available for making decisions.

Revising Probabilities Based on Sample Information

Continuing with our Tastree Partner manager, let us assume that the manager has obtained one additional piece of information about attendance at the Big Theater. Table 17-7 presents this information along with the historical probabilities that were given in Table 17-3.

TABLE 17-7. *Big Theater Data.*

PROPORTION OF SEASON TICKET HOLDERS ATTENDING	DOZENS OF TASTREES DEMANDED	PROBABILITY OF SALES
0.05	30	0.10
0.15	35	0.35
0.30	40	0.30
0.60	45	0.15
0.90	50	0.10

The additional information is contained in the first column of Table 17-7 and is the probability of a season ticket holder attending a particular performance. These season ticket holder probabilities are associated with the Tastrees demanded in that the higher probabilities of a season ticket holder attending occur as demand becomes greater. Thus, we are saying when more season ticket holders attend a performance, overall attendance is likely to be higher, which means more Tastrees should be sold. Realizing this, the manager has decided to randomly sample the list of season ticket holders to find out what their attendance plans are.

To demonstrate, let us suppose that she takes a sample of 10 and learns that 4 are planning to attend the next performance. Under the assumption of sampling from a binomial distribution, let us see how the sample information can be used to revise the historical probabilities.

First, we can use Appendix D to find the probabilities of a sample of 10 showing that four will attend using the historical probabilities of season ticket holder attendance.   In other words, the probability of finding 4 yes answers in a sample of 10 if, in fact, only 5 percent of the season ticket holders attend the performance, and so forth. As read from the appendix, the probabilities are presented below.

$$P \, (X = 4) \text{ if } n = 10 \text{ and } \pi = 0.05 = 0.001$$

$$P \, (X = 4) \text{ if } n = 10 \text{ and } \pi = 0.15 = 0.0401$$

$$P \, (X = 4) \text{ if } n = 10 \text{ and } \pi = 0.30 = 0.2001$$

$$P \, (X = 4) \text{ if } n = 10 \text{ and } \pi = 0.60 = P \, (n - X) =$$
$$P \, (10 - 4) \text{ if } n = 10 \text{ and } (1 - \pi) = 0.40 = 0.1115$$

$$P \, (X = 4) \text{ if } n = 10 \text{ and } \pi = 0.90 = P \, (n - X) =$$
$$P \, (10 - 4) \text{ if } n = 10 \text{ and } (1 - \pi) = 0.10 = 0.0001$$

You will also find these probabilities in column two of Table 17-8.

TABLE 17-8.   *Revision of Probabilities: Sampling From a Binomial Distribution.*

(1) PROPORTION (π) OF SEASON TICKET HOLDERS ATTENDING	(2) $P \, (X = 4)$ IF $n = 10$ AND π IS GIVEN	(3) (PRIOR) HISTORICAL PROBABILITIES	(4) = (2) (3) JOINT PROBABILITIES	(5) = $\frac{(4)}{\Sigma \, (4)}$ REVISED PROBABILITIES (POSTERIOR)
0.05	0.001	0.10	0.000100	0.0011
0.15	0.0401	0.35	0.014035	0.1544
0.30	0.2001	0.30	0.060030	0.6604
0.60	0.1115	0.15	0.016725	0.1840
0.90	0.0001	0.10	0.000010	0.0001
			0.090900	1.0000

The next step is to develop joint probabilities using the probabilities in column two and the historical probabilities for Tastree demand (Column 3 of the table).   For example, (0.2001) (0.30) = 0.06003 as shown in Column 4.

The results of the final step in revising the probabilities based on the information from the sample are shown in Column 5 of Table 17-8.   This step consists of dividing each of the joint probabilities by the sum of all the joint probabilities, or (0.0001) ÷ (0.0909) = 0.0011, etc.

In summary, now that we know that 4 out of a sample of 10 season ticket holders plan to attend the performance, the revised probability of selling 30 Tastrees is now 0.0011 instead of the original 0.10. This, and all other revisions, are shown in Table 17-9. Using the improved information, we can now recompute the expected money values (EMV). In our example, the action that will yield the highest average profit is now the preparation of 40 dozen Tastrees.

TABLE 17-9. *Payoff Table and Calculations Using Revised Probabilities.*

	EVENTS (i)	ACTIONS (j)				
PROBABILITY	DOZENS OF	DOZENS OF TASTREES PREPARED				
OF DEMAND	TASTREES DEMANDED	30	35	40	45	50
0.0011	30	60	40	20	0	−20
0.1544	35	60	70	50	30	10
0.6604	40	60	70	80	60	40
0.1840	45	60	70	80	90	70
0.0001	50	60	70	80	90	100
Expected Money Value		$60.00	$69.97	$75.30	$60.83	$40.83

Best Action: Prepare 40 dozen

Expected Profit Under Certainty = (0.0011) ($60) + (0.1544) ($70) + (0.6604) ($80)
    + (0.1840) ($90) + (0.0001) ($100)

    = 0.066 + 10.808 + 52.832 + 16.56 + 0.01

    = $80.28

Expected Value of Perfect Information = $80.28 − $75.30

EVPI = $4.98

Continuing with the revised probabilities, the expected profit under certainty for the next performance is $80.28. And by comparing the expected profit under certainty and the maximum expected money value, we find that the expected value of perfect information (EVPI) has been reduced to $4.98. In other words, the sample gave information that will result in a higher average profit, but we could still increase average profits. However, we should not spend more than $4.98 doing so.

It should be pointed out that the new information gained from the sample does not change the average expected profit under certainty (78.00) shown in Table 17-6. This amount ($78.00) is the maximum average profit possible if we never make an error. The data in Table 17-9 are for *the next performance* and should be interpreted to mean that the average expected profit under certainty for those performances where four in a sample of 10 season ticket holders have said they will attend is $80.28.

At this point, you may wonder about the value of sample information as compared to its cost. In our example, it is obvious that the sample data increaséd the expected money value from $67.00 to $75.30, or by $8.30. If the sample costs less than $8.30, then deciding to "improve" the probabilities increased the firm's profits. But, is there a way we could have known this prior to taking the sample? To answer this question, you must know the *expected value of sample information* (EVSI).

Expected Value of Sample Information

The expected value of sample information is the difference between the expected return using sample information and the expected return under risk. The expected return under risk is the maximum profit (minimum cost) obtained from payoff table analysis (see $67.00 in Table 17-5) and is obtained using prior probabilities. The expected return using sample information is obtained using the revised probabilities. Then a comparison of the return (with and without sampling) is made to determine EVSI.

Earlier in this chapter you learned that the expected value of perfect information was the difference between expected profit under certainty and the maximum expected money value. In the Tastree Partner example, the expected value of perfect information is $11.00. Thus, the decision maker would not pay more than $11.00 for information. If additional information, perhaps from a sample, is not available at a cost equal to or less than the expected value of perfect information, then the decision maker cannot improve the average profit of his firm.

Returning to our example, assume that it costs 50¢ per sample unit to sample the theater's season ticket holders. With an expected value of perfect information of $11.00, the manager could conceivably have a sample of 22 theater customers rather than 10. How, then, did she arrive at a sample size of 10 instead of 22, or 12, or perhaps only one? Once again, to answer this question you must know the expected value of sample information, but you also have to know the cost of each size sample.

Let's examine how to calculate EVSI. Actually, what is required is the construction of Tables 17-8 and 17-9 for all possible responses for each of the possible sample sizes from 1 to 22. To serve as an example, with the sample of 10, we need to determine the total of the joint probabilities for every value of X (just like it was done for $X = 4$ in Table 17-8) and the maximum expected money value. The next step requires that each joint probability (0.1373 for $X = 0$) be multiplied by the corresponding EMV ($60.00 when $X = 0$) as shown in Table 17-10. These products are then summed and the original EMV is subtracted from this sum to yield the expected value of sample information. Thus, in Table 17-10, $67.00 is subtracted from $72.80 to give EVSI = $5.80. Since the sample of 10 would cost $5.00 (50¢/unit), it is an economical decision to take the sample.

TABLE 17-10. *Determination of the Expected Value of Sample Information* ($n = 10$).

	(1)	(2)	(3)	(4) = (1) (3)
			MAXIMUM	
	$P (X = x)$		EXPECTED MONEY	
x	PROBABILITY	DECISION	VALUE	$P (X = x) \cdot$ (Max EMV)
0	0.1373	30	$60.00	$8.24
1	0.1897	35	65.02	12.33
2	0.1757	35	68.73	12.08
3	0.1329	35	69.76	9.27
4	0.0909[a]	40[b]	75.30[b]	6.85
5	0.0641	40	78.60	5.04
6	0.0502	45	82.88	4.16
7	0.0407	45	88.01	3.58
8	0.0379	45	89.67	3.40
9	0.0448	50	95.91	4.30
10	0.0358	50	99.24	3.55
	1.0000[c]			$72.80

EVSI = E (Max EMV) − Max (EMV under risk)
  = $P (X = x)$ (Max EMV) − Max (EMV under risk)
EVSI = $72.80 − 67.00[d] = $5.80

[a]From Table 17-8, the probability of obtaining 4 yes responses.
[b]From Table 17-9.
[c]The probability of all possible outcomes from sample, $n = 10$.
[d]From Table 17-5.

Actually, we still do not know if this is the "best" size sample to take. To answer this question, we must now repeat the above process for all possible sample sizes. The optimum sample size is the one with the maximum difference between EVSI and the cost of sampling. We will not pursue this point any further except to say that it could be done easily with the aid of a computer. However, before completing the discussion, let us look at one more example.

▶ **Example 17.5**   The executives of Supercard are considering accepting a proposal from a travel agency. The travel agency has proposed that Supercard offer all of its cardholders the opportunity to participate in a variety of vacation trips sponsored by the travel agency. The promotion and mailing expense incurred by Supercard would be $20,000. In return for this participation, Supercard would receive $5.00 per participant for the first 4,000 participants and $2.50 per participant for all participants above 4,000. At present, Supercard has 40,000 cardholders, and the executives have established the following anticipated participation rates and their probabilities.

PROPORTION OF CARDHOLDERS PARTICIPATING	PROBABILITY OF PARTICIPATION
0.05	0.2
0.10	0.4
0.15	0.3
0.20	0.1

a. Construct the appropriate payoff table.

First let us calculate the payoff for each participation rate

    (1) (40,000) (0.05) = 2,000
       (2,000) ($5) − $20,000 = −$10,000
    (2) (40,000) (0.10) = 4,000
       (4,000) ($5) − $20,000 = 0
    (3) (40,000) (0.15) = 6,000
       (4,000) ($5) + ($2,000) ($2.50) − $20,000 =
       $20,000 + $5,000 − $20,000 = $5,000
    (4) (40,000) (0.20) = 8,000
       (4,000) ($5) + (4,000) ($2.50) − $20,000 =
       $20,000 + $10,000 − $20,000 = $10,000

TABLE 17-11. *Payoff Table for Marketing Example.*

PROBABILITY OF PARTICIPATION	PROPORTION OF CARDHOLDERS PARTICIPATING	POSSIBLE ACTIONS ACCEPT PROPOSAL	REJECT PROPOSAL
0.2	0.05	−$10,000	0
0.4	0.10	0	0
0.3	0.15	5,000	0
0.1	0.20	10,000	0
Expected Money Value		$500	0

b. Determine the expected profit under certainty (EPUC).

    EPUC = (0.2) (0) + (0.4) (0) + (0.3) (5) + (0.10) (10)
    EPUC = 0 + 0 + 1.5 + 1
    EPUC = $2.5 (thousands)

c. Determine the expected value of perfect information (EVPI).

EVPI = EPUC − Max (EMV)

= $2,500 − $500

EVPI = $2,000

d. Determine the expected value of sample information (EVSI) using a sample size of three.

Obtaining the solution to EVSI necessitates the determination of the revised probabilities for each possible set of responses that might be obtained from the sample. That is, of the three people interviewed, 0, 1, 2, or 3 people might respond yes when asked if they would take one of the vacation trips. Tables 17-12 through 17-15 show the probability revisions and calculation of the expected money values.

TABLE 17-12. *Probability Revisions, n = 3, X = 0.*

(1)	(2)	(3)	(4) = (2) (3)	$(5) = \dfrac{(4)}{\Sigma\,(4)}$
PROPORTION OF CARDHOLDERS (π)	$P\,(X = 0)$ IF $n = 3$ AND π IS GIVEN	PRIOR PROBABILITIES	JOINT PROBABILITIES	REVISED PROBABILITIES
0.05	0.8574	0.20	0.17480	0.24549
0.10	0.7290	0.40	0.29160	0.41746
0.15	0.6141	0.30	0.18423	0.26375
0.20	0.5120	0.10	0.05120	0.07330
			0.69851	

EMV (Accept Proposal) = 0.24549 (−$10,000) + 0.41746 (0) + 0.26375 ($5,000) +
 0.07330 ($10,000)
 = −$2454.90 + $1318.75 + $733
 = −$403.15

EMV (Reject Proposal) = 0.24549 (0) + 0.41746 (0) + 0.26375 (0) + 0.07330 (0)
 = 0

TABLE 17-13. *Probability Revisions, n = 3, X = 1.*

(1)	(2)	(3)	(4) = (2) (3)	$(5) = \dfrac{(4)}{\Sigma\,(4)}$
PROPORTION OF CARDHOLDERS (π)	$P\,(X = 1)$ IF $n = 3$ AND π IS GIVEN	PRIOR PROBABILITIES	JOINT PROBABILITIES	REVISED PROBABILITIES
0.05	0.1354	0.20	0.02708	0.10407
0.10	0.2430	0.40	0.09720	0.37354
0.15	0.3251	0.30	0.09753	0.37481
0.20	0.3840	0.10	0.03840	0.14757
			0.26021	

EMV (Accept Proposal) = 0.10407 (−$10,000) + 0.37354 (0) + 0.37481 ($5,000) +
0.14757 ($10,000)
= −$1040.70 + 0 + $1874.05 + $1475.70
= $2309.05

EMV (Reject Proposal) = 0.10407 (0) + 0.37354 (0) + 0.37481 (0) + 0.14757 (0)
= 0

TABLE 17-14. *Probability Revisions, n = 3, X = 2.*

(1)	(2)	(3)	(4) = (2) (3)	(5) = $\frac{(4)}{\Sigma\,(4)}$
PROPORTION OF CARDHOLDERS (π)	$P\,(X = 2)$ IF $n = 3$ AND π IS GIVEN	PRIOR PROBABILITIES	JOINT PROBABILITIES	REVISED PROBABILITIES
0.05	0.0071	0.20	0.00142	0.03637
0.10	0.0270	0.40	0.01080	0.27664
0.15	0.0574	0.30	0.01722	0.44109
0.20	0.0960	0.10	0.00960	0.24590
			0.03904	

EMV (Accept Proposal) = 0.03637 (−$10,000) + 0.27664 (0) + 0.44109 ($5,000) +
0.24590 ($10,000)
= −$363.70 + 0 + $2205.45 + $2459.00
= $4300.75

EMV (Reject Proposal) = 0.03637 (0) + 0.27664 (0) + 0.44109 (0) + 0.24590 (0)
= 0

TABLE 17-15. *Probability Revisions, n = 3, X = 3.*

(1)	(2)	(3)	(4) = (2) (3)	(5) = $\frac{(4)}{\Sigma\,(4)}$
PROPORTION OF CARDHOLDERS (π)	$P\,(X = 3)$ IF $n = 3$ AND π IS GIVEN	PRIOR PROBABILITIES	JOINT PROBABILITIES	REVISED PROBABILITIES
0.05	0.0001	0.20	0.00002	0.00893
0.10	0.0010	0.40	0.00040	0.17857
0.15	0.0034	0.30	0.00102	0.45536
0.20	0.0080	0.10	0.00080	0.35714
			0.00224	

EMV (Accept Proposal) = 0.00893 (−$10,000) + 0.17857 (0) + 0.45536 ($5,000) +
0.35714 ($10,000)
= −$89.30 + 0 + $2276.80 + $3571.40
= $5758.90

EMV (Reject Proposal) = 0.00893 (0) + 0.17857 (0) + 0.45536 (0) + 0.35714 (0)
= 0

Now, using the expected money values and the sum of the joint probabilities from the preceding tables and the original maximum expected money value from Table 17-11, we can calculate the expected value of sample information (EVSI).

EVSI = $P (X = x)$ (maximum EMV) − Max (EMV under risk)

$$= [0.69851 (0) + 0.26021 (\$2309.05) + 0.03904 (\$4300.75)$$
$$+ 0.00224 (\$5758.90)] - \$500$$
$$= [0 + \$600.84 + 167.90 \quad \$12.90] - \$500$$
$$= \$781.64 - \$500$$

EVSI = $281.64

Thus, if taking the sample of size three costs less than $281.64, then it is advantageous to take the sample. Some other sample size, however, may be a more optimum sample size. In order to decide on the optimum size, it would be necessary to find EVSI for all samples that would cost less than the EVPI ($2,000), and select the sample with the maximum difference between EVSI and cost of sampling.

SUMMARY OF EQUATIONS

17.1 $E (X) = \Sigma [x \cdot P (X = x)]$,   *page 415*

17.2 $payoff_{ij} = (P - C) \cdot j$,   *page 418*

17.3 $payoff_{ij} = (P - C) i + (S - C) (j - i)$,   *page 419*

REVIEW QUESTIONS

1. Explain the meaning of each equation at the end of the chapter.
2. Define
   a. certainty
   b. risk
   c. uncertainty
   d. payoff table
   e. expected value of perfect information
3. Discuss the relationship between expected profit under certainty, expected money value, and expected value of perfect information.
4. What is the expected value of sample information and how is it used?

1. A study by the manager of a large store determined that the checkout times had the probability distribution shown below. Use the probability distribution to determine the average checkout time for a customer.

CHECKOUT TIME (MINUTES)	PROBABILITY
2	0.30
4	0.45
6	0.15
8	0.10

2. A retailer has maintained the following records on weekly sales for Product M. Based on these records, estimate the average weekly sales for this product.

WEEKLY SALES	FREQUENCY
10	60
11	100
12	150
13	120
14	70

3. A Christmas tree lot owner anticipates the demand distribution for a particular size tree. If a tree costs $4.00 and is sold for $7.00:

NUMBER OF TREES	PROBABILITY OF SALE
100	0.30
110	0.40
120	0.20
130	0.10

   a. Construct the appropriate payoff table.
   b. Calculate the expected money value for each action.
   c. Determine the best stocking action.
   d. Calculate the expected profit under certainty.
   e. Calculate EVPI.

4. The Film Shop has determined the following probability distribution for a particular film:

DEMAND	PROBABILITY
20	0.20
25	0.30
30	0.40
35	0.10

If each roll costs $3.00 and sells for $4.00:
a. Construct the appropriate payoff table.
b. Determine the best stocking action.
c. Determine the expected profit uncertainty.
d. Calculate EVPI.

5. The Book Store expects the following demand distribution to exist for a certain book. If the book costs $10.00 per copy and sells for $12.00 per copy:

NUMBER OF BOOKS	PROBABILITY OF SALE
10	0.30
20	0.50
30	0.20

a. Construct the appropriate payoff table.
b. Calculate the expected money value for each action and determine the best stocking action.
c. Calculate the expected profit under certainty.
d. Calculate EVPI.

6. Repeat Problem 5 if the books can be salvaged for $6.00 per copy.

7. Use the results of Problem 5(a) to construct a loss table by subtracting every element in each row from the largest element in each row. (Hint: 0's should appear on the diagonal.)

8. Use the results of Problem 7 to
a. calculate the expected loss for each action.
b. determine the optimum stocking action by locating the minimum expected loss.
c. compare the minimum expected loss with the EVPI of Problem 1.

9. A store owner has determined the following probability distribution for calendars. The calendars sell for $8.00 per dozen and cost $5.00 per dozen.

CALENDARS (DOZENS)	PROBABILITY
15	0.25
20	0.45
25	0.30

   a. Construct the appropriate payoff table.
   b. Determine the best stocking action.
   c. Determine the expected value of perfect information.

10. Repeat Problem 9 if the calendars can be salvaged for $2.00 per dozen.

11. Use the results of Problem 10 to establish a loss table by subtracting every element in each row from the largest element in each row. (Hint: The diagonal elements will be 0.)

12. Use the results of Problem 11 to
   a. calculate the expected loss for each action.
   b. determine the optimum stocking by locating the minimum expected loss.
   c. compare your answer in Problem 12(b) with your answer in Problem 10(c).

13. A florist has recorded the following information concerning the demand for her famous Easter bouquets.

BOUQUETS	PROBABILITY
200	0.10
210	0.40
220	0.30
230	0.20

If the Bouquets sell for $10.00 and cost $5.00 to make,
   a. construct the payoff table for the possible action–event combinations.
   b. determine the best stocking action.
   c. determine the expected profit under certainty.
   d. determine the EVPI.
   e. determine the expected demand.

14. Repeat Problem 13 if the salvage value of each bouquet is $3.00.

15. Assume that the florist of Problem 13 has acquired additional information as shown below:

PROBABILITY OF REPEAT CUSTOMERS	BOUQUETS	PROBABILITY
0.05	200	0.10
0.25	210	0.40
0.35	220	0.30
0.50	230	0.20

   a. Based on a sample size of 2 previous customers who responded that they intended to purchase an Easter bouquet again this year, revise the prior probabilities.
   b. Determine the stocking action based on the revised probabilities.

16. Calculate the expected value of sample information for the sample size of 2 in Problem 15.

17. Manufacturing, Inc., receives a particular raw material in lot sizes of 1,000 units. The quality of previous shipments has had the following distribution.

FRACTION DEFECTIVE	PROBABILITY
0.05	0.65
0.10	0.30
0.15	0.05

The company must decide between two procedures for accepting the materials. The first procedure involves 100 percent inspection at a fixed cost of $400.00. The second procedure is to place received lots directly into inventory and repair the defective parts as they are used at a cost of $5.00 per unit.

a. Construct a table showing the cost of each action–event combination. (Hint: Do not use payoff table equations.)

b. Determine which procedure is better by selecting the procedure with the lowest expected cost under risk.

c. Determine the expected cost under certainty.

d. Determine the expected value of perfect information.

18. Based on a sample of two items, both of which are perfect, revise the prior probability distribution of Problem 17. Is it better to inspect 100 percent or use the material without inspection?

19. Determine the expected value of sample information for the situation in Problems 17 and 18. If sampling costs are 20 cents per unit, is it profitable to sample?

20. Suppose EVSI = $75.00 and sampling costs are 30¢ per unit; what is the expected net gain from sampling (ENGS)? Assume sample size of 150.

21. A manufacturer of model airplanes is considering a new model. Market research has indicated the following possible annual demands. If the model cost $3.00 per unit and sells for $6.00 per unit:

MODEL AIRPLANES (1000's)	PROBABILITY
0	0.15
5	0.40
10	0.35
15	0.10

a. Construct the appropriate payoff table.

b. Determine the best production quantity.

c. Determine the EVPI.

d. Determine the expected demand for the new model.

22. An outboard motor distributor has established the following probability distribution of sales for the upcoming year:

OUTBOARD MOTORS	PROBABILITY
20	0.10
21	0.25
22	0.40
23	0.25

a. What is the expected demand for outboard motors?

b. If the profit per motor is $150.00, what is the expected profit from the sale of outboard motors?

23. Big Breakfast, Inc., is ready to introduce a new cereal, Toothbreakers. Market research has established the following demand information:

DEMAND (NUMBER OF BOXES)	PROBABILITY
75,000	0.19
100,000	0.28
125,000	0.25
150,000	0.17
175,000	0.11

If the cost of producing each box is 45¢, and each box sells for 75¢:

a. Construct the appropriate payoff table.

b. Determine the best quantity to produce.

c. Calculate EVPI.

Appendixes

APPENDIX A
TABLE OF COMMON LOGARITHMS

Common logs are exponents of the base 10. In other words, the common log of a number is that value such that when 10 is raised to that power, it is equal to the number. As a result, the common log of 100 is equal to 2.00000, since $10^{2.00000}$ is equal to 100. Similarly,

log 1000 = 3.00000, because $10^{3.00000}$ = 1000

log 10   = 1.00000, because $10^{1.00000}$ = 10

log 0.1   = −1.00000, since 10^{-1} = 0.1 (this would
                     normally be written as 9.00000 − 10.00000)

The part of the common log to the left of the decimal point, 2, 3, 1, and −1 in the above examples, is called the characteristic of the log.

The part of the log to the right of the decimal point is the mantissa. It can be found from the tables in this appendix. For example, suppose we needed the common log of 4.325. Since there is one digit to the left of the decimal point, the characteristic of this number is equal to 0. Moving to the table we can find the mantissa of the log 4.325. We first move down the column headed by N until we find the number 432. Then the column is found having the value of 5 at its top, the 4th digit of the number. Reading from the column titled 5, we see the mantissa of the number is equal to 0.63599. Therefore,

log 4.325 = 0.63599

Similarly,

log 86.05 = 1.93475

Due to the mathematical properties of logs, their use can be most helpful. Since $X^a \cdot X^b = X^{a+b}$, to find the product of two numbers we merely need to add their common logs and find the number having the resulting log. This process is called finding the antilog, that number which has a log of a certain value.

For example, to find the product of 4.325 × 86.05, we need to add their common logs, 0.63599 + 1.93475 = 2.57074, and then find the antilog. Looking up 0.57074 in the body of the log table, we see the first four digits of the number would be 3722. Since a characteristic of 2 means there are 3 digits to the left of the decimal point, we see that the product of 4.325 × 86.05 = 372.2.

The mathematical property $X^a/X^b = X^{a-b}$ is also useful in working with logs. For example, suppose we needed to find the quotient of 86.05/4.325. Using logs we would subtract the log of 4.325 from the log of 86.05, and find the antilog of the resulting value. This process is shown on p. 440.

$$1.93475 - 0.63599 = 1.29876$$

$$\text{antilog } 1.29876 = 19.90$$

Thus the quotient of 86.05/4.325 is equal to 19.90.

Logs can also be used to raise numbers to powers (square, cube, etc.) and to find the roots of numbers.

power $X^a = \text{antilog } (a \cdot \log X)$

root $X^{1/a} = \text{antilog } \dfrac{\log X}{a}$

As an illustration, suppose you want to raise 5 to the 6th power (5^6) and you also want to find the 3rd root of 9 or $9^{1/3}$.

$$5^6 = \text{antilog } (6) (0.69897)$$

$$= \text{antilog } 4.19382$$

$$5^6 = 15625$$

$$9^{1/3} = \text{antilog } \frac{0.95424}{3}$$

$$= \text{antilog } 0.31808$$

$$9^{1/3} = 2.0801$$

TABLE OF COMMON LOGARITHMS.

N.	0	1	2	3	4	5	6	7	8	9
100	00 000	043	087	130	173	217	260	303	346	389
101	432	475	518	561	604	647	689	732	775	817
102	860	903	945	988	*030	*072	*115	*157	*199	*242
103	01 284	326	368	410	452	494	536	578	620	662
104	703	745	787	828	870	912	953	995	*036	*078
105	02 119	160	202	243	284	325	366	407	449	490
106	531	572	612	653	694	735	776	816	857	898
107	938	979	*019	*060	*100	*141	*181	*222	*262	*302
108	03 342	383	423	463	503	543	583	623	663	703
109	743	782	822	862	902	941	981	*021	*060	*100
110	04 139	179	218	258	297	336	376	415	454	493
111	532	571	610	650	689	727	766	805	844	883
112	922	961	999	*038	*077	*115	*154	*192	*231	*269
113	05 308	346	385	423	461	500	538	576	614	652
114	690	729	767	805	843	881	918	956	994	*032
115	06 070	108	145	183	221	258	296	333	371	408
116	446	483	521	558	595	633	670	707	744	781
117	819	856	893	930	967	*004	*041	*078	*115	*151
118	07 188	225	262	298	335	372	408	445	482	518
119	555	591	628	664	700	737	773	809	846	882
120	918	954	990	*027	*063	*099	*135	*171	*207	*243
121	08 279	314	350	386	422	458	493	529	565	600
122	636	672	707	743	778	814	849	884	920	955
123	991	*026	*061	*096	*132	*167	*202	*237	*272	*307
124	09 342	377	412	447	482	517	552	587	621	656
125	691	726	760	795	830	864	899	934	968	*003
126	10 037	072	106	140	175	209	243	278	312	346
127	380	415	449	483	517	551	585	619	653	687
128	721	755	789	823	857	890	924	958	992	*025
129	11 059	093	126	160	193	227	261	294	327	361
130	394	428	461	494	528	561	594	628	661	694
131	727	760	793	826	860	893	926	959	992	*024
132	12 057	090	123	156	189	222	254	287	320	352
133	385	418	450	483	516	548	581	613	646	678
134	710	743	775	808	840	872	905	937	969	*001
135	13 033	066	098	130	162	194	226	258	290	322
136	354	386	418	450	481	513	545	577	609	640
137	672	704	735	767	799	830	862	893	925	956
138	988	*019	*051	*082	*114	*145	*176	*208	*239	*270
139	14 301	333	364	395	426	457	489	520	551	582
140	613	644	675	706	737	768	799	829	860	891
141	922	953	983	*014	*045	*076	*106	*137	*168	*198
142	15 229	259	290	320	351	381	412	442	473	503
143	534	564	594	625	655	685	715	746	776	806
144	836	866	897	927	957	987	*017	*047	*077	*107
145	16 137	167	197	227	256	286	316	346	376	406
146	435	465	495	524	554	584	613	643	673	702
147	732	761	791	820	850	879	909	938	967	997
148	17 026	056	085	114	143	173	202	231	260	289
149	319	348	377	406	435	464	493	522	551	580
150	609	638	667	696	725	754	782	811	840	869
N.	0	1	2	3	4	5	6	7	8	9

Proportional parts

	44	43	42
1	4.4	4.3	4.2
2	8.8	8.6	8.4
3	13.2	12.9	12.6
4	17.6	17.2	16.8
5	22.0	21.5	21.0
6	26.4	25.8	25.2
7	30.8	30.1	29.4
8	35.2	34.4	33.6
9	39.6	38.7	37.8

	41	40	39
1	4.1	4.0	3.9
2	8.2	8.0	7.8
3	12.3	12.0	11.7
4	16.4	16 0	15.6
5	20.5	20.0	19.5
6	24.6	24.0	23 4
7	28.7	28.0	27.3
8	32.8	32.0	31.2
9	36.9	36.0	35.1

	38	37	36
1	3.8	3.7	3.6
2	7 6	7.4	7.2
3	11.4	11.1	10.8
4	15.2	14.8	14.4
5	19.0	18.5	18.0
6	22.8	22.2	21.6
7	26.6	25.9	25.2
8	30.4	29.6	28.8
9	34.2	33.3	32.4

	35	34	33
1	3.5	3.4	3.3
2	7.0	6.8	6.6
3	10.5	10.2	9.9
4	14.0	13.6	13.2
5	17.5	17.0	16.5
6	21.0	20.4	19.8
7	24.5	23.8	23.1
8	28.0	27.2	26.4
9	31.5	30.6	29.7

	32	31	30
1	3.2	3.1	3.0
2	6.4	6.2	6.0
3	9.6	9.3	9.0
4	12.8	12.4	12.0
5	16.0	15.5	15.0
6	19.2	18.6	18.0
7	22.4	21.7	21.0
8	25.6	24.8	24.0
9	28.8	27.9	27.0

Proportional parts

Source: From *Handbook of Mathematical Tables and Formulas* by Burlington and May. Copyright © 1973 by McGraw-Hill, Inc. Used with permission of McGraw-Hill Book Company.

TABLE OF COMMON LOGARITHMS.

N.	0	1	2	3	4	5	6	7	8	9
150	17 609	638	667	696	725	754	782	811	840	869
151	898	926	955	984	*013	*041	*070	*099	*127	*156
152	18 184	213	241	270	298	327	355	384	412	441
153	469	498	526	554	583	611	639	667	696	724
154	752	780	808	837	865	893	921	949	977	*005
155	19 033	061	089	117	145	173	201	229	257	285
156	312	340	368	396	424	451	479	507	535	562
157	590	618	645	673	700	728	756	783	811	838
158	866	893	921	948	976	*003	*030	*058	*085	*112
159	20 140	167	194	222	249	276	303	330	358	385
160	412	439	466	493	520	548	575	602	629	656
161	683	710	737	763	790	817	844	871	898	925
162	952	978	*005	*032	*059	*085	*112	*139	*165	*192
163	21 219	245	272	299	325	352	378	405	431	458
164	484	511	537	564	590	617	643	669	696	722
165	748	775	801	827	854	880	906	932	958	985
166	22 011	037	063	089	115	141	167	194	220	246
167	272	298	324	350	376	401	427	453	479	505
168	531	557	583	608	634	660	686	712	737	763
169	789	814	840	866	891	917	943	968	994	*019
170	23 045	070	096	121	147	172	198	223	249	274
171	300	325	350	376	401	426	452	477	502	528
172	553	578	603	629	654	679	704	729	754	779
173	805	830	855	880	905	930	955	980	*005	*030
174	24 055	080	105	130	155	180	204	229	254	279
175	304	329	353	378	403	428	452	477	502	527
176	551	576	601	625	650	674	699	724	748	773
177	797	822	846	871	895	920	944	969	993	*018
178	25 042	066	091	115	139	164	188	212	237	261
179	285	310	334	358	382	406	431	455	479	503
180	527	551	575	600	624	648	672	696	720	744
181	768	792	816	840	864	888	912	935	959	983
182	26 007	031	055	079	102	126	150	174	198	221
183	245	269	293	316	340	364	387	411	435	458
184	482	505	529	553	576	600	623	647	670	694
185	717	741	764	788	811	834	858	881	905	928
186	951	975	998	*021	*045	*068	*091	*114	*138	*161
187	27 184	207	231	254	277	300	323	346	370	393
188	416	439	462	485	508	531	554	577	600	623
189	646	669	692	715	738	761	784	807	830	852
190	875	898	921	944	967	989	*012	*035	*058	*081
191	28 103	126	149	171	194	217	240	262	285	307
192	330	353	375	398	421	443	466	488	511	533
193	556	578	601	623	646	668	691	713	735	758
194	780	803	825	847	870	892	914	937	959	981
195	29 003	026	048	070	092	115	137	159	181	203
196	226	248	270	292	314	336	358	380	403	425
197	447	469	491	513	535	557	579	601	623	645
198	667	688	710	732	754	776	798	820	842	863
199	885	907	929	951	973	994	*016	*038	*060	*081
200	30 103	125	146	168	190	211	233	255	276	298
N.	0	1	2	3	4	5	6	7	8	9

Proportional parts

	29	28
1	2.9	2.8
2	5.8	5.6
3	8.7	8.4
4	11.6	11.2
5	14.5	14.0
6	17.4	16.8
7	20.3	19.6
8	23.2	22.4
9	26.1	25.2

	27	26
1	2.7	2.6
2	5.4	5.2
3	8.1	7.8
4	10.8	10.4
5	13.5	13.0
6	16.2	15.6
7	18.9	18.2
8	21.6	20.8
9	24.3	23.4

	25
1	2.5
2	5.0
3	7.5
4	10.0
5	12.5
6	15.0
7	17.5
8	20.0
9	22.5

	24	23
1	2.4	2.3
2	4.8	4.6
3	7.2	6.9
4	9.6	9.2
5	12.0	11.5
6	14.4	13.8
7	16.8	16.1
8	19.2	18.4
9	21.6	20.7

	22	21
1	2.2	2.1
2	4.4	4.2
3	6.6	6.3
4	8.8	8.4
5	11.0	10.5
6	13.2	12.6
7	15.4	14.7
8	17.6	16.8
9	19.8	18.9

Proportional parts

TABLE OF COMMON LOGARITHMS.

N.	0	1	2	3	4	5	6	7	8	9	Proportional parts	
200	30 103	125	146	168	190	211	233	255	276	298		
201	320	341	363	384	406	428	449	471	492	514	**22**	**21**
202	535	557	578	600	621	643	664	685	707	728	1 2.2	2.1
203	750	771	792	814	835	856	878	899	920	942	2 4.4	4.2
204	963	984	*006	*027	*048	*069	*091	*112	*133	*154	3 6.6	6.3
											4 8.8	8.4
205	31 175	197	218	239	260	281	302	323	345	366	5 11.0	10.5
206	387	408	429	450	471	492	513	534	555	576	6 13.2	12.6
207	597	618	639	660	681	702	723	744	765	785	7 15.4	14.7
208	806	827	848	869	890	911	931	952	973	994	8 17.6	16.8
209	32 015	035	056	077	098	118	139	160	181	201	9 19.8	18.9
210	222	243	263	284	305	325	346	366	387	408		
211	428	449	469	490	510	531	552	572	593	613	**20**	
212	634	654	675	695	715	736	756	777	797	818	1 2.0	
213	838	858	879	899	919	940	960	980	*001	*021	2 4.0	
214	33 041	062	082	102	122	143	163	183	203	224	3 6.0	
											4 8.0	
215	244	264	284	304	325	345	365	385	405	425	5 10.0	
216	445	465	486	506	526	546	566	586	606	626	6 12.0	
217	646	666	686	706	726	746	766	786	806	826	7 14.0	
218	846	866	885	905	925	945	965	985	*005	*025	8 16.0	
219	34 044	064	084	104	124	143	163	183	203	223	9 18.0	
220	242	262	282	301	321	341	361	380	400	420		
221	439	459	479	498	518	537	557	577	596	616	**19**	
222	635	655	674	694	713	733	753	772	792	811	1 1.9	
223	830	850	869	889	908	928	947	967	986	*005	2 3.8	
224	35 025	044	064	083	102	122	141	160	180	199	3 5.7	
											4 7.6	
225	218	238	257	276	295	315	334	353	372	392	5 9.5	
226	411	430	449	468	488	507	526	545	564	583	6 11.4	
227	603	622	641	660	679	698	717	736	755	774	7 13.3	
228	793	813	832	851	870	889	908	927	946	965	8 15.2	
229	984	*003	*021	*040	*059	*078	*097	*116	*135	*154	9 17.1	
230	36 173	192	211	229	248	267	286	305	324	342		
231	361	380	399	418	436	455	474	493	511	530	**18**	
232	549	568	586	605	624	642	661	680	698	717	1 1.8	
233	736	754	773	791	810	829	847	866	884	903	2 3.6	
234	922	940	959	977	996	*014	*033	*051	*070	*088	3 5.4	
											4 7.2	
235	37 107	125	144	162	181	199	218	236	254	273	5 9.0	
236	291	310	328	346	365	383	401	420	438	457	6 10.8	
237	475	493	511	530	548	566	585	603	621	639	7 12.6	
238	658	676	694	712	731	749	767	785	803	822	8 14.4	
239	840	858	876	894	912	931	949	967	985	*003	9 16.2	
240	38 021	039	057	075	093	112	130	148	166	184		
241	202	220	238	256	274	292	310	328	346	364		
242	382	399	417	435	453	471	489	507	525	543	**17**	
243	561	578	596	614	632	650	668	686	703	721	1 1.7	
244	739	757	775	792	810	828	846	863	881	899	2 3.4	
											3 5.1	
245	917	934	952	970	987	*005	*023	*041	*058	*076	4 6.8	
246	39 094	111	129	146	164	182	199	217	235	252	5 8.5	
247	27C	287	305	322	340	358	375	393	410	428	6 10.2	
248	445	463	480	498	515	533	550	568	585	602	7 11.9	
249	620	637	655	672	690	707	724	742	759	777	8 13.6	
											9 15.3	
250	794	811	829	846	863	881	898	915	933	950		
N.	0	1	2	3	4	5	6	7	8	9	Proportional parts	

TABLE OF COMMON LOGARITHMS.

N.	0	1	2	3	4	5	6	7	8	9
250	39 794	811	829	846	863	881	898	915	933	950
251	967	985	*002	*019	*037	*054	*071	*088	*106	*123
252	40 140	157	175	192	209	226	243	261	278	295
253	312	329	346	364	381	398	415	432	449	466
254	483	500	518	535	552	569	586	603	620	637
255	654	671	688	705	722	739	756	773	790	807
256	824	841	858	875	892	909	926	943	960	976
257	993	*010	*027	*044	*061	*078	*095	*111	*128	*145
258	41 162	179	196	212	229	246	263	280	296	313
259	330	347	363	380	397	414	430	447	464	481
260	497	514	531	547	564	581	597	614	631	647
261	664	681	697	714	731	747	764	780	797	814
262	830	847	863	880	896	913	929	946	963	979
263	996	*012	*029	*045	*062	*078	*095	*111	*127	*144
264	42 160	177	193	210	226	243	259	275	292	308
265	325	341	357	374	390	406	423	439	455	472
266	488	504	521	537	553	570	586	602	619	635
267	651	667	684	700	716	732	749	765	781	797
268	813	830	846	862	878	894	911	927	943	959
269	975	991	*008	*024	*040	*056	*072	*088	*104	*120
270	43 136	152	169	185	201	217	233	249	265	281
271	297	313	329	345	361	377	393	409	425	441
272	457	473	489	505	521	537	553	569	584	600
273	616	632	648	664	680	696	712	727	743	759
274	775	791	807	823	838	854	870	886	902	917
275	*933	949	965	981	996	*012	*028	*044	*059	*075
276	44 091	107	122	138	154	170	185	201	217	232
277	248	264	279	295	311	326	342	358	373	389
278	404	420	436	451	467	483	498	514	529	545
279	560	576	592	607	623	638	654	669	685	700
280	716	731	747	762	778	793	809	824	840	855
281	871	886	902	917	932	948	963	979	994	*010
282	45 025	040	056	071	086	102	117	133	148	163
283	179	194	209	225	240	255	271	286	301	317
284	332	347	362	378	393	408	423	439	454	469
285	484	500	515	530	545	561	576	591	606	621
286	637	652	667	682	697	712	728	743	758	773
287	788	803	818	834	849	864	879	894	909	924
288	939	954	969	984	*000	*015	*030	*045	*060	*075
289	46 090	105	120	135	150	165	180	195	210	225
290	240	255	270	285	300	315	330	345	359	374
291	389	404	419	434	449	464	479	494	509	523
292	538	553	568	583	598	613	627	642	657	672
293	687	702	716	731	746	761	776	790	805	820
294	835	850	864	879	894	909	923	938	953	967
295	982	997	*012	*026	*041	*056	*070	*085	*100	*114
296	47 129	144	159	173	188	202	217	232	246	261
297	276	290	305	319	334	349	363	378	392	407
298	422	436	451	465	480	494	509	524	538	553
299	567	582	596	611	625	640	654	669	683	698
300	712	727	741	756	770	784	799	813	828	842
N.	0	1	2	3	4	5	6	7	8	9

Proportional parts

18	17	16	15	14
1 1.8	1 1.7	1 1.6	1 1.5	1 1.4
2 3.6	2 3.4	2 3.2	2 3.0	2 2.8
3 5.4	3 5.1	3 4.8	3 4.5	3 4.2
4 7.2	4 6.8	4 6.4	4 6.0	4 5.6
5 9.0	5 8.5	5 8.0	5 7.5	5 7.0
6 10.8	6 10.2	6 9.6	6 9.0	6 8.4
7 12.6	7 11.9	7 11.2	7 10.5	7 9.8
8 14.4	8 13.6	8 12.8	8 12.0	8 11.2
9 16.2	9 15.3	9 14.4	9 13.5	9 12.6

$\log e = 0.43429$

Proportional parts

TABLE OF COMMON LOGARITHMS.

N.	0	1	2	3	4	5	6	7	8	9
300	47 712	727	741	756	770	784	799	813	828	842
301	857	871	885	900	914	929	943	958	972	986
302	48 001	015	029	044	058	073	087	101	116	130
303	144	159	173	187	202	216	230	244	259	273
304	287	302	316	330	344	359	373	387	401	416
305	430	444	458	473	487	501	515	530	544	558
306	572	586	601	615	629	643	657	671	686	700
307	714	728	742	756	770	785	799	813	827	841
308	855	869	883	897	911	926	940	954	968	982
309	996	*010	*024	*038	*052	*066	*080	*094	*108	*122
310	49 136	150	164	178	192	206	220	234	248	262
311	276	290	304	318	332	346	360	374	388	402
312	415	429	443	457	471	485	499	513	527	541
313	554	568	582	596	610	624	638	651	665	679
314	693	707	721	734	748	762	776	790	803	817
315	831	845	859	872	886	900	914	927	941	955
316	969	982	996	*010	*024	*037	*051	*065	*079	*092
317	50 106	120	133	147	161	174	188	202	215	229
318	243	256	270	284	297	311	325	338	352	365
319	379	393	406	420	433	447	461	474	488	501
320	515	529	542	556	569	583	596	610	623	637
321	651	664	678	691	705	718	732	745	759	772
322	786	799	813	826	840	853	866	880	893	907
323	920	934	947	961	974	987	*001	*014	*028	*041
324	51 055	068	081	095	108	121	135	148	162	175
325	188	202	215	228	242	255	268	282	295	308
326	322	335	348	362	375	388	402	415	428	441
327	455	468	481	495	508	521	534	548	561	574
328	587	601	614	627	640	654	667	680	693	706
329	720	733	746	759	772	786	799	812	825	838
330	851	865	878	891	904	917	930	943	957	970
331	983	996	*009	*022	*035	*048	*061	*075	*088	*101
332	52 114	127	140	153	166	179	192	205	218	231
333	244	257	270	284	297	310	323	336	349	362
334	375	388	401	414	427	440	453	466	479	492
335	504	517	530	543	556	569	582	595	608	621
336	634	647	660	673	686	699	711	724	737	750
337	763	776	789	802	815	827	840	853	866	879
338	892	905	917	930	943	956	969	982	994	*007
339	53 020	033	046	058	071	084	097	110	122	135
340	148	161	173	186	199	212	224	237	250	263
341	275	288	301	314	326	339	352	364	377	390
342	403	415	428	441	453	466	479	491	504	517
343	529	542	555	567	580	593	605	618	631	643
344	656	668	681	694	706	719	732	744	757	769
345	782	794	807	820	832	845	857	870	882	895
346	908	920	933	945	958	970	983	995	*008	*020
347	54 033	045	058	070	083	095	108	120	133	145
348	158	170	183	195	208	220	233	245	258	270
349	283	295	307	320	332	345	357	370	382	394
350	407	419	432	444	456	469	481	494	506	518
N.	0	1	2	3	4	5	6	7	8	9

Proportional parts

	15		14		13		12
1	1.5	1	1.4	1	1.3	1	1.2
2	3.0	2	2.8	2	2.6	2	2.4
3	4.5	3	4.2	3	3.9	3	3.6
4	6.0	4	5.6	4	5.2	4	4.8
5	7.5	5	7.0	5	6.5	5	6.0
6	9.0	6	8.4	6	7.8	6	7.2
7	10.5	7	9.8	7	9.1	7	8.4
8	12.0	8	11.2	8	10.4	8	9.6
9	13.5	9	12.6	9	11.7	9	10.8

$\log \pi = 0.49715$

TABLE OF COMMON LOGARITHMS.

N.	0	1	2	3	4	5	6	7	8	9
350	54 407	419	432	444	456	469	481	494	506	518
351	531	543	555	568	580	593	605	617	630	642
352	654	667	679	691	704	716	728	741	753	765
353	777	790	802	814	827	839	851	864	876	888
354	900	913	925	937	949	962	974	986	998	*011
355	55 023	035	047	060	072	084	096	108	121	133
356	145	157	169	182	194	206	218	230	242	255
357	267	279	291	303	315	328	340	352	364	376
358	388	400	413	425	437	449	461	473	485	497
359	509	522	534	546	558	570	582	594	606	618
360	630	642	654	666	678	691	703	715	727	739
361	751	763	775	787	799	811	823	835	847	859
362	871	883	895	907	919	931	943	955	967	979
363	991	*003	*015	*027	*038	*050	*062	*074	*086	*098
364	56 110	122	134	146	158	170	182	194	205	217
365	229	241	253	265	277	289	301	312	324	336
366	348	360	372	384	396	407	419	431	443	455
367	467	478	490	502	514	526	538	549	561	573
368	585	597	608	620	632	644	656	667	679	691
369	703	714	726	738	750	761	773	785	797	808
370	820	832	844	855	867	879	891	902	914	926
371	937	949	961	972	984	996	*008	*019	*031	*043
372	57 054	066	078	089	101	113	124	136	148	159
373	171	183	194	206	217	229	241	252	264	276
374	287	299	310	322	334	345	357	368	380	392
375	403	415	426	438	449	461	473	484	496	507
376	519	530	542	553	565	576	588	600	611	623
377	634	646	657	669	680	692	703	715	726	738
378	749	761	772	784	795	807	818	830	841	852
379	864	875	887	898	910	921	933	944	955	967
380	978	990	*001	*013	*024	*035	*047	*058	*070	*081
381	58 092	104	115	127	138	149	161	172	184	195
382	206	218	229	240	252	263	274	286	297	309
383	320	331	343	354	365	377	388	399	410	422
384	433	444	456	467	478	490	501	512	524	535
385	546	557	569	580	591	602	614	625	636	647
386	659	670	681	692	704	715	726	737	749	760
387	771	782	794	805	816	827	838	850	861	872
388	883	894	906	917	928	939	950	961	973	984
389	995	*006	*017	*028	*040	*051	*062	*073	*084	*095
390	59 106	118	129	140	151	162	173	184	195	207
391	218	229	240	251	262	273	284	295	306	318
392	329	340	351	362	373	384	395	406	417	428
393	439	450	461	472	483	494	506	517	528	539
394	550	561	572	583	594	605	616	627	638	649
395	660	671	682	693	704	715	726	737	748	759
396	770	780	791	802	813	824	835	846	857	868
397	879	890	901	912	923	934	945	956	966	977
398	988	999	*010	*021	*032	*043	*054	*065	*076	*086
399	60 097	108	119	130	141	152	163	173	184	195
400	206	217	228	239	249	260	271	282	293	304
N.	0	1	2	3	4	5	6	7	8	9

Proportional parts

	13		12		11		10
1	1.3	1	1.2	1	1.1	1	1.0
2	2.6	2	2.4	2	2.2	2	2.0
3	3.9	3	3.6	3	3.3	3	3.0
4	5.2	4	4.8	4	4.4	4	4.0
5	6.5	5	6.0	5	5.5	5	5.0
6	7.8	6	7.2	6	6.6	6	6.0
7	9.1	7	8.4	7	7.7	7	7.0
8	10.4	8	9.6	8	8.8	8	8.0
9	11.7	9	10.8	9	9.9	9	9.0

TABLE OF COMMON LOGARITHMS.

N.	0	1	2	3	4	5	6	7	8	9
400	60 206	217	228	239	249	260	271	282	293	304
401	314	325	336	347	358	369	379	390	401	412
402	423	433	444	455	466	477	487	498	509	520
403	531	541	552	563	574	584	595	606	617	627
404	638	649	660	670	681	692	703	713	724	735
405	746	756	767	778	788	799	810	821	831	842
406	853	863	874	885	895	906	917	927	938	949
407	959	970	981	991	*002	*013	*023	*034	*045	*055
408	61 066	077	087	098	109	119	130	140	151	162
409	172	183	194	204	215	225	236	247	257	268
410	278	289	300	310	321	331	342	352	363	374
411	384	395	405	416	426	437	448	458	469	479
412	490	500	511	521	532	542	553	563	574	584
413	595	606	616	627	637	648	658	669	679	690
414	700	711	721	731	742	752	763	773	784	794
415	805	815	826	836	847	857	868	878	888	899
416	909	920	930	941	951	962	972	982	993	*003
417	62 014	024	034	045	055	066	076	086	097	107
418	118	128	138	149	159	170	180	190	201	211
419	221	232	242	252	263	273	284	294	304	315
420	325	335	346	356	366	377	387	397	408	418
421	428	439	449	459	469	480	490	500	511	521
422	531	542	552	562	572	583	593	603	613	624
423	634	644	655	665	675	685	696	706	716	726
424	737	747	757	767	778	788	798	808	818	829
425	839	849	859	870	880	890	900	910	921	931
426	941	951	961	972	982	992	*002	*012	*022	*033
427	63 043	053	063	073	083	094	104	114	124	134
428	144	155	165	175	185	195	205	215	225	236
429	246	256	266	276	286	296	306	317	327	337
430	347	357	367	377	387	397	407	417	428	438
431	448	458	468	478	488	498	508	518	528	538
432	548	558	568	579	589	599	609	619	629	639
433	649	659	669	679	689	699	709	719	729	739
434	749	759	769	779	789	799	809	819	829	839
435	849	859	869	879	889	899	909	919	929	939
436	949	959	969	979	988	998	*008	*018	*028	*038
437	64 048	058	068	078	088	098	108	118	128	137
438	147	157	167	177	187	197	207	217	227	237
439	246	256	266	276	286	296	306	316	326	335
440	345	355	365	375	385	395	404	414	424	434
441	444	454	464	473	483	493	503	513	523	532
442	542	552	562	572	582	591	601	611	621	631
443	640	650	660	670	680	689	699	709	719	729
444	738	748	758	768	777	787	797	807	816	826
445	836	846	856	865	875	885	895	904	914	924
446	933	943	953	963	972	982	992	*002	*011	*021
447	65 031	040	050	060	070	079	089	099	108	118
448	128	137	147	157	167	176	186	196	205	215
449	225	234	244	254	263	273	283	292	302	312
450	321	331	341	350	360	369	379	389	398	408
N.	0	1	2	3	4	5	6	7	8	9

Proportional parts

	11
1	1.1
2	2.2
3	3.3
4	4.4
5	5.5
6	6.6
7	7.7
8	8.8
9	9.9

	10
1	1.0
2	2.0
3	3.0
4	4.0
5	5.0
6	6.0
7	7.0
8	8.0
9	9.0

	9
1	0.9
2	1.8
3	2.7
4	3.6
5	4.5
6	5.4
7	6.3
8	7.2
9	8.1

TABLE OF COMMON LOGARITHMS.

N.	0	1	2	3	4	5	6	7	8	9	Proportional parts
450	65 321	331	341	350	360	369	379	389	398	408	
451	418	427	437	447	456	466	475	485	495	504	
452	514	523	533	543	552	562	571	581	591	600	
453	610	619	629	639	648	658	667	677	686	696	
454	706	715	725	734	744	753	763	772	782	792	
455	801	811	820	830	839	849	858	868	877	887	
456	896	906	916	925	935	944	954	963	973	982	
457	992	*001	*011	*020	*030	*039	*049	*058	*068	*077	
458	66 087	096	106	115	124	134	143	153	162	172	
459	181	191	200	210	219	229	238	247	257	266	
460	276	285	295	304	314	323	332	342	351	361	
461	370	380	389	398	408	417	427	436	445	455	
462	464	474	483	492	502	511	521	530	539	549	
463	558	567	577	586	596	605	614	624	633	642	
464	652	661	671	680	689	699	708	717	727	736	
465	745	755	764	773	783	792	801	811	820	829	
466	839	848	857	867	876	885	894	904	913	922	
467	932	941	950	960	969	978	987	997	*006	*015	
468	67 025	034	043	052	062	071	080	089	099	108	
469	117	127	136	145	154	164	173	182	191	201	
470	210	219	228	237	247	256	265	274	284	293	
471	302	311	321	330	339	348	357	367	376	385	
472	394	403	413	422	431	440	449	459	468	477	
473	486	495	504	514	523	532	541	550	560	569	
474	578	587	596	605	614	624	633	642	651	660	
475	669	679	688	697	706	715	724	733	742	752	
476	761	770	779	788	797	806	815	825	834	843	
477	852	861	870	879	888	897	906	916	925	934	
478	943	952	961	970	979	988	997	*006	*015	*024	
479	68 034	043	052	061	070	079	088	097	106	115	
480	124	133	142	151	160	169	178	187	196	205	
481	215	224	233	242	251	260	269	278	287	296	
482	305	314	323	332	341	350	359	368	377	386	
483	395	404	413	422	431	440	449	458	467	476	
484	485	494	502	511	520	529	538	547	556	565	
485	574	583	592	601	610	619	628	637	646	655	
486	664	673	681	690	699	708	717	726	735	744	
487	753	762	771	780	789	797	806	815	824	833	
488	842	851	860	869	878	886	895	904	913	922	
489	931	940	949	958	966	975	984	993	*002	*011	
490	69 020	028	037	046	055	064	073	082	090	099	
491	108	117	126	135	144	152	161	170	179	188	
492	197	205	214	223	232	241	249	258	267	276	
493	285	294	302	311	320	329	338	346	355	364	
494	373	381	390	399	408	417	425	434	443	452	
495	461	469	478	487	496	504	513	522	531	539	
496	548	557	566	574	583	592	601	609	618	627	
497	636	644	653	662	671	679	688	697	705	714	
498	723	732	740	749	758	767	775	784	793	801	
499	810	819	827	836	845	854	862	871	880	888	
500	897	906	914	923	932	940	949	958	966	975	
N.	0	1	2	3	4	5	6	7	8	9	Proportional parts

Proportional parts

10
1	1.0
2	2.0
3	3.0
4	4.0
5	5.0
6	6.0
7	7.0
8	8.0
9	9.0

9
1	0.9
2	1.8
3	2.7
4	3.6
5	4.5
6	5.4
7	6.3
8	7.2
9	8.1

8
1	0.8
2	1.6
3	2.4
4	3.2
5	4.0
6	4.8
7	5.6
8	6.4
9	7.2

TABLE OF COMMON LOGARITHMS.

N.	0	1	2	3	4	5	6	7	8	9	Proportional parts
500	69 897	906	914	923	932	940	949	958	966	975	
501	984	992	*001	*010	*018	*027	*036	*044	*053	*062	
502	70 070	079	088	096	105	114	122	131	140	148	
503	157	165	174	183	191	200	209	217	226	234	
504	243	252	260	269	278	286	295	303	312	321	
505	329	338	346	355	364	372	381	389	398	406	
506	415	424	432	441	449	458	467	475	484	492	
507	501	509	518	526	535	544	552	561	569	578	
508	586	595	603	612	621	629	638	646	655	663	
509	672	680	689	697	706	714	723	731	740	749	
510	757	766	774	783	791	800	808	817	825	834	
511	842	851	859	868	876	885	893	902	910	919	
512	927	935	944	952	961	969	978	986	995	*003	
513	71 012	020	029	037	046	054	063	071	079	088	
514	096	105	113	122	130	139	147	155	164	172	
515	181	189	198	206	214	223	231	240	248	257	
516	265	273	282	290	299	307	315	324	332	341	
517	349	357	366	374	383	391	399	408	416	425	
518	433	441	450	458	466	475	483	492	500	508	
519	517	525	533	542	550	559	567	575	584	592	
520	600	609	617	625	634	642	650	659	667	675	
521	684	692	700	709	717	725	734	742	750	759	
522	767	775	784	792	800	809	817	825	834	842	
523	850	858	867	875	883	892	900	908	917	925	
524	933	941	950	958	966	975	983	991	999	*008	
525	72 016	024	032	041	049	057	066	074	082	090	
526	099	107	115	123	132	140	148	156	165	173	
527	181	189	198	206	214	222	230	239	247	255	
528	263	272	280	288	296	304	313	321	329	337	
529	346	354	362	370	378	387	395	403	411	419	
530	428	436	444	452	460	469	477	485	493	501	
531	509	518	526	534	542	550	558	567	575	583	
532	591	599	607	616	624	632	640	648	656	665	
533	673	681	689	697	705	713	722	730	738	746	
534	754	762	770	779	787	795	803	811	819	827	
535	835	843	852	860	868	876	884	892	900	908	
536	916	925	933	941	949	957	965	973	981	989	
537	997	*006	*014	*022	*030	*038	*046	*054	*062	*070	
538	73 078	086	094	102	111	119	127	135	143	151	
539	159	167	175	183	191	199	207	215	223	231	
540	239	247	255	263	272	280	288	296	304	312	
541	320	328	336	344	352	360	368	376	384	392	
542	400	408	416	424	432	440	448	456	464	472	
543	480	488	496	504	512	520	528	536	544	552	
544	560	568	576	584	592	600	608	616	624	632	
545	640	648	656	664	672	679	687	695	703	711	
546	719	727	735	743	751	759	767	775	783	791	
547	799	807	815	823	830	838	846	854	862	870	
548	878	886	894	902	910	918	926	933	941	949	
549	957	965	973	981	989	997	*005	*013	*020	*028	
550	74 036	044	052	060	068	076	084	092	099	107	
N.	0	1	2	3	4	5	6	7	8	9	Proportional parts

Proportional parts:

	9
1	0.9
2	1.8
3	2.7
4	3.6
5	4.5
6	5.4
7	6.3
8	7.2
9	8.1

	8
1	0.8
2	1.6
3	2.4
4	3.2
5	4.0
6	4.8
7	5.6
8	6.4
9	7.2

	7
1	0.7
2	1.4
3	2.1
4	2.8
5	3.5
6	4.2
7	4.9
8	5.6
9	6.3

TABLE OF COMMON LOGARITHMS.

N.	0	1	2	3	4	5	6	7	8	9	Proportional parts
550	74 036	044	052	060	068	076	084	092	099	107	
551	115	123	131	139	147	155	162	170	178	186	
552	194	202	210	218	225	233	241	249	257	265	
553	273	280	288	296	304	312	320	327	335	343	
554	351	359	367	374	382	390	398	406	414	421	
555	429	437	445	453	461	468	476	484	492	500	
556	507	515	523	531	539	547	554	562	570	578	
557	586	593	601	609	617	624	632	640	648	656	
558	663	671	679	687	695	702	710	718	726	733	
559	741	749	757	764	772	780	788	796	803	811	
560	819	827	834	842	850	858	865	873	881	889	
561	896	904	912	920	927	935	943	950	958	966	**8**
562	974	981	989	997	*005	*012	*020	*028	*035	*043	1 | 0.8
563	75 051	059	066	074	082	089	097	105	113	120	2 | 1.6
564	128	136	143	151	159	166	174	182	189	197	3 | 2.4
											4 | 3.2
565	205	213	220	228	236	243	251	259	266	274	5 | 4.0
566	282	289	297	305	312	320	328	335	343	351	6 | 4.8
567	358	366	374	381	389	397	404	412	420	427	7 | 5.6
568	435	442	450	458	465	473	481	488	496	504	8 | 6.4
569	511	519	526	534	542	549	557	565	572	580	9 | 7.2
570	587	595	603	610	618	626	633	641	648	656	
571	664	671	679	686	694	702	709	717	724	732	
572	740	747	755	762	770	778	785	793	800	808	
573	815	823	831	838	846	853	861	868	876	884	
574	891	899	906	914	921	929	937	944	952	959	
575	967	974	982	989	997	*005	*012	*020	*027	*035	
576	76 042	050	057	065	072	080	087	095	103	110	
577	118	125	133	140	148	155	163	170	178	185	
578	193	200	208	215	223	230	238	245	253	260	
579	268	275	283	290	298	305	313	320	328	335	
580	343	350	358	365	373	380	388	395	403	410	
581	418	425	433	440	448	455	462	470	477	485	**7**
582	492	500	507	515	522	530	537	545	552	559	1 | 0.7
583	567	574	582	589	597	604	612	619	626	634	2 | 1.4
584	641	649	656	664	671	678	686	693	701	708	3 | 2.1
											4 | 2.8
585	716	723	730	738	745	753	760	768	775	782	5 | 3.5
586	790	797	805	812	819	827	834	842	849	856	6 | 4.2
587	864	871	879	886	893	901	908	916	923	930	7 | 4.9
588	938	945	953	960	967	975	982	989	997	*004	8 | 5.6
589	77 012	019	026	034	041	048	056	063	070	078	9 | 6.3
590	085	093	100	107	115	122	129	137	144	151	
591	159	166	173	181	188	195	203	210	217	225	
592	232	240	247	254	262	269	276	283	291	298	
593	305	313	320	327	335	342	349	357	364	371	
594	379	386	393	401	408	415	422	430	437	444	
595	' 452	459	466	474	481	488	495	503	510	517	
596	525	532	539	546	554	561	568	576	583	590	
597	597	605	612	619	627	634	641	648	656	663	
598	670	677	685	692	699	706	714	721	728	735	
599	743	750	757	764	772	779	786	793	801	808	
600	815	822	830	837	844	851	859	866	873	880	
N.	0	1	2	3	4	5	6	7	8	9	Proportional parts

TABLE OF COMMON LOGARITHMS.

N.	0	1	2	3	4	5	6	7	8	9
600	77 815	822	830	837	844	851	859	866	873	880
601	887	895	902	909	916	924	931	938	945	952
602	960	967	974	981	988	996	*003	*010	*017	*025
603	78 032	039	046	053	061	068	075	082	089	097
604	104	111	118	125	132	140	147	154	161	168
605	176	183	190	197	204	211	219	226	233	240
606	247	254	262	269	276	283	290	297	305	312
607	319	326	333	340	347	355	362	369	376	383
608	390	398	405	412	419	426	433	440	447	455
609	462	469	476	483	490	497	504	512	519	526
610	533	540	547	554	561	569	576	583	590	597
611	604	611	618	625	633	640	647	654	661	668
612	675	682	689	696	704	711	718	725	732	739
613	746	753	760	767	774	781	789	796	803	810
614	817	824	831	838	845	852	859	866	873	880
615	888	895	902	909	916	923	930	937	944	951
616	958	965	972	979	986	993	*000	*007	*014	*021
617	79 029	036	043	050	057	064	071	078	085	092
618	099	106	113	120	127	134	141	148	155	162
619	169	176	183	190	197	204	211	218	225	232
620	239	246	253	260	267	274	281	288	295	302
621	309	316	323	330	337	344	351	358	365	372
622	379	386	393	400	407	414	421	428	435	442
623	449	456	463	470	477	484	491	498	505	511
624	518	525	532	539	546	553	560	567	574	581
625	588	595	602	609	616	623	630	637	644	650
626	657	664	671	678	685	692	699	706	713	720
627	727	734	741	748	754	761	768	775	782	789
628	796	803	810	817	824	831	837	844	851	858
629	865	872	879	886	893	900	906	913	920	927
630	934	941	948	955	962	969	975	982	989	996
631	80 003	010	017	024	030	037	044	051	058	065
632	072	079	085	092	099	106	113	120	127	134
633	140	147	154	161	168	175	182	188	195	202
634	209	216	223	229	236	243	250	257	264	271
635	277	284	291	298	305	312	318	325	332	339
636	346	353	359	366	373	380	387	393	400	407
637	414	421	428	434	441	448	455	462	468	475
638	482	489	496	502	509	516	523	530	536	543
639	550	557	564	570	577	584	591	598	604	611
640	618	625	632	638	645	652	659	665	672	679
641	686	693	699	706	713	720	726	733	740	747
642	754	760	767	774	781	787	794	801	808	814
643	821	828	835	841	848	855	862	868	875	882
644	889	895	902	909	916	922	929	936	943	949
645	956	963	969	976	983	990	996	*003	*010	*017
646	81 023	030	037	043	050	057	064	070	077	084
647	090	097	104	111	117	124	131	137	144	151
648	158	164	171	178	184	191	198	204	211	218
649	224	231	238	245	251	258	265	271	278	285
650	291	298	305	311	318	325	331	338	345	351
N.	0	1	2	3	4	5	6	7	8	9

Proportional parts

	8		7		6
1	0.8	1	0.7	1	0.6
2	1.6	2	1.4	2	1.2
3	2.4	3	2.1	3	1.8
4	3.2	4	2.8	4	2.4
5	4.0	5	3.5	5	3.0
6	4.8	6	4.2	6	3.6
7	5.6	7	4.9	7	4.2
8	6.4	8	5.6	8	4.8
9	7.2	9	6.3	9	5.4

TABLE OF COMMON LOGARITHMS.

N.	0	1	2	3	4	5	6	7	8	9
650	81 291	298	305	311	318	325	331	338	345	351
651	358	365	371	378	385	391	398	405	411	418
652	425	431	438	445	451	458	465	471	478	485
653	491	498	505	511	518	525	531	538	544	551
654	558	564	571	578	584	591	598	604	611	617
655	624	631	637	644	651	657	664	671	677	684
656	690	697	704	710	717	723	730	737	743	750
657	757	763	770	776	783	790	796	803	809	816
658	823	829	836	842	849	856	862	869	875	882
659	889	895	902	908	915	921	928	935	941	948
660	954	961	968	974	981	987	994	*000	*007	*014
661	82 020	027	033	040	046	053	060	066	073	079
662	086	092	099	105	112	119	125	132	138	145
663	151	158	164	171	178	184	191	197	204	210
664	217	223	230	236	243	249	256	263	269	276
665	282	289	295	302	308	315	321	328	334	341
666	347	354	360	367	373	380	387	393	400	406
667	413	419	426	432	439	445	452	458	465	471
668	478	484	491	497	504	510	517	523	530	536
669	543	549	556	562	569	575	582	588	595	601
670	607	614	620	627	633	640	646	653	659	666
671	672	679	685	692	698	705	711	718	724	730
672	737	743	750	756	763	769	776	782	789	795
673	802	808	814	821	827	834	840	847	853	860
674	866	872	879	885	892	898	905	911	918	924
675	930	937	943	950	956	963	969	975	982	988
676	995	*001	*008	*014	*020	*027	*033	*040	*046	*052
677	83 059	065	072	078	085	091	097	104	110	117
678	123	129	136	142	149	155	161	168	174	181
679	187	193	200	206	213	219	225	232	238	245
680	251	257	264	270	276	283	289	296	302	308
681	315	321	327	334	340	347	353	359	366	372
682	378	385	391	398	404	410	417	423	429	436
683	442	448	455	461	467	474	480	487	493	499
684	506	512	518	525	531	537	544	550	556	563
685	569	575	582	588	594	601	607	613	620	626
686	632	639	645	651	658	664	670	677	683	689
687	696	702	708	715	721	727	734	740	746	753
688	759	765	771	778	784	790	797	803	809	816
689	822	828	835	841	847	853	860	866	872	879
690	885	891	897	904	910	916	923	929	935	942
691	948	954	960	967	973	979	985	992	998	*004
692	84 011	017	023	029	036	042	048	055	061	067
693	073	080	086	092	098	105	111	117	123	130
694	136	142	148	155	161	167	173	180	186	192
695	198	205	211	217	223	230	236	242	248	255
696	261	267	273	280	286	292	298	305	311	317
697	323	330	336	342	348	354	361	367	373	379
698	386	392	398	404	410	417	423	429	435	442
699	448	454	460	466	473	479	485	491	497	504
700	510	516	522	528	535	541	547	553	559	566
N.	0	1	2	3	4	5	6	7	8	9

Proportional parts

7

1	0.7
2	1.4
3	2.1
4	2.8
5	3.5
6	4.2
7	4.9
8	5.6
9	6 3

6

1	0.6
2	1.2
3	1.8
4	2.4
5	3.0
6	3.6
7	4.2
8	4.8
9	5.4

TABLE OF COMMON LOGARITHMS.

N.	0	1	2	3	4	5	6	7	8	9	Proportional parts
700	84 510	516	522	528	535	541	547	553	559	566	
701	572	578	584	590	597	603	609	615	621	628	
702	634	640	646	652	658	665	671	677	683	689	
703	696	702	708	714	720	726	733	739	745	751	
704	757	763	770	776	782	788	794	800	807	813	
705	819	825	831	837	844	850	856	862	868	874	
706	880	887	893	899	905	911	917	924	930	936	
707	942	948	954	960	967	973	979	985	991	997	
708	85 003	009	016	022	028	034	040	046	052	058	
709	065	071	077	083	089	095	101	107	114	120	
710	126	132	138	144	150	156	163	169	175	181	
711	187	193	199	205	211	217	224	230	236	242	
712	248	254	260	266	272	278	285	291	297	303	
713	309	315	321	327	333	339	345	352	358	364	
714	370	376	382	388	394	400	406	412	418	425	
715	431	437	443	449	455	461	467	473	479	485	
716	491	497	503	509	516	522	528	534	540	546	
717	552	558	564	570	576	582	588	594	600	606	
718	612	618	625	631	637	643	649	655	661	667	
719	673	679	685	691	697	703	709	715	721	727	
720	733	739	745	751	757	763	769	775	781	788	
721	794	800	806	812	818	824	830	836	842	848	
722	854	860	866	872	878	884	890	896	902	908	
723	914	920	926	932	938	944	950	956	962	968	
724	974	980	986	992	998	*004	*010	*016	*022	*028	
725	86 034	040	046	052	058	064	070	076	082	088	
726	094	100	106	112	118	124	130	136	141	147	
727	153	159	165	171	177	183	189	195	201	207	
728	213	219	225	231	237	243	249	255	261	267	
729	273	279	285	291	297	303	308	314	320	326	
730	332	338	344	350	356	362	368	374	380	386	
731	392	398	404	410	415	421	427	433	439	445	
732	451	457	463	469	475	481	487	493	499	504	
733	510	516	522	528	534	540	546	552	558	564	
734	570	576	581	587	593	599	605	611	617	623	
735	629	635	641	646	652	658	664	670	676	682	
736	688	694	700	705	711	717	723	729	735	741	
737	747	753	759	764	770	776	782	788	794	800	
738	806	812	817	823	829	835	841	847	853	859	
739	864	870	876	882	888	894	900	906	911	917	
740	923	929	935	941	947	953	958	964	970	976	
741	982	988	994	999	*005	*011	*017	*023	*029	*035	
742	87 040	046	052	058	064	070	075	081	087	093	
743	099	105	111	116	122	128	134	140	146	151	
744	157	163	169	175	181	186	192	198	204	210	
745	216	221	227	233	239	245	251	256	262	268	
746	274	280	286	291	297	303	309	315	320	326	
747	332	338	344	349	355	361	367	373	379	384	
748	390	396	402	408	413	419	425	431	437	442	
749	448	454	460	466	471	477	483	489	495	500	
750	506	512	518	523	529	535	541	547	552	558	
N.	0	1	2	3	4	5	6	7	8	9	Proportional parts

Proportional parts:

	7
1	0.7
2	1.4
3	2.1
4	2.8
5	3.5
6	4.2
7	4.9
8	5.6
9	6.3

	6
1	0.6
2	1.2
3	1.8
4	2.4
5	3.0
6	3.6
7	4.2
8	4.8
9	5.4

	5
1	0.5
2	1.0
3	1.5
4	2.0
5	2.5
6	3.0
7	3.5
8	4.0
9	4.5

TABLE OF COMMON LOGARITHMS.

N.	0	1	2	3	4	5	6	7	8	9	Proportional parts
750	87 506	512	518	523	529	535	541	547	552	558	
751	564	570	576	581	587	593	599	604	610	616	
752	622	628	633	639	645	651	656	662	668	674	
753	679	685	691	697	703	708	714	720	726	731	
754	737	743	749	754	760	766	772	777	783	789	
755	795	800	806	812	818	823	829	835	841	846	
756	852	858	864	869	875	881	887	892	898	904	
757	910	915	921	927	933	938	944	950	955	961	
758	967	973	978	984	990	996	*001	*007	*013	*018	
759	88 024	030	036	041	047	053	058	064	070	076	
760	081	087	093	098	104	110	116	121	127	133	
761	138	144	150	156	161	167	173	178	184	190	
762	195	201	207	213	218	224	230	235	241	247	
763	252	258	264	270	275	281	287	292	298	304	
764	309	315	321	326	332	338	343	349	355	360	
765	366	372	377	383	389	395	400	406	412	417	
766	423	429	434	440	446	451	457	463	468	474	
767	480	485	491	497	502	508	513	519	525	530	
768	536	542	547	553	559	564	570	576	581	587	
769	593	598	604	610	615	621	627	632	638	643	
770	649	655	660	666	672	677	683	689	694	700	
771	705	711	717	722	728	734	739	745	750	756	
772	762	767	773	779	784	790	795	801	807	812	
773	818	824	829	835	840	846	852	857	863	868	
774	874	880	885	891	897	902	908	913	919	925	
775	930	936	941	947	953	958	964	969	975	981	
776	986	992	997	*003	*009	*014	*020	*025	*031	*037	
777	89 042	048	053	059	064	070	076	081	087	092	
778	098	104	109	115	120	126	131	137	143	148	
779	154	159	165	170	176	182	187	193	198	204	
780	209	215	221	226	232	237	243	248	254	260	
781	265	271	276	282	287	293	298	304	310	315	
782	321	326	332	337	343	348	354	360	365	371	
783	376	382	387	393	398	404	409	415	421	426	
784	432	437	443	448	454	459	465	470	476	481	
785	487	492	498	504	509	515	520	526	531	537	
786	542	548	553	559	564	570	575	581	586	592	
787	597	603	609	614	620	625	631	636	642	647	
788	653	658	664	669	675	680	686	691	697	702	
789	708	713	719	724	730	735	741	746	752	757	
790	763	768	774	779	785	790	796	801	807	812	
791	818	823	829	834	840	845	851	856	862	867	
792	873	878	883	889	894	900	905	911	916	922	
793	927	933	938	944	949	955	960	966	971	977	
794	982	988	993	998	*004	*009	*015	*020	*026	*031	
795	90 037	042	048	053	059	064	069	075	080	086	
796	091	097	102	108	113	119	124	129	135	140	
797	146	151	157	162	168	173	179	184	189	195	
798	200	206	211	217	222	227	233	238	244	249	
799	255	260	266	271	276	282	287	293	298	304	
800	309	314	320	325	331	336	342	347	352	358	
N.	0	1	2	3	4	5	6	7	8	9	Proportional parts

Proportional parts:

6

1	0.6
2	1.2
3	1.8
4	2.4
5	3.0
6	3.6
7	4.2
8	4.8
9	5.4

5

1	0.5
2	1.0
3	1.5
4	2.0
5	2.5
6	3.0
7	3.5
8	4.0
9	4.5

TABLE OF COMMON LOGARITHMS.

N.	0	1	2	3	4	5	6	7	8	9	Proportional parts
800	90 309	314	320	325	331	336	342	347	352	358	
801	363	369	374	380	385	390	396	401	407	412	
802	417	423	428	434	439	445	450	455	461	466	
803	472	477	482	488	493	499	504	509	515	520	
804	526	531	536	542	547	553	558	563	569	574	
805	580	585	590	596	601	607	612	617	623	628	
806	634	639	644	650	655	660	666	671	677	682	
807	687	693	698	703	709	714	720	725	730	736	
808	741	747	752	757	763	768	773	779	784	789	
809	795	800	806	811	816	822	827	832	838	843	
810	849	854	859	865	870	875	881	886	891	897	**6**
811	902	907	913	918	924	929	934	940	945	950	1 0.6
812	956	961	966	972	977	982	988	993	998	*004	2 1.2
813	91 009	014	020	025	030	036	041	046	052	057	3 1.8
814	062	068	073	078	084	089	094	100	105	110	4 2.4 / 5 3.0
815	116	121	126	132	137	142	148	153	158	164	6 3.6
816	169	174	180	185	190	196	201	206	212	217	7 4.2
817	222	228	233	238	243	249	254	259	265	270	8 4.8
818	275	281	286	291	297	302	307	312	318	323	9 5.4
819	328	334	339	344	350	355	360	365	371	376	
820	381	387	392	397	403	408	413	418	424	429	
821	434	440	445	450	455	461	466	471	477	482	
822	487	492	498	503	508	514	519	524	529	535	
823	540	545	551	556	561	566	572	577	582	587	
824	593	598	603	609	614	619	624	630	635	640	
825	645	651	656	661	666	672	677	682	687	693	
826	698	703	709	714	719	724	730	735	740	745	
827	751	756	761	766	772	777	782	787	793	798	
828	803	808	814	819	824	829	834	840	845	850	
829	855	861	866	871	876	882	887	892	897	903	
830	908	913	918	924	929	934	939	944	950	955	**5**
831	960	965	971	976	981	986	991	997	*002	*007	1 0.5
832	92 012	018	023	028	033	038	044	049	054	059	2 1.0
833	065	070	075	080	085	091	096	101	106	111	3 1.5
834	117	122	127	132	137	143	148	153	158	163	4 2.0 / 5 2.5
835	169	174	179	184	189	195	200	205	210	215	6 3.0
836	221	226	231	236	241	247	252	257	262	267	7 3.5
837	273	278	283	288	293	298	304	309	314	319	8 4.0
838	324	330	335	340	345	350	355	361	366	371	9 4.5
839	376	381	387	392	397	402	407	412	418	423	
840	428	433	438	443	449	454	459	464	469	474	
841	480	485	490	495	500	505	511	516	521	526	
842	531	536	542	547	552	557	562	567	572	578	
843	583	588	593	598	603	609	614	619	624	629	
844	634	639	645	650	655	660	665	670	675	681	
845	686	691	696	701	706	711	716	722	727	732	
846	737	742	747	752	758	763	768	773	778	783	
847	788	793	799	804	809	814	819	824	829	834	
848	840	845	850	855	860	865	870	875	881	886	
849	891	896	901	906	911	916	921	927	932	937	
850	942	947	952	957	962	967	973	978	983	988	
N.	0	1	2	3	4	5	6	7	8	9	Proportional parts

Appendix A

TABLE OF COMMON LOGARITHMS.

N.	0	1	2	3	4	5	6	7	8	9	Proportional parts
850	92 942	947	952	957	962	967	973	978	983	988	
851	993	998	*003	*008	*013	*018	*024	*029	*034	*039	
852	93 044	049	054	059	064	069	075	080	085	090	
853	095	100	105	110	115	120	125	131	136	141	
854	146	151	156	161	166	171	176	181	186	192	
855	197	202	207	212	217	222	227	232	237	242	
856	247	252	258	263	268	273	278	283	288	293	
857	298	303	308	313	318	323	328	334	339	344	
858	349	354	359	364	369	374	379	384	389	394	
859	399	404	409	414	420	425	430	435	440	445	
860	450	455	460	465	470	475	480	485	490	495	
861	500	505	510	515	520	526	531	536	541	546	
862	551	556	561	566	571	576	581	586	591	596	
863	601	606	611	616	621	626	631	636	641	646	
864	651	656	661	666	671	676	682	687	692	697	
865	702	707	712	717	722	727	732	737	742	747	
866	752	757	762	767	772	777	782	787	792	797	
867	802	807	812	817	822	827	832	837	842	847	
868	852	857	862	867	872	877	882	887	892	897	
869	902	907	912	917	922	927	932	937	942	947	
870	952	957	962	967	972	977	982	987	992	997	
871	94 002	007	012	017	022	027	032	037	042	047	
872	052	057	062	067	072	077	082	086	091	096	
873	101	106	111	116	121	126	131	136	141	146	
874	151	156	161	166	171	176	181	186	191	196	
875	201	206	211	216	221	226	231	236	240	245	
876	250	255	260	265	270	275	280	285	290	295	
877	300	305	310	315	320	325	330	335	340	345	
878	349	354	359	364	369	374	379	384	389	394	
879	399	404	409	414	419	424	429	433	438	443	
880	448	453	458	463	468	473	478	483	488	493	
881	498	503	507	512	517	522	527	532	537	542	
882	547	552	557	562	567	571	576	581	586	591	
883	596	601	606	611	616	621	626	630	635	640	
884	645	650	655	660	665	670	675	680	685	689	
885	694	699	704	709	714	719	724	729	734	738	
886	743	748	753	758	763	768	773	778	783	787	
887	792	797	802	807	812	817	822	827	832	836	
888	841	846	851	856	861	866	871	876	880	885	
889	890	895	900	905	910	915	919	924	929	934	
890	939	944	949	954	959	963	968	973	978	983	
891	988	993	998	*002	*007	*012	*017	*022	*027	*032	
892	95 036	041	046	051	056	061	066	071	075	080	
893	085	090	095	100	105	109	114	119	124	129	
894	134	139	143	148	153	158	163	168	173	177	
895	182	187	192	197	202	207	211	216	221	226	
896	231	236	240	245	250	255	260	265	270	274	
897	279	284	289	294	299	303	308	313	318	323	
898	328	332	337	342	347	352	357	361	366	371	
899	376	381	386	390	395	400	405	410	415	419	
900	424	429	434	439	444	448	453	458	463	468	
N.	0	1	2	3	4	5	6	7	8	9	Proportional parts

Proportional parts:

6
1	0.6
2	1.2
3	1.8
4	2.4
5	3.0
6	3.6
7	4.2
8	4.8
9	5.4

5
1	0.5
2	1.0
3	1.5
4	2.0
5	2.5
6	3.0
7	3.5
8	4.0
9	4.5

4
1	0.4
2	0.8
3	1.2
4	1.6
5	2.0
6	2.4
7	2.8
8	3.2
9	3.6

Table of Common Logarithms.

N.	0	1	2	3	4	5	6	7	8	9
900	95 424	429	434	439	444	448	453	458	463	468
901	472	477	482	487	492	497	501	506	511	516
902	521	525	530	535	540	545	550	554	559	564
903	569	574	578	583	588	593	598	602	607	612
904	617	622	626	631	636	641	646	650	655	660
905	665	670	674	679	684	689	694	698	703	708
906	713	718	722	727	732	737	742	746	751	756
907	761	766	770	775	780	785	789	794	799	804
908	809	813	818	823	828	832	837	842	847	852
909	856	861	866	871	875	880	885	890	895	899
910	904	909	914	918	923	928	933	938	942	947
911	952	957	961	966	971	976	980	985	990	995
912	999	*004	*009	*014	*019	*023	*028	*033	*038	*042
913	96 047	052	057	061	066	071	076	080	085	090
914	095	099	104	109	114	118	123	128	133	137
915	142	147	152	156	161	166	171	175	180	185
916	190	194	199	204	209	213	218	223	227	232
917	237	242	246	251	256	261	265	270	275	280
918	284	289	294	298	303	308	313	317	322	327
919	332	336	341	346	350	355	360	365	369	374
920	379	384	388	393	398	402	407	412	417	421
921	426	431	435	440	445	450	454	459	464	468
922	473	478	483	487	492	497	501	506	511	515
923	520	525	530	534	539	544	548	553	558	562
924	567	572	577	581	586	591	595	600	605	609
925	614	619	624	628	633	638	642	647	652	656
926	661	666	670	675	680	685	689	694	699	703
927	708	713	717	722	727	731	736	741	745	750
928	755	759	764	769	774	778	783	788	792	797
929	802	806	811	816	820	825	830	834	839	844
930	848	853	858	862	867	872	876	881	886	890
931	895	900	904	909	914	918	923	928	932	937
932	942	946	951	956	960	965	970	974	979	984
933	988	993	997	*002	*007	*011	*016	*021	*025	*030
934	97 035	039	044	049	053	058	063	067	072	077
935	081	086	090	095	100	104	109	114	118	123
936	128	132	137	142	146	151	155	160	165	169
937	174	179	183	188	192	197	202	206	211	216
938	220	225	230	234	239	243	248	253	257	262
939	267	271	276	280	285	290	294	299	304	308
940	313	317	322	327	331	336	340	345	350	354
941	359	364	368	373	377	382	387	391	396	400
942	405	410	414	419	424	428	433	437	442	447
943	451	456	460	465	470	474	479	483	488	493
944	497	502	506	511	516	520	525	529	534	539
945	543	548	552	557	562	566	571	575	580	585
946	589	594	598	603	607	612	617	621	626	630
947	635	640	644	649	653	658	663	667	672	676
948	681	685	690	695	699	704	708	713	717	722
949	727	731	736	740	745	749	754	759	763	768
950	772	777	782	786	791	795	800	804	809	813
N.	0	1	2	3	4	5	6	7	8	9

Proportional parts

	5
1	0.5
2	1.0
3	1.5
4	2.0
5	2.5
6	3.0
7	3.5
8	4.0
9	4.5

	4
1	0.4
2	0.8
3	1.2
4	1.6
5	2.0
6	2.4
7	2.8
8	3.2
9	3.6

TABLE OF COMMON LOGARITHMS.

N.	0	1	2	3	4	5	6	7	8	9	Proportional parts
950	97 772	777	782	786	791	795	800	804	809	813	
951	818	823	827	832	836	841	845	850	855	859	
952	864	868	873	877	882	886	891	896	900	905	
953	909	914	918	923	928	932	937	941	946	950	
954	955	959	964	968	973	978	982	987	991	996	
955	98 000	005	009	014	019	023	028	032	037	041	
956	046	050	055	059	064	068	073	078	082	087	
957	091	096	100	105	109	114	118	123	127	132	
958	137	141	146	150	155	159	164	168	173	177	
959	182	186	191	195	200	204	209	214	218	223	
960	227	232	236	241	245	250	254	259	263	268	**5**
961	272	277	281	286	290	295	299	304	308	313	1 — 0.5
962	318	322	327	331	336	340	345	349	354	358	2 — 1.0
963	363	367	372	376	381	385	390	394	399	403	3 — 1.5
964	408	412	417	421	426	430	435	439	444	448	4 — 2.0 / 5 — 2.5
965	453	457	462	466	471	475	480	484	489	493	6 — 3.0
966	498	502	507	511	516	520	525	529	534	538	7 — 3.5
967	543	547	552	556	561	565	570	574	579	583	8 — 4.0
968	588	592	597	601	605	610	614	619	623	628	9 — 4.5
969	632	637	641	646	650	655	659	664	668	673	
970	677	682	686	691	695	700	704	709	713	717	
971	722	726	731	735	740	744	749	753	758	762	
972	767	771	776	780	784	789	793	798	802	807	
973	811	816	820	825	829	834	838	843	847	851	
974	856	860	865	869	874	878	883	887	892	896	
975	900	905	909	914	918	923	927	932	936	941	
976	945	949	954	958	963	967	972	976	981	985	
977	989	994	998	*003	*007	*012	*016	*021	*025	*029	
978	99 034	038	043	047	052	056	061	065	069	074	
979	078	083	087	092	096	100	105	109	114	118	
980	123	127	131	136	140	145	149	154	158	162	**4**
981	167	171	176	180	185	189	193	198	202	207	1 — 0.4
982	211	216	220	224	229	233	238	242	247	251	2 — 0.8
983	255	260	264	269	273	277	282	286	291	295	3 — 1.2
984	300	304	308	313	317	322	326	330	335	339	4 — 1.6 / 5 — 2.0
985	344	348	352	357	361	366	370	374	379	383	6 — 2.4
986	388	392	396	401	405	410	414	419	423	427	7 — 2.8
987	432	436	441	445	449	454	458	463	467	471	8 — 3.2
988	476	480	484	489	493	498	502	506	511	515	9 — 3.6
989	520	524	528	533	537	542	546	550	555	559	
990	564	568	572	577	581	585	590	594	599	603	
991	607	612	616	621	625	629	634	638	642	647	
992	651	656	660	664	669	673	677	682	686	691	
993	695	699	704	708	712	717	721	726	730	734	
994	739	743	747	752	756	760	765	769	774	778	
995	782	787	791	795	800	804	808	813	817	822	
996	826	830	835	839	843	848	852	856	861	865	
997	870	874	878	883	887	891	896	900	904	909	
998	913	917	922	926	930	935	939	944	948	952	
999	957	961	965	970	974	978	983	987	991	996	
1000	00 000	004	009	013	017	022	026	030	035	039	
N.	0	1	2	3	4	5	6	7	8	9	Proportional parts

APPENDIX B
TABLE OF SQUARES AND SQUARE ROOTS

This appendix can be most useful whenever an electronic calculator is not available. It gives N^2, $\sqrt{N}$, and $\sqrt{10N}$ for various values of N. If your value of N is an integer between 1 and 1000, merely find that number under the N column. Then by reading across the row you obtain the values of N^2, $\sqrt{N}$, and $\sqrt{10N}$. For example, 50^2 is equal to 2500, $\sqrt{50}$ is equal to 7.071, and $\sqrt{10\,(50)}$ or $\sqrt{500}$ is equal to 22.36.

By the movement of the decimal point similar information can be obtained for many other numbers. For example, 0.50^2 is equal to 0.2500, $\sqrt{.50}$ is equal to 0.7071, and $\sqrt{10\,(0.5)}$ or $\sqrt{5}$ is equal to 2.236. Likewise, 5000^2 is equal to 25,000,000, $\sqrt{5000}$ is equal to 70.71, and $\sqrt{10\,(5000)}$ or $\sqrt{50,000}$ is equal to 223.607.

TABLE OF SQUARES AND SQUARE ROOTS.

N	N^2	$\sqrt{N}$	$\sqrt{10N}$	N	N^2	$\sqrt{N}$	$\sqrt{10N}$
1	1	1.00 000	3.16 228	50	2 500	7.07 107	22.36 07
2	4	1.41 421	4.47 214	51	2 601	7.14 143	22.58 32
3	9	1.73 205	5.47 723	52	2 704	7.21 110	22.80 35
4	16	2.00 000	6.32 456	53	2 809	7.28 011	23.02 17
5	25	2.23 607	7.07 107	54	2 916	7.34 847	23.23 79
				55	3 025	7.41 620	23.45 21
6	36	2.44 949	7.74 597	56	3 136	7.48 331	23.66 43
7	49	2.64 575	8.36 660	57	3 249	7.54 983	23.87 47
8	64	2.82 843	8.94 427	58	3 364	7.61 577	24.08 32
9	81	3.00 000	9.48 683	59	3 481	7.68 115	24.28 99
10	100	3.16 228	10.00 00	60	3 600	7.74 597	24.49 49
11	121	3.31 662	10.48 81	61	3 721	7.81 025	24.69 82
12	144	3.46 410	10.95 45	62	3 844	7.87 401	24.89 98
13	169	3.60 555	11.40 18	63	3 969	7.93 725	25.09 98
14	196	3.74 166	11.83 22	64	4 096	8.00 000	25.29 82
15	225	3.87 298	12.24 74	65	4 225	8.06 226	25.49 51
16	256	4.00 000	12.64 91	66	4 356	8.12 404	25.69 05
17	289	4.12 311	13.03 84	67	4 489	8.18 535	25.88 44
18	324	4.24 264	13.41 64	68	4 624	8.24 621	26.07 68
19	361	4.35 890	13.78 40	69	4 761	8.30 662	26.26 79
20	400	4.47 214	14.14 21	70	4 900	8.36 660	26.45 75
21	441	4.58 258	14.49 14	71	5 041	8.42 615	26.64 58
22	484	4.69 042	14.83 24	72	5 184	8.48 528	26.83 28
23	529	4.79 583	15.16 58	73	5 329	8.54 400	27.01 85
24	576	4.89 898	15.49 19	74	5 476	8.60 233	27.20 29
25	625	5.00 000	15.81 14	75	5 625	8.66 025	27.38 61
26	676	5.09 902	16.12 45	76	5 776	8.71 780	27.56 81
27	729	5.19 615	16.43 17	77	5 929	8.77 496	27.74 89
28	784	5.29 150	16.73 32	78	6 084	8.83 176	27.92 85
29	841	5.38 516	17.02 94	79	6 241	8.88 819	28.10 69
30	900	5.47 723	17.32 05	80	6 400	8.94 427	28.28 43
31	961	5.56 776	17.60 68	81	6 561	9.00 000	28.46 05
32	1 024	5.65 685	17.88 85	82	6 724	9.05 539	28.63 56
33	1 089	5.74 456	18.16 59	83	6 889	9.11 043	28.80 97
34	1 156	5.83 095	18.43 91	84	7 056	9.16 515	28.98 28
35	1 225	5.91 608	18.70 83	85	7 225	9.21 954	29.15 48
36	1 296	6.00 000	18.97 37	86	7 396	9.27 362	29.32 58
37	1 369	6.08 276	19.23 54	87	7 569	9.32 738	29.49 58
38	1 444	6.16 441	19.49 36	88	7 744	9.38 083	29.66 48
39	1 521	6.24 500	19.74 84	89	7 921	9.43 398	29.83 29
40	1 600	6.32 456	20.00 00	90	8 100	9.48 683	30.00 00
41	1 681	6.40 312	20.24 85	91	8 281	9.53 939	30.16 62
42	1 764	6.48 074	20.49 39	92	8 464	9.59 166	30.33 15
43	1 849	6.55 744	20.73 64	93	8 649	9.64 365	30.49 59
44	1 936	6.63 325	20.97 62	94	8 836	9.69 536	30.65 94
45	2 025	6.70 820	21.21 32	95	9 025	9.74 679	30.82 21
46	2 116	6.78 233	21.44 76	96	9 216	9.79 796	30.98 39
47	2 209	6.85 565	21.67 95	97	9 409	9.84 886	31.14 48
48	2 304	6.92 820	21.90 89	98	9 604	9.89 949	31.30 50
49	2 401	7.00 000	22.13 59	99	9 801	9.94 987	31.46 43
50	2 500	7.07 107	22.36 07	100	10 000	10.00 000	31.62 28
N	N^2	$\sqrt{N}$	$\sqrt{10N}$	N	N^2	$\sqrt{N}$	$\sqrt{10N}$

Source: From *Statistical Analysis: With Business and Economic Applications,* second edition, by Ya-lun Chou. Copyright © 1969, 1975 by Holt, Rinehart and Winston. Reprinted by permission of Holt, Rinehart and Winston.

TABLE OF SQUARES AND SQUARE ROOTS.

N	N²	√N	√10N
100	10 000	10.00 00	31.62 28
101	10 201	10.04 99	31.78 05
102	10 404	10.09 95	31.93 74
103	10 609	10.14 89	32.09 36
104	10 816	10.19 80	32.24 90
105	11 025	10.24 70	32.40 37
106	11 236	10.29 56	32.55 76
107	11 449	10.34 41	32.71 09
108	11 664	10.39 23	32.86 34
109	11 881	10.44 03	33.01 51
110	12 100	10.48 81	33.16 62
111	12 321	10.53 57	33.31 67
112	12 544	10.58 30	33.46 64
113	12 769	10.63 01	33.61 55
114	12 996	10.67 71	33.76 39
115	13 225	10.72 38	33.91 16
116	13 456	10.77 03	34.05 88
117	13 689	10.81 67	34.20 53
118	13 924	10.86 28	34.35 11
119	14 161	10.90 87	54.49 64
120	14 400	10.95 45	34.64 10
121	14 641	11.00 00	34.78 51
122	14 884	11.04 54	34.92 85
123	15 129	11.09 05	35.07 14
124	15 376	11.13 55	35.21 36
125	15 625	11.18 03	35.35 53
126	15 876	11.22 50	35.49 65
127	16 129	11.26 94	35.63 71
128	16 384	11.31 37	35.77 71
129	16 641	11.35 78	35.91 66
130	16 900	11.40 18	36.05 55
131	17 161	11.44 55	36.19 39
132	17 424	11.48 91	36.33 18
133	17 689	11.53 26	36.46 92
134	17 956	11.57 58	36.60 60
135	18 225	11.61 90	36.74 23
136	18 496	11.66 19	36.87 82
137	18 769	11.70 47	37.01 35
138	19 044	11.74 73	37.14 84
139	19 321	11.78 98	37.28 27
140	19 600	11.83 22	37.41 66
141	19 881	11.87 43	37.55 00
142	20 164	11.91 64	37.68 29
143	20 449	11.95 83	37.81 53
144	20 736	12.00 00	37.94 73
145	21 025	12.04 16	38.07 89
146	21 316	12.08 30	38.20 99
147	21 609	12.12 44	38.34 06
148	21 904	12.16 55	38.47 08
149	22 201	12.20 66	38.60 05
150	22 500	12.24 74	38.72 98
N	N²	√N	√10N

N	N²	√N	√10N
150	22 500	12.24 74	38.72 98
151	22 801	12.28 82	38.85 87
152	23 104	12.32 88	38.98 72
153	23 409	12.36 93	39.11 52
154	23 716	12.40 97	39.24 28
155	24 025	12.44 99	39.37 00
156	24 336	12.49 00	39.49 68
157	24 649	12.53 00	39.62 32
158	24 964	12.56 98	39.74 92
159	25 281	12.60 95	39.87 48
160	25 600	12.64 91	40.00 00
161	25 921	12.68 86	40.12 48
162	26 244	12.72 79	40.24 92
163	26 569	12.76 71	40.37 33
164	26 896	12.80 62	40.49 69
165	27 225	12.84 52	40.62 02
166	27 556	12.88 41	40.74 31
167	27 889	12.92 28	40.86 56
168	28 224	12.96 15	40.98 78
169	28 561	13.00 00	41.10 96
170	28 900	13.03 84	41.23 11
171	29 241	13.07 67	41.35 21
172	29 584	13.11 49	41.47 29
173	29 929	13.15 29	41.59 33
174	30 276	13.19 09	41.71 33
175	30 625	13.22 88	41.83 30
176	30 976	13.26 65	41.95 24
177	31 329	13.30 41	42.07 14
178	31 684	13.34 17	42.19 00
179	32 041	13.37 91	42.30 84
180	32 400	13.41 64	42.42 64
181	32 761	13.45 36	42.54 41
182	33 124	13.49 07	42.66 15
183	33 489	13.52 77	42.77 85
184	33 856	13.56 47	42.89 52
185	34 225	13.60 15	43.01 16
186	34 596	13.63 82	43.12 77
187	34 969	13.67 48	43.24 35
188	35 344	13.71 13	43.35 90
189	35 721	13.74 77	43.47 41
190	36 100	13.78 40	43.58 90
191	36 481	13.82 03	43.70 35
192	36 864	13.85 64	43.81 78
193	37 249	13.89 24	43.93 18
194	37 636	13.92 84	44.04 54
195	38 025	13.96 42	44.15 88
196	38 416	14.00 00	44.27 19
197	38 809	14.03 57	44.38 47
198	39 204	14.07 12	44.49 72
199	39 601	14.10 67	44.60 94
200	40 000	14.14 21	44.72 14
N	N²	√N	√10N

TABLE OF SQUARES AND SQUARE ROOTS.

N	N²	√N	√10N
200	40 000	14.14 21	44.72 14
201	40 401	14.17 74	44.83 30
202	40 804	14.21 27	44.94 44
203	41 209	14.24 78	45.05 55
204	41 616	14.28 29	45.16 64
205	42 025	14.31 78	45.27 69
206	42 436	14.35 27	45.38 72
207	42 849	14.38 75	45.49 73
208	43 264	14.42 22	45.60 70
209	43 681	14.45 68	45.71 65
210	44 100	14.49 14	45.82 58
211	44 521	14.52 58	45.93 47
212	44 944	14.56 02	46.04 35
213	45 369	14.59 45	46.15 19
214	45 796	14.62 87	46.26 01
215	46 225	14.66 29	46.36 81
216	46 656	14.69 69	46.47 58
217	47 089	14.73 09	46.58 33
218	47 524	14.76 48	46.69 05
219	47 961	14.79 86	46.79 74
220	48 400	14.83 24	46.90 42
221	48 841	14.86 61	47.01 06
222	49 284	14.89 97	47.11 69
223	49 729	14.93 32	47.22 29
224	50 176	14.96 66	47.32 86
225	50 625	15.00 00	47.43 42
226	51 076	15.03 33	47.53 95
227	51 529	15.06 65	47.64 45
228	51 984	15.09 97	47.74 93
229	52 441	15.13 27	47.85 39
230	52 900	15.16 58	47.95 83
231	53 361	15.19 87	48.06 25
232	53 824	15.23 15	48.16 64
233	54 289	15.26 43	48.27 01
234	54 756	15.29 71	48.37 35
235	55 225	15.32 97	48.47 68
236	55 696	15.36 23	48.57 98
237	56 169	15.39 48	48.68 26
238	56 644	15.42 72	48.78 52
239	57 121	15.45 96	48.88 76
240	57 600	15.49 19	48.98 98
241	58 081	15.52 42	49.09 18
242	58 564	15.55 63	49.19 35
243	59 049	15.58 85	49.29 50
244	59 536	15.62 05	49.39 64
245	60 025	15.65 25	49.49 75
246	60 516	15.68 44	49.59 84
247	61 009	15.71 62	49.69 91
248	61 504	15.74 80	49.79 96
249	62 001	15.77 97	49.89 99
250	62 500	15.81 14	50.00 00
N	N²	√N	√10N

N	N²	√N	√10N
250	62 500	15.81 14	50.00 00
251	63 001	15.84 30	50.09 99
252	63 504	15.87 45	50.19 96
253	64 009	15.90 60	50.29 91
254	64 516	15.93 74	50.39 84
255	65 025	15.96 87	50.49 75
256	65 536	16.00 00	50.59 64
257	66 049	16.03 12	50.69 52
258	66 564	16.06 24	50.79 37
259	67 081	16.09 35	50.89 20
260	67 600	16.12 45	50.99 02
261	68 121	16.15 55	51.08 82
262	68 644	16.18 64	51.18 59
263	69 169	16.21 73	51.28 35
264	69 696	16.24 81	51.38 09
265	70 225	16.27 88	51.47 82
266	70 756	16.30 95	51.57 52
267	71 289	16.34 01	51.67 20
268	71 824	16.37 07	51.76 87
269	72 361	16.40 12	51.86 52
270	72 900	16.43 17	51.96 15
271	73 441	16.46 21	52.05 77
272	73 984	16.49 24	52.15 36
273	74 529	16.52 27	52.24 94
274	75 076	16.55 29	52.34 50
275	75 625	16.58 31	52.44 04
276	76 176	16.61 32	52.53 57
277	76 729	16.64 33	52.63 08
278	77 284	16.67 33	52.72 57
279	77 841	16.70 33	52.82 05
280	78 400	16.73 32	52.91 50
281	78 961	16.76 31	53.00 94
282	79 524	16.79 29	53.10 37
283	80 089	16.82 26	53.19 77
284	80 656	16.85 23	53.29 17
285	81 225	16.88 19	53.38 54
286	81 796	16.91 15	53.47 90
287	82 369	16.94 11	53.57 24
288	82 944	16.97 06	53.66 56
289	83 521	17.00 00	53.75 87
290	84 100	17.02 94	53.85 16
291	84 681	17.05 87	53.94 44
292	85 264	17.08 80	54.03 70
293	85 849	17.11 72	54.12 95
294	86 436	17.14 64	54.22 18
295	87 025	17.17 56	54.31 39
296	87 616	17.20 47	54.40 59
297	88 209	17.23 37	54.49 77
298	88 804	17.26 27	54.58 94
299	89 401	17.29 16	54.68 09
300	90 000	17.32 05	54.77 23
N	N²	√N	√10N

TABLE OF SQUARES AND SQUARE ROOTS.

N	N²	√N	√10N	N	N²	√N	√10N
300	90 000	17.32 05	54.77 23	**350**	122 500	18.70 83	59.16 08
301	90 601	17.34 94	54.86 35	351	123 201	18.73 50	59.24 53
302	91 204	17.37 81	54.95 45	352	123 904	18.76 17	59.32 96
303	91 809	17.40 69	55.04 54	353	124 609	18.78 83	59.41 38
304	92 416	17.43 56	55.13 62	354	125 316	18.81 49	59.49 79
305	93 025	17.46 42	55.22 68	355	126 025	18.84 14	59.58 19
306	93 636	17.49 29	55.31 73	**356**	126 736	18.86 80	59.66 57
307	94 249	17.52 14	55.40 76	357	127 449	18.89 44	59.74 95
308	94 864	17.54 99	55.49 77	358	128 164	18.92 09	59.83 31
309	95 481	17.57 84	55.58 78	359	128 881	18.94 73	59.91 66
310	96 100	17.60 68	55.67 76	360	129 600	18.97 37	60.00 00
311	96 721	17.63 52	55.76 74	**361**	130 321	19.00 00	60.08 33
312	97 344	17.66 35	55.85 70	362	131 044	19.02 63	60.16 64
313	97 969	17.69 18	55.94 64	363	131 769	19.05 26	60.24 95
314	98 596	17.72 00	56.03 57	364	132 496	19.07 88	60.33 24
315	99 225	17.74 82	56.12 49	365	133 225	19.10 50	60.41 52
316	99 856	17.77 64	56.21 39	**366**	133 956	19.13 11	60.49 79
317	100 489	17.80 45	56.30 28	367	134 689	19.15 72	60.58 05
318	101 124	17.83 26	56.39 15	368	135 424	19.18 33	60.66 30
319	101 761	17.86 06	56.48 01	369	136 161	19.20 94	60.74 54
320	102 400	17.88 85	56.56 85	370	136 900	19.23 54	60.82 76
321	103 041	17.91 65	56.65 69	**371**	137 641	19.26 14	60.90 98
322	103 684	17.94 44	56.74 50	372	138 384	19.28 73	60.99 18
323	104 329	17.97 22	56.83 31	373	139 129	19.31 32	61.07 37
324	104 976	18.00 00	56.92 10	374	139 876	19.33 91	61.15 55
325	105 625	18.02 78	57.00 88	375	140 625	19.36 49	61.23 72
326	106 276	18.05 55	57.09 64	**376**	141 376	19.39 07	61.31 88
327	106 929	18.08 31	57.18 39	377	142 129	19.41 65	61.40 03
328	107 584	18.11 08	57.27 13	378	142 884	19.44 22	61.48 17
329	108 241	18.13 84	57.35 85	379	143 641	19.46 79	61.56 30
330	108 900	18.16 59	57.44 56	380	144 400	19.49 36	61.64 41
331	109 561	18.19 34	57.53 26	**381**	145 161	19.51 92	61.72 52
332	110 224	18.22 09	57.61 94	382	145 924	19.54 48	61.80 61
333	110 889	18.24 83	57.70 62	383	146 689	19.57 04	61.88 70
334	111 556	18.27 57	57.79 27	384	147 456	19.59 59	61.96 77
335	112 225	18.30 30	57.87 92	385	148 225	19.62 14	62.04 84
336	112 896	18.33 03	57.96 55	**386**	148 996	19.64 69	62.12 89
337	113 569	18.35 76	58.05 17	387	149 769	19.67 23	62.20 93
338	114 244	18.38 48	58.13 78	388	150 544	19.69 77	62.28 96
339	114 921	18.41 20	58.22 37	389	151 321	19.72 31	62.36 99
340	115 600	18 43 91	58.30 95	390	152 100	19.74 84	62.45 00
341	116 281	18.46 62	58.39 52	**391**	152 881	19.77 37	62.53 00
342	116 964	18.49 32	58.48 08	392	153 664	19.79 90	62.60 99
343	117 649	18.52 03	58.56 62	393	154 449	19.82 42	62.68 97
344	118 336	18.54 72	58.65 15	394	155 236	19.84 94	62.76 94
345	119 025	18.57 42	58.73 67	395	156 025	19.87 46	62.84 90
346	119 716	18.60 11	58.82 18	**396**	156 816	19.89 97	62.92 85
347	120 409	18.62 79	58.90 67	397	157 609	19.92 49	63.00 79
348	121 104	18.65 48	58.99 15	398	158 404	19.94 99	63.08 72
349	121 801	18.68 15	59.07 62	399	159 201	19.97 50	63.16 64
350	122 500	18.70 83	59.16 08	400	160 000	20.00 00	63.24 56
N	N²	√N	√10N	N	N²	√N	√10N

TABLE OF SQUARES AND SQUARE ROOTS.

N	N²	√N	√10N
400	160 000	20.00 00	63.24 56
401	160 801	20.02 50	63.32 46
402	161 604	20.04 99	63.40 35
403	162 409	20.07 49	63.48 23
404	163 216	20.09 98	63.56 10
405	164 025	20.12 46	63.63 96
406	164 836	20.14 94	63.71 81
407	165 649	20.17 42	63.79 66
408	166 464	20.19 90	63.87 49
409	167 281	20.22 37	63.95 31
410	168 100	20.24 85	64.03 12
411	168 921	20.27 31	64.10 93
412	169 744	20.29 78	64.18 72
413	170 569	20.32 24	64.26 51
414	171 396	20.34 70	64.34 28
415	172 225	20.37 15	64.42 05
416	173 056	20.39 61	64.49 81
417	173 889	20.42 06	64.57 55
418	174 724	20.44 50	64.65 29
419	175 561	20.46 95	64.73 02
420	176 400	20.49 39	64.80 74
421	177 241	20.51 83	64.88 45
422	178 084	20.54 26	64.96 15
423	178 929	20.56 70	65.03 85
424	179 776	20.59 13	65.11 53
425	180 625	20.61 55	65.19 20
426	181 476	20.63 98	65.26 87
427	182 329	20.66 40	65.34 52
428	183 184	20.68 82	65.42 17
429	184 041	20.71 23	65.49 81
430	184 900	20.73 64	65.57 44
431	185 761	20.76 05	65.65 06
432	186 624	20.78 46	65.72 67
433	187 489	20.80 87	65.80 27
434	188 356	20.83 27	65.87 87
435	189 225	20.85 67	65.95 45
436	190 096	20.88 06	66.03 03
437	190 969	20.90 45	66.10 60
438	191 844	20.92 84	66.18 16
439	192 721	20.95 23	66.25 71
440	193 600	20.97 62	66.33 25
441	194 481	21.00 00	66.40 78
442	195 364	21.02 38	66.48 31
443	196 249	21.04 76	66.55 82
444	197 136	21.07 13	66.63 33
445	198 025	21.09 50	66.70 83
446	198 916	21.11 87	66.78 32
447	199 809	21.14 24	66.85 81
448	200 704	21.16 60	66.93 28
449	201 601	21.18 96	67.00 75
450	202 500	21.21 32	67.08 20
N	N²	√N	√10N

N	N²	√N	√10N
450	202 500	21.21 32	67.08 20
451	203 401	21.23 68	67.15 65
452	204 304	21.26 03	67.23 09
453	205 209	21.28 38	67.30 53
454	206 116	21.30 73	67.37 95
455	207 025	21.33 07	67.45 37
456	207 936	21.35 42	67.52 78
457	208 849	21.37 76	67.60 18
458	209 764	21.40 09	67.67 57
459	210 681	21.42 43	67.74 95
460	211 600	21.44 76	67.82 33
461	212 521	21.47 09	67.89 70
462	213 444	21.49 42	67.97 06
463	214 369	21.51 74	68.04 41
464	215 296	21.54 07	68.11 75
465	216 225	21.56 39	68.19 09
466	217 156	21.58 70	68.26 42
467	218 089	21.61 02	68.33 74
468	219 024	21.63 33	68.41 05
469	219 961	21.65 64	68.48 36
470	220 900	21.67 95	68.55 65
471	221 841	21.70 25	68.62 94
472	222 784	21.72 56	68.70 23
473	223 729	21.74 86	68.77 50
474	224 676	21.77 15	68.84 77
475	225 625	21.79 45	68.92 02
476	226 576	21.81 74	68.99 28
477	227 529	21.84 03	69.06 52
478	228 484	21.86 32	69.13 75
479	229 441	21.88 61	69.20 98
480	230 400	21.90 89	69.28 20
481	231 361	21.93 17	69.35 42
482	232 324	21.95 45	69.42 62
483	233 289	21.97 73	69.49 82
484	234 256	22.00 00	69.57 01
485	235 225	22.02 27	69.64 19
486	236 196	22.04 54	69.71 37
487	237 169	22.06 81	69.78 54
488	238 144	22.09 07	69.85 70
489	239 121	22.11 33	69.92 85
490	240 100	22.13 59	70.00 00
491	241 081	22.15 85	70.07 14
492	242 064	22.18 11	70.14 27
493	243 049	22.20 36	70.21 40
494	244 036	22.22 61	70.28 51
495	245 025	22.24 86	70.35 62
496	246 016	22.27 11	70.42 73
497	247 009	22.29 35	70.49 82
498	248 004	22.31 59	70.56 91
499	249 001	22.33 83	70.63 99
500	250 000	22.36 07	70.71 07
N	N²	√N	√10N

TABLE OF SQUARES AND SQUARE ROOTS.

N	N²	√N	√10N	N	N²	√N	√10N
500	250 000	22.36 07	70.71 07	**550**	302 500	23.45 21	74.16 20
501	251 001	22.38 30	70.78 14	551	303 601	23.47 34	74.22 94
502	252 004	22.40 54	70.85 20	552	304 704	23.49 47	74.29 67
503	253 009	22.42 77	70.92 25	553	305 809	23.51 60	74.36 40
504	254 016	22.44 99	70.99 30	554	306 916	23.53 72	74.43 12
505	255 025	22.47 22	71.06 34	555	308 025	23.55 84	74.49 83
506	256 036	22.49 44	71.13 37	**556**	309 136	23.57 97	74.56 54
507	257 049	22.51 67	71.20 39	557	310 249	23.60 08	74.63 24
508	258 064	22.53 89	71.27 41	558	311 364	23.62 20	74.69 94
509	259 081	22.56 10	71.34 42	559	312 481	23.64 32	74.76 63
510	260 100	22.58 32	71.41 43	560	313 600	23.66 43	74.83 31
511	261 121	22.60 53	71.48 43	**561**	314 721	23.68 54	74.89 99
512	262 144	22.62 74	71.55 42	562	315 844	23.70 65	74.96 67
513	263 169	22.64 95	71.62 40	563	316 969	23.72 76	75.03 33
514	264 196	22.67 16	71.69 38	564	318 096	23.74 87	75.09 99
515	265 225	22.69 36	71.76 35	565	319 225	23.76 97	75.16 65
516	266 256	22.71 56	71.83 31	**566**	320 356	23.79 08	75.23 30
517	267 289	22.73 76	71.90 27	567	321 489	23.81 18	75.29 94
518	268 324	22.75 96	71.97 22	568	322 624	23.83 28	75.36 58
519	269 361	22.78 16	72.04 17	569	323 761	23.85 37	75.43 21
520	270 400	22.80 35	72.11 10	570	324 900	23.87 47	75.49 83
521	271 441	22.82 54	72.18 03	**571**	326 041	23.89 56	75.56 45
522	272 484	22.84 73	72.24 96	572	327 184	23.91 65	75.63 07
523	273 529	22.86 92	72.31 87	573	328 329	23.93 74	75.69 68
524	274 576	22.89 10	72.38 78	574	329 476	23.95 83	75.76 28
525	275 625	22.91 29	72.45 69	575	330 625	23.97 92	75.82 88
526	276 676	22.93 47	72.52 59	**576**	331 776	24.00 00	75.89 47
527	277 729	22.95 65	72.59 48	577	332 929	24.02 08	75.96 05
528	278 784	22.97 83	72.66 36	578	334 084	24.04 16	76.02 63
529	279 841	23.00 00	72.73 24	579	335 241	24.06 24	76.09 20
530	280 900	23.02 17	72.80 11	580	336 400	24.08 32	76.15 77
531	281 961	23.04 34	72.86 97	**581**	337 561	24.10 39	76.22 34
532	283 024	23.06 51	72.93 83	582	338 724	24.12 47	76.28 89
533	284 089	23.08 68	73.00 68	583	339 889	24.14 54	76.35 44
534	285 156	23.10 84	73.07 53	584	341 056	24.16 61	76.41 99
535	286 225	23.13 01	73.14 37	585	342 225	24.18 68	76.48 53
536	287 296	23.15 17	73.21 20	**586**	343 396	24.20 74	76.55 06
537	288 369	23.17 33	73.28 03	587	344 569	24.22 81	76.61 59
538	289 444	23.19 48	73.34 85	588	345 744	24.24 87	76.68 12
539	290 521	23.21 64	73.41 66	589	346 921	24.26 93	76.74 63
540	291 600	23.23 79	73.48 47	590	348 100	24.28 99	76.81 15
541	292 681	23.25 94	73.55 27	**591**	349 281	24.31 05	76.87 65
542	293 764	23.28 09	73.62 06	592	350 464	24.33 11	76.94 15
543	294 849	23.30 24	73.68 85	593	351 649	24.35 16	77.00 65
544	295 936	23.32 38	73.75 64	594	352 836	24.37 21	77.07 14
545	297 025	23.34 52	73.82 41	595	354 025	24.39 26	77.13 62
546	298 116	23.36 66	73.89 18	**596**	355 216	24.41 31	77.20 10
547	299 209	23.38 80	73.95 94	597	356 409	24.43 36	77.26 58
548	300 304	23.40 94	74.02 70	598	357 604	24.45 40	77.33 05
549	301 401	23.43 07	74.09 45	599	358 801	24.47 45	77.39 51
550	302 500	23.45 21	74.16 20	600	360 000	24.49 49	77.45 97
N	N²	√N	√10N	N	N²	√N	√10N

TABLE OF SQUARES AND SQUARE ROOTS.

N	N²	√N	√10N		N	N²	√N	√10N
600	360 000	24.49 49	77.45 97		**650**	422 500	25.49 51	80.62 26
601	361 201	24.51 53	77.52 42		651	423 801	25.51 47	80.68 46
602	362 404	24.53 57	77.58 87		652	425 104	25.53 43	80.74 65
603	363 609	24.55 61	77.65 31		653	426 409	25.55 39	80.80 84
604	364 816	24.57 64	77.71 74		654	427 716	25.57 34	80.87 03
605	366 025	24.59 67	77.78 17		655	429 025	25.59 30	80.93 21
606	367 236	24.61 71	77.84 60		**656**	430 336	25.61 25	80.99 38
607	368 449	24.63 74	77.91 02		657	431 649	25.63 20	81.05 55
608	369 664	24.65 77	77.97 44		658	432 964	25.65 15	81.11 72
609	370 881	24.67 79	78.03 85		659	434 281	25.67 10	81.17 88
610	372 100	24.69 82	78.10 25		660	435 600	25.69 05	81.24 04
611	373 321	24.71 84	78.16 65		**661**	436 921	25.70 99	81.30 19
612	374 544	24.73 86	78.23 04		662	438 244	25.72 94	81.36 34
613	375 769	24.75 88	78.29 43		663	439 569	25.74 88	81.42 48
614	376 996	24.77 90	78.35 82		664	440 896	25.76 82	81.48 62
615	378 225	24.79 92	78.42 19		665	442 225	25.78 76	81.54 75
616	379 456	24.81 93	78.48 57		**666**	443 556	25.80 70	81.60 88
617	380 689	24.83 95	78.54 93		667	444 889	25.82 63	81.67 01
618	381 924	24.85 96	78.61 30		668	446 224	25.84 57	81.73 13
619	383 161	24.87 97	78.67 66		669	447 561	25.86 50	81.79 24
620	384 400	24.89 98	78.74 01		670	448 900	25.88 44	81.85 35
621	385 641	24.91 99	78.80 36		**671**	450 241	25.90 37	81.91 46
622	386 884	24.93 99	78.86 70		672	451 584	25.92 30	81.97 56
623	388 129	24.96 00	78.93 03		673	452 929	25.94 22	82.03 66
624	389 376	24.98 00	78.99 37		674	454 276	25.96 15	82.09 75
625	390 625	25.00 00	79.05 69		675	455 625	25.98 08	82.15 84
626	391 876	25.02 00	79.12 02		**676**	456 976	26.00 00	82.21 92
627	393 129	25.04 00	79.18 33		677	458 329	26.01 92	82.28 00
628	394 384	25.05 99	79.24 65		678	459 684	26.03 84	82.34 08
629	395 641	25.07 99	79.30 95		679	461 041	26.05 76	82.40 15
630	396 900	25.09 98	79.37 25		680	462 400	26.07 68	82.46 21
631	398 161	25.11 97	79.43 55		**681**	463 761	26.09 60	82.52 27
632	399 424	25.13 96	79.49 84		682	465 124	26.11 51	82.58 33
633	400 689	25.15 95	79.56 13		683	466 489	26.13 43	82.64 38
634	401 956	25.17 94	79.62 41		684	467 856	26.15 34	82.70 43
635	403 225	25.19 92	79.68 69		685	469 225	26.17 25	82.76 47
636	404 496	25.21 90	79.74 96		**686**	470 596	26.19 16	82.82 51
637	405 769	25.23 89	79.81 23		687	471 969	26.21 07	82.88 55
638	407 044	25.25 87	79.87 49		688	473 344	26.22 98	82.94 58
639	408 321	25.27 84	79.93 75		689	474 721	26.24 88	83.00 60
640	409 600	25.29 82	80.00 00		690	476 100	26.26 79	83.06 62
641	410 881	25.31 80	80.06 25		**691**	477 481	26.28 69	83.12 64
642	412 164	25.33 77	80.12 49		692	478 864	26.30 59	83.18 65
643	413 449	25.35 74	80.18 73		693	480 249	26.32 49	83.24 66
644	414 736	25.37 72	80.24 96		694	481 636	26.34 39	83.30 67
645	416 025	25.39 69	80.31 19		695	483 025	26.36 29	83.36 67
646	417 316	25.41 65	80.37 41		**696**	484 416	26.38 18	83.42 66
647	418 609	25.43 62	80.43 63		697	485 809	26.40 08	83.48 65
648	419 904	25.45 58	80.49 84		698	487 204	26.41 97	83.54 64
649	421 201	25.47 55	80.56 05		699	488 601	26.43 86	83.60 62
650	422 500	25.49 51	80.62 26		700	490 000	26.45 75	83.66 60
N	N²	√N	√10N		N	N²	√N	√10N

TABLE OF SQUARES AND SQUARE ROOTS.

N	N²	√N	√10N
700	490 000	26.45 75	83.66 60
701	491 401	26.47 64	83.72 57
702	492 804	26.49 53	83.78 54
703	494 209	26.51 41	83.84 51
704	495 616	26.53 30	83.90 47
705	497 025	26.55 18	83.96 43
706	498 436	26.57 07	84.02 38
707	499 849	26.58 95	84.08 33
708	501 264	26.60 83	84.14 27
709	502 681	26.62 71	84.20 21
710	504 100	26.64 58	84.26 15
711	505 521	26.66 46	84.32 08
712	506 944	26.68 33	84.38 01
713	508 369	26.70 21	84.43 93
714	509 796	26.72 08	84.49 85
715	511 225	26.73 95	84.55 77
716	512 656	26.75 82	84.61 68
717	514 089	26.77 69	84.67 59
718	515 524	26.79 55	84.73 49
719	516 961	26.81 42	84.79 39
720	518 400	26.83 28	84.85 28
721	519 841	26.85 14	84.91 17
722	521 284	26.87 01	84.97 06
723	522 729	26.88 87	85.02 94
724	524 176	26.90 72	85.08 82
725	525 625	26.92 58	85.14 69
726	527 076	26.94 44	85.20 56
727	528 529	26.96 29	85.26 43
728	529 984	26.98 15	85.32 29
729	531 441	27.00 00	85.38 15
730	532 900	27.01 85	85.44 00
731	534 361	27.03 70	85.49 85
732	535 824	27.05 55	85.55 70
733	537 289	27.07 40	85.61 54
734	538 756	27.09 24	85.67 38
735	540 225	27.11 09	85.73 21
736	541 696	27.12 93	85.79 04
737	543 169	27.14 77	85.84 87
738	544 644	27.16 62	85.90 69
739	546 121	27.18 46	85.96 51
740	547 600	27.20 29	86.02 33
741	549 081	27.22 13	86.08 14
742	550 564	27.23 97	86.13 94
743	552 049	27.25 80	86.19 74
744	553 536	27.27 64	86.25 54
745	555 025	27.29 47	86.31 34
746	556 516	27.31 30	86.37 13
747	558 009	27.33 13	86.42 92
748	559 504	27.34 96	86.48 70
749	561 001	27.36 79	86.54 48
750	562 500	27.38 61	86.60 25
N	N²	√N	√10N

N	N²	√N	√10N
750	562 500	27.38 61	86.60 25
751	564 001	27.40 44	86.66 03
752	565 504	27.42 26	86.71 79
753	567 009	27.44 08	86.77 56
754	568 516	27.45 91	86.83 32
755	570 025	27.47 73	86.89 07
756	571 536	27.49 55	86.94 83
757	573 049	27.51 36	87.00 57
758	574 564	27.53 18	87.06 32
759	576 081	27.55 00	87.12 06
760	577 600	27.56 81	87.17 80
761	579 121	27.58 62	87.23 53
762	580 644	27.60 43	87.29 26
763	582 169	27.62 25	87.34 99
764	583 696	27.64 05	87.40 71
765	585 225	27.65 86	87.46 43
766	586 756	27.67 67	87.52 14
767	588 289	27.69 48	87.57 85
768	589 824	27.71 28	87.63 56
769	591 361	27.73 08	87.69 26
770	592 900	27.74 89	87.74 96
771	594 441	27.76 69	87.80 66
772	595 984	27.78 49	87.86 35
773	597 529	27.80 29	87.92 04
774	599 076	27.82 09	87.97 73
775	600 625	27.83 88	88.03 41
776	602 176	27.85 68	88.09 09
777	603 729	27.87 47	88.14 76
778	605 284	27.89 27	88.20 43
779	606 841	27.91 06	88.26 10
780	608 400	27.92 85	88.31 76
781	609 961	27.94 64	88.37 42
782	611 524	27.96 43	88.43 08
783	613 089	27.98 21	88.48 73
784	614 656	28.00 00	88.54 38
785	616 225	28.01 79	88.60 02
786	617 796	28.03 57	88.65 66
787	619 369	28.05 35	88.71 30
788	620 944	28.07 13	88.76 94
789	622 521	28.08 91	88.82 57
790	624 100	28.10 69	88.88 19
791	625 681	28.12 47	88.93 82
792	627 264	28.14 25	88.99 44
793	628 849	28.16 03	89.05 05
794	630 436	28.17 80	89.10 67
795	632 025	28.19 57	89.16 28
796	633 616	28.21 35	89.21 88
797	635 209	28.23 12	89.27 49
798	636 804	28.24 89	89.33 08
799	638 401	28.26 66	89.38 68
800	640 000	28.28 43	89.44 27
N	N²	√N	√10N

TABLE OF SQUARES AND SQUARE ROOTS.

N	N²	√N	√10N
800	640 000	28.28 43	89.44 27
801	641 601	28.30 19	89.49 86
802	643 204	28.31 96	89.55 45
803	644 809	28.33 73	89.61 03
804	646 416	28.35 49	89.66 60
805	648 025	28.37 25	89.72 18
806	649 636	28.39 01	89.77 75
807	651 249	28.40 77	89.83 32
808	652 864	28.42 53	89.88 88
809	654 481	28.44 29	89.94 44
810	656 100	28.46 05	90.00 00
811	657 721	28.47 81	90.05 55
812	659 344	28.49 56	90.11 10
813	660 969	28.51 32	90.16 65
814	662 596	28.53 07	90.22 19
815	664 225	28.54 82	90.27 74
816	665 856	28.56 57	90.33 27
817	667 489	28.58 32	90.38 81
818	669 124	28.60 07	90.44 34
819	670 761	28.61 82	90.49 86
820	672 400	28.63 56	90.55 39
821	674 041	28.65 31	90.60 91
822	675 684	28.67 05	90.66 42
823	677 329	28.68 80	90.71 93
824	678 976	28.70 54	90.77 44
825	680 625	28.72 28	90.82 95
826	682 276	28.74 02	90.88 45
827	683 929	28.75 76	90.93 95
828	685 584	28.77 50	90.99 45
829	687 241	28.79 24	91.04 94
830	688 900	28.80 97	91.10 43
831	690 561	28.82 71	91.15 92
832	692 224	28.84 44	91.21 40
833	693 889	28.86 17	91.26 88
834	695 556	28.87 91	91.32 36
835	697 225	28.89 64	91.37 83
836	698 896	28.91 37	91.43 30
837	700 569	28.93 10	91.48 77
838	702 244	28.94 82	91.54 23
839	703 921	28.96 55	91.59 69
840	705 600	28.98 28	91.65 15
841	707 281	29.00 00	91.70 61
842	708 964	29.01 72	91.76 06
843	710 649	29.03 45	91.81 50
844	712 336	29.05 17	91.86 95
845	714 025	29.06 89	91.92 39
846	715 716	29.08 61	91.97 83
847	717 409	29.10 33	92.03 26
848	719 104	29.12 04	92.08 69
849	720 801	29.13 76	92.14 12
850	722 500	29.15 48	92.19 54
N	N²	√N	√10N

N	N²	√N	√10N
850	722 500	29.15 48	92.19 54
851	724 201	29.17 19	92.24 97
852	725 904	29.18 90	92.30 38
853	727 609	29.20 62	92.35 80
854	729 316	29.22 33	92.41 21
855	731 025	29.24 04	92.46 62
856	732 736	29.25 75	92.52 03
857	734 449	29.27 46	92.57 43
858	736 164	29.29 16	92.62 83
859	737 881	29.30 87	92.68 23
860	739 600	29.32 58	92.73 62
861	741 321	29.34 28	92.79 01
862	743 044	29.35 98	92.84 40
863	744 769	29.37 69	92.89 78
864	746 496	29.39 39	92.95 16
865	748 225	29.41 09	93.00 54
866	749 956	29.42 79	93.05 91
867	751 689	29.44 49	93.11 28
868	753 424	29.46 18	93.16 65
869	755 161	29.47 88	93.22 02
870	756 900	29.49 58	93.27 38
871	758 641	29.51 27	93.32 74
872	760 384	29.52 96	93.38 09
873	762 129	29.54 66	93.43 45
874	763 876	29.56 35	93.48 80
875	765 625	29.58 04	93.54 14
876	767 376	29.59 73	93.59 49
877	769 129	29.61 42	93.64 83
878	770 884	29.63 11	93.70 17
879	772 641	29.64 79	93.75 50
880	774 400	29.66 48	93.80 83
881	776 161	29.68 16	93.86 16
882	777 924	29.69 85	93.91 49
883	779 689	29.71 53	93.96 81
884	781 456	29.73 21	94.02 13
885	783 225	29.74 89	94.07 44
886	784 996	29.76 58	94.12 76
887	786 769	29.78 25	94.18 07
888	788 544	29.79 93	94.23 38
889	790 321	29.81 61	94.28 68
890	792 100	29.83 29	94.33 98
891	793 881	29.84 96	94.39 28
892	795 664	29.86 64	94.44 58
893	797 449	29.88 31	94.49 87
894	799 236	29.89 98	94.55 16
895	801 025	29.91 66	94.60 44
896	802 816	29.93 33	94.65 73
897	804 609	29.95 00	94.71 01
898	806 404	29.96 66	94.76 29
899	808 201	29.98 33	94.81 56
900	810 000	30.00 00	94.86 83
N	N²	√N	√10N

TABLE OF SQUARES AND SQUARE ROOTS.

N	N²	√N	√10N
900	810 000	30.00 00	94.86 83
901	811 801	30.01 67	94.92 10
902	813 604	30.03 33	94.97 37
903	815 409	30.05 00	95.02 63
904	817 216	30.06 66	95.07 89
905	819 025	30.08 32	95.13 15
906	820 836	30.09 98	95.18 40
907	822 649	30.11 64	95.23 65
908	824 464	30.13 30	95.28 90
909	826 281	30.14 96	95.34 15
910	828 100	30.16 62	95.39 39
911	829 921	30.18 28	95.44 63
912	831 744	30.19 93	95.49 87
913	833 569	30.21 59	95.55 10
914	835 396	30.23 24	95.60 33
915	837 225	30.24 90	95.65 56
916	839 056	30.26 55	95.70 79
917	840 889	30.28 20	95.76 01
918	842 724	30.29 85	95.81 23
919	844 561	30.31 50	95.86 45
920	846 400	30.33 15	95.91 66
921	848 241	30.34 80	95.96 87
922	850 084	30.36 45	96.02 08
923	851 929	30.38 09	96.07 29
924	853 776	30.39 74	96.12 49
925	855 625	30.41 38	96.17 69
926	857 476	30.43 02	96.22 89
927	859 329	30.44 67	96.28 08
928	861 184	30.46 31	96.33 28
929	863 041	30.47 95	96.38 46
930	864 900	30.49 59	96.43 65
931	866 761	30.51 23	96.48 83
932	868 624	30.52 87	96.54 01
933	870 489	30.54 50	96.59 19
934	872 356	30.56 14	96.64 37
935	874 225	30.57 78	96.69 54
936	876 096	30.59 41	96.74 71
937	877 969	30.61 05	96.79 88
938	879 844	30.62 68	96.85 04
939	881 721	30.64 31	96.90 20
940	883 600	30.65 94	96.95 36
941	885 481	30.67 57	97.00 52
942	887 364	30.69 20	97.05 67
943	889 249	30.70 83	97.10 82
944	891 136	30.72 46	97.15 97
945	893 025	30.74 09	97.21 11
946	894 916	30.75 71	97.26 25
947	896 809	30.77 34	97.31 39
948	898 704	30.78 96	97.36 53
949	900 601	30.80 58	97.41 66
950	902 500	30.82 21	97.46 79
N	N²	√N	√10N

N	N²	√N	√10N
950	902 500	30.82 21	97.46 79
951	904 401	30.83 83	97.51 92
952	906 304	30.85 45	97.57 05
953	908 209	30.87 07	97.62 17
954	910 116	30.88 69	97.67 29
955	912 025	30.90 31	97.72 41
956	913 936	30.91 92	97.77 53
957	915 849	30.93 54	97.82 64
958	917 764	30.95 16	97.87 75
959	919 681	30.96 77	97.92 85
960	921 600	30.98 39	97.97 96
961	923 521	31.00 00	98.03 06
962	925 444	31.01 61	98.08 16
963	927 369	31.03 22	98.13 26
964	929 296	31.04 83	98.18 35
965	931 225	31.06 44	98.23 44
966	933 156	31.08 05	98.28 53
967	935 089	31.09 66	98.33 62
968	937 024	31.11 27	98.38 70
969	938 961	31.12 88	98.43 78
970	940 900	31.14 48	98.48 86
971	942 841	31.16 09	98.53 93
972	944 784	31.17 69	98.59 01
973	946 729	31.19 29	98.64 08
974	948 676	31.20 90	98.69 14
975	950 625	31.22 50	98.74 21
976	952 576	31.24 10	98.79 27
977	954 529	31.25 70	98.84 33
978	956 484	31.27 30	98.89 39
979	958 441	31.28 90	98.94 44
980	960 400	31.30 50	98.99 49
981	962 361	31.32 09	99.04 54
982	964 324	31.33 69	99.09 59
983	966 289	31.35 28	99.14 64
984	968 256	31.36 88	99.19 68
985	970 225	31.38 47	99.24 72
986	972 196	31.40 06	99.29 75
987	974 169	31.41 66	99.34 79
988	976 144	31.43 25	99.39 82
989	978 121	31.44 84	99.44 85
990	980 100	31.46 43	99.49 87
991	982 081	31.48 02	99.54 90
992	984 064	31.49 60	99.59 92
993	986 049	31.51 19	99.64 94
994	988 036	31.52 78	99.69 95
995	990 025	31.54 36	99.74 97
996	992 016	31.55 95	99.79 98
997	994 009	31.57 53	99.84 99
998	996 004	31.59 11	99.89 99
999	998 001	31.60 70	99.95 00
1000	1000 000	31.62 28	100.00 00
N	N²	√N	√10N

APPENDIX C
TABLE OF COMBINATORIALS

The table in this appendix gives the values for the combination of n items taken x at a time up to n and x equal to 20. It should be remembered that to use combinations that the order in which the items are chosen is not of importance, and no item can be chosen more than once.

To illustrate the use of this table, suppose we had the following combination problem.

$$C_4^{15}$$

where 15 is n (the total number of items) and 4 is x (the number of items drawn). To find the total number of possible combinations, first find the correct column, in this case the column headed by $\binom{n}{4}$. Next, this column is moved down until the row is found which represents the value of n, in this case 15. As seen from the table, the C_4^{15} is equal to 1365.

For values of x which are greater than 10, the identity

$$C\frac{n}{x} = C\frac{n}{n-x}$$

can be used to determine the total number of combinations. For example, the determination of the total number of possible combinations for n equal to 15 and x equal to 11 is given below.

$$C_{11}^{15} = C_4^{15}$$

which as seen earlier is equal to 1365.

TABLE OF COMBINATORIALS.

n	$C\frac{n}{0}$	$C\frac{n}{1}$	$C\frac{n}{2}$	$C\frac{n}{3}$	$C\frac{n}{4}$	$C\frac{n}{5}$	$C\frac{n}{6}$	$C\frac{n}{7}$	$C\frac{n}{8}$	$C\frac{n}{9}$	$C\frac{n}{10}$
0	1										
1	1	1									
2	1	2	1								
3	1	3	3	1							
4	1	4	6	4	1						
5	1	5	10	10	5	1					
6	1	6	15	20	15	6	1				
7	1	7	21	35	35	21	7	1			
8	1	8	28	56	70	56	28	8	1		
9	1	9	36	84	126	126	84	36	9	1	
10	1	10	45	120	210	252	210	120	45	10	1
11	1	11	55	165	330	462	462	330	165	55	11
12	1	12	66	220	495	792	924	792	495	220	66
13	1	13	78	286	715	1287	1716	1716	1287	715	286
14	1	14	91	364	1001	2002	3003	3432	3003	2002	1001
15	1	15	105	455	1365	3003	5005	6435	6435	5005	3003
16	1	16	120	560	1820	4368	8008	11440	12870	11440	8008
17	1	17	136	680	2380	6188	12376	19448	24310	24310	19448
18	1	18	153	816	3060	8568	18564	31824	43758	48620	43758
19	1	19	171	969	3876	11628	27132	50388	75582	92378	92378
20	1	20	190	1140	4845	15504	38760	77520	125970	167960	184756

For $x > 10$ it may be necessary to make use of the identity $C\frac{n}{x} = C\frac{n}{n-x}$.

Source: John E. Freund, *Modern Elementary Statistics*, 3rd ed., © 1967, p. 392. By permission of Prentice-Hall, Inc., Englewood Cliffs, N.J.

APPENDIX D
THE BINOMIAL PROBABILITY DISTRIBUTION

This appendix is actually a collection of the values of several binomial probability distributions. There is a distribution included for each value of n from 1 to 20.

As an illustration of how the tables are used, consider the following problem:

$$C_2^3 \, (0.50)^2 \, (0.50)^1$$

Remember that the general form of this equation is:

$$C_x^n \, \pi^x \, (1 - \pi)^{n-x}$$

Since $n = 3$, refer to the section of the table headed by $n = 3$. Across the top of the table are a series of column headings for specific values of π. Find the one labeled 0.50. Now, referring to the column headed by x, find the row designating $x = 2$. Reading in that row of the 0.50 column, you will find the value 0.3750. This is the solution to the problem.

Notice that the values of π terminate at 0.5 in the table. To determine probabilities associated with values of π greater than 0.5, simply subtract π from one $(1 - \pi)$, and let $x = (n - x)$ and proceed as before. For example,

$$C_4^5 \, (0.6)^4 \, (0.4)^1$$

Using the table for $n = 5$, look under the column headed 0.4 $(1 - 0.6)$ to the row corresponding with $x = 1$ $(5 - 4)$. The answer to the above problem is 0.2592.

THE BINOMIAL PROBABILITY DISTRIBUTION.

							π				
n	x	.05	.10	.15	.20	.25	.30	.35	.40	.45	.50
1	0	.9500	.9000	.8500	.8000	.7500	.7000	.6500	.6000	.5500	.5000
	1	.0500	.1000	.1500	.2000	.2500	.3000	.3500	.4000	.4500	.5000
2	0	.9025	.8100	.7225	.6400	.5625	.4900	.4225	.3600	.3025	.2500
	1	.0950	.1800	.2550	.3200	.3750	.4200	.4550	.4800	.4950	.5000
	2	.0025	.0100	.0225	.0400	.0625	.0900	.1225	.1600	.2025	.2500
3	0	.8574	.7290	.6141	.5120	.4219	.3430	.2746	.2160	.1664	.1250
	1	.1354	.2430	.3251	.3840	.4219	.4410	.4436	.4320	.4084	.3750
	2	.0071	.0270	.0574	.0960	.1406	.1890	.2389	.2880	.3341	.3750
	3	.0001	.0010	.0034	.0080	.0156	.0270	.0429	.0640	.0911	.1250
4	0	.8145	.6561	.5220	.4096	.3164	.2401	.1785	.1296	.0915	.0625
	1	.1715	.2916	.3685	.4096	.4219	.4116	.3845	.3456	.2995	.2500
	2	.0135	.0486	.0975	.1536	.2109	.2646	.3105	.3456	.3675	.3750
	3	.0005	.0036	.0115	.0256	.0469	.0756	.1115	.1536	.2005	.2500
	4	.0000	.0001	.0005	.0016	.0039	.0081	.0150	.0256	.0410	.0625
5	0	.7738	.5905	.4437	.3277	.2373	.1681	.1160	.0778	.0503	.0312
	1	.2036	.3280	.3915	.4096	.3955	.3602	.3124	.2592	.2059	.1562
	2	.0214	.0729	.1382	.2048	.2637	.3087	.3364	.3456	.3369	.3125
	3	.0011	.0081	.0244	.0512	.0879	.1323	.1811	.2304	.2757	.3125
	4	.0000	.0004	.0022	.0064	.0146	.0284	.0488	.0768	.1128	.1562
	5	.0000	.0000	.0001	.0003	.0010	.0024	.0053	.0102	.0185	.0312
6	0	.7351	.5314	.3771	.2621	.1780	.1176	.0754	.0467	.0277	.0156
	1	.2321	.3543	.3993	.3932	.3560	.3025	.2437	.1866	.1359	.0938
	2	.0305	.0984	.1762	.2458	.2966	.3241	.3280	.3110	.2780	.2344
	3	.0021	.0146	.0415	.0819	.1318	.1852	.2355	.2765	.3032	.3125
	4	.0001	.0012	.0055	.0154	.0330	.0595	.0951	.1382	.1861	.2344
	5	.0000	.0001	.0004	.0015	.0044	.0102	.0205	.0369	.0609	.0938
	6	.0000	.0000	.0000	.0001	.0002	.0007	.0018	.0041	.0083	.0156
7	0	.6983	.4783	.3206	.2097	.1335	.0824	.0490	.0280	.0152	.0078
	1	.2573	.3720	.3960	.3670	.3115	.2471	.1848	.1306	.0872	.0547
	2	.0406	.1240	.2097	.2753	.3115	.3177	.2985	.2613	.2140	.1641
	3	.0036	.0230	.0617	.1147	.1730	.2269	.2679	.2903	.2918	.2734
	4	.0002	.0026	.0109	.0287	.0577	.0972	.1442	.1935	.2388	.2734
	5	.0000	.0002	.0012	.0043	.0115	.0250	.0466	.0774	.1172	.1641
	6	.0000	.0000	.0001	.0004	.0013	.0036	.0084	.0172	.0320	.0547
	7	.0000	.0000	.0000	.0000	.0001	.0002	.0006	.0016	.0037	.0078

Source: From *Handbook of Probability and Statistics with Tables* by Burlington & May. Copyright © 1970 by McGraw-Hill, Inc. Used with permission of McGraw-Hill Book Company.

THE BINOMIAL PROBABILITY DISTRIBUTION.

n	x	.05	.10	.15	.20	.25	.30	.35	.40	.45	.50
8	0	.6634	.4305	.2725	.1678	.1001	.0576	.0319	.0168	.0084	.0039
	1	.2793	.3826	.3847	.3355	.2760	.1977	.1373	.0896	.0548	.0312
	2	.0515	.1488	.2376	.2936	.3115	.2965	.2587	.2090	.1569	.1094
	3	.0054	.0331	.0839	.1468	.2076	.2541	.2786	.2787	.2568	.2188
	4	.0004	.0046	.0185	.0459	.0865	.1361	.1875	.2322	.2627	.2734
	5	.0000	.0004	.0026	.0092	.0231	.0467	.0808	.1239	.1719	.2188
	6	.0000	.0000	.0002	.0011	.0038	.0100	.0217	.0413	.0703	.1094
	7	.0000	.0000	.0000	.0001	.0004	.0012	.0033	.0079	.0164	.0312
	8	.0000	.0000	.0000	.0000	.0000	.0001	.0002	.0007	.0017	.0039
9	0	.6302	.3874	.2316	.1342	.0751	.0404	.0277	.0101	.0046	.0020
	1	.2985	.3874	.3679	.3020	.2253	.1556	.1004	.0605	.0339	.0176
	2	.0629	.1722	.2597	.3020	.3003	.2668	.2162	.1612	.1110	.0703
	3	.0077	.0446	.1069	.1762	.2336	.2668	.2716	.2508	.2119	.1641
	4	.0006	.0074	.0283	.0661	.1168	.1715	.2194	.2508	.2600	.2461
	5	.0000	.0008	.0050	.0165	.0389	.0735	.1181	.1672	.2128	.2461
	6	.0000	.0001	.0006	.0028	.0087	.0210	.0424	.0743	.1160	.1641
	7	.0000	.0000	.0000	.0003	.0012	.0039	.0098	.0212	.0407	.0703
	8	.0000	.0000	.0000	.0000	.0001	.0004	.0013	.0035	.0083	.0176
	9	.0000	.0000	.0000	.0000	.0000	.0000	.0001	.0003	.0008	.0020
10	0	.5987	.3487	.1969	.1074	.0563	.0282	.0135	.0060	.0025	.0010
	1	.3151	.3874	.3474	.2684	.1877	.1211	.0725	.0403	.0207	.0098
	2	.0746	.1937	.2759	.3020	.2816	.2335	.1757	.1209	.0763	.0439
	3	.0105	.0574	.1298	.2013	.2503	.2668	.2522	.2150	.1665	.1172
	4	.0010	.0112	.0401	.0881	.1460	.2001	.2377	.2508	.2384	.2051
	5	.0001	.0015	.0085	.0264	.0584	.1029	.1536	.2007	.2340	.2461
	6	.0000	.0001	.0012	.0055	.0162	.0368	.0689	.1115	.1596	.2051
	7	.0000	.0000	.0001	.0008	.0031	.0090	.0212	.0425	.0746	.1172
	8	.0000	.0000	.0000	.0001	.0004	.0014	.0043	.0106	.0229	.0439
	9	.0000	.0000	.0000	.0000	.0000	.0001	.0005	.0016	.0042	.0098
	10	.0000	.0000	.0000	.0000	.0000	.0000	.0000	.0001	.0003	.0010
11	0	.5688	.3138	.1673	.0859	.0422	.0198	.0088	.0036	.0014	.0005
	1	.3293	.3835	.3248	.2362	.1549	.0932	.0518	.0266	.0125	.0054
	2	.0867	.2131	.2866	.2953	.2581	.1998	.1395	.0887	.0513	.0269
	3	.0137	.0710	.1517	.2215	.2581	.2568	.2254	.1774	.1259	.0806
	4	.0014	.0158	.0536	.1107	.1721	.2201	.2428	.2365	.2060	.1611
	5	.0001	.0025	.0132	.0388	.0803	.1231	.1830	.2207	.2360	.2256
	6	.0000	.0003	.0023	.0097	.0268	.0566	.0985	.1471	.1931	.2256
	7	.0000	.0000	.0003	.0017	.0064	.0173	.0379	.0701	.1128	.1611
	8	.0000	.0000	.0000	.0002	.0011	.0037	.0102	.0234	.0462	.0806
	9	.0000	.0000	.0000	.0000	.0001	.0005	.0018	.0052	.0126	.0269
	10	.0000	.0000	.0000	.0000	.0000	.0000	.0002	.0007	.0021	.0054
	11	.0000	.0000	.0000	.0000	.0000	.0000	.0000	.0000	.0002	.0005
12	0	.5404	.2824	.1422	.0687	.0317	.0138	.0057	.0022	.0008	.0002
	1	.3413	.3766	.3012	.2062	.1267	.0712	.0368	.0174	.0075	.0029
	2	.0988	.2301	.2924	.2835	.2323	.1678	.1088	.0639	.0339	.0161
	3	.0173	.0852	.1720	.2362	.2581	.2397	.1954	.1419	.0923	.0537
	4	.0021	.0213	.0683	.1329	.1936	.2311	.2367	.2128	.1700	.1208
	5	.0002	.0038	.0193	.0532	.1032	.1585	.2039	.2270	.2225	.1934
	6	.0000	.0005	.0040	.0155	.0401	.0792	.1281	.1766	.2124	.2256

THE BINOMIAL PROBABILITY DISTRIBUTION.

n	x	.05	.10	.15	.20	.25	.30	.35	.40	.45	.50
12	7	.0000	.0000	.0006	.0033	.0115	.0291	.0591	.1009	.1489	.1934
	8	.0000	.0000	.0001	.0005	.0024	.0078	.0199	.0420	.0762	.1208
	9	.0000	.0000	.0000	.0001	.0004	.0015	.0048	.0125	.0277	.0537
	10	.0000	.0000	.0000	.0000	.0000	.0002	.0008	.0025	.0068	.0161
	11	.0000	.0000	.0000	.0000	.0000	.0000	.0001	.0003	.0010	.0029
	12	.0000	.0000	.0000	.0000	.0000	.0000	.0000	.0000	.0001	.0002
13	0	.5133	.2542	.1209	.0550	.0238	.0097	.0037	.0013	.0004	.0001
	1	.3512	.3672	.2774	.1787	.1029	.0540	.0259	.0113	.0045	.0016
	2	.1109	.2448	.2937	.2680	.2059	.1388	.0836	.0453	.0220	.0095
	3	.0214	.0997	.1900	.2457	.2517	.2181	.1651	.1107	.0660	.0349
	4	.0028	.0277	.0838	.1535	.2097	.2337	.2222	.1845	.1350	.0873
	5	.0003	.0055	.0266	.0691	.1258	.1803	.2154	.2214	.1989	.1571
	6	.0000	.0008	.0063	.0230	.0559	.1030	.1546	.1968	.2169	.2095
	7	.0000	.0001	.0011	.0058	.0186	.0442	.0833	.1312	.1775	.2095
	8	.0000	.0000	.0001	.0011	.0047	.0142	.0336	.0656	.1089	.1571
	9	.0000	.0000	.0000	.0001	.0009	.0034	.0101	.0243	.0495	.0873
	10	.0000	.0000	.0000	.0000	.0001	.0006	.0022	.0065	.0162	.0349
	11	.0000	.0000	.9000	.0000	.0000	.0001	.0003	.0012	.0036	.0095
	12	.0000	.0000	.0000	.0000	.0000	.0000	.0000	.0001	.0005	.0016
	13	.0000	.0000	.0000	.0000	.0000	.0000	.0000	.0000	.0000	.0001
14	0	.4877	.2288	.1028	.0440	.0178	.0068	.0024	.0008	.0002	.0001
	1	.3593	.3559	.2539	.1539	.0832	.0407	.0181	.0073	.0027	.0009
	2	.1229	.2570	.2912	.2501	.1802	.1134	.0634	.0317	.0141	.0056
	3	.0259	.1142	.2056	.2501	.2402	.1943	.1366	.0845	.0462	.0222
	4	.0037	.0349	.0998	.1720	.2202	.2290	.2022	.1549	.1040	.0611
	5	.0004	.0078	.0352	.0860	.1468	.1963	.2178	.2066	.1701	.1222
	6	.0000	.0013	.0093	.0322	.0734	.1262	.1759	.2066	.2088	.1833
	7	.0000	.0002	.0019	.0092	.0280	.0618	.1082	.1574	.1952	.2095
	8	.0000	.0000	.0003	.0020	.0082	.0232	.0510	.0918	.1398	.1833
	9	.0000	.0000	.0000	.0003	.0018	.0066	.0183	.0408	.0762	.1222
	10	.0000	.0000	.0000	.0000	.0003	.0014	.0049	.0136	.0312	.0611
	11	.0000	.0000	.0000	.0000	.0000	.0002	.0010	.0033	.0093	.0222
	12	.0000	.0000	.0000	.0000	.0000	.0000	.0001	.0005	.0019	.0056
	13	.0000	.0000	.0000	.0000	.0000	.0000	.0000	.0001	.0002	.0009
	14	.0000	.0000	.0000	.0000	.0000	.0000	.0000	.0000	.0000	.0001
15	0	.4633	.2059	.0874	.0352	.0134	.0047	.0016	.0005	.0001	.0000
	1	.3658	.3432	.2312	.1319	.0668	.0305	.0126	.0047	.0016	.0005
	2	.1348	.2669	.2856	.2309	.1559	.0916	.0476	.0219	.0090	.0032
	3	.0307	.1285	.2184	.2501	.2252	.1700	.1110	.0634	.0318	.0139
	4	.0049	.0428	.1156	.1876	.2252	.2186	.1792	.1268	.0780	.0417
	5	.0006	.0105	.0449	.1032	.1651	.2061	.2123	.1859	.1404	.0916
	6	.0000	.0019	.0132	.0430	.0917	.1472	.1906	.2066	.1914	.1527
	7	.0000	.0003	.0030	.0138	.0393	.0811	.1319	.1771	.2013	.1964
	8	.0000	.0000	.0005	.0035	.0131	.0348	.0710	.1181	.1647	.1964
	9	.0000	.0000	.0001	.0007	.0034	.0116	.0298	.0612	.1048	.1527
	10	.0000	.0000	.0000	.0001	.0007	.0030	.0096	.0245	.0515	.0916
	11	.0000	.0000	.0000	.0000	.0001	.0006	.0024	.0074	.0191	.0417
	12	.0000	.0000	.0000	.0000	.0000	.0001	.0004	.0016	.0052	.0139
	13	.0000	.0000	.0000	.0000	.0000	.0000	.0001	.0003	.0010	.0032
	14	.0000	.0000	.0000	.0000	.0000	.0000	.0000	.0000	.0001	.0005
	15	.0000	.0000	.0000	.0000	.0000	.0000	.0000	.0000	.0000	.0000

THE BINOMIAL PROBABILITY DISTRIBUTION.

n	x	π									
		.05	.10	.15	.20	.25	.30	.35	.40	.45	.50
16	0	.4401	.1853	.0743	.0281	.0100	.0033	.0010	.0003	.0001	.0000
	1	.3706	.3294	.2097	.1126	.0535	.0228	.0087	.0030	.0009	.0002
	2	.1463	.2745	.2775	.2111	.1336	.0732	.0353	.0150	.0056	.0018
	3	.0359	.1423	.2285	.2463	.2079	.1465	.0888	.0468	.0215	.0085
	4	.0061	.0514	.1311	.2001	.2252	.2040	.1553	.1014	.0572	.0278
	5	.0008	.0137	.0555	.1201	.1802	.2099	.2008	.1623	.1123	.0667
	6	.0001	.0028	.0180	.0550	.1101	.1649	.1982	.1983	.1684	.1222
	7	.0000	.0004	.0045	.0197	.0524	.1010	.1524	.1889	.1969	.1746
	8	.0000	.0001	.0009	.0055	.0197	.0487	.0923	.1417	.1812	.1964
	9	.0000	.0000	.0001	.0012	.0058	.0185	.0442	.0840	.1318	.1746
	10	.0000	.0000	.0000	.0002	.0014	.0056	.0167	.0392	.0755	.1222
	11	.0000	.0000	.0000	.0000	.0002	.0013	.0049	.0142	.0337	.0667
	12	.0000	.0000	.0000	.0000	.0000	.0002	.0011	.0040	.0115	.0278
	13	.0000	.0000	.0000	.0000	.0000	.0000	.0002	.0008	.0029	.0085
	14	.0000	.0000	.0000	.0000	.0000	.0000	.0000	.0001	.0005	.0018
	15	.0000	.0000	.0000	.0000	.0000	.0000	.0000	.0000	.0001	.0002
	16	.0000	.0000	.0000	.0000	.0000	.0000	.0000	.0000	.0000	.0000
17	0	.4181	.1668	.0631	.0225	.0075	.0023	.0007	.0002	.0000	.0000
	1	.3741	.3150	.1893	.0957	.0426	.0169	.0060	.0019	.0005	.0001
	2	.1575	.2800	.2673	.1914	.1136	.0581	.0260	.0102	.0035	.0010
	3	.0415	.1556	.2359	.2393	.1893	.1245	.0701	.0341	.0144	.0052
	4	.0076	.0605	.1457	.2093	.2209	.1868	.1320	.0796	.0411	.0182
	5	.0010	.0175	.0668	.1361	.1914	.2081	.1849	.1379	.0875	.0472
	6	.0001	.0039	.0236	.0680	.1276	.1784	.1991	.1839	.1432	.0944
	7	.0000	.0007	.0065	.0267	.0668	.1201	.1685	.1927	.1841	.1484
	8	.0000	.0001	.0014	.0084	.0279	.0644	.1143	.1606	.1883	.1855
	9	.0000	.0000	.0003	.0021	.0093	.0276	.0611	.1070	.1540	.1855
	10	.0000	.0000	.0000	.0004	.0025	.0095	.0263	.0571	.1008	.1434
	11	.0000	.0000	.0000	.0001	.0005	.0026	.0090	.0242	.0525	.0944
	12	.0000	.0000	.0000	.0000	.0001	.0006	.0024	.0081	.0215	.0472
	13	.0000	.0000	.0000	.0000	.0000	.0001	.0005	.0021	.0068	.0182
	14	.0000	.0000	.0000	.0000	.0000	.0000	.0001	.0004	.0016	.0052
	15	.0000	.0000	.0000	.0000	.0000	.0000	.0000	.0001	.0003	.0010
	16	.0000	.0000	.0000	.0000	.0000	.0000	.0000	.0000	.0000	.0001
	17	.0000	.0000	.0000	.0000	.0000	.0000	.0000	.0000	.0000	.0000
18	0	.3972	.1501	.0536	.0180	.0056	.0016	.0004	.0001	.0000	.0000
	1	.3763	.3002	.1704	.0811	.0338	.0126	.0042	.0012	.0003	.0001
	2	.1683	.2835	.2556	.1723	.0958	.0458	.0190	.0069	.0022	.0006
	3	.0473	.1680	.2406	.2297	.1704	.1046	.0547	.0246	.0095	.0031
	4	.0093	.0700	.1592	.2153	.2130	.1681	.1104	.0614	.0291	.0117
	5	.0014	.0218	.0787	.1507	.1988	.2017	.1664	.1146	.0666	.0327
	6	.0002	.0052	.0310	.0816	.1436	.1873	.1941	.1655	.1181	.0708
	7	.0000	.0010	.0091	.0350	.0820	.1376	.1792	.1892	.1657	.1214
	8	.0000	.0002	.0022	.0120	.0376	.0811	.1327	.1734	.1864	.1669
	9	.0000	.0000	.0004	.0033	.0139	.0386	.0794	.1284	.1694	.1855
	10	.0000	.0000	.0001	.0008	.0042	.0149	.0385	.0771	.1248	.1669
	11	.0000	.0000	.0000	.0001	.0010	.0046	.0151	.0374	.0742	.1214
	12	.0000	.0000	.0000	.0000	.0002	.0012	.0047	.0145	.0354	.0708

THE BINOMIAL PROBABILITY DISTRIBUTION.

n	x	.05	.10	.15	.20	.25	.30	.35	.40	.45	.50
						π					
18	13	.0000	.0000	.0000	.0000	.0000	.0002	.0012	.0045	.0134	.0327
	14	.0000	.0000	.0000	.0000	.0000	.0000	.0002	.0011	.0039	.0117
	15	.0000	.0000	.0000	.0000	.0000	.0000	.0000	.0002	.0009	.0031
	16	.0000	.0000	.0000	.0000	.0000	.0000	.0000	.0000	.0001	.0006
	17	.0000	.0000	.0000	.0000	.0000	.0000	.0000	.0000	.0000	.0001
	18	.0000	.0000	.0000	.0000	.0000	.0000	.0000	.0000	.0000	.0000
19	0	.3774	.1351	.0456	.0144	.0042	.0011	.0003	.0001	.0000	.0000
	1	.3774	.2852	.1529	.0685	.0268	.0093	.0029	.0008	.0002	.0000
	2	.1787	.2852	.2428	.1540	.0803	.0358	.0138	.0046	.0013	.0003
	3	.0533	.1796	.2428	.2182	.1517	.0869	.0422	.0175	.0062	.0018
	4	.0112	.0798	.1714	.2182	.2023	.1491	.0909	.0467	.0203	.0074
	5	.0018	.0266	.0907	.1636	.2023	.1916	.1468	.0933	.0497	.0222
	6	.0002	.0069	.0374	.0955	.1574	.1916	.1844	.1451	.0949	.0518
	7	.0000	.0014	.0122	.0443	.0974	.1525	.1844	.1797	.1443	.0961
	8	.0000	.0002	.0032	.0166	.0487	.0981	.1489	.1797	.1771	.1442
	9	.0000	.0000	.0007	.0051	.0198	.0514	.0980	.1464	.1771	.1762
	10	.0000	.0000	.0001	.0013	.0066	.0220	.0528	.0976	.1449	.1762
	11	.0000	.0000	.0000	.0003	.0018	.0077	.0233	.0532	.0970	.1442
	12	.0000	.0000	.0000	.0000	.0004	.0022	.0083	.0237	.0529	.0961
	13	.0000	.0000	.0000	.0000	.0001	.0005	.0024	.0085	.0233	.0518
	14	.0000	.0000	.0000	.0000	.0000	.0001	.0006	.0024	.0082	.0222
	15	.0000	.0000	.0000	.0000	.0000	.0000	.0001	.0005	.0022	.0074
	16	.0000	.0000	.0000	.0000	.0000	.0000	.0000	.0001	.0005	.0018
	17	.0000	.0000	.0000	.0000	.0000	.0000	.0000	.0000	.0001	.0003
	18	.0000	.0000	.0000	.0000	.0000	.0000	.0000	.0000	.0000	.0000
	19	.0000	.0000	.0000	.0000	.0000	.0000	.0000	.0000	.0000	.0000
20	0	.3585	.1216	.0388	.0115	.0032	.0008	.0002	.0000	.0000	.0000
	1	.3774	.2702	.1368	.0576	.0211	.0068	.0020	.0005	.0001	.0000
	2	.1887	.2852	.2293	.1369	.0669	.0278	.0100	.0031	.0008	.0002
	3	.0596	.1901	.2428	.2054	.1339	.0716	.0323	.0123	.0040	.0011
	4	.0133	.0898	.1821	.2182	.1897	.1304	.0738	.0350	.0139	.0046
	5	.0022	.0319	.1028	.1746	.2023	.1789	.1272	.0746	.0365	.0148
	6	.0003	.0089	.0454	.1091	.1686	.1916	.1712	.1244	.0746	.0370
	7	.0000	.0020	.0160	.0545	.1124	.1643	.1844	.1659	.1221	.0739
	8	.0000	.0004	.0046	.0222	.0609	.1144	.1614	.1797	.1623	.1201
	9	.0000	.0001	.0011	.0074	.0271	.0654	.1158	.1597	.1771	.1602
	10	.0000	.0000	.0002	.0020	.0099	.0308	.0686	.1171	.1593	.1762
	11	.0000	.0000	.0000	.0005	.0030	.0120	.0336	.0710	.1185	.1602
	12	.0000	.0000	.0000	.0001	.0008	.0039	.0136	.0355	.0727	.1201
	13	.0000	.0000	.0000	.0000	.0002	.0010	.0045	.0146	.0366	.0739
	14	.0000	.0000	.0000	.0000	.0000	.0002	.0012	.0049	.0150	.0370
	15	.0000	.0000	.0000	.0000	.0000	.0000	.0003	.0013	.0049	.0148
	16	.0000	.0000	.0000	.0000	.0000	.0000	.0000	.0003	.0013	.0046
	17	.0000	.0000	.0000	.0000	.0000	.0000	.0000	.0000	.0002	.0011
	18	.0000	.0000	.0000	.0000	.0000	.0000	.0000	.0000	.0000	.0002
	19	.0000	.0000	.0000	.0000	.0000	.0000	.0000	.0000	.0000	.0000
	20	.0000	.0000	.0000	.0000	.0000	.0000	.0000	.0000	.0000	.0000

APPENDIX E
TABLE OF CUMULATIVE BINOMIAL

This appendix is actually a series of tables which give the cumulative values of the binomial probability distribution. It should be remembered that it is appropriate to use the binomial distribution whenever the simple events are independent, the outcome of a simple event can be classified into a dichotomy, and the number of successful simple events is a discrete value.

To illustrate the use of these tables, suppose we have a problem in which the probability of success on any simple event π is equal to 0.4, the number of simple events observed n is equal to 5, and the number of successful simple events X is equal to either 0, 1, 2, or 3. In other words, we want to find

$$\sum_{x=0}^{3} C_x^5 \, (0.4)^x \, (0.6)^{5-x}$$

First move to the table headed by $n = 5$. Upon finding this table, locate the column with a π of 0.4. Moving down this column, find the row representing 3 successful simple events. Reading from this intersection, we see that the cumulative probability of obtaining a 0, 1, 2, or 3 in this problem is equal to 0.9130. The probability of obtaining 4 or 5 successful simple events in this problem is equal to $1.0000 - 0.9130$ or 0.0870, since a probability distribution sums to one.

If a problem is encountered where $\pi > 0.50$, this table can also be used in its solution. For example, suppose we have a problem in which $n = 5$, $\pi = 0.6$, and $X \leq 2$. In this case we must focus on the $(1 - \pi)$ portion of Equation 5.2. For our illustration, the probability of 0, 1 or 2 successes is the same as the probability of 3, 4, or 5 failures. In the table, we can find the probability of $X \leq 2$ failures and subtract from 1.0 to determine the probability of $X > 2$. From the table headed $n = 5$ under the 0.40 column,

$$P\,(X \leq 2) = 0.6826$$

then

$$P\,(X > 2) = 1 - 0.6826 = 0.3174$$

Thus, the probability of $X \leq 2$ when $\pi = 0.60$ is 0.3174, which is the same as the probability of $X > 2$ when $(1 - \pi) = 0.40$.

TABLE OF CUMULATIVE BINOMIAL.

$n = 1$

π x	.01	.05	.10	.20	.30	.40	.50
0	0.9900	0.9500	0.9000	0.8000	0.7000	0.6000	0.5000
1	1.0000	1.0000	1.0000	1.0000	1.0000	1.0000	1.0000

$n = 2$

π x	.01	.05	.10	.20	.30	.40	.50
0	0.9801	0.9025	0.8100	0.6400	0.4900	0.3600	0.2500
1	0.9999	0.9975	0.9900	0.9600	0.9100	0.8400	0.7500
2	1.0000	1.0000	1.0000	1.0000	1.0000	1.0000	1.0000

$n = 3$

π x	.01	.05	.10	.20	.30	.40	.50
0	0.9703	0.8574	0.7290	0.5120	0.3430	0.2160	0.1250
1	0.9997	0.9927	0.9720	0.8960	0.7840	0.6480	0.5000
2	1.0000	0.9999	0.9990	0.9920	0.9730	0.9360	0.8750
3	1.0000	1.0000	1.0000	1.0000	1.0000	1.0000	1.0000

$n = 4$

π x	.01	.05	.10	.20	.30	.40	.50
0	0.9606	0.8145	0.6561	0.4096	0.2401	0.1296	0.0625
1	0.9994	0.9860	0.9477	0.8192	0.6517	0.4752	0.3125
2	1.0000	0.9995	0.9963	0.9728	0.9163	0.8208	0.6875
3	1.0000	1.0000	0.9999	0.9984	0.9919	0.9744	0.9375
4	1.0000	1.0000	1.0000	1.0000	1.0000	1.0000	1.0000

$n = 5$

π x	.01	.05	.10	.20	.30	.40	.50
0	0.9510	0.7738	0.5905	0.3277	0.1681	0.0778	0.0313
1	0.9990	0.9774	0.9185	0.7373	0.5282	0.3370	0.1875
2	1.0000	0.9988	0.9914	0.9421	0.8369	0.6826	0.5000
3	1.0000	1.0000	0.9995	0.9933	0.9692	0.9130	0.8125
4	1.0000	1.0000	1.0000	0.9997	0.9976	0.9898	0.9688
5				1.0000	1.0000	1.0000	1.0000

Source: From *Statistics: Meaning and Method* by Lawrence L. Lapin, copyright © 1975 by Harcourt Brace Jovanovich, Inc., and reprinted with their permission.

TABLE OF CUMULATIVE BINOMIAL.

n = 6

π x	.01	.05	.10	.20	.30	.40	.50
0	0.9415	0.7351	0.5314	0.2621	0.1176	0.0467	0.0156
1	0.9985	0.9672	0.8857	0.6554	0.4202	0.2333	0.1094
2	1.0000	0.9978	0.9841	0.9011	0.7443	0.5443	0.3438
3	1.0000	0.9999	0.9987	0.9830	0.9295	0.8208	0.6563
4	1.0000	1.0000	0.9999	0.9984	0.9891	0.9590	0.8906
5	1.0000	1.0000	1.0000	0.9999	0.9993	0.9959	0.9844
6				1.0000	1.0000	1.0000	1.0000

n = 7

π x	.01	.05	.10	.20	.30	.40	.50
0	0.9321	0.6983	0.4783	0.2097	0.0824	0.0280	0.0078
1	0.9980	0.9556	0.8503	0.5767	0.3294	0.1586	0.0625
2	1.0000	0.9962	0.9743	0.8520	0.6471	0.4199	0.2266
3	1.0000	0.9998	0.9973	0.9667	0.8740	0.7102	0.5000
4	1.0000	1.0000	0.9998	0.9953	0.9712	0.9037	0.7734
5	1.0000	1.0000	1.0000	0.9996	0.9962	0.9812	0.9375
6				1.0000	0.9998	0.9984	0.9922
7					1.0000	1.0000	1.0000

n = 8

π x	.01	.05	.10	.20	.30	.40	.50
0	0.9227	0.6634	0.4305	0.1678	0.0576	0.0168	0.0039
1	0.9973	0.9428	0.8131	0.5033	0.2553	0.1064	0.0352
2	0.9999	0.9942	0.9619	0.7969	0.5518	0.3154	0.1445
3	1.0000	0.9996	0.9950	0.9437	0.8059	0.5941	0.3633
4	1.0000	1.0000	0.9996	0.9896	0.9420	0.8263	0.6367
5	1.0000	1.0000	1.0000	0.9988	0.9887	0.9502	0.8555
6				0.9999	0.9987	0.9915	0.9648
7				1.0000	0.9999	0.9993	0.9961
8					1.0000	1.0000	1.0000

n = 9

π x	.01	.05	.10	.20	.30	.40	.50
0	0.9135	0.6302	0.3874	0.1342	0.0404	0.0101	0.0020
1	0.9966	0.9288	0.7748	0.4362	0.1960	0.0705	0.0195
2	0.9999	0.9916	0.9470	0.7382	0.4628	0.2318	0.0898
3	1.0000	0.9994	0.9917	0.9144	0.7297	0.4826	0.2539
4	1.0000	1.0000	0.9991	0.9804	0.9012	0.7334	0.5000
5	1.0000	1.0000	0.9999	0.9969	0.9747	0.9006	0.7461

TABLE OF CUMULATIVE BINOMIAL.

$n = 9$

π x	.01	.05	.10	.20	.30	.40	.50
6	1.0000	1.0000	1.0000	0.9997	0.9957	0.9750	0.9102
7				1.0000	0.9996	0.9962	0.9805
8					1.0000	0.9997	0.9980
9						1.0000	1.0000

$n = 10$

π x	.01	.05	.10	.20	.30	.40	.50
0	0.9044	0.5987	0.3487	0.1074	0.0282	0.0060	0.0010
1	0.9957	0.9139	0.7361	0.3758	0.1493	0.0464	0.0107
2	0.9999	0.9885	0.9298	0.6778	0.3828	0.1673	0.0547
3	1.0000	0.9990	0.9872	0.8791	0.6496	0.3823	0.1719
4	1.0000	0.9999	0.9984	0.9672	0.8497	0.6331	0.3770
5	1.0000	1.0000	0.9999	0.9936	0.9526	0.8338	0.6230
6	1.0000	1.0000	1.0000	0.9991	0.9894	0.9452	0.8281
7				0.9999	0.9999	0.9877	0.9453
8				1.0000	1.0000	0.9983	0.9893
9						0.9999	0.9990
10						1.0000	1.0000

$n = 11$

π x	.01	.05	.10	.20	.30	.40	.50
0	0.8953	0.5688	0.3138	0.0859	0.0198	0.0036	0.0005
1	0.9948	0.8981	0.6974	0.3221	0.1130	0.0302	0.0059
2	0.9998	0.9848	0.9104	0.6174	0.3127	0.1189	0.0327
3	1.0000	0.9984	0.9815	0.8369	0.5696	0.2963	0.1133
4	1.0000	0.9999	0.9972	0.9496	0.7897	0.5328	0.2744
5	1.0000	1.0000	0.9997	0.9883	0.9218	0.7535	0.5000
6	1.0000	1.0000	1.0000	0.9980	0.9784	0.9006	0.7256
7				0.9998	0.9957	0.9707	0.8867
8				1.0000	0.9994	0.9941	0.9673
9					1.0000	0.9993	0.9941
10						1.0000	0.9995
11							1.0000

TABLE OF CUMULATIVE BINOMIAL.

$n = 12$

π x	.01	.05	.10	.20	.30	.40	.50
0	0.8864	0.5404	0.2824	0.0687	0.0138	0.0022	0.0002
1	0.9938	0.8816	0.6590	0.2749	0.0850	0.0196	0.0032
2	0.9998	0.9804	0.8891	0.5583	0.2528	0.0834	0.0193
3	1.0000	0.9978	0.9744	0.7946	0.4925	0.2253	0.0730
4	1.0000	0.9998	0.9957	0.9274	0.7237	0.4382	0.1938
5	1.0000	1.0000	0.9995	0.9806	0.8821	0.6652	0.3872
6	1.0000	1.0000	0.9999	0.9961	0.9614	0.8418	0.6128
7	1.0000	1.0000	1.0000	0.9994	0.9905	0.9427	0.8062
8				0.9999	0.9983	0.9847	0.9270
9				1.0000	0.9998	0.9972	0.9807
10					1.0000	0.9997	0.9968
11						1.0000	0.9998
12							1.0000

$n = 13$

π x	.01	.05	.10	.20	.30	.40	.50
0	0.8775	0.5133	0.2542	0.0550	0.0097	0.0013	0.0001
1	0.9928	0.8646	0.6213	0.2336	0.0637	0.0126	0.0017
2	0.9997	0.9755	0.8661	0.5017	0.2025	0.0579	0.0112
3	1.0000	0.9969	0.9658	0.7473	0.4206	0.1686	0.0461
4	1.0000	0.9997	0.9935	0.9009	0.6543	0.3530	0.1334
5	1.0000	1.0000	0.9991	0.9700	0.8346	0.5744	0.2905
6	1.0000	1.0000	0.9999	0.9930	0.9376	0.7712	0.5000
7	1.0000	1.0000	1.0000	0.9988	0.9818	0.9023	0.7095
8				0.9998	0.9960	0.9679	0.8666
9				1.0000	0.9993	0.9922	0.9539
10					0.9999	0.9987	0.9888
11					1.0000	0.9999	0.9983
12						1.0000	0.9999
13							1.0000

TABLE OF CUMULATIVE BINOMIAL.

$n = 14$

π x	.01	.05	.10	.20	.30	.40	.50
0	0.8687	0.4877	0.2288	0.0440	0.0068	0.0008	0.0001
1	0.9916	0.8470	0.5846	0.1979	0.0475	0.0081	0.0009
2	0.9997	0.9699	0.8416	0.4481	0.1608	0.0398	0.0065
3	1.0000	0.9958	0.9559	0.6982	0.3552	0.1243	0.0287
4	1.0000	0.9996	0.9908	0.8702	0.5842	0.2793	0.0898
5	1.0000	1.0000	0.9985	0.9561	0.7805	0.4859	0.2120
6	1.0000	1.0000	0.9998	0.9884	0.9067	0.6925	0.3953
7	1.0000	1.0000	1.0000	0.9976	0.9685	0.8499	0.6047
8				0.9996	0.9917	0.9417	0.7880
9				1.0000	0.9983	0.9825	0.9102
10					0.9998	0.9961	0.9713
11					1.0000	0.9994	0.9935
12						0.9999	0.9991
13						1.0000	0.9999
14							1.0000

$n = 15$

π x	.01	.05	.10	.20	.30	.40	.50
0	0.8601	0.4633	0.2059	0.0352	0.0047	0.0005	0.0000
1	0.9904	0.8290	0.5490	0.1671	0.0353	0.0052	0.0005
2	0.9996	0.9638	0.8159	0.3980	0.1268	0.0271	0.0037
3	1.0000	0.9945	0.9444	0.6482	0.2969	0.0905	0.0176
4	1.0000	0.9994	0.9873	0.8358	0.5155	0.2173	0.0592
5	1.0000	0.9999	0.9978	0.9389	0.7216	0.4032	0.1509
6	1.0000	1.0000	0.9997	0.9819	0.8689	0.6098	0.3036
7	1.0000	1.0000	1.0000	0.9958	0.9500	0.7869	0.5000
8				0.9992	0.9848	0.9050	0.6964
9				0.9999	0.9963	0.9662	0.8491
10				1.0000	0.9993	0.9907	0.9408
11					0.9999	0.9981	0.9824
12					1.0000	0.9997	0.9963
13						1.0000	0.9995
14							1.0000

TABLE OF CUMULATIVE BINOMIAL.

$n = 16$

π	.01	.05	.10	.20	.30	.40	.50
x							
0	0.8515	0.4401	0.1853	0.0281	0.0033	0.0003	0.0000
1	0.9891	0.8108	0.5147	0.1407	0.0261	0.0033	0.0003
2	0.9995	0.9571	0.7892	0.3518	0.0994	0.0183	0.0021
3	1.0000	0.9930	0.9316	0.5981	0.2459	0.0651	0.0106
4	1.0000	0.9991	0.9830	0.7982	0.4499	0.1666	0.0384
5	1.0000	0.9999	0.9967	0.9183	0.6598	0.3288	0.1051
6	1.0000	1.0000	0.9995	0.9733	0.8247	0.5272	0.2272
7	1.0000	1.0000	0.9999	0.9930	0.9256	0.7161	0.4018
8	1.0000	1.0000	1.0000	0.9985	0.9743	0.8577	0.5982
9				0.9998	0.9929	0.9417	0.7728
10				1.0000	0.9984	0.9809	0.8949
11					0.9997	0.9951	0.9616
12					1.0000	0.9991	0.9894
13						0.9999	0.9979
14						1.0000	0.9997
15							1.0000

$n = 17$

π	.01	.05	.10	.20	.30	.40	.50
x							
0	0.8429	0.4181	0.1668	0.0225	0.0023	0.0002	0.0000
1	0.9877	0.7922	0.4818	0.1182	0.0193	0.0021	0.0001
2	0.9994	0.9497	0.7618	0.3096	0.0774	0.0123	0.0012
3	1.0000	0.9912	0.9174	0.5489	0.2019	0.0464	0.0064
4	1.0000	0.9988	0.9779	0.7582	0.3887	0.1260	0.0245
5	1.0000	0.9999	0.9953	0.8943	0.5968	0.2639	0.0717
6	1.0000	1.0000	0.9992	0.9623	0.7752	0.4478	0.1662
7	1.0000	1.0000	0.9999	0.9891	0.8954	0.6405	0.3145
8	1.0000	1.0000	1.0000	0.9974	0.9597	0.8011	0.5000
9				0.9995	0.9873	0.9081	0.6855
10				0.9999	0.9968	0.9652	0.8338
11				1.0000	0.9993	0.9894	0.9283
12					0.9999	0.9975	0.9755
13					1.0000	0.9995	0.9936
14						0.9999	0.9988
15						1.0000	0.9999
16							1.0000

TABLE OF CUMULATIVE BINOMIAL.

$n = 18$

π	.01	.05	.10	.20	.30	.40	.50
x							
0	0.8345	0.3972	0.1501	0.0180	0.0016	0.0001	0.0000
1	0.9862	0.7735	0.4503	0.0991	0.0142	0.0013	0.0001
2	0.9993	0.9419	0.7338	0.2713	0.0600	0.0082	0.0007
3	1.0000	0.9891	0.9018	0.5010	0.1646	0.0328	0.0038
4	1.0000	0.9985	0.9718	0.7164	0.3327	0.0942	0.0154
5	1.0000	0.9998	0.9936	0.8671	0.5344	0.2088	0.0481
6	1.0000	1.0000	0.9988	0.9487	0.7217	0.3743	0.1189
7	1.0000	1.0000	0.9998	0.9837	0.8593	0.5634	0.2403
8	1.0000	1.0000	1.0000	0.9957	0.9404	0.7368	0.4073
9				0.9991	0.9790	0.8653	0.5927
10				0.9998	0.9939	0.9424	0.7597
11				1.0000	0.9986	0.9797	0.8811
12					0.9997	0.9942	0.9519
13					1.0000	0.9987	0.9846
14						0.9998	0.9962
15						1.0000	0.9993
16							0.9999
17							1.0000

$n = 19$

π	.01	.05	.10	.20	.30	.40	.50
x							
0	0.8262	0.3774	0.1351	0.0144	0.0011	0.0001	0.0000
1	0.9847	0.7547	0.4203	0.0829	0.0104	0.0008	0.0000
2	0.9991	0.9335	0.7054	0.2369	0.0462	0.0055	0.0004
3	1.0000	0.9868	0.8850	0.4551	0.1332	0.0230	0.0022
4	1.0000	0.9980	0.9648	0.6733	0.2822	0.0696	0.0096
5	1.0000	0.9998	0.9914	0.8369	0.4739	0.1629	0.0318
6	1.0000	1.0000	0.9983	0.9324	0.6655	0.3081	0.0835
7	1.0000	1.0000	0.9997	0.9767	0.8180	0.4878	0.1796
8	1.0000	1.0000	1.0000	0.9933	0.9161	0.6675	0.3238
9				0.9984	0.9674	0.8139	0.5000
10				0.9997	0.9895	0.9115	0.6762

TABLE OF CUMULATIVE BINOMIAL.

$n = 19$

π x	.01	.05	.10	.20	.30	.40	.50
11				0.9999	0.9972	0.9648	0.8204
12				1.0000	0.9994	0.9884	0.9165
13					0.9999	0.9969	0.9682
14					1.0000	0.9994	0.9904
15						0.9999	0.9978
16						1.0000	0.9996
17							1.0000

$n = 20$

π x	.01	.05	.10	.20	.30	.40	.50
0	0.8179	0.3585	0.1216	0.0115	0.0008	0.0000	0.0000
1	0.9831	0.7358	0.3917	0.0692	0.0076	0.0005	0.0000
2	0.9990	0.9245	0.6769	0.2061	0.0355	0.0036	0.0002
3	1.0000	0.9841	0.8670	0.4114	0.1071	0.0160	0.0013
4	1.0000	0.9974	0.9568	0.6296	0.2375	0.0510	0.0059
5	1.0000	0.9997	0.9887	0.8042	0.4164	0.1256	0.0207
6	1.0000	1.0000	0.9976	0.9133	0.6080	0.2500	0.0577
7	1.0000	1.0000	0.9996	0.9679	0.7723	0.4159	0.1316
8	1.0000	1.0000	0.9999	0.9900	0.8867	0.5956	0.2517
9	1.0000	1.0000	1.0000	0.9974	0.9520	0.7553	0.4119
10				0.9994	0.9829	0.8725	0.5881
11				0.9999	0.9949	0.9435	0.7483
12				1.0000	0.9987	0.9790	0.8684
13					0.9997	0.9935	0.9423
14					1.0000	0.9984	0.9793
15						0.9997	0.9941
16						1.0000	0.9987
17							0.9998
18							1.0000

APPENDIX F
POISSON DISTRIBUTION

As pointed out in the text, the Poisson distribution can be most useful in determining certain probabilities whenever the probability, π, for a simple event is relatively small. The table in this appendix gives the probability of various values of x and μ, with the Poisson formula being

$$P(x) = \frac{\mu^x e^{-\mu}}{x!}$$

Suppose we have a problem in which the probability of the simple event occurring is 0.05 and we are to observe 10 such simple events. In addition, suppose we are interested in finding the probability of 4 of these simple events being successful. Since the mean μ of this problem is equal to $(0.05)(10) = 0.5$, we first locate the column headed by 0.5 in the table. Moving down the column, we find the row representing the number of successful simple events in which we are interested, in this case 4. Reading from the intersection of the 0.5 column and 4 row, we see that the answer to this problem is 0.0016.

POISSON DISTRIBUTION.

					μ					
x	0.1	0.2	0.3	0.4	0.5	0.6	0.7	0.8	0.9	1.0
0	.9048	.8187	.7408	.6703	.6065	.5488	.4966	.4493	.4066	.3679
1	.0905	.1637	.2222	.2681	.3033	.3293	.3476	.3595	.3659	.3679
2	.0045	.0164	.0333	.0536	.0758	.0988	.1217	.1438	.1647	.1839
3	.0002	.0011	.0033	.0072	.0126	.0198	.0284	.0383	.0494	.0613
4	.0000	.0001	.0002	.0007	.0016	.0030	.0050	.0077	.0111	.0153
5	.0000	.0000	.0000	.0001	.0002	.0004	.0007	.0012	.0020	.0031
6	.0000	.0000	.0000	.0000	.0000	.0000	.0001	.0002	.0003	.0005
7	.0000	.0000	.0000	.0000	.0000	.0000	.0000	.0000	.0000	.0001

					μ					
x	1.1	1.2	1.3	1.4	1.5	1.6	1.7	1.8	1.9	2.0
0	.3329	.3012	.2725	.2466	.2231	.2019	.1827	.1653	.1496	.1353
1	.3662	.3614	.3543	.3452	.3347	.3230	.3106	.2975	.2842	.2707
2	.2014	.2169	.2303	.2417	.2510	.2584	.2640	.2678	.2700	.2707
3	.0738	.0867	.0998	.1128	.1255	.1378	.1496	.1607	.1710	.1804
4	.0203	.0260	.0324	.0395	.0471	.0551	.0636	.0723	.0812	.0902
5	.0045	.0062	.0084	.0111	.0141	.0176	.0216	.0260	.0309	.0361
6	.0008	.0012	.0018	.0026	.0035	.0047	.0061	.0078	.0098	.0120
7	.0001	.0002	.0003	.0005	.0008	.0011	.0015	.0020	.0027	.0034
8	.0000	.0000	.0001	.0001	.0001	.0002	.0003	.0005	.0006	.0009
9	.0000	.0000	.0000	.0000	.0000	.0000	.0001	.0001	.0001	.0002

					μ					
x	2.1	2.2	2.3	2.4	2.5	2.6	2.7	2.8	2.9	3.0
0	.1225	.1108	.1003	.0907	.0821	.0743	.0672	.0608	.0550	.0498
1	.2572	.2438	.2306	.2177	.2052	.1931	.1815	.1703	.1596	.1494
2	.2700	.2681	.2652	.2613	.2565	.2510	.2450	.2384	.2314	.2240
3	.1890	.1966	.2033	.2090	.2176	.2205	.2205	.2225	.2237	.2240
4	.0992	.1082	.1169	.1254	.1336	.1414	.1488	.1557	.1622	.1680
5	.0417	.0476	.0538	.0602	.0668	.0735	.0804	.0872	.0940	.1008
6	.0146	.0174	.0206	.0241	.0278	.0319	.0362	.0407	.0455	.0540
7	.0044	.0055	.0068	.0083	.0099	.0118	.0139	.0163	.0188	.0216
8	.0011	.0015	.0019	.0025	.0031	.0038	.0047	.0057	.0068	.0081
9	.0003	.0004	.0005	.0007	.0009	.0011	.0014	.0018	.0022	.0027
10	.0001	.0001	.0001	.0002	.0002	.0003	.0004	.0005	.0006	.0008
11	.0000	.0000	.0000	.0000	.0000	.0001	.0001	.0001	.0002	.0002
12	.0000	.0000	.0000	.0000	.0000	.0000	.0000	.0000	.0000	.0001

Source: From *Handbook of Probability and Statistics with Tables* by Burlington & May. Copyright © 1970 by McGraw-Hill, Inc. Used with permission of McGraw-Hill Book Company.

POISSON DISTRIBUTION.

	μ									
x	3.1	3.2	3.3	3.4	3.5	3.6	3.7	3.8	3.9	4.0
0	.0450	.0408	.0369	.0344	.0302	.0273	.0247	.0224	.0202	.0183
1	.1397	.1304	.1217	.1135	.1057	.0984	.0915	.0850	.0789	.0733
2	.2165	.2087	.2008	.1929	.1850	.1771	.1692	.1615	.1539	.1465
3	.2237	.2226	.2209	.2186	.2158	.2125	.2087	.2046	.2001	.1954
4	.1734	.1781	.1823	.1858	.1888	.1912	.1931	.1944	.1951	.1954
5	.1075	.1140	.1203	.1264	.1322	.1377	.1429	.1477	.1522	.1563
6	.0555	.0608	.0662	.0716	.0771	.0826	.0881	.0936	.0989	.1042
7	.0246	.0278	.0312	.0348	.0385	.0425	.0466	.0508	.0551	.0595
8	.0095	.0111	.0129	.0148	.0169	.0191	.0215	.0241	.0269	.0298
9	.0033	.0040	.0047	.0056	.0066	.0076	.0089	.0102	.0116	.0132
10	.0010	.0013	.0016	.0019	.0023	.0028	.0033	.0039	.0045	.0053
11	.0003	.0004	.0005	.0006	.0007	.0009	.0011	.0013	.0016	.0019
12	.0001	.0001	.0001	.0002	.0002	.0003	.0003	.0004	.0005	.0006
13	.0000	.0000	.0000	.0000	.0001	.0001	.0001	.0001	.0002	.0002
14	.0000	.0000	.0000	.0000	.0000	.0000	.0000	.0000	.0000	.0001

	μ									
x	4.1	4.2	4.3	4.4	4.5	4.6	4.7	4.8	4.9	5.0
0	.0166	.0150	.0136	.0123	.0111	.0101	.0091	.0082	.0074	.0067
1	.0679	.0630	.0583	.0540	.0500	.0462	.0427	.0395	.0365	.0337
2	.1393	.1323	.1254	.1188	.1125	.1063	.1005	.0948	.0894	.0842
3	.1904	.1852	.1798	.1743	.1687	.1631	.1574	.1517	.1460	.1404
4	.1951	.1944	.1933	.1917	.1898	.1875	.1849	.1820	.1789	.1755
5	.1600	.1633	.1662	.1687	.1708	.1725	.1738	.1747	.1753	.1755
6	.1093	.1143	.1191	.1237	.1281	.1323	.1362	.1398	.1432	.1462
7	.0640	.0686	.0732	.0778	.0824	.0869	.0914	.0959	.1002	.1044
8	.0328	.0360	.0393	.0428	.0463	.0500	.0537	.0575	.0614	.0653
9	.0150	.0168	.0188	.0209	.0232	.0255	.0280	.0307	.0334	.0363
10	.0061	.0071	.0081	.0092	.0104	.0118	.0132	.0147	.0164	.0181
11	.0023	.0027	.0032	.0037	.0043	.0049	.0056	.0064	.0073	.0082
12	.0008	.0009	.0011	.0014	.0016	.0019	.0022	.0026	.0030	.0034
13	.0002	.0003	.0004	.0005	.0006	.0007	.0008	.0009	.0011	.0013
14	.0001	.0001	.0001	.0001	.0002	.0002	.0003	.0003	.0004	.0005
15	.0000	.0000	.0000	.0000	.0001	.0001	.0001	.0001	.0001	.0002

	μ									
x	5.1	5.2	5.3	5.4	5.5	5.6	5.7	5.8	5.9	6.0
0	.0061	.0055	.0050	.0045	.0041	.0037	.0033	.0030	.0027	.0025
1	.0311	.0287	.0265	.0244	.0225	.0207	.0191	.0176	.0162	.0149
2	.0793	.0746	.0701	.0659	.0618	.0580	.0544	.0509	.0477	.0446
3	.1348	.1293	.1239	.1185	.1133	.1082	.1033	.0985	.0938	.0892
4	.1719	.1681	.1641	.1600	.1558	.1515	.1472	.1428	.1383	.1339
5	.1753	.1748	.1740	.1728	.1714	.1697	.1678	.1656	.1632	.1606
6	.1490	.1515	.1537	.1555	.1571	.1584	.1594	.1601	.1605	.1606
7	.1086	.1125	.1163	.1200	.1234	.1267	.1298	.1326	.1353	.1377
8	.0692	.0731	.0771	.0810	.0849	.0887	.0925	.0962	.0998	.1033
9	.0392	.0423	.0454	.0486	.0519	.0552	.0586	.0620	.0654	.0688

POISSON DISTRIBUTION.

μ

x	5.1	5.2	5.3	5.4	5.5	5.6	5.7	5.8	5.9	6.0
10	.0200	.0220	.0241	.0262	.0285	.0309	.0334	.0359	.0386	.0413
11	.0093	.0104	.0116	.0129	.0143	.0157	.0173	.0190	.0207	.0225
12	.0039	.0045	.0051	.0058	.0065	.0073	.0082	.0092	.0102	.0113
13	.0015	.0018	.0021	.0024	.0028	.0032	.0036	.0041	.0046	.0052
14	.0006	.0007	.0008	.0009	.0011	.0013	.0015	.0017	.0019	.0022
15	.0002	.0002	.0003	.0003	.0004	.0005	.0006	.0007	.0008	.0009
16	.0001	.0001	.0001	.0001	.0001	.0002	.0002	.0002	.0003	.0003
17	.0000	.0000	.0000	.0000	.0000	.0001	.0001	.0001	.0001	.0001

μ

x	6.1	6.2	6.3	6.4	6.5	6.6	6.7	6.8	6.9	7.0
0	.0022	.0020	.0018	.0017	.0015	.0014	.0012	.0011	.0010	.0009
1	.0137	.0126	.0116	.0106	.0098	.0090	.0082	.0076	.0070	.0064
2	.0417	.0390	.0364	.0340	.0318	.0296	.0276	.0258	.0240	.0223
3	.0848	.0806	.0765	.0726	.0688	.0652	.0617	.0584	.0552	.0521
4	.1294	.1249	.1205	.1162	.1118	.1076	.1034	.0992	.0952	.0912
5	.1579	.1549	.1519	.1487	.1454	.1420	.1385	.1349	.1314	.1277
6	.1605	.1601	.1595	.1586	.1575	.1562	.1546	.1529	.1511	.1490
7	.1399	.1418	.1435	.1450	.1462	.1472	.1480	.1486	.1489	.1490
8	.1066	.1099	.1130	.1160	.1188	.1215	.1240	.1263	.1284	.1304
9	.0723	.0757	.0791	.0825	.0858	.0891	.0923	.0954	.0985	.1014
10	.0441	.0469	.0498	.0528	.0558	.0588	.0618	.0649	.0679	.0710
11	.0245	.0265	.0285	.0307	.0330	.0353	.0377	.0401	.0426	.0452
12	.0124	.0137	.0150	.0164	.0179	.0194	.0210	.0227	.0245	.0264
13	.0058	.0065	.0073	.0081	.0089	.0098	.0108	.0119	.0130	.0142
14	.0025	.0029	.0033	.0037	.0041	.0046	.0052	.0058	.0064	.0071
15	.0010	.0012	.0014	.0016	.0018	.0020	.0023	.0026	.0029	.0033
16	.0004	.0005	.0005	.0006	.0007	.0008	.0010	.0011	.0013	.0014
17	.0001	.0002	.0002	.0002	.0003	.0003	.0004	.0004	.0005	.0006
18	.0000	.0001	.0001	.0001	.0001	.0001	.0001	.0002	.0002	.0002
19	.0000	.0000	.0000	.0000	.0000	.0000	.0000	.0001	.0001	.0001

μ

x	7.1	7.2	7.3	7.4	7.5	7.6	7.7	7.8	7.9	8.0
0	.0008	.0007	.0007	.0006	.0006	.0005	.0005	.0004	.0004	.0003
1	.0059	.0054	.0049	.0045	.0041	.0038	.0035	.0032	.0029	.0027
2	.0208	.0194	.0180	.0167	.0156	.0145	.0134	.0125	.0116	.0107
3	.0492	.0464	.0438	.0413	.0389	.0366	.0345	.0324	.0305	.0286
4	.0874	.0836	.0799	.0764	.0729	.0696	.0663	.0632	.0602	.0573
5	.1241	.1204	.1167	.1130	.1094	.1057	.1021	.0986	.0951	.0916
6	.1468	.1445	.1420	.1394	.1367	.1339	.1311	.1282	.1252	.1221
7	.1489	.1486	.1481	.1474	.1465	.1454	.1442	.1428	.1413	.1396
8	.1321	.1337	.1351	.1363	.1373	.1382	.1388	.1392	.1395	.1396
9	.1042	.1070	.1096	.1121	.1144	.1167	.1187	.1207	.1224	.1241
10	.0740	.0770	.0800	.0829	.0858	.0887	.0914	.0941	.0967	.0993
11	.0478	.0504	.0531	.0558	.0585	.0613	.0640	.0667	.0695	.0722

POISSON DISTRIBUTION.

μ

x	7.1	7.2	7.3	7.4	7.5	7.6	7.7	7.8	7.9	8.0
12	.0283	.0303	.0323	.0344	.0366	.0388	.0411	.0434	.0457	.0481
13	.0154	.0168	.0181	.0196	.0211	.0227	.0243	.0260	.0278	.0296
14	.0078	.0086	.0095	.0104	.0113	.0123	.0134	.0145	.0157	.0169
15	.0037	.0041	.0046	.0051	.0057	.0062	.0069	.0075	.0083	.0090
16	.0016	.0019	.0021	.0024	.0026	.0030	.0033	.0037	.0041	.0045
17	.0007	.0008	.0009	.0010	.0012	.0013	.0015	.0017	.0019	.0021
18	.0003	.0003	.0004	.0004	.0005	.0006	.0006	.0007	.0008	.0009
19	.0001	.0001	.0001	.0002	.0002	.0002	.0003	.0003	.0003	.0004
20	.0000	.0000	.0001	.0001	.0001	.0000	.0001	.0001	.0001	.0002
21	.0000	.0000	.0000	.0000	.0000	.0000	.0000	.0000	.0001	.0001

μ

x	8.1	8.2	8.3	8.4	8.5	8.6	8.7	8.8	8.9	9.0
0	.0003	.0003	.0002	.0002	.0002	.0002	.0002	.0002	.0001	.0001
1	.0025	.0023	.0021	.0019	.0017	.0016	.0014	.0013	.0012	.0011
2	.0100	.0092	.0086	.0079	.0074	.0068	.0063	.0058	.0054	.0050
3	.0269	.0252	.0237	.0222	.0208	.0195	.0183	.0171	.0160	.0150
4	.0544	.0517	.0491	.0466	.0443	.0420	.0398	.0377	.0357	.0337
5	.0882	.0849	.0816	.0784	.0752	.0722	.0692	.0663	.0635	.0607
6	.1191	.1160	.1128	.1097	.1066	.1034	.1003	.0972	.0941	.0911
7	.1378	.1358	.1338	.1317	.1294	.1271	.1247	.1222	.1197	.1171
8	.1395	.1392	.1388	.1382	.1375	.1366	.1356	.1344	.1332	.1318
9	.1256	.1269	.1280	.1290	.1299	.1306	.1311	.1315	.1317	.1318
10	.1017	.1040	.1063	.1084	.1104	.1123	.1140	.1157	.1172	.1186
11	.0749	.0776	.0802	.0828	.0853	.0878	.0902	.0925	.0948	.0970
12	.0505	.0530	.0555	.0579	.0604	.0629	.0654	.0679	.0703	.0728
13	.0315	.0334	.0354	.0374	.0395	.0416	.0438	.0459	.0481	.0504
14	.0182	.0196	.0210	.0225	.0240	.0256	.0272	.0289	.0306	.0324
15	.0098	.0107	.0116	.0126	.0136	.0147	.0158	.0169	.0182	.0194
16	.0050	.0055	.0060	.0066	.0072	.0079	.0086	.0093	.0101	.0109
17	.0024	.0026	.0029	.0033	.0036	.0040	.0044	.0048	.0053	.0058
18	.0011	.0012	.0014	.0015	.0017	.0019	.0021	.0024	.0026	.0029
19	.0005	.0005	.0006	.0007	.0008	.0009	.0010	.0011	.0012	.0014
20	.0002	.0002	.0002	.0003	.0003	.0004	.0004	.0005	.0005	.0006
21	.0001	.0001	.0001	.0001	.0001	.0002	.0002	.0002	.0002	.0003
22	.0000	.0000	.0000	.0000	.0001	.0001	.0001	.0001	.0001	.0001

μ

x	9.1	9.2	9.3	9.4	9.5	9.6	9.7	9.8	9.9	10
0	.0001	.0001	.0001	.0001	.0001	.0001	.0001	.0001	.0001	.0000
1	.0010	.0009	.0009	.0008	.0007	.0007	.0006	.0005	.0005	.0005
2	.0046	.0043	.0040	.0037	.0034	.0031	.0029	.0027	.0025	.0023
3	.0140	.0131	.0123	.0115	.0107	.0100	.0093	.0087	.0081	.0076
4	.0319	.0302	.0285	.0269	.0254	.0240	.0226	.0213	.0201	.0189

POISSON DISTRIBUTION.

						μ					
x	9.1	9.2	9.3	9.4	9.5	9.6	9.7	9.8	9.9	10
5	.0581	.0555	.0530	.0506	.0483	.0460	.0439	.0418	.0398	.0378
6	.0881	.0851	.0822	.0793	.0764	.0736	.0709	.0682	.0656	.0631
7	.1145	.1118	.1091	.1064	.1037	.1010	.0982	.0955	.0928	.0901
8	.1302	.1286	.1269	.1251	.1232	.1212	.1191	.1170	.1148	.1126
9	.1317	.1315	.1311	.1306	.1300	.1293	.1284	.1274	.1263	.1251
10	.1198	.1210	.1219	.1228	.1235	.1241	.1245	.1249	.1250	.1251
11	.0991	.1012	.1031	.1049	.1067	.1083	.1098	.1112	.1125	.1137
12	.0752	.0776	.0799	.0822	.0844	.0866	.0888	.0908	.0928	.0948
13	.0526	.0549	.0572	.0594	.0617	.0640	.0662	.0685	.0707	.0729
14	.0342	.0361	.0380	.0399	.0419	.0439	.0459	.0479	.0500	.0521
15	.0208	.0221	.0235	.0250	.0265	.0281	.0297	.0313	.0330	.0347
16	.0118	.0127	.0137	.0147	.0157	.0168	.0180	.0192	.0204	.0217
17	.0063	.0069	.0075	.0081	.0088	.0095	.0103	.0111	.0119	.0128
18	.0032	.0035	.0039	.0042	.0046	.0051	.0055	.0060	.0065	.0071
19	.0015	.0017	.0019	.0021	.0023	.0026	.0028	.0031	.0034	.0037
20	.0007	.0008	.0009	.0010	.0011	.0012	.0014	.0015	.0017	.0019
21	.0003	.0003	.0004	.0004	.0005	.0006	.0006	.0007	.0008	.0009
22	.0001	.0001	.0002	.0002	.0002	.0002	.0003	.0003	.0004	.0004
23	.0000	.0001	.0001	.0001	.0001	.0001	.0001	.0001	.0002	.0002
24	.0000	.0000	.0000	.0000	.0000	.0000	.0000	.0001	.0001	.0001

					μ					
x	11	12	13	14	15	16	17	18	19	20
0	.0000	.0000	.0000	.0000	.0000	.0000	.0000	.0000	.0000	.0000
1	.0002	.0001	.0000	.0000	.0000	.0000	.0000	.0000	.0000	.0000
2	.0010	.0004	.0002	.0001	.0000	.0000	.0000	.0000	.0000	.0000
3	.0037	.0018	.0008	.0004	.0002	.0001	.0000	.0000	.0000	.0000
4	.0102	.0053	.0027	.0013	.0006	.0003	.0001	.0001	.0000	.0000
5	.0224	.0127	.0070	.0037	.0019	.0010	.0005	.0002	.0001	.0001
6	.0411	.0255	.0152	.0087	.0048	.0026	.0014	.0007	.0004	.0002
7	.0646	.0437	.0281	.0174	.0104	.0060	.0034	.0018	.0010	.0005
8	.0888	.0655	.0457	.0304	.0194	.0120	.0072	.0042	.0024	.0013
9	.1085	.0874	.0661	.0473	.0324	.0213	.0135	.0083	.0050	.0029
10	.1194	.1048	.0859	.0663	.0486	.0341	.0230	.0150	.0095	.0058
11	.1194	.1144	.1015	.0844	.0663	.0496	.0355	.0245	.0164	.0106
12	.1094	.1144	.1099	.0984	.0829	.0661	.0504	.0368	.0259	.0176
13	.0926	.1056	.1099	.1060	.0956	.0814	.0658	.0509	.0378	.0271
14	.0728	.0905	.1021	.1060	.1024	.0930	.0800	.0655	.0541	.0387
15	.0534	.0724	.0885	.0989	.1024	.0992	.0906	.0786	.0650	.0516
16	.0367	.0543	.0719	.0866	.0960	.0992	.0963	.0884	.0772	.0646
17	.0237	.0383	.0550	.0713	.0847	.0934	.0963	.0936	.0863	.0760
18	.0145	.0256	.0397	.0554	.0706	.0830	.0909	.0936	.0911	.0844
19	.0084	.0161	.0272	.0409	.0557	.0699	.0814	.0887	.0911	.0888
20	.0046	.0097	.0177	.0286	.0418	.0559	.0692	.0798	.0866	.0888
21	.0024	.0055	.0109	.0191	.0299	.0426	.0560	.0684	.0783	.0846
22	.0012	.0030	.0065	.0121	.0204	.0310	.0433	.0560	.0676	.0769
23	.0006	.0016	.0037	.0074	.0133	.0216	.0320	.0438	.0559	.0669
24	.0003	.0008	.0020	.0043	.0083	.0144	.0226	.0328	.0442	.0557

POISSON DISTRIBUTION.

					μ					
x	11	12	13	14	15	16	17	18	19	20
25	.0001	.0004	.0010	.0024	.0050	.0092	.0154	.0237	.0336	.0446
26	.0000	.0002	.0005	.0013	.0029	.0057	.0101	.0164	.9246	.0343
27	.0000	.0001	.0002	.0007	.0016	.0034	.0063	.0109	.0173	.0254
28	.0000	.0000	.0001	.0003	.0009	.0019	.0038	.0070	.0117	.0181
29	.0000	.0000	.0001	.0002	.0004	.0011	.0023	.0044	.0077	.0125
30	.0000	.0000	.0000	.0001	.0002	.0006	.0013	.0026	.0049	.0083
31	.0000	.0000	.0000	.0000	.0001	.0003	.0007	.0015	.0030	.0054
32	.0000	.0000	.0000	.0000	.0001	.0001	.0004	.0009	.0018	.0034
33	.0000	.0000	.0000	.0000	.0000	.0001	.0002	.0005	.0010	.0020
34	.0000	.0000	.0000	.0000	.0000	.0000	.0001	.0002	.0006	.0012
35	.0000	.0000	.0000	.0000	.0000	.0000	.0000	.0001	.0003	.0007
36	.0000	.0000	.0000	.0000	.0000	.0000	.0000	.0001	.0002	.0004
37	.0000	.0000	.0000	.0000	.0000	.0000	.0000	.0000	.0001	.0002
38	.0000	.0000	.0000	.0000	.0000	.0000	.0000	.0000	.0000	.0001
39	.0000	.0000	.0000	.0000	.0000	.0000	.0000	.0000	.0000	.0001

APPENDIX G
ORDINATES OF THE NORMAL CURVE

The ordinate of the normal curve, Y, is the vertical distance from the X-axis to the normal curve. The table in this appendix gives this height for various points along the curve. The first column is the number of standard deviations, Z, you are away from the mean. The second column gives the ordinate for this point.

As seen from the table, the ordinate at the mean (or 0σ from the mean) of the normal curve is equal to 0.3989. On the other hand, the ordinate at the point 1.00σ away from the mean is equal to 0.2420.

It should be remembered that the normal curve is symmetrical. As a result, to find the ordinate at a negative value of Z, merely look up the positive value of Z. For example, the ordinate of -1.00σ is equal to 0.2420.

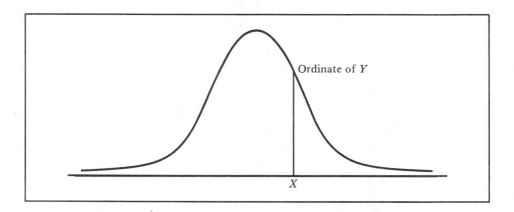

ORDINATES OF THE NORMAL CURVE.

Z	Second decimal place in Z									
	0.00	0.01	0.02	0.03	0.04	0.05	0.06	0.07	0.08	0.09
0.0	0.3989	0.3989	0.3989	0.3988	0.3986	0.3984	0.3982	0.3980	0.3977	0.3973
0.1	.3970	.3965	.3961	.3956	.3951	.3945	.3939	.3932	.3925	.3918
0.2	.3910	.3902	.3894	.3885	.3876	.3867	.3857	.3847	.3836	.3825
0.3	.3814	.3802	.3790	.3778	.3765	.3752	.3739	.3725	.3712	.3697
0.4	.3683	.3668	.3653	.3637	.3621	.3605	.3589	.3572	.3555	.3538
0.5	.3521	.3503	.3485	.3467	.3448	.3429	.3410	.3391	.3372	.3352
0.6	.3332	.3312	.3292	.3271	.3251	.3230	.3209	.3187	.3166	.3144
0.7	.3123	.3101	.3079	.3056	.3034	.3011	.2989	.2966	.2943	.2920
0.8	.2897	.2874	.2850	.2827	.2803	.2780	.2756	.2732	.2709	.2685
0.9	.2661	.2637	.2613	.2589	.2565	.2541	.2516	.2492	.2468	.2444
1.0	.2420	.2396	.2371	.2347	.2323	.2299	.2275	.2251	.2227	.2203
1.1	.2179	.2155	.2131	.2107	.2083	.2059	.2036	.2012	.1989	.1965
1.2	.1942	.1919	.1895	.1872	.1849	.1826	.1804	.1781	.1758	.1736
1.3	.1714	.1691	.1669	.1647	.1626	.1604	.1582	.1561	.1539	.1518
1.4	.1497	.1476	.1456	.1435	.1415	.1394	.1374	.1354	.1334	.1315
1.5	.1295	.1276	.1257	.1238	.1219	.1200	.1182	.1163	.1145	.1127
1.6	.1109	.1092	.1074	.1057	.1040	.1023	.1006	.0989	.0973	.0957
1.7	.0940	.0925	.0909	.0893	.0878	.0863	.0848	.0833	.0818	.0804
1.8	.0790	.0775	.0761	.0748	.0734	.0721	.0707	.0694	.0681	.0669
1.9	.0656	.0644	.0632	.0620	.0608	.0596	.0584	.0573	.0562	.0551
2.0	.0540	.0529	.0519	.0508	.0498	.0488	.0478	.0468	.0459	.0449
2.1	.0440	.0431	.0422	.0413	.0404	.0396	.0387	.0379	.0371	.0363
2.2	.0355	.0347	.0339	.0332	.0325	.0317	.0310	.0303	.0297	.0290
2.3	.0283	.0277	.0270	.0264	.0258	.0252	.0246	.0241	.0235	.0229
2.4	.0224	.0219	.0213	.0208	.0203	.0198	.0194	.0189	.0184	.0180
2.5	.0175	.0171	.0167	.0163	.0158	.0154	.0151	.0147	.0143	.0139
2.6	.0136	.0132	.0129	.0126	.0122	.0119	.0116	.0113	.0110	.0107
2.7	.0104	.0101	.0099	.0096	.0093	.0091	.0088	.0086	.0084	.0081
2.8	.0079	.0077	.0075	.0073	.0071	.0069	.0067	.0065	.0063	.0061
2.9	.0060	.0058	.0056	.0055	.0053	.0051	.0050	.0048	.0047	.0046

Z	First decimal place in Z									
	0.0	0.1	0.2	0.3	0.4	0.5	0.6	0.7	0.8	0.9
3	0.0044	0.0033	0.0024	0.0017	0.0012	0.0009	0.0006	0.0004	0.0003	0.0002
4	.0001	.0001	.0001	.0000	.0000	.0000	.0000	.0000	.0000	.0000

Source: Reprinted by permission from *Statistical Methods* by George W. Snedecor and William G. Cochran, sixth edition © 1967 by Iowa State University Press, Ames, Iowa.

APPENDIX H
AREAS UNDER THE NORMAL CURVE FROM 0 TO Z

The table in this appendix gives areas under the unit normal curve, this curve having a mean equal to 0 and a standard deviation equal to 1. It is constructed such that it will give the area between the mean of the normal curve and a point represented by Z, the number of standard deviations the point is away from the mean. For example, to find the area under the unit normal curve between the mean and a point 1.96σ to the right of the mean, the column on the left side of the table is moved down to the row having a Z of 1.9. Then this row is moved across to the 0.06 column, thus representing the point 1.96σ from the mean. As read from the intersection of the 1.9 row and the 0.06 column, the area between the mean of the unit normal curve and the point 1.96σ to the right of the mean is 0.4750.

Due to the fact that the area under the normal curve is equal to 1, and the normal curve is symmetrical, this table can be used to solve other types of problems than that outlined above. For example, due to the symmetry of the distribution, the area between the mean and $-Z\sigma$ is equal to the area between the mean and $+Z\sigma$. To illustrate, the area between the mean and $+1.96\sigma$ is equal to the area between the mean and -1.96σ, or 0.4750.

A second type of problem which can be solved is the determination of the area in the tail of the distribution. Since the area on either side of the mean is equal to .5000, the area to the right of 1.96σ is equal to $0.5000 - 0.4750$, or 0.0250.

Other problems involving the normal curve can be solved by similar methods.

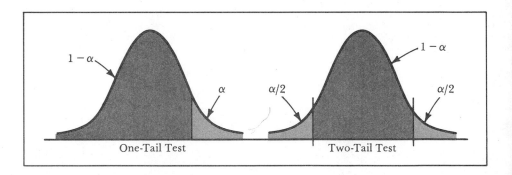

AREAS UNDER THE NORMAL CURVE FROM 0 TO Z.

Z	0.00	0.01	0.02	0.03	0.04	0.05	0.06	0.07	0.08	0.09
0.0	0.0000	0.0040	0.0080	0.0120	0.0160	0.0199	0.0239	0.0279	0.0319	0.0359
0.1	.0398	.0438	.0478	.0517	.0557	.0596	.0636	.0675	.0714	.0753
0.2	.0793	.0832	.0871	.0910	.0948	.0987	.1026	.1064	.1103	.1141
0.3	.1179	.1217	.1255	.1293	.1331	.1368	.1406	.1443	.1480	.1517
0.4	.1554	.1591	.1628	.1664	.1700	.1736	.1772	.1808	.1844	.1879
0.5	.1915	.1950	.1985	.2019	.2054	.2088	.2123	.2157	.2190	.2224
0.6	.2257	.2291	.2324	.2357	.2389	.2422	.2454	.2486	.2517	.2549
0.7	.2580	.2611	.2642	.2673	.2704	.2734	.2764	.2794	.2823	.2852
0.8	.2881	.2910	.2939	.2967	.2995	.3023	.3051	.3078	.3106	.3133
0.9	.3159	.3186	.3212	.3238	.3264	.3289	.3315	.3340	.3365	.3389
1.0	.3413	.3438	.3461	.3485	.3508	.3531	.3554	.3577	.3599	.3621
1.1	.3643	.3665	.3686	.3708	.3729	.3749	.3770	.3790	.3810	.3830
1.2	.3849	.3869	.3888	.3907	.3925	.3944	.3962	.3980	.3997	.4015
1.3	.4032	.4049	.4066	.4082	.4099	.4115	.4131	.4147	.4162	.4177
1.4	.4192	.4207	.4222	.4236	.4251	.4265	.4279	.4292	.4306	.4319
1.5	.4332	.4345	.4357	.4370	.4382	.4394	.4406	.4418	.4429	.4441
1.6	.4452	.4463	.4474	.4484	.4495	.4505	.4515	.4525	.4535	.4545
1.7	.4554	.4564	.4573	.4582	.4591	.4599	.4608	.4616	.4625	.4633
1.8	.4641	.4649	.4656	.4664	.4671	.4678	.4686	.4693	.4699	.4706
1.9	.4713	.4719	.4726	.4732	.4738	.4744	.4750	.4756	.4761	.4767
2.0	.4772	.4778	.4783	.4788	.4793	.4798	.4803	.4808	.4812	.4817
2.1	.4821	.4826	.4830	.4834	.4838	.4842	.4846	.4850	.4854	.4857
2.2	.4861	.4864	.4868	.4871	.4875	.4878	.4881	.4884	.4887	.4890
2.3	.4893	.4896	.4898	.4901	.4904	.4906	.4909	.4911	.4913	.4916
2.4	.4918	.4920	.4922	.4925	.4927	.4929	.4931	.4932	.4934	.4936
2.5	.4938	.4940	.4941	.4943	.4945	.4946	.4948	.4949	.4951	.4952
2.6	.4953	.4955	.4956	.4957	.4959	.4960	.4961	.4962	.4963	.4964
2.7	.4965	.4966	.4967	.4968	.4969	.4970	.4971	.4972	.4973	.4974
2.8	.4974	.4975	.4976	.4977	.4977	.4978	.4979	.4979	.4980	.4981
2.9	.4981	.4982	.4982	.4983	.4984	.4984	.4985	.4985	.4986	.4986
3.0	.4987	.4987	.4987	.4988	.4988	.4989	.4989	.4989	.4990	.4990
3.1	.4990	.4991	.4991	.4991	.4992	.4992	.4992	.4992	.4993	.4993
3.2	.4993	.4993	.4994	.4994	.4994	.4994	.4994	.4995	.4995	.4995
3.3	.4995	.4995	.4995	.4996	.4996	.4996	.4996	.4996	.4996	.4997
3.4	.4997	.4997	.4997	.4997	.4997	.4997	.4997	.4997	.4997	.4998
3.6	.4998	.4998	.4999	.4999	.4999	.4999	.4999	.4999	.4999	.4999
3.9	.5000									

Source: Reprinted by permission from *Statistical Methods* by George W. Snedecor and William G. Cochran, sixth edition © 1967 by Iowa State University Press, Ames, Iowa.

APPENDIX I
VALUES OF e^{-x}

The table in this appendix provides values for the expression e^{-x}. To use the table, locate x on the left side of the column, and then opposite the value of x is the value of e^{-x}. For instance, if x equals 0.40, then e^{-x} equals 0.67032.

Values of x between 1 and 10 not given in the table can be obtained in the following manner.

$$e^{-1.3} = (e^{-1})(e^{-0.3}) = (0.36788)(0.74082)$$

$$e^{-1.3} = 0.27253$$

VALUES OF e^{-x}.

x	e^{-x}	x	e^{-x}	x	e^{-x}	x	e^{-x}
.00	1.00000	.30	.74082	.60	.54881	.90	.40657
.01	.99005	.31	.73345	.61	.54335	.91	.40252
.02	.98020	.32	.72615	.62	.53794	.92	.39852
.03	.97045	.33	.71892	.63	.53259	.93	.39455
.04	.96079	.34	.71177	.64	.52729	.94	.39063
.05	.95123	.35	.70469	.65	.52205	.95	.38674
.06	.94176	.36	.69768	.66	.51865	.96	.38289
.07	.93239	.37	.69073	.67	.51171	.97	.37908
.08	.92312	.38	.68386	.68	.50662	.98	.37531
.09	.91393	.39	.67706	.69	.50158	.99	.37158
.10	.90484	.40	.67032	.70	.49659	1.00	.36788
.11	.89583	.41	.66365	.71	.49164		
.12	.88692	.42	.65705	.72	.48675	2.00	.13534
.13	.87810	.43	.65051	.73	.48191		
.14	.86936	.44	.64404	.74	.47711	3.00	.04979
.15	.86071	.45	.63763	.75	.47237		
.16	.85214	.46	.63128	.76	.46767	4.00	.01832
.17	.84366	.47	.62500	.77	.46301		
.18	.83527	.48	.61878	.78	.45841	5.00	.00674
.19	.82696	.49	.61263	.79	.45384		
.20	.81873	.50	.60653	.80	.44933	6.00	.00248
.21	.81058	.51	.60050	.81	.44486		
.22	.80252	.52	.59452	.82	.44043	7.00	.00091
.23	.79453	.53	.58860	.83	.43605		
.24	.78663	.54	.58275	.84	.43171	8.00	.00034
.25	.77880	.55	.57695	.85	.42741		
.26	.77105	.56	.57121	.86	.42316	9.00	.00012
.27	.76338	.57	.56553	.87	.41895		
.28	.75578	.58	.55990	.88	.41478	10.00	.00005
.29	.74826	.59	.55433	.89	.41066		

Source: *Study Guide/Workbook for Basic Statistics* by Donald W. Satterfield and Douglas E. Huffman, © 1977 by West Publishing Co., St. Paul, Minn. Used with permission of the publisher.

APPENDIX J
STUDENT *t* DISTRIBUTION

This appendix gives values for *t* of the Student's distribution, with the value being determined by the number of degrees of freedom and the level of confidence being used. The Student's distribution is used whenever the standard deviation of a sample is used as an estimator for the standard deviation of the population, and the sample size is less than 31. It should be remembered that in order to use this distribution the population from which the sample is taken must be assumed to be normally distributed.

To illustrate the use of the Student's distribution, suppose there was a sample of size 25 whose standard deviation was to be used as an estimator of the population's standard deviation. In addition, suppose that a confidence interval was being constructed at a 95 percent level of confidence. To determine the appropriate *t* value, the column for a two-tail 95 percent level of confidence situation is located, representing an α of 0.05. This column is then moved down until the row representing 24 degrees of freedom is found, since the number of degrees of freedom is equal to the sample size minus one. As read from the table, the appropriate *t* value for this situation is 2.064.

If this same sample was to be used in a one-tail hypothesis testing situation, the 95 percent one-tail column would instead be used. The resulting *t* value would be 1.711.

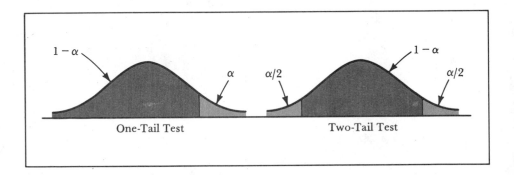

STUDENT t DISTRIBUTION.

	Level of Confidence $(1-\alpha)$													
	.10	.20	.30	.40	.50	.60	.70	.80	.90	.95	.98	.99	.999	Two-Tail Test
$n-1$ Degrees of Freedom	Level of Significance (α)													
	.45	.4	.35	.3	.25	.2	.15	.1	.05	.025	.01	.005	.0005	One-Tail Test
	.9	.8	.7	.6	.5	.4	.3	.2	.1	.05	.02	.01	.001	Two-Tail Test
1	.158	.325	.510	.727	1.000	1.376	1.963	3.078	6.314	12.706	31.821	63.657	636.619	
2	.142	.289	.445	.617	.816	1.061	1.386	1.886	2.910	4.303	6.965	9.925	31.598	
3	.137	.277	.424	.584	.765	.978	1.250	1.638	2.353	3.182	4.541	5.841	12.941	
4	.134	.271	.414	.569	.741	.941	1.190	1.533	2.132	2.776	3.747	4.604	8.610	
5	.132	.267	.408	.559	.727	.920	1.156	1.476	2.015	2.571	3.365	4.032	6.859	
6	.131	.265	.404	.553	.718	.906	1.134	1.440	1.943	2.447	3.143	3.707	5.959	
7	.130	.263	.402	.549	.711	.896	1.119	1.415	1.895	2.365	2.998	3.499	5.405	
8	.130	.262	.399	.546	.706	.889	1.108	1.397	1.860	2.306	2.896	3.355	5.041	
9	.129	.261	.398	.543	.703	.883	1.100	1.383	1.833	2.262	2.821	3.250	4.781	
10	.129	.260	.397	.542	.700	.879	1.093	1.372	1.812	2.228	2.764	3.169	4.587	
11	.129	.260	.396	.540	.697	.876	1.088	1.363	1.796	2.201	2.718	3.106	4.437	
12	.128	.259	.395	.539	.695	.873	1.083	1.356	1.782	2.179	2.681	3.055	4.318	
13	.128	.259	.394	.538	.694	.870	1.079	1.350	1.771	2.160	2.650	3.012	4.221	
14	.128	.258	.393	.537	.692	.868	1.076	1.345	1.761	2.145	2.624	2.977	4.140	
15	.128	.258	.393	.536	.691	.866	1.074	1.341	1.753	2.131	2.602	2.947	4.073	
16	.128	.258	.392	.535	.690	.865	1.071	1.337	1.746	2.120	2.583	2.921	4.015	
17	.128	.257	.392	.534	.689	.863	1.069	1.333	1.740	2.110	2.567	2.898	3.965	
18	.127	.257	.392	.534	.688	.862	1.067	1.330	1.734	2.101	2.552	2.878	3.922	
19	.127	.257	.391	.533	.688	.861	1.066	1.328	1.729	2.093	2.539	2.861	3.883	
20	.127	.257	.391	.533	.687	.860	1.064	1.325	1.725	2.086	2.528	2.845	3.850	
21	.127	.257	.391	.532	.686	.859	1.063	1.323	1.721	2.080	2.518	2.831	3.819	
22	.127	.256	.390	.532	.686	.858	1.061	1.321	1.717	2.074	2.508	2.819	3.792	
23	.127	.256	.390	.532	.685	.858	1.060	1.319	1.714	2.069	2.500	2.807	3.767	
24	.127	.256	.390	.531	.685	.857	1.059	1.318	1.711	2.064	2.492	2.797	3.745	
25	.127	.256	.390	.531	.684	.856	1.058	1.316	1.708	2.060	2.485	2.788	3.725	
26	.127	.256	.390	.531	.684	.856	1.058	1.315	1.706	2.056	2.479	2.779	3.707	
27	.127	.256	.389	.531	.684	.855	1.057	1.314	1.703	2.052	2.473	2.771	3.690	
28	.127	.256	.389	.530	.683	.855	1.056	1.313	1.701	2.048	2.467	2.763	3.674	
29	.127	.256	.389	.530	.683	.854	1.055	1.311	1.699	2.045	2.462	2.756	3.659	
30	.127	.256	.389	.530	.683	.854	1.055	1.310	1.697	2.042	2.457	2.750	3.646	
40	.126	.255	.388	.529	.681	.851	1.050	1.303	1.684	2.021	2.423	2.704	3.551	
60	.126	.254	.387	.527	.679	.848	1.046	1.296	1.671	2.000	2.390	2.660	3.460	
120	.126	.254	.386	.526	.677	.845	1.041	1.289	1.658	1.980	2.358	2.617	3.373	
∞	.126	.253	.385	.524	.674	.842	1.036	1.282	1.645	1.960	2.326	2.576	3.291	

Source: From *Statistics for Business Decision Making* by I. Robert Parket. Reprinted with permission of Random House, Inc. Numerical contents originally from Table III of Fisher and Yates: *Statistical Tables for Biological, Agricultural and Medical Research,* Sixth Edition, 1974, published by Longman Group, Ltd., London, and reprinted by permission of the publishers.

APPENDIX K
F DISTRIBUTION

The table in this appendix gives the values of F for various degrees of freedom and α's of 0.01 and 0.05. The F values for an α of 0.05 are given in light face type, while the F values for an α of 0.01 are given in the bold face type.

To illustrate the use of this table, suppose three samples of size 8 each were chosen, and an α value 0.05 is to be used. The column corresponding to the number of degrees of freedom in the numerator is located. For the numerator the number of degrees of freedom is equal to the number of samples minus one. In this problem the number of degrees of freedom in the numerator is $3 - 1 = 2$. After the correct column is located, it is moved down until the row is reached which represents the number of degrees of freedom in the denominator. For the denominator the number of degrees of freedom is equal to the combined sample size minus the number of samples. In this problem the number of degrees of freedom in the denominator is $24 - 3 = 21$. Reading from the table, with an α of 0.05 the appropriate F value is 3.47. If an α of 0.01 was to be used the appropriate F value would be 5.78.

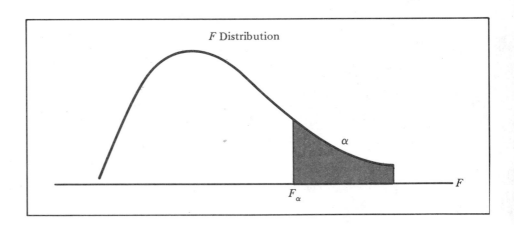

F DISTRIBUTION.

5% (Roman Type) and 1% (Bold Face Type) Points for the Distribution of F

f_1, Degrees of Freedom (for greater mean square)

f_2	1	2	3	4	5	6	7	8	9	10	11	12	14	16	20	24	30	40	50	75	100	200	500	∞
1	161 **4,052**	200 **4,999**	216 **5,403**	225 **5,625**	230 **5,764**	234 **5,859**	237 **5,928**	239 **5,981**	241 **6,022**	242 **6,056**	243 **6,082**	244 **6,106**	245 **6,142**	246 **6,169**	248 **6,208**	249 **6,234**	250 **6,261**	251 **6,286**	252 **6,302**	253 **6,323**	253 **6,334**	254 **6,352**	254 **6,361**	254 **6,366**
2	18.51 **98.49**	19.00 **99.00**	19.16 **99.17**	19.25 **99.25**	19.30 **99.30**	19.33 **99.33**	19.36 **99.36**	19.37 **99.37**	19.38 **99.39**	19.39 **99.40**	19.40 **99.41**	19.41 **99.42**	19.42 **99.43**	19.43 **99.44**	19.44 **99.45**	19.45 **99.46**	19.46 **99.47**	19.47 **99.48**	19.47 **99.48**	19.48 **99.49**	19.49 **99.49**	19.49 **99.49**	19.50 **99.50**	19.50 **99.50**
3	10.13 **34.12**	9.55 **30.82**	9.28 **29.46**	9.12 **28.71**	9.01 **28.24**	8.94 **27.91**	8.88 **27.67**	8.84 **27.49**	8.81 **27.34**	8.78 **27.23**	8.76 **27.13**	8.74 **27.05**	8.71 **26.92**	8.69 **26.83**	8.66 **26.69**	8.64 **26.60**	8.62 **26.50**	8.60 **26.41**	8.58 **26.35**	8.57 **26.27**	8.56 **26.23**	8.54 **26.18**	8.54 **26.14**	8.53 **26.12**
4	7.71 **21.20**	6.94 **18.00**	6.59 **16.69**	6.39 **15.98**	6.26 **15.52**	6.16 **15.21**	6.09 **14.98**	6.04 **14.80**	6.00 **14.66**	5.96 **14.54**	5.93 **14.45**	5.91 **14.37**	5.87 **14.24**	5.84 **14.15**	5.80 **14.02**	5.77 **13.93**	5.74 **13.83**	5.71 **13.74**	5.70 **13.69**	5.68 **13.61**	5.66 **13.57**	5.65 **13.52**	5.64 **13.48**	5.63 **13.46**
5	6.61 **16.26**	5.79 **13.27**	5.41 **12.06**	5.19 **11.39**	5.05 **10.97**	4.95 **10.67**	4.88 **10.45**	4.82 **10.29**	4.78 **10.15**	4.74 **10.05**	4.70 **9.96**	4.68 **9.89**	4.64 **9.77**	4.60 **9.68**	4.56 **9.55**	4.53 **9.47**	4.50 **9.38**	4.46 **9.29**	4.44 **9.24**	4.42 **9.17**	4.40 **9.13**	4.38 **9.07**	4.37 **9.04**	4.36 **9.02**
6	5.99 **13.74**	5.14 **10.92**	4.76 **9.78**	4.53 **9.15**	4.39 **8.75**	4.28 **8.47**	4.21 **8.26**	4.15 **8.10**	4.10 **7.98**	4.06 **7.87**	4.03 **7.79**	4.00 **7.72**	3.96 **7.60**	3.92 **7.52**	3.87 **7.39**	3.84 **7.31**	3.81 **7.23**	3.77 **7.14**	3.75 **7.09**	3.72 **7.02**	3.71 **6.99**	3.69 **6.94**	3.68 **6.90**	3.67 **6.88**
7	5.59 **12.25**	4.74 **9.55**	4.35 **8.45**	4.12 **7.85**	3.97 **7.46**	3.87 **7.19**	3.79 **7.00**	3.73 **6.84**	3.68 **6.71**	3.63 **6.62**	3.60 **6.54**	3.57 **6.47**	3.52 **6.35**	3.49 **6.27**	3.44 **6.15**	3.41 **6.07**	3.38 **5.98**	3.34 **5.90**	3.32 **5.85**	3.29 **5.78**	3.28 **5.75**	3.25 **5.70**	3.24 **5.67**	3.23 **5.65**
8	5.32 **11.26**	4.46 **8.65**	4.07 **7.59**	3.84 **7.01**	3.69 **6.63**	3.58 **6.37**	3.50 **6.19**	3.44 **6.03**	3.39 **5.91**	3.34 **5.82**	3.31 **5.74**	3.28 **5.67**	3.23 **5.56**	3.20 **5.48**	3.15 **5.36**	3.12 **5.28**	3.08 **5.20**	3.05 **5.11**	3.03 **5.06**	3.00 **5.00**	2.98 **4.96**	2.96 **4.91**	2.94 **4.88**	2.93 **4.86**
9	5.12 **10.56**	4.26 **8.02**	3.86 **6.99**	3.63 **6.42**	3.48 **6.06**	3.37 **5.80**	3.29 **5.62**	3.23 **5.47**	3.18 **5.35**	3.13 **5.26**	3.10 **5.18**	3.07 **5.11**	3.02 **5.00**	2.98 **4.92**	2.93 **4.80**	2.90 **4.73**	2.86 **4.64**	2.82 **4.56**	2.80 **4.51**	2.77 **4.45**	2.76 **4.41**	2.73 **4.36**	2.72 **4.33**	2.71 **4.31**
10	4.96 **10.04**	4.10 **7.56**	3.71 **6.55**	3.48 **5.99**	3.33 **5.64**	3.22 **5.39**	3.14 **5.21**	3.07 **5.06**	3.02 **4.95**	2.97 **4.85**	2.94 **4.78**	2.91 **4.71**	2.86 **4.60**	2.82 **4.52**	2.77 **4.41**	2.74 **4.33**	2.70 **4.25**	2.67 **4.17**	2.64 **4.12**	2.61 **4.05**	2.59 **4.01**	2.56 **3.96**	2.55 **3.93**	2.54 **3.91**
11	4.84 **9.65**	3.98 **7.20**	3.59 **6.22**	3.36 **5.67**	3.20 **5.32**	3.09 **5.07**	3.01 **4.88**	2.95 **4.74**	2.90 **4.63**	2.86 **4.54**	2.82 **4.46**	2.79 **4.40**	2.74 **4.29**	2.70 **4.21**	2.65 **4.10**	2.61 **4.02**	2.57 **3.94**	2.53 **3.86**	2.50 **3.80**	2.47 **3.74**	2.45 **3.70**	2.42 **3.66**	2.41 **3.62**	2.40 **3.60**
12	4.75 **9.33**	3.88 **6.93**	3.49 **5.95**	3.26 **5.41**	3.11 **5.06**	3.00 **4.82**	2.92 **4.65**	2.85 **4.50**	2.80 **4.39**	2.76 **4.30**	2.72 **4.22**	2.69 **4.16**	2.64 **4.05**	2.60 **3.98**	2.54 **3.86**	2.50 **3.78**	2.46 **3.70**	2.42 **3.61**	2.40 **3.56**	2.36 **3.49**	2.35 **3.46**	2.32 **3.41**	2.31 **3.38**	2.30 **3.36**
13	4.67 **9.07**	3.80 **6.70**	3.41 **5.74**	3.18 **5.20**	3.02 **4.86**	2.92 **4.62**	2.84 **4.44**	2.77 **4.30**	2.72 **4.19**	2.67 **4.10**	2.63 **4.02**	2.60 **3.96**	2.55 **3.85**	2.51 **3.78**	2.46 **3.67**	2.42 **3.59**	2.38 **3.51**	2.34 **3.42**	2.32 **3.37**	2.28 **3.30**	2.26 **3.27**	2.24 **3.21**	2.22 **3.18**	2.21 **3.16**

Source: Reprinted by permission from *Statistical Methods* by George W. Snedecor and William G. Cochran, sixth edition © 1967 by Iowa State University Press, Ames, Iowa.

F Distribution.

f_1, Degrees of Freedom (for greater mean square)

f_2	1	2	3	4	5	6	7	8	9	10	11	12	14	16	20	24	30	40	50	75	100	200	500	∞
14	4.60 **8.86**	3.74 **6.51**	3.34 **5.56**	3.11 **5.03**	2.96 **4.69**	2.85 **4.46**	2.77 **4.28**	2.70 **4.14**	2.65 **4.03**	2.60 **3.94**	2.56 **3.86**	2.53 **3.80**	2.48 **3.70**	2.44 **3.62**	2.39 **3.51**	2.35 **3.43**	2.31 **3.34**	2.27 **3.26**	2.24 **3.21**	2.21 **3.14**	2.19 **3.11**	2.16 **3.06**	2.14 **3.02**	2.13 **3.00**
15	4.54 **8.68**	3.68 **6.36**	3.29 **5.42**	3.06 **4.89**	2.90 **4.56**	2.79 **4.32**	2.70 **4.14**	2.64 **4.00**	2.59 **3.89**	2.55 **3.80**	2.51 **3.73**	2.48 **3.67**	2.43 **3.56**	2.39 **3.48**	2.33 **3.36**	2.29 **3.29**	2.25 **3.20**	2.21 **3.12**	2.18 **3.07**	2.15 **3.00**	2.12 **2.97**	2.10 **2.92**	2.08 **2.89**	2.07 **2.87**
16	4.49 **8.53**	3.63 **6.23**	3.24 **5.29**	3.01 **4.77**	2.85 **4.44**	2.74 **4.20**	2.66 **4.03**	2.59 **3.89**	2.54 **3.78**	2.49 **3.69**	2.45 **3.61**	2.42 **3.55**	2.37 **3.45**	2.33 **3.37**	2.28 **3.25**	2.24 **3.18**	2.20 **3.10**	2.16 **3.01**	2.13 **2.96**	2.09 **2.98**	2.07 **2.86**	2.04 **2.80**	2.02 **2.77**	2.01 **2.75**
17	4.45 **8.40**	3.59 **6.11**	3.20 **5.18**	2.96 **4.67**	2.81 **4.34**	2.70 **4.10**	2.62 **3.93**	2.55 **3.79**	2.50 **3.68**	2.45 **3.59**	2.41 **3.52**	2.38 **3.45**	2.33 **3.35**	2.29 **3.27**	2.23 **3.16**	2.19 **3.08**	2.15 **3.00**	2.11 **2.92**	2.08 **2.86**	2.04 **2.79**	2.02 **2.76**	1.99 **2.70**	1.97 **2.67**	1.96 **2.65**
18	4.41 **8.28**	3.55 **6.01**	3.16 **5.09**	2.93 **4.58**	2.77 **4.25**	2.66 **4.01**	2.58 **3.85**	2.51 **3.71**	2.46 **3.60**	2.41 **3.51**	2.37 **3.44**	2.34 **3.37**	2.29 **3.27**	2.25 **3.19**	2.19 **3.07**	2.15 **3.00**	2.11 **2.91**	2.07 **2.83**	2.04 **2.78**	2.00 **2.71**	1.98 **2.68**	1.95 **2.62**	1.93 **2.59**	1.92 **2.57**
19	4.38 **8.18**	3.52 **5.93**	3.13 **5.01**	2.90 **4.50**	2.74 **4.17**	2.63 **3.94**	2.55 **3.77**	2.48 **3.63**	2.43 **3.52**	2.38 **3.43**	2.34 **3.36**	2.31 **3.30**	2.26 **3.19**	2.21 **3.12**	2.15 **3.00**	2.11 **2.92**	2.07 **2.84**	2.02 **2.76**	2.00 **2.70**	1.96 **2.63**	1.94 **2.60**	1.91 **2.54**	1.90 **2.51**	1.88 **2.49**
20	4.35 **8.10**	3.49 **5.85**	3.10 **4.94**	2.87 **4.43**	2.71 **4.10**	2.60 **3.87**	2.52 **3.71**	2.45 **3.56**	2.40 **3.45**	2.35 **3.37**	2.31 **3.30**	2.28 **3.23**	2.23 **3.13**	2.18 **3.05**	2.12 **2.94**	2.08 **2.86**	2.04 **2.77**	1.99 **2.69**	1.96 **2.63**	1.92 **2.56**	1.90 **2.53**	1.87 **2.47**	1.85 **2.44**	1.84 **2.42**
21	4.32 **8.02**	3.47 **5.78**	3.07 **4.87**	2.84 **4.37**	2.68 **4.04**	2.57 **3.81**	2.49 **3.65**	2.42 **3.51**	2.37 **3.40**	2.32 **3.31**	2.28 **3.24**	2.25 **3.17**	2.20 **3.07**	2.15 **2.99**	2.09 **2.88**	2.05 **2.80**	2.00 **2.72**	1.96 **2.63**	1.93 **2.58**	1.89 **2.51**	1.87 **2.47**	1.84 **2.42**	1.82 **2.38**	1.81 **2.36**
22	4.30 **7.94**	3.44 **5.72**	3.05 **4.82**	2.82 **4.31**	2.66 **3.99**	2.55 **3.76**	2.47 **3.59**	2.40 **3.45**	2.35 **3.35**	2.30 **3.26**	2.26 **3.18**	2.23 **3.12**	2.18 **3.02**	2.13 **2.94**	2.07 **2.83**	2.03 **2.75**	1.98 **2.67**	1.93 **2.58**	1.91 **2.53**	1.87 **2.46**	1.84 **2.42**	1.81 **2.37**	1.80 **2.33**	1.78 **2.31**
23	4.28 **7.88**	3.42 **5.66**	3.03 **4.76**	2.80 **4.26**	2.64 **3.94**	2.53 **3.71**	2.45 **3.54**	2.38 **3.41**	2.32 **3.30**	2.28 **3.21**	2.24 **3.14**	2.20 **3.07**	2.14 **2.97**	2.10 **2.89**	2.04 **2.78**	2.00 **2.70**	1.96 **2.62**	1.91 **2.53**	1.88 **2.48**	1.84 **2.41**	1.82 **2.37**	1.79 **2.32**	1.77 **2.28**	1.76 **2.26**
24	4.26 **7.82**	3.40 **5.61**	3.01 **4.72**	2.78 **4.22**	2.62 **3.90**	2.51 **3.67**	2.43 **3.50**	2.36 **3.36**	2.30 **3.25**	2.26 **3.17**	2.22 **3.09**	2.18 **3.03**	2.13 **2.93**	2.09 **2.85**	2.02 **2.74**	1.98 **2.66**	1.94 **2.58**	1.89 **2.49**	1.86 **2.44**	1.82 **2.36**	1.80 **2.33**	1.76 **2.27**	1.74 **2.23**	1.73 **2.21**
25	4.24 **7.77**	3.38 **5.57**	2.99 **4.68**	2.76 **4.18**	2.60 **3.86**	2.49 **3.63**	2.41 **3.46**	2.34 **3.32**	2.28 **3.21**	2.24 **3.13**	2.20 **3.05**	2.16 **2.99**	2.11 **2.89**	2.06 **2.81**	2.00 **2.70**	1.96 **2.62**	1.92 **2.54**	1.87 **2.45**	1.84 **2.40**	1.80 **2.32**	1.77 **2.29**	1.74 **2.23**	1.72 **2.19**	1.71 **2.17**
26	4.22 **7.72**	3.37 **5.53**	2.98 **4.64**	2.74 **4.14**	2.59 **3.82**	2.47 **3.59**	2.39 **3.42**	2.32 **3.29**	2.27 **3.17**	2.22 **3.09**	2.18 **3.02**	2.15 **2.96**	2.10 **2.86**	2.05 **2.77**	1.99 **2.66**	1.95 **2.58**	1.90 **2.50**	1.85 **2.41**	1.82 **2.36**	1.78 **2.28**	1.76 **2.25**	1.72 **2.19**	1.70 **2.15**	1.69 **2.13**

The function, $F = e$ with exponent $2z$, is computed in part from Fisher's table VI (7). Additional entries are by interpolation, mostly graphical.

F DISTRIBUTION.

f_1, Degrees of Freedom (for greater mean square)

f_2	1	2	3	4	5	6	7	8	9	10	11	12	14	16	20	24	30	40	50	75	100	200	500	∞
27	4.21	3.35	2.96	2.73	2.57	2.46	2.37	2.30	2.25	2.20	2.16	2.13	2.08	2.03	1.97	1.93	1.88	1.84	1.80	1.76	1.74	1.71	1.68	1.67
	7.68	5.49	4.60	4.11	3.79	3.56	3.39	3.26	3.14	3.06	2.98	2.93	2.83	2.74	2.63	2.55	2.47	2.38	2.33	2.25	2.21	2.16	2.12	2.10
28	4.20	3.34	2.95	2.71	2.56	2.44	2.36	2.29	2.24	2.19	2.15	2.12	2.06	2.02	1.96	1.91	1.87	1.81	1.78	1.75	1.72	1.69	1.67	1.65
	7.64	5.45	4.57	4.07	3.76	3.53	3.36	3.23	3.11	3.03	2.95	2.90	2.80	2.71	2.60	2.52	2.44	2.35	2.30	2.22	2.18	2.13	2.09	2.06
29	4.18	3.33	2.93	2.70	2.54	2.43	2.35	2.28	2.22	2.18	2.14	2.10	2.05	2.00	1.94	1.90	1.85	1.80	1.77	1.73	1.71	1.68	1.65	1.64
	7.60	5.42	4.54	4.04	3.73	3.50	3.33	3.20	3.08	3.00	2.92	2.87	2.77	2.68	2.57	2.49	2.41	2.32	2.27	2.19	2.15	2.10	2.06	2.03
30	4.17	3.32	2.92	2.69	2.53	2.42	2.34	2.27	2.21	2.16	2.12	2.09	2.04	1.99	1.93	1.89	1.84	1.79	1.76	1.72	1.69	1.66	1.64	1.62
	7.56	5.39	4.51	4.02	3.70	3.47	3.30	3.17	3.06	2.98	2.90	2.84	2.74	2.66	2.55	2.47	2.38	2.29	2.24	2.16	2.13	2.07	2.03	2.01
32	4.15	3.30	2.90	2.67	2.51	2.40	2.32	2.25	2.19	2.14	2.10	2.07	2.02	1.97	1.91	1.86	1.82	1.76	1.74	1.69	1.67	1.64	1.61	1.59
	7.50	5.34	4.46	3.97	3.66	3.42	3.25	3.12	3.01	2.94	2.86	2.80	2.70	2.62	2.51	2.42	2.34	2.25	2.20	2.12	2.08	2.02	1.98	1.96
34	4.13	3.28	2.88	2.65	2.49	2.38	2.30	2.23	2.17	2.12	2.08	2.05	2.00	1.95	1.89	1.84	1.80	1.74	1.71	1.67	1.64	1.61	1.59	1.57
	7.44	5.29	4.42	3.93	3.61	3.38	3.21	3.08	2.97	2.89	2.82	2.76	2.66	2.58	2.47	2.38	2.30	2.21	2.15	2.08	2.04	1.98	1.94	1.91
36	4.11	3.26	2.86	2.63	2.48	2.36	2.28	2.21	2.15	2.10	2.06	2.03	1.98	1.93	1.87	1.82	1.78	1.72	1.69	1.65	1.62	1.59	1.56	1.55
	7.39	5.25	4.38	3.89	3.58	3.35	3.18	3.04	2.94	2.86	2.78	2.72	2.62	2.54	2.43	2.35	2.26	2.17	2.12	2.04	2.00	1.94	1.90	1.87
38	4.10	3.25	2.85	2.62	2.46	2.35	2.26	2.19	2.14	2.09	2.05	2.02	1.96	1.92	1.85	1.80	1.76	1.71	1.67	1.63	1.60	1.57	1.54	1.53
	7.35	5.21	4.34	3.86	3.54	3.32	3.15	3.02	2.91	2.82	2.75	2.69	2.59	2.51	2.40	2.32	2.22	2.14	2.08	2.00	1.97	1.90	1.86	1.84
40	4.08	3.23	2.84	2.61	2.45	2.34	2.25	2.18	2.12	2.07	2.04	2.00	1.95	1.90	1.84	1.79	1.74	1.69	1.66	1.61	1.59	1.55	1.53	1.51
	7.31	5.18	4.31	3.83	3.51	3.29	3.12	2.99	2.88	2.80	2.73	2.66	2.56	2.49	2.37	2.29	2.20	2.11	2.05	1.97	1.94	1.88	1.84	1.81
42	4.07	3.22	2.83	2.59	2.44	2.32	2.24	2.17	2.11	2.06	2.02	1.99	1.94	1.89	1.82	1.78	1.73	1.68	1.64	1.60	1.57	1.54	1.51	1.49
	7.27	5.15	4.29	3.80	3.49	3.26	3.10	2.96	2.86	2.77	2.70	2.64	2.54	2.46	2.35	2.26	2.17	2.08	2.02	1.94	1.91	1.85	1.80	1.78
44	4.06	3.21	2.82	2.58	2.43	2.31	2.23	2.16	2.10	2.05	2.01	1.98	1.92	1.88	1.81	1.76	1.72	1.66	1.63	1.58	1.56	1.52	1.50	1.48
	7.24	5.12	4.26	3.78	3.46	3.24	3.07	2.94	2.84	2.75	2.68	2.62	2.52	2.44	2.32	2.24	2.15	2.06	2.00	1.92	1.88	1.82	1.78	1.75
46	4.05	3.20	2.81	2.57	2.42	2.30	2.22	2.14	2.09	2.04	2.00	1.97	1.91	1.87	1.80	1.75	1.71	1.65	1.62	1.57	1.54	1.51	1.48	1.46
	7.21	5.10	4.24	3.76	3.44	3.22	3.05	2.92	2.82	2.73	2.66	2.60	2.50	2.42	2.30	2.22	2.13	2.04	1.98	1.90	1.86	1.80	1.76	1.72
48	4.04	3.19	2.80	2.56	2.41	2.30	2.21	2.14	2.08	2.03	1.99	1.96	1.90	1.86	1.79	1.74	1.70	1.64	1.61	1.56	1.53	1.50	1.47	1.45
	7.19	5.08	4.22	3.74	3.42	3.20	3.04	2.90	2.80	2.71	2.64	2.58	2.48	2.40	2.28	2.20	2.11	2.02	1.96	1.88	1.84	1.78	1.73	1.70

F DISTRIBUTION.

f_1 Degrees of Freedom (for greater mean square)

f_2	1	2	3	4	5	6	7	8	9	10	11	12	14	16	20	24	30	40	50	75	100	200	500	∞	f_2
50	4.03 / 7.17	3.18 / 5.06	2.79 / 4.20	2.56 / 3.72	2.40 / 3.41	2.29 / 3.18	2.20 / 3.02	2.13 / 2.88	2.07 / 2.78	2.02 / 2.70	1.98 / 2.62	1.95 / 2.56	1.90 / 2.46	1.85 / 2.39	1.78 / 2.26	1.74 / 2.18	1.69 / 2.10	1.63 / 2.00	1.60 / 1.94	1.55 / 1.86	1.52 / 1.82	1.48 / 1.76	1.46 / 1.71	1.44 / 1.68	50
55	4.02 / 7.12	3.17 / 5.01	2.78 / 4.16	2.54 / 3.68	2.38 / 3.37	2.27 / 3.15	2.18 / 2.98	2.11 / 2.85	2.05 / 2.75	2.00 / 2.66	1.97 / 2.59	1.93 / 2.53	1.88 / 2.43	1.83 / 2.35	1.76 / 2.23	1.72 / 2.15	1.67 / 2.06	1.61 / 1.96	1.58 / 1.90	1.52 / 1.82	1.50 / 1.78	1.46 / 1.71	1.43 / 1.66	1.41 / 1.64	55
60	4.00 / 7.08	3.15 / 4.98	2.76 / 4.13	2.52 / 3.65	2.37 / 3.34	2.25 / 3.12	2.17 / 2.95	2.10 / 2.82	2.04 / 2.72	1.99 / 2.63	1.95 / 2.56	1.92 / 2.50	1.86 / 2.40	1.81 / 2.32	1.75 / 2.20	1.70 / 2.12	1.65 / 2.03	1.59 / 1.93	1.56 / 1.87	1.50 / 1.79	1.48 / 1.74	1.44 / 1.68	1.41 / 1.63	1.39 / 1.60	60
65	3.99 / 7.04	3.14 / 4.95	2.75 / 4.10	2.51 / 3.62	2.36 / 3.31	2.24 / 3.09	2.15 / 2.93	2.08 / 2.79	2.02 / 2.70	1.98 / 2.61	1.94 / 2.54	1.90 / 2.47	1.85 / 2.37	1.80 / 2.30	1.73 / 2.18	1.68 / 2.09	1.63 / 2.00	1.57 / 1.90	1.54 / 1.84	1.49 / 1.76	1.46 / 1.71	1.42 / 1.64	1.39 / 1.60	1.37 / 1.56	65
70	3.98 / 7.01	3.13 / 4.92	2.74 / 4.08	2.50 / 3.60	2.35 / 3.29	2.23 / 3.07	2.14 / 2.91	2.07 / 2.77	2.01 / 2.67	1.97 / 2.59	1.93 / 2.51	1.89 / 2.45	1.84 / 2.35	1.79 / 2.28	1.72 / 2.15	1.67 / 2.07	1.62 / 1.98	1.56 / 1.88	1.53 / 1.82	1.47 / 1.74	1.45 / 1.69	1.40 / 1.62	1.37 / 1.56	1.35 / 1.53	70
80	3.96 / 6.96	3.11 / 4.88	2.72 / 4.04	2.48 / 3.56	2.33 / 3.25	2.21 / 3.04	2.12 / 2.87	2.05 / 2.74	1.99 / 2.64	1.95 / 2.55	1.91 / 2.48	1.88 / 2.41	1.82 / 2.32	1.77 / 2.24	1.70 / 2.11	1.65 / 2.03	1.60 / 1.94	1.54 / 1.84	1.51 / 1.78	1.45 / 1.70	1.42 / 1.65	1.38 / 1.57	1.35 / 1.52	1.32 / 1.49	80
100	3.94 / 6.90	3.09 / 4.82	2.70 / 3.98	2.46 / 3.51	2.30 / 3.20	2.19 / 2.99	2.10 / 2.82	2.03 / 2.69	1.97 / 2.59	1.92 / 2.51	1.88 / 2.43	1.85 / 2.36	1.79 / 2.26	1.75 / 2.19	1.68 / 2.06	1.63 / 1.98	1.57 / 1.89	1.51 / 1.79	1.48 / 1.73	1.42 / 1.64	1.39 / 1.59	1.34 / 1.51	1.30 / 1.46	1.28 / 1.43	100
125	3.92 / 6.84	3.07 / 4.78	2.68 / 3.94	2.44 / 3.47	2.29 / 3.17	2.17 / 2.95	2.08 / 2.79	2.01 / 2.65	1.95 / 2.56	1.90 / 2.47	1.86 / 2.40	1.83 / 2.33	1.77 / 2.23	1.72 / 2.15	1.65 / 2.03	1.60 / 1.94	1.55 / 1.85	1.49 / 1.75	1.45 / 1.68	1.39 / 1.59	1.36 / 1.54	1.31 / 1.46	1.27 / 1.40	1.25 / 1.37	125
150	3.91 / 6.81	3.06 / 4.75	2.67 / 3.91	2.43 / 3.44	2.27 / 3.14	2.16 / 2.92	2.07 / 2.76	2.00 / 2.62	1.94 / 2.53	1.89 / 2.44	1.85 / 2.37	1.82 / 2.30	1.76 / 2.20	1.71 / 2.12	1.64 / 2.00	1.59 / 1.91	1.54 / 1.83	1.47 / 1.72	1.44 / 1.66	1.37 / 1.56	1.34 / 1.51	1.29 / 1.43	1.25 / 1.37	1.22 / 1.33	150
200	3.89 / 6.76	3.04 / 4.71	2.65 / 3.88	2.41 / 3.41	2.26 / 3.11	2.14 / 2.90	2.05 / 2.73	1.98 / 2.60	1.92 / 2.50	1.87 / 2.41	1.83 / 2.34	1.80 / 2.28	1.74 / 2.17	1.69 / 2.09	1.62 / 1.97	1.57 / 1.88	1.52 / 1.79	1.45 / 1.69	1.42 / 1.62	1.35 / 1.53	1.32 / 1.48	1.26 / 1.39	1.22 / 1.33	1.19 / 1.28	200
400	3.86 / 6.70	3.02 / 4.66	2.62 / 3.83	2.39 / 3.36	2.23 / 3.06	2.12 / 2.85	2.03 / 2.69	1.96 / 2.55	1.90 / 2.46	1.85 / 2.37	1.81 / 2.29	1.78 / 2.23	1.72 / 2.12	1.67 / 2.04	1.60 / 1.92	1.54 / 1.84	1.49 / 1.74	1.42 / 1.64	1.38 / 1.57	1.32 / 1.47	1.28 / 1.42	1.22 / 1.32	1.16 / 1.24	1.13 / 1.19	400
1000	3.85 / 6.66	3.00 / 4.62	2.61 / 3.80	2.38 / 3.34	2.22 / 3.04	2.10 / 2.82	2.02 / 2.66	1.95 / 2.53	1.89 / 2.43	1.84 / 2.34	1.80 / 2.26	1.76 / 2.20	1.70 / 2.09	1.65 / 2.01	1.58 / 1.89	1.53 / 1.81	1.46· / 1.69	1.41 / 1.61	1.36 / 1.54	1.30 / 1.44	1.26 / 1.38	1.19 / 1.28	1.13 / 1.19	1.08 / 1.11	1000
∞	3.84 / 6.63	2.99 / 4.60	2.60 / 3.78	2.37 / 3.32	2.21 / 3.02	2.09 / 2.80	2.01 / 2.64	1.94 / 2.51	1.88 / 2.41	1.83 / 2.32	1.79 / 2.24	1.75 / 2.18	1.69 / 2.07	1.64 / 1.99	1.57 / 1.87	1.52 / 1.79	1.46· / 1.69	1.40 / 1.59	1.35 / 1.52	1.28 / 1.41	1.24 / 1.36	1.17 / 1.25	1.11 / 1.15	1.00 / 1.00	∞

APPENDIX L
CHI SQUARE DISTRIBUTION

The table in this appendix gives values for χ^2 for various degrees of freedom and levels of confidence. The values down the left side of the table represent the number of degrees of freedom in the problem. The values across the top of this table represent the area in the tail of the distribution.

To illustrate the use of this table, suppose we have an experiment in which the observations are classified into a table of 4 rows and 3 columns. In addition, a 95% level of confidence is to be used in the problem. The number of degrees of freedom in this type of problem is equal to $(r-1)(c-1)$, where r is the number of rows in the table of observations, and c is the number of columns in the table of observations. To find the appropriate χ^2 value, the column headed by 0.05 is located since the level of confidence is 0.95. This column is then moved down until the row representing the correct number of degrees of freedom is located, $(4-1)(3-1) = 6$ in this case. Reading from the intersection of this column and row, the appropriate χ^2 value is seen to be 12.5916.

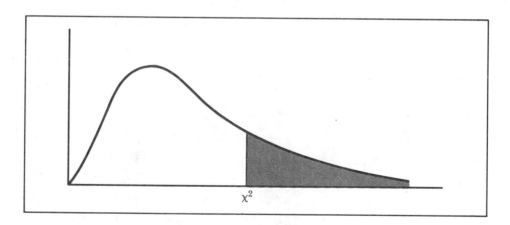

$$\chi^2$$

CHI SQUARE DISTRIBUTION.

d.f. \ α	0.995	0.975	0.050	0.025	0.010	0.005
1	0.0^43927	0.0^39821	3.84146	5.02389	6.63490	7.87944
2	0.010025	0.050636	5.99147	7.37776	9.21034	10.5966
3	0.071721	0.215795	7.81473	9.34840	11.3449	12.8381
4	0.206990	0.484419	9.48773	11.1433	13.2767	14.8602
5	0.411740	0.831211	11.0705	12.8325	15.0863	16.7496
6	0.675727	1.237347	12.5916	14.4494	16.8119	18.5476
7	0.989265	1.68987	14.0671	16.0128	18.4753	20.2777
8	1.344419	2.17973	15.5073	17.5346	20.0902	21.9550
9	1.734926	2.70039	16.9190	19.0228	21.6660	23.5893
10	2.15585	3.24697	18.3070	20.4831	23.2093	25.1882
11	2.60321	3.81575	19.6751	21.9200	24.7250	26.7569
12	3.07382	4.40379	21.0261	23.3367	26.2170	28.2995
13	3.56503	5.00874	22.3621	24.7356	27.6883	29.8194
14	4.07468	5.62872	23.6848	26.1190	29.1413	31.3193
15	4.60094	6.26214	24.9958	27.4884	30.5779	32.8013
16	5.14224	6.90766	26.2962	28.8454	31.9999	34.2672
17	5.69724	7.56418	27.5871	30.1910	33.4087	35.7185
18	6.26481	8.23075	28.8693	31.5264	34.8053	37.1564
19	6.84398	8.90655	30.1435	32.8523	36.1908	38.5822
20	7.43386	9.59083	31.4104	34.1696	37.5662	39.9968
21	8.03366	10.28293	32.6705	35.4789	38.9321	41.4010
22	8.64272	10.9823	33.9244	36.7807	40.2894	42.7956
23	9.26042	11.6885	35.1725	30.0757	41.6384	44.1813
24	9.88623	12.4001	36.4151	39.3641	42.9798	45.5585
25	10.5197	13.1197	37.6525	40.6465	44.3141	46.9278
26	11.1603	13.8439	38.8852	41.9232	45.6417	48.2899
27	11.8076	14.5733	40.1133	43.1944	46.9630	49.6449
28	12.4613	15.3079	41.3372	44.4607	48.2782	50.9933
29	13.1211	16.0471	42.5569	45.7222	49.5879	52.3356
30	13.7867	16.7908	43.7729	46.9792	50.8922	53.6720
40	20.7065	24.4331	55.7585	59.3417	63.6907	66.7659
50	27.9907	32.3574	67.5048	71.4202	76.1539	79.4900
60	35.5346	40.4817	79.0819	83.2976	88.3794	91.9517
70	43.2752	48.7576	90.5312	95.0231	100.425	104.215
80	51.1720	57.1532	101.879	106.629	112.329	116.321
90	59.1963	65.6466	113.145	118.136	124.116	128.299
100	67.3276	74.2219	124.342	129.561	135.807	140.169

Source: From *Statistics for Business and Economics* by Paul Hoel and Raymond Jessen. Copyright © 1971 by John Wiley & Sons, Inc. Reprinted with permission of John Wiley & Sons, Inc. Numerical contents originally from "Percentage Points of χ^2 Distribution," *Biometrika*, Vol. 32, edited by E. S. Pearson. Used with permission.

APPENDIX M
TABLE OF RANDOM DIGITS

TABLE OF RANDOM DIGITS.

	00–04	05–09	10–14	15–19	20–24	25–29	30–34	35–39	40–44	45–49
00	54463	22662	65905	70639	79365	67382	29085	69831	47058	08186
01	15389	85205	18850	39226	42249	90669	96325	23248	60933	26927
02	85941	40756	82414	02015	13858	78030	16269	65978	01385	15345
03	61149	69440	11286	88218	58925	03638	52862	62733	33451	77455
04	05219	81619	10651	67079	92511	59888	84502	72095	83463	75577
05	41417	98326	87719	92294	46614	50948	64886	20002	97365	30976
06	28357	94070	20652	35774	16249	75019	21145	05217	47286	76305
07	17783	00015	10806	83091	91530	36466	39981	62481	49177	75779
08	40950	84820	29881	85966	62800	70326	84740	62660	77379	90279
09	82995	64157	66164	41180	10089	41757	78258	96488	88629	37231
10	96754	17676	55659	44105	47361	34833	86679	23930	53249	27083
11	34357	88040	53364	71726	45690	66334	60332	22554	90600	71113
12	06318	37403	49927	57715	50423	67372	63116	48888	21505	80182
13	62111	52820	07243	79931	89292	84767	85693	73947	22278	11551
14	47534	09243	67879	00544	23410	12740	02540	54440	32949	13491
15	98614	75993	84460	62846	59844	14922	48730	73443	48167	34770
16	24856	03648	44898	09351	98795	18644	39765	71058	90368	44104
17	96887	12479	80621	66223	86085	78285	02432	53342	42846	94771
18	90801	21472	42815	77408	37390	76766	52615	32141	30268	18106
19	55165	77312	83666	36028	28420	70219	81369	41943	47366	41067
20	75884	12952	84318	95108	72305	64620	91318	89872	45375	85436
21	16777	37116	58550	42958	21460	43910	01175	87894	81378	10620
22	46230	43877	80207	88877	89380	32992	91380	03164	98656	59337
23	42902	66892	46134	01432	94710	23474	20423	60137	60609	13119
24	81007	00333	39693	28039	10154	95425	39220	19774	31782	49037
25	68089	01122	51111	72373	06902	74373	96199	97017	41273	21546
26	20411	67081	89950	16944	93054	87687	96693	87236	77054	33848
27	58212	13160	06468	15718	82627	76999	05999	58680	96739	63700
28	70577	42866	24969	61210	76046	67699	42054	12696	93758	03283
29	94522	74358	71659	62038	79643	79169	44741	05437	39038	13163
30	42626	86819	85651	88678	17401	03252	99547	32404	17918	62880
31	16051	33763	57194	16752	54450	19031	58580	47629	54132	60631
32	08244	27647	33851	44705	94211	46716	11738	55784	95374	72655
33	59497	04392	09419	89964	51211	04894	72882	17805	21896	83864
34	97155	13428	40293	09985	58434	01412	69124	8217i	59058	82859
35	98409	66162	95763	47420	20792	61527	20441	39435	11859	41567
36	45476	84882	65109	96597	25930	66790	65706	61203	53634	22557
37	89300	69700	50741	30329	11658	23166	05400	66669	48708	03887
38	50051	95137	91631	66315	91428	12275	24816	68091	71710	33258
39	31753	85178	31310	89642	98364	02306	24617	09609	83942	22716
40	79152	53829	77250	20190	56535	18760	69942	77448	33278	48805
41	44560	38750	83635	56540	64900	42912	13953	79149	18710	68618
42	68328	83378	63369	71381	39564	05615	42451	64559	97501	65747
43	46939	38689	58625	08342	30459	85863	20781	09284	26333	91777
44	83544	86141	15707	96256	23068	13782	08467	89469	93842	55349
45	91621	00881	04900	54224	46177	55309	17852	27491	89415	23466
46	91896	67126	04151	03795	59077	11848	12630	98375	52068	60142
47	55751	62515	21108	80830	02263	29303	37204	96926	30506	09808
48	85156	87689	95493	88842	00664	55017	55539	17771	69448	87530
49	07521	56898	12236	60277	39102	62315	12239	07105	11844	01117

Source: Reprinted by permission from *Statistical Methods* by George W. Snedecor and William G. Cochran, sixth edition © 1967 by Iowa State University Press, Ames, Iowa.

Appendix M

TABLE OF RANDOM DIGITS.

	50–54	55–59	60–64	65–69	70–74	75–79	80–84	85–89	90–94	95–99
00	59391	58030	52098	82718	87024	82848	04190	96574	90464	29065
01	99567	76364	77204	04615	27062	96621	43918	01896	83991	51141
02	10363	97518	51400	25670	98342	61891	27101	37855	06235	33316
03	86859	19558	64432	16706	99612	59798	32803	67708	15297	28612
04	11258	24591	36863	55368	31721	94335	34936	02566	80972	08188
05	95068	88628	35911	14530	33020	80428	39936	31855	34334	64865
06	54463	47237	73800	91017	36239	71824	83671	39892	60518	37092
07	16874	62677	57412	13215	31389	62233	80827	73917	82802	84420
08	92494	63157	76593	91316	03505	72389	96363	52887	01087	66091
09	15669	56689	35682	40844	53256	81872	35213	09840	34471	74441
10	99116	75486	84989	23476	52967	67104	39495	39100	17217	74073
11	15696	10703	65178	90637	63110	17622	53988	71087	84148	11670
12	97720	15369	51269	69620	03388	13699	33423	67453	43269	56720
13	11666	13841	71681	98000	35979	39719	81899	07449	47985	46967
14	71628	73130	78783	75691	41632	09847	61547	18707	85489	69944
15	40501	51089	99943	91843	41995	88931	73631	69361	05375	15417
16	22518	55576	98215	82068	10798	86211	36584	67466	69373	40054
17	75112	30485	62173	02132	14878	92879	22281	16783	86352	00077
18	80327	02671	98191	84342	90813	49268	95441	15496	20168	09271
19	60251	45548	02146	05597	48228	81366	34598	72856	66762	17002
20	57430	82270	10421	00540	43648	75888	66049	21511	47676	33444
21	73528	39559	34434	88596	54086	71693	43132	14414	79949	85193
22	25991	65959	70769	64721	86413	33475	42740	06175	82758	66248
23	78388	16638	09134	59980	63806	48472	39318	35434	24057	74739
24	12477	09965	96657	57994	59439	76330	24596	77515	09577	91871
25	83266	32883	42451	15579	38155	29793	40914	65990	16255	17777
26	76970	80876	10237	39515	79152	74798	39357	09054	73579	92359
27	37074	65198	44785	68624	98336	84481	97610	78735	46703	98265
28	83712	06514	30101	78295	54656	85417	43189	60048	72781	72606
29	20287	56862	69727	94443	64936	08366	27227	05158	50326	59566
30	74261	32592	86538	27041	65172	85532	07571	80609	39285	65340
31	64081	49863	08478	96001	18888	14810	70545	89755	59064	07210
32	05617	75818	47750	67814	29575	10526	66192	44464	27058	40467
33	26793	74951	95466	74307	13330	42664	85515	20632	05497	33625
34	65988	72850	48737	54719	52056	01596	03845	35067	03134	70322
35	27366	42271	44300	73399	21105	03280	73457	43093	05192	48657
36	56760	10909	98147	34736	33863	95256	12731	66598	50771	83665
37	72880	43338	93643	58904	59543	23943	11231	83268	65938	81581
38	77888	38100	03062	58103	47961	83841	25878	23746	55903	44115
39	28440	07819	21580	51459	47971	29882	13990	29226	23608	15872
40	63525	94441	77033	12147	51054	49955	58312	76923	96071	05813
41	47606	93410	16359	89033	89696	47231	64498	31776	05383	39902
42	52669	45030	96279	14709	52372	87832	02735	50803	72744	88208
43	16738	60159	07425	62369	07515	82721	37875	71153	21315	00132
44	59348	11695	45751	15865	74739	05572	32688	20271	65128	14551
45	12900	71775	29845	60774	94924	21810	38636	33717	67598	82521
46	75086	23537	49939	33595	13484	97588	28617	17979	70749	35234
47	99495	51434	29181	09993	38190	42553	68922	52125	91077	40197
48	26075	31671	45386	36583	93459	48599	52022	41330	60651	91321
49	13636	93596	23377	51133	95126	61496	42474	45141	46660	42338

Answers to Odd-Numbered Problems

CHAPTER 1

3. (a)

SECTORS	GROSS NATIONAL PRODUCT (BILLIONS OF DOLLARS)	
	1975	1976
Durable Goods	$ 258.2	$ 303.4
Nondurable Goods	428.0	460.9
Services	699.2	782.0
Structures	143.5	160.2
Total	$1528.9	$1706.5

Source: *Survey of Current Business*, October 1977, p. 8

15. (a)

ACCOUNT SIZE	FREQUENCY
50 and under 250	7
250 and under 450	9
450 and under 650	14
650 and under 850	10
850 and under 1050	6
1050 and under 1250	4
1250 and under 1450	10

(b)

ACCOUNT SIZE	FREQUENCY
50 or more	60
250 or more	53
450 or more	44
650 or more	30
850 or more	20
1050 or more	14
1250 or more	10
1450 or more	0

17. (a)

AGES	f
18 or more	50
24 or more	48
30 or more	44
36 or more	36
42 or more	24
48 or more	14
54 or more	7
60 or more	2
66 or more	0

(c)

AGES	f
less than 18	0
less than 24	2
less than 30	6
less than 36	14
less than 42	26
less than 48	36
less than 54	43
less than 60	48
less than 66	50

19. (a) $(60,000)(0.10) = \$6,000$ (b) $(\$60,000)(0.05) = \$3,000$
(c) $\$70,000 - \$40,000 = \$30,000$

(d) $\dfrac{\$70,000 - \$40,000}{\$40,000}(100) = 75\%$

CHAPTER 2

1. $\overline{X} = 26.575$
3. $\overline{X} = 5.1$;   Median $= 5$;   Mode $= 6$
    Half of the employes work 5 or more hours overtime with the most frequent number of overtime hours equaling 6.
5. (a) $\overline{X} = 148.57$ (b) Median $= 125$ (c) There is no mode
7. $\overline{X} = 24.83$;   Median $= 26$;   Mode $= 26$;   Yes, Sam may make it
9. (a) Median, because of extreme value ($\$100,000$)
    (b) Median $= \$7800$
11. $\overline{X} = 6.88$
13. (a) $\overline{X} = 53.4$, Md $= 52.14$
15. (a) $\overline{X} = \$7.833$ (b) Md $= \$8.20$ (c) Modal class is 9–11
17. $X_w = 5.83\%$
19. $\overline{X}_w = 34.4¢$
21. Average annual rate of return $= 3.6\%$

CHAPTER 3

1. (a) Range = 24   (b) $Q_1 = 3, Q_3 = 8$; Half the workers drive between
     3 and 8 miles
    (c) $s = 6.94$
3. (a) $\overline{X} = 15.5$ (b) $s = 9.16$ (c) $s = 9.16$
    (d) identical,   (e) variance = $s^2 = 83.83$
5. (a) $\mu = 72.67$ (b) $Q_2 = 70$ (c) using $\mu = 72.67$; $\sigma = 39.58$
    (d) $\sigma^2 = 1566.22$
7. (a) $\overline{X} = 53.52$ (b) $s = 23.64$ (c) $s^2 = 558.76$
9. (a) largest $\Sigma X = \$2,250,000$
    (b) smallest coefficient of variation, $V_1 = 9.04\%$
    (c) Plant #2 with the largest standard deviation
    (d) High = \$14,800; low = \$6,050
11. (a) $\overline{X} = 31$ (b) $s = 11.45$ (c) $s = 11.45$ (d) same
    (e) $s^2 = 131.19$ (f) $Q_1 = 24$ (g) $Q_2 = 30.833$ (h) $Q_3 = 39.167$
    (i) $V = 36.94\%$ (j) = +0.045
13. V (Assistant Prof.) = 33.3%
15. (a) $V_m = 6.67\%$, Grower M has the more uniform fruit
    (b) Grower N, Largest = 20 ounces
17. (a) ΣX will be larger for Battery I
    (b) $V_{II} = 5.56\%$, Since Battery II has least relative variation, II should
     be more uniform.

anticipates normal distribution

CHAPTER 4

1. 21
3. 336
5. 17,576,000
7. (a) 42/90   (b) 27/90
9. 20
11. 15
13. 0.75
15. 2/10
17. 0.5303
19. (a) 0.125   (b) 0.083   (c) 0.375
21. (a) 0.49   (b) 0.42
23. 0.382
25. 0.0036
27. 0.923
29. (a) 0.585   (b) 0.433   (c) No
31. (a) 20/100   (b) 55/100   (c) 5/21   (d) 3/20   (e) No
33. (a) 0.30   (b) 0.90
35. 0.67

CHAPTER 5

1. (a)

SALES ($1000)	PROB. DIST.
50	0.04
100	0.30
150	0.34
200	0.26
250	0.06

(b)

SALES ($1000)	P (X ≤ x)
50	0.04
100	0.34
150	0.68
200	0.94
250	1.000

3.

NUMBER OF HEADS	f	PROB.
3	1	0.125
2	3	0.375
1	3	0.375
0	1	0.125

5.

NUMBER OF HEADS	P (X ≤ x) (a)	(b)
0	0.125	0.064
1	0.500	0.352
2	0.875	0.784
3	1.000	1.000

7. (a) no   (b) yes   (c) yes   (d) yes
9. (a) 0.2508   (b) 0.6177   (c) 0.0123   (d) 0.0464
11. 0.0574
13. (a) 0.0734   (b) 0.9883   (c) 0.7037   (d) 0.0559   (e) 0.5202
      (f) 0.7821
15. (a) 0.0498   (b) 0.2240   (c) 0.0337
17. (a) 0.1465   (b) 0.1221   (c) 0.7852
19. 0.0467
21. 0.24303
23. 0.2119   (b) 0.2743   (c) 0.1464   (d) 0.0228   (e) 0.1151
      (f) 0.3811   (g) 0.2620
25. 17.048
27. 641
29. 0.0223
31. 0.2231
33. (a) 0.03995   (b) 0.00006265

CHAPTER 6

1. (a) $5^2 = 25$ (b) $P_2^5 = 20$ (c) $C_2^5 = 10$

3. Possible Samples $(2, 2)$, $(2, 4)$, $(2, 6)$, $(2, 8)$, $(4, 2)$, $(4, 4)$, $(4, 6)$, $(4, 8)$, $(6, 2)$, $(6, 4)$, $(6, 6)$, $(6, 8)$, $(8, 2)$, $(8, 4)$, $(8, 6)$, $(8, 8)$

Sample means for the preceding samples: 2, 3, 4, 5, 3, 4, 5, 6, 4, 5, 6, 7, 5, 6, 7, 8

5. (a) 5   (b) $\sigma_{\bar{x}} = 1.58$ (c) $\sigma_{\bar{x}} = 1.58$

7.

$\overline{X}$	f
3	2
4	2
5	4
6	2
7	2

9. (a) $P_2^4 = 12$

(b)

$\overline{X}$	f
7.5	2
9.0	2
10.5	4
12.0	2
13.5	2

(c) $\overline{\overline{X}} = 10.5$
(d) $\sigma_{\bar{x}} = 1.936$
(e) $\sigma_{\bar{x}} = 1.936$

11. $\pi = \frac{1}{4}$

13. (a) 0.6826   (b) 0.2514   (c) 0.9544   (d) 0.0475

15. 181.066

17. (a) $p = 0.40$ (b) $s_p = 0.0693$

19. $\sigma_p = 0.03456$

21. (a) $s_{\bar{x}} = 1$ (b) yes

23. (a) $p = 0.167$ (b) $s_p = 0.03004$

25. 0.0409

CHAPTER 7

1. $35.08 \leq \mu \leq 42.92$

3. $211.58 \leq \mu \leq 238.42$

5. $41.54 \leq \mu \leq 45.46$

7. $6.383 \leq \mu \leq 7.617$

9. $15.226 \leq \mu \leq 16.774$

11. $557.22 \leq \mu \leq 602.78$

13. $23.9468 \leq \mu \leq 28.0532$

15. $80.821 \leq \mu \leq 119.179$

17. $316.12 \leq \mu \leq 483.88$ 3016 - 4084

19. $0.56 \leq \pi \leq 0.64$

21. $88.032 \leq \mu \leq 91.968$

23. $0.546 \leq \pi \leq 0.734$

25. $535.3 \leq \mu \leq 564.7$

27. $0.564 \leq \pi \leq 0.676$

29. $0.02 \leq \pi \leq 0.08$

CHAPTER 8

1. $Z_{test} = -2$, Reject H_0; Reject H_0 if $Z_{test} < -1.96$ or $Z_{test} > 1.96$
3. $Z_{test} = 1.33$, Accept (do not reject) H_0; Reject H_0 if $t_{test} > 1.64$
5. $Z_{test} = -5$, Reject H_0; Reject H_0 if $Z_{test} < -2.33$
7. $Z_{test} = 13.33$, Reject H_0; Conclude battery life has improved; Reject H_0 if $Z_{test} > 1.64$
9. $t_{test} = 1.33$, Accept (do not reject) H_0 and discontinue prizes. Assume sales are normally distributed.   Reject H_0 if $t_{test} > 1.341$
11. (a) If fills are normally distributed   (b) $t_{test} = -1.14$, so accept (do not reject) H_0.   Reject H_0 if $t_{test} < -2.131$ or $t_{test} > 2.131$
13. $Z_{test} = -2$, Accept H_0; Reject H_0 if $Z_{test} < -2.33$
15. $Z_{test} = 2.5$, Reject H_0; The sample results do not support the company's hypotheses, $H_0: \mu = 15$.   Reject H_0 if $Z_{test} > 1.64$
17. $Z_{test} = 1.51$, Accept (do not reject) H_0; Reject H_0 if $Z_{test} < -1.96$ or $Z_{test} > 1.96$
19. $Z_{test} = -1.75$, Accept (do not reject) H_0.   Conclude shipment is acceptable; Reject H_0 if $Z_{test} < -2.33$
21. $Z_{test} = -0.52$, Accept (do not reject) H_0.   Conclude that the manager's claim has not been refuted.   Reject H_0 if $Z_{test} < -1.64$.
23. $Z_{test} = -3.0$, Reject H_0; Conclude that the bags are light.   Reject H_0 if $Z_{test} < -1.64$
25. $t_{test} = -2.195$; Reject H_0; Conclude that process average changed.   Reject H_0 if $t_{test} < -2.069$ or $t_{test} > 2.069$
27. $\beta = 0.4840$
29. $\beta = 0.7352$
31. $Z_{test} = -3$; Reject H_0; Claim is not supported.   Reject H_0 if $Z_{test} < -2.33$

CHAPTER 9

1. A partial listing of the 81 sample pairs is given below:

SAMPLES FROM N_1	$\overline{X}_1$	SAMPLES FROM N_2	$\overline{X}_2$	$\overline{X}_1 - \overline{X}_2$
4,4	4.0	1,1	1.0	3.0
4,4	4.0	1,2	1.5	2.5
4,4	4.0	1,3	2.0	2.0
4,4	4.0	2,1	1.5	2.5
4,4	4.0	2,2	2.0	2.0
4,4	4.0	2,3	2.5	1.5
4,4	4.0	3,1	2.0	2.0

(cont'd)

SAMPLES FROM N_1	$\overline{X}_1$	SAMPLES FROM N_2	$\overline{X}_2$	$\overline{X}_1 - \overline{X}_2$
4,4	4.0	3,2	2.5	1.5
4,4	4.0	3,3	3.0	1.0
4,5	4.5	1,1	1.0	3.5
4,5	4.5	1,2	1.5	3.0
4,5	4.5	1,3	2.0	2.5
.	.	.	.	.
.	.	.	.	.
.	.	.	.	.
6,6	6.0	1,1	1.0	5.0
6,6	6.0	1,2	1.5	4.5
6,6	6.0	1,3	2.0	4.0
6,6	6.0	2,1	1.5	4.5
6,6	6.0	2,2	2.0	4.0
6,6	6.0	2,3	2.5	3.5
6,6	6.0	3,1	2.0	4.0
6,6	6.0	3,2	2.5	3.5
6,6	6.0	3,3	3.0	3.0

3. (a) 3   (b) 0.816, yes

5. $Z_{test} = 0.775$, Accept (do not reject) H_0; Reject H_0 if $Z_{test} < -1.96$ or $Z_{test} > 1.96$

7. $t_{test} = -2.05$, Accept (do not reject) H_0; Conclude the operations are not different.   Reject H_0 if $t_{test} < -2.807$ or $t_{test} > 2.807$

9. $t_{test} = -3.25$, Reject H_0; Reject H_0 if $t_{test} < -2.306$ or $t_{test} > 2.306$

11. $Z_{test} = 12.20$, Accept (do not reject) H_0; Reject H_0 if $Z_{test} < -1.28$

13. $Z_{test} = 4.76$, Reject H_0, Conclude SE method is superior; Reject H_0 if $Z_{test} > 1.64$

15. $Z_{test} = -1.32$, Accept (do not reject) H_0.   Reject H_0 if $Z_{test} < -1.64$

17. $t_{test} = 2.90$, Reject H_0; Conclude that the Penguins are bigger spenders. Reject H_0 if $t_{test} > 2.5$

19. $Z_{test} = -0.817$, Accept (do not reject) H_0; Reject H_0 if $Z_{test} < -1.64$ or $Z_{test} > 1.64$

21. $Z_{test} = -0.472$, Accept (do not reject) H_0; Reject H_0 if $Z_{test} < -2.58$ or $Z_{test} > 2.58$

23. $F_{test} = 1.96$, Reject H_0; Conclude that the variances are not equal.   Reject H_0 if $F_{test} > 1.50$

25. $t_{test} = 5.40$, Reject H_0; Conclude scores are improved.   Reject H_0 if $t_{test} < -2.861$ or $t_{test} > 2.861$

27. $F_{test} = 7.553$, Reject H_0; Reject H_0 if $F_{test} > 5.29$

29. (a) $t_{test} = -0.557$, Accept, (do not reject) H_0.   Reject H_0 if $t_{test} < -2.306$ or $t_{test} > 2.306$

(b) $t_{test} = -0.779$, Accept (do not reject) H_0.   Reject H_0 if $t_{test} < -3.355$ or $t_{test} > 3.355$

(c) $F_{test} = 0.863$, Accept (do not reject) H_0; Reject H_0 if $F_{test} > 6.93$

CHAPTER 10

1. 51, 89, 06, 24, 71, 85, 57, 33, 09, 40
3. (a) $\overline{X} = 3$ (b) $\overline{X} = 5.5$
5. $n_1 = 159$, $n_2 = 79$, $n_3 = 40$, $n_4 = 22$
7. $n_1 = 67$, $n_2 = 133$
9. (a) $n = 59.9076$ (b) $n = 2.6896$ (c) $n = 10.7584$
   (d) $n = 14.9769$
11. $n = 61.4656$
13. $n = 1068$
15. $n = 90.39$

CHAPTER 11

1. $\chi^2_{test} = 39.90$
3. $\chi^2_{test} = 0.78$
5. $\chi^2_{test} = 38.283$
7. $\chi^2_{test} = 0.73$
9. $\chi^2_{test} = 403.412$
11. $Z_{test} = 1.12$
13. $Z_{test} = -0.13$
15. $H' = 6.90$

CHAPTER 12

1. (b) $Y_c = 3.7501 + 3.0729X$
   (c) $t_{test} = 3.998$
   (d) $Y_c = 40.6249$
   (e) 95% confidence interval for $Y_c = 40.6249 \pm 12.0703$
   (f) 95% confidence interval for $Y = 40.6249 \pm 24.1387$
3. (b) $Y_c = 26.762 - 0.2024X$
   (c) $t_{test} = -16.73$
5. $r^2 = 0.9726$
7. (a) $Y_c = \$3139.30$
   (b) 95% confidence interval for $Y_c = 3.1393 \pm 0.1808$
   (c) 95% confidence interval for $Y = 3.1393 \pm 0.4206$
9. (a) $Y_c = \$1,304.20$
   (b) 90% confidence interval for $Y_c = 1.3042 \pm 0.1089$
   (c) 90% confidence interval for $Y = 1.3042 \pm 0.2534$
11. (a) $Y_c = 17.0816 + 0.2398X$
    (b) $t_{test} = 1.332$
13. (a) $r = 0.9354$
    (b) $s_{y.x} = 3.1944$

15. (b) $Y_c = -7.5533 + 2.3355X$
    (c) $t_{test} = 3.4124$
    (d) $r^2 = 0.5927$
17. (a) $t_{test} = 15.314$
    (b) $r^2 = 0.9671$
    (c) 95% confidence interval for $Y_c = 13.9884 \pm 1.0605$
    (d) 95% confidence for the upper limit for $Y_c = 10.2824 + 1.9247$

CHAPTER 13

1. (a) $\log Y_c = -1.67821 + 0.03936X$ (b) $Y_c = 28.15$
   (c) $r = 0.993$ (d) 9.50% increased for each additional service call
3. (a) $Y_c = 654.545 - 79.318x$
   (b) $\log Y_c = 2.78042 - 0.0563x$
   (c) $Y_c = 636.77155 - 79.31818x + 1.77739x^2$
   (d) $Y_c = 86.16$ using the equation from (b). The other two equations show negative employment.
   (e) $r^2 = 0.9793$
   (f) $r^2 = 0.98755$
5. Price: $\log Y_c = 0.03973 + X (1.99956 - 2)$
   Cost: $\log Y_c = (9.65116 - 10) + 0.00048X$
   Profit = \$38,220
7. (a) $Y_c = 32.516 + 3.027X_1 - 0.360X_2$
   X_1 = Dog licenses X_2 = supermarkets X_1 $t_{test} = 3.40$
        X_2 $t_{test} = 0.43$
   (b) No
9. $Y_c = -715870.896 + 19788.200X_2 + 56.558X_2$
            $R^2 = 0.975$
   $Y_c = -627720.386 + 57.221$
            $R^2 = 0.962$
   X_1 = unemployment X_2 = population
11. (a) $Y_c = -27.697 + 0.785X_1 + 0.069X_2$
    X_1 = temperature X_2 = traffic count
    (b) X_1 $t_{test} = 2.21$ X_2 $t_{test} = 2.12$
    (c) $R = 0.897$

CHAPTER 14

1. (a) 100, 101.9, 101.0, 97.8, 99.5
   (b) 99.0, 100.9, 100, 96.8, 98.5
   (c) 100, 99.9, 100.6, 101.9, 101.7
   (d) 98.8, 100, 100.6, 100.5, 101.5
3. (a) 5.2%   (b) 0.7%

5. 100, 112.4, 101.1
7. 100, 111.5, 112.3
9. (a)   100, 112.1, 101.4
   (b)   100, 112.14, 101.0
   (c)   The Laspeyres index represents the increase in expenditures necessary to maintain the same buying power as 1973, whereas the Paasche index represents the increase in expenditures necessary to purchase the current market basket of goods
11. (a)   100, 109.3, 122.5
   (b)   100, 109.3, 122.0
13. (a)   100, 110.86, 118.29   (b)   18.29%   (c)   6.7%
15. (a)   116.37   (b)   The manufacturer has to spend a little over $1.16 in 1978 to buy what $1.00 purchased in 1977
   (c)   116.03   (d)   The quantities purchased in 1978 cost 16.03% more than they would have if purchased at 1977 prices.
17. (a)   113.3, 115   (b)   134, 137.5, 144.5
19. No, his real earnings are higher in present job.
21. $3.22
23. $13750

CHAPTER 15

1. (b) $Y_c = 220.71 + 7.86x$, $x = 1$ year, $Y =$ annual yield (bushels), origin = July 1, 1975
   (c) $Y_{1979} = 252.15$ bushels
3. (a)   (I) $Y_c = 401 + 14.64x$, $x = 1$ year, $Y =$ annual sales ($1,000), origin = July 1, 1975
      (II) $Y_c = 445.71 + 8.18x$, $x = 1$ year, $Y =$ annual sales ($1,000), origin = July 1, 1975
   (b)   I sales will exceed II sales in 1982
5. (a) $Y_c = 467.94 + 33.50x$, $x = 1$ year, $Y =$ annual sales (billions), origin = Apr. 1, 1972 (Apr. origin because of fiscal year from Oct. 1 to Sept. 30 of following year)
   (b) $Y_{1978} = 668.94 billions. Y_{1978} estimate is below actual 1977 value. Thus, linear trend may not be very good estimator.
7. (a) $Y_c = 106 + 5.96x$, $x = 1$ year, $Y = $1,000$, origin = July 1, 1975
   (b) $\log Y_c = 1.72627 + 0.07100x$, $x = 1$ year, $Y = $1,000$, origin = July 1, 1975
   (c) $Y_c = 167.20 thousands
9. (a)   17 years
   (b) $500 thousand
   (c) $14.4 thousand
11. (a)   10 years   (b) $12 thousand per year   (c) $140 thousand
13. $Y_c = 3900.02 + 32.14x - 89.29x^2$, $x = 1$ year, $Y =$ annual employment, origin = July 1, 1971
15. $\log Y_c = 1.00327 + 0.07209x$, $x = 1$ year, $Y =$ annual sales (billions), origin = Dec. 31, 1974
17. 70,760,000 pounds
19. (a) $Y_c = 327.2 + 14.4x$ (b) $Y_c = 23.67 + 0.1x$ (c)   38.32

CHAPTER 16

1.

	J	F	M	A	M	J	J	A	S	O	N	D
1972	—	—	—	—	—	—	112.15	139.37	125.44	147.22	110.47	87.67
1973	80.54	78.60	67.24	59.10	81.79	101.59	130.82	149.09	116.17	139.31	100.00	90.24
1974	82.42	82.51	75.00	69.53	66.67	104.66	128.44	134.38	123.74	139.80	104.85	78.76
1975	85.14	91.08	78.43	60.00	66.24	104.44	122.80	139.66	132.50	126.76	108.15	88.36
1976	79.80	83.39	80.74	70.18	79.34	101.96	112.80	136.65	124.20	121.59	111.28	89.98
1977	78.54	90.98	94.33	68.62	86.16	101.48	105.19	127.77	113.06	118.98	105.66	95.54
1978	92.24	94.95	89.43	62.70	94.46	98.35						
Modified Total	327.9	347.96	323.60	260.85	313.96	409.47	476.19	550.06	489.55	527.46	429.13	356.25
Average Specific Seasonal (Unadjusted)	81.98	86.99	80.90	65.21	78.49	102.37	119.05	137.52	122.39	131.87	107.28	89.06 = 1203.1
Seasonal Index	81.77	86.77	80.69	65.04	78.29	102.11	118.74	137.16	122.07	131.53	107.00	88.83 = 1200

$$\text{Adjustment multiplier} = \frac{1200}{1203.11} = 0.99741503$$

3.

	J	F	M	A	M	J	J	A	S	O	N	D
1971	—	—	—	—	—	—	90.33	94.86	93.49	93.22	85.28	84.91
1972	90.84	122.76	136.60	110.51	110.17	103.39	83.31	83.33	94.12	93.36	74.28	78.34
1973	96.35	123.91	139.22	115.19	107.94	101.89	89.33	87.87	87.62	96.46	80.29	79.69
1974	99.63	123.22	127.26	114.97	104.53	103.42	87.96	92.75	95.49	94.32	79.29	78.25
1975	95.51	112.02	138.47	119.35	109.90	104.99	86.85	89.58	93.11	84.18	83.51	79.03
1976	102.50	113.38	139.27	110.23	104.79	99.44	91.73	96.32	96.22	97.16	93.87	86.50
1977	94.37	110.64	127.51	104.11	100.45	100.51	89.53	97.12	96.44	99.82	83.98	84.03
1978	91.47	111.27	139.44	109.04	104.58	100.64	—	—	—	—	—	—
Modified Total	477.33	582.65	681.07	559.94	531.74	509.85	444.00	461.38	472.43	474.52	412.35	406.00
Average Specific Seasonal (Unadjusted)	95.47	116.53	136.21	111.99	106.35	101.97	88.80	92.28	94.49	94.90	82.47	81.20 = 1202.66
Seasonal Index Adjusted	95.26	116.27	135.91	111.74	106.11	101.74	88.60	92.08	94.28	94.69	82.29	81.02 = 1199.99

Multiplier for Adjustment

$$\frac{1200}{1202.66} = 0.99778824$$

5.

	I	II	III	IV
1974	—	—	108.44	102.04
1975	89.57	99.24	108.82	101.49
1976	88.45	103.41	106.54	102.42
1977	88.67	101.68	107.18	103.35
1978	87.62	99.76	—	—
Modified Total	177.12	201.44	215.62	204.46
Unadjusted Index	88.56	100.72	107.81	102.23
Adjusted Index	88.71	100.89	107.99	102.40

7. $222,455, $216,932, $255,933, $137,615
9. (a) $96,000   (b) $125,000
11. $2,800,000
13. January = 90.31,   February = 73.12,   March = 82.02,   April = 96.50, May = 100.76, June = 106.32, July = 126.57, August = 116.65, September = 115.79, October = 107.81, November = 94.46, December = 89.69

CHAPTER 17

1. 4.1

3. (a)

		Actions			
		100	110	120	130
Events	100	300	260	220	180
	110	300	330	290	250
	120	300	330	360	320
	130	300	330	360	390

(b)   EMV (100) = $300,   EMV (110) = $309,   EMV (120) = $290, EMV (130) = $257

(c) Stock 110   (d) EPUC = $333   (e) EVPI = $24

5. (a)

		Actions		
		10	20	30
Events	10	20	−80	−180
	20	20	40	−60
	30	20	40	60

(b)   EMV (10) = $20, EMV (20) = $4, EMV (30) = −$72; Stock 10 books

(c) EPUC = $38   (d) EVPI = $18

7.

		Actions		
		10	20	30
Events	10	0	100	200
	20	20	0	100
	30	40	20	0

9. (a)

		Actions		
		15	20	25
Events	15	45	20	−5
	20	45	60	35
	25	45	60	75

(b)   EMV (15) = $45, EMV (20) = $50, EMV (25) = $37; Stock 20 calendars

(c) EVPI = $10.75

11.

	Actions		
	15	20	25
Events 15	0	15	30
20	15	0	15
25	30	15	0

13. (a)

	Actions			
	200	210	220	230
200	1000	950	900	850
Events 210	1000	1050	1000	950
220	1000	1050	1100	1050
230	1000	1050	1100	1150

(b) EMV (200) = \$1,000; EMV (210) = \$1,040; EMV (220) = \$1,040; EMV (230) = \$1,010; Stock 210 or 220 bouquets

(c) EPUC = \$1080   (d) EVPI = \$40   (e) 216

15. (a)

PROB. OF REPEAT	(P (X = 2, n = 2)	PRIOR	JOINT	REVISED
0.05	0.0025	0.10	0.00025	0.00223
0.25	0.0625	0.40	0.02500	0.22321
0.35	0.1225	0.30	0.03675	0.32813
0.50	0.2500	0.20	0.05000	0.44643
			0.11200	1.00000

(b) EMV (200) = \$1,000;   EMV (210) = \$1,049.777;   EMV (220) = \$1077.233; EMV (230) = \$1071.876; Stock 220

17. (a)

FRACTION DEFECTIVE	100% INSPECT	ACCEPT WITHOUT INSPECTION
0.05	400	250
0.10	400	500
0.15	400	750

(b) EMV (100% Inspect) = \$400, EMV (accept without inspection) = \$350; Accept without inspection.

(c) ECUC = \$302.50   (d) EVPI = \$47.50

19.

w/o sampling   with sampling

EVSI = Min (EMV)   —   E (min EMV)

= \$350 − [0.86576 (341.035) + 0.12850 (400) + 0.00576 (400)]

= \$350 − [\$295.25 + \$51.40 + \$2.30]

EVSI = \$1.05

at \$0.20 per unit, cost of sampling = \$0.40, thus sampling is worth \$1.05 − 0.40 = \$0.65

21. (a)

		Actions 0	5	10	15
Events	0	0	−15	−30	−45
	5	0	15	0	−15
	10	0	15	30	15
	15	0	15	30	45

(b) EMV (0) = 0, EMV (5) = $10.50, EMV (10) = $9.00, EMV (15) = −$3.00; Stock 5 planes

(c) $10.50 (d) 7

23. (a)

		Actions (1000) 75	100	125	150	175
Events	75	22500	11250	0	−11250	−22500
	100	22500	30000	18750	7500	−3750
	125	22500	30000	37500	26250	15000
	150	22500	30000	37500	45000	33750
	175	22500	30000	37500	45000	52500

(b) EMV (75,000) = $22,500; EMV (100,000) = $26437.50; EMV (125,000) = $25125; EMV (150,000) = $19125; EMV (175,000) = $9937.50; Stock 100,000 boxes

(c) EVPI = $9037.50

Index